TABLE A Fundamental constants†

1. Speed of light
   $c = (2.9979250 \pm 0.0000010) \times 10^{10}$ cm s^{-1}

2. Electronic Charge
   $e = (4.803250 \pm 0.000021) \times 10^{-10}$ esu

3. Electron rest mass
   $m = (9.109558 \pm 0.000054) \times 10^{-28}$ g

4. Proton rest mass
   $m_p = (1.672614 \pm 0.000011) \times 10^{-24}$ g

5. Planck's constant
   $h = (6.626196 \pm 0.000050) \times 10^{-27}$ erg s
   $\hbar = h/2\pi = (1.0545919 \pm 0.0000080) \times 10^{-27}$ erg s

6. Bohr magneton
   $$\beta = \frac{e\hbar}{2mc} = (9.274096 \pm 0.000065) \times 10^{-21} \text{ erg G}^{-1}$$

7. Electron free-spin g-factor
   $g_e = 2.0023192778 \pm 0.0000000062$

8. Electron magnetic moment
   $$\mu_e = \frac{g_e\beta}{2} = (9.284851 \pm 0.000065) \times 10^{-21} \text{ erg G}^{-1}$$

9. Nuclear magneton
   $$\beta_N = \frac{e\hbar}{2m_pc} = (5.050951 \pm 0.000050) \times 10^{-24} \text{ erg G}^{-1}$$

10. $\mu_p = (1.4106203 \pm 0.0000099) \times 10^{-23}$ erg G^{-1}

11. Magnetogyric ratio of proton in spherical water samples
    γ_p(corrected for diamagnetism of host compound) $= (2.6751965 \pm 0.0000082)$
    $$\times 10^4 \text{ rad s}^{-1} \text{ G}^{-1}$$

12. g_p(corrected for diamagnetism) $= \dfrac{2\mu_p}{\beta_N} = 5.585564 \pm 0.000017$

13. Magnetogyric ratio of a free electron
    $$\gamma_e = \frac{|\mu_e|}{S\hbar} = -(1.7608425 \pm 0.0000010) \times 10^7 \text{ rad s}^{-1} \text{ G}^{-1}$$

14. Boltzmann's constant
    $k = 1.380622 \pm 0.000059 \times 10^{-16}$ erg K^{-1}

---

† Taken from B. N. Taylor, W. H. Parker and D. N. Langenberg, *Rev. Mod. Phys.*, **41**, 375 (1969).

TABLE B Useful conversion factors

1. Magnetic field, H (gauss), to electron resonant frequency, ν_{el} (MHz) and $\bar{\nu}_{el}$ (cm^{-1})

$$\nu_{el} = \left(\frac{g_e\beta H}{h}\right)\left(\frac{g}{g_e}\right) \doteq 2.80247\left(\frac{g}{g_e}\right)H$$

$$H = 0.356828\left(\frac{g_e}{g}\right)\nu_{el}$$

ν_{el} (MHz) $= c \times 10^{-6}\bar{\nu}_{el}$ (cm^{-1}) $= 2.99793 \times 10^4\bar{\nu}_{el}$
$\bar{\nu}_{el}$ (cm^{-1}) $= 0.333564 \times 10^{-4}\nu_{el}$ (MHz)

2. Magnetic field, H (gauss), to proton resonant frequency, ν_p (MHz)

$\nu_p = 4.257708 \times 10^{-3}H$
$H = 234.868\nu_p$

3. Ratio of proton to electron resonant frequency

$$\nu_p/\nu_{el} = 1.51927 \times 10^{-3}\left(\frac{g_e}{g}\right)$$

4. Calculation of g factors

$$g = \frac{h\nu_{el}}{\beta H} = 0.714484\frac{\nu_{el}\ (\text{MHz})}{H\ (\text{G})}$$

$$= \frac{2\mu_p}{\beta}\frac{\nu_{el}}{\nu_p} = 3.042065 \times 10^{-3}\frac{\nu_{el}}{\nu_p}$$

5. Hyperfine couplings and hyperfine splittings

$$A\ (\text{MHz}) = 2.80247\left(\frac{g}{g_e}\right)a\ (\text{gauss})$$

$$a\ (\text{G}) = 0.356828\left(\frac{g_e}{g}\right)A\ (\text{MHz})$$

A/c (cm^{-1}) $= 0.333564 \times 10^{-4}A$ (MHz)

$$A_0\ (\text{MHz}) = \frac{8\pi}{3}\frac{g_e\beta}{h}\frac{\mu_N}{I}|\psi_s(0)|^2 = 23.4779\frac{\mu_N}{I}|\psi_s(0)|^2$$

ELECTRON SPIN RESONANCE
Elementary Theory and Practical Applications

John E. Wertz

Professor of Chemistry
University of Minnesota

James R. Bolton

Professor of Chemistry
University of Western Ontario

Chapman and Hall
New York • London

First edition published 1972 by
McGraw-Hill Book Company
This reprint published 1986 by
Chapman and Hall
29 West 35th St. New York, N.Y. 10001
Published in Great Britain by
Chapman and Hall Ltd
11 New Fetter Lane, London EC4P 4EE

© 1986 Chapman and Hall

Printed in the United States of America

ISBN 0 412 01181 6(cloth)
ISBN 0 412 01161 1(paper)

Congress Cataloging-in-Publication Data

Wertz, John E., 1916—
    Electron spin resonance.

    Reprint. Originally published: New York: McGraw-Hill,
1972.
    Bibliography: p.
    Includes index.
    1. Electron paramagnetic resonance. I. Bolton,
James R., 1937—   II. Title.
[QC762.W47 1986]   538'.36   85-29080
ISBN 0-412-01181-6
ISBN 0-412-01161-1

Contents

Preface

In the twenty-five years since its discovery by Zavoiskii, the technique of electron spin resonance (ESR) spectroscopy has provided detailed structural information on a variety of paramagnetic organic and inorganic systems. It is doubtful that even much later than 1945 any chemist would have been so bold as to predict the great diversity of systems which have proved amenable to study by ESR spectroscopy. In this book we have attempted to provide numerous examples of actual ESR spectra to illustrate the wide scope of application. No attempt has been made to present a comprehensive coverage of the literature in any field, but references to reviews and key articles are given throughout the book.

This introductory textbook had its origin in lecture notes prepared for an American Chemical Society short course on electron spin resonance. The present version is the result of extensive revision and expansion of the original notes. Experience with such courses has convinced us that there are large numbers of chemists, physicists, and biologists who have a strong interest in electron spin resonance. The mathematical training of most of the short-course students is limited to calculus. Their contact with theories of molecular structure is largely limited to that obtained in an elementary physical chemistry course. It is to an audience of such background that this book is directed. Numerous advanced undergraduates and beginning graduate students have also found that the level of presentation in this book is appropriate. A number of industrial chemists have found the notes useful in a self-study program.

Many students find it useful to have available in one text a large body of supplementary material on mathematical techniques and elementary quantum mechanics, especially that of angular momentum. Some of this material is readily found in standard texts; however, especially for the quantum mechanics of angular momentum, one finds few treatments at the level which is helpful to the greatest number of students. Even students with advanced preparation find it advantageous to have this supplementary material self-contained in the Appendices.

The starting level of this book is low; however, as the student acquires skill in the application of spin operators and matrix techniques, he advances to a respectable level of understanding. Much of this is gained through solution of the many problems given at the end of each chapter. The level of presentation reaches its peak in Chapters 11 and 12 on transition-metal ions. The student who has progressed through this level should find it possible to read with comprehension most of the ESR literature.

It is suggested that on first reading of a chapter the student should confine his attention to the main body of the text, omitting indented sections. At such points where further background material is helpful, we give references to the Appendices.

We are deeply grateful to our many colleagues, students and friends who read and criticized parts of the manuscript. Special thanks are due to H. Beinert, A. Carrington, K. Cost, J. J. Davies, J. S. Hyde, C. A. Mc-Dowell, B. R. McGarvey, A. D. McLachlan, C. A. Mead, J. R. Morton, S. Prager, W. L. Reynolds, I. C. P. Smith, S. R. P. Smith, P. D. Sullivan, J. H. van der Waals, G. Vincow, and D. H. Whiffen. We are also grateful to T. Cole and J. dos Santos Veiga for their collaboration in the preparation and presentation of the first ESR short course. Carolyn Warden kindly helped to prepare the indices. We are indebted to the many workers who have provided us with original figures or detailed data. This material is individually acknowledged when presented. If it had not been for the invitation by M. Passer of the American Chemical Society to present a series of short courses on ESR spectroscopy, it is unlikely that this project would have been undertaken. Finally, we are grateful for the unceasing patience and selflessness of our wives, Florence Wertz and Wilma Bolton, during the preparation of this book.

John E. Wertz
James R. Bolton

1

Basic Principles of
Electron Spin Resonance

1-1 INTRODUCTION

The technique of electron spin resonance may be regarded as a fascinating extension of the Stern-Gerlach experiment. In one of the most fundamental experiments on the structure of matter, Stern and Gerlach showed that an atom with a net electron magnetic moment can take up only discrete orientations in a magnetic field. Subsequently, Uhlenbeck and Goudsmit linked the electron magnetic moment with the radical idea of electron spin. Whether or not the reader has an immediate interest in the multitude of systems to which the technique is applicable, he can ill afford to ignore the insights which it can provide. Further, there is hardly another application from which one can gain a clearer insight into many of the fundamental concepts of quantum mechanics. Much of our knowledge of the structure of molecules has been obtained from the analysis of molecular absorption spectra. Such spectra are obtained by measuring the attenuation vs. wavelength (or frequency) of a beam of electromagnetic radiation as it passes through a sample of matter. Lines or bands in a spectrum represent transitions between energy levels of a molecule. Hence the frequency of each line or band measures the energy separation of two levels. Given enough data and some guidance from theory, one may construct an energy-level diagram from a spectrum. Comparison of an energy-level diagram and an observed spectrum shows clearly that of the many transitions which may be drawn between levels, only a few "allowed" transitions are observed. Hence the prediction of allowed transitions requires a knowledge of selection rules.

Electromagnetic radiation may be regarded as coupled electric and magnetic fields oscillating perpendicular to one another and to the direction of propagation (Fig. 1-1). In most cases it is the electric field component which interacts with molecules. For absorption to occur, two conditions must be fulfilled: (1) The energy of a quantum must correspond to the separation between energy levels in the molecule. (2) The oscillating *electric* field component must be able to stimulate an oscillating electric dipole in the molecule. For example, electromagnetic radiation in the microwave region will interact with molecules which have a permanent electric dipole moment; an example is HCl. Molecular rotation creates the required oscillating electric dipole. Likewise, infrared radiation will interact with molecules in vibrational modes which give rise to a change in the electric dipole moment.

Similarly a molecule containing *magnetic* dipoles might be expected to interact with the *magnetic* component of microwave radiation. When such a molecule is irradiated over a wide range of spectral frequencies, one normally finds no absorption attributable to a magnetic interaction. If, however, the sample of interest is placed in a static magnetic field, absorption attributable to magnetic dipole transitions may occur at one or more characteristic frequencies.

A possible experimental arrangement for the detection of magnetic dipolar transitions is the electron spin resonance microwave spectrometer shown in Fig. 1-2b; an optical spectrometer in Fig. 1-2a is shown to suggest by analogy the function of components in the electron spin resonance spectrometer. In either case, monochromatic radiation falls upon a sample in an appropriate cell, and one looks for changes in the intensity of the transmitted radiation by means of the appropriate detector. Absorption will

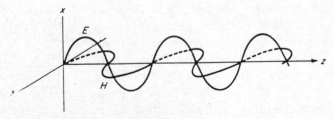

Fig. 1-1 Instantaneous amplitudes of electric and magnetic field components in a propagating electromagnetic beam.

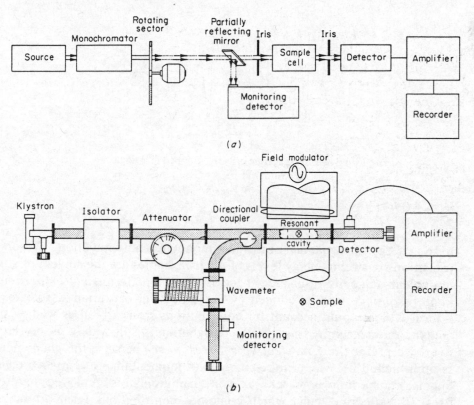

Fig. 1-2 (*a*) Block diagram of an optical spectrometer. (*b*) Block diagram of an analogous electron spin resonance microwave spectrometer.

occur when the energy of a quantum of incident radiation matches the energy-level separation. The requirement of a static magnetic field† is the unique aspect of magnetic dipolar transitions. In the absence of a magnetic field the energy levels are coincident. Permanent magnetic dipoles in a molecule are associated either with electrons or with nuclei. The magnetic dipoles arise from net electronic or nuclear angular momentum. The fundamental phenomenon to be understood is thus the quantization of angular momentum.

Magnetic dipoles attributable to electrons arise from net spin or net orbital angular momenta or from a combination of these. In the overwhelming majority of cases encountered, over 99 percent of the magnetic dipole is due to spin angular momentum, with but a small orbital contribution. Resonant absorption of radiation in a static magnetic field by such systems is variously called "paramagnetic resonance," "electron paramagnetic resonance," or "electron spin resonance." The term "resonance" is appropriate, since the well-defined separation of energy levels is matched

† The static field may be contributed internally by the nuclei of a molecule.

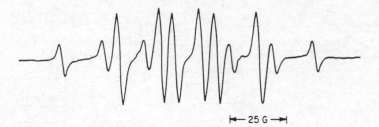

Fig. 1-3 First-derivative ESR spectrum of NH_3^+ trapped in a single crystal of NH_4ClO_4. This spectrum will be discussed in Sec. 4-6. [*Spectrum taken from J. S. Hyde and E. S. Freeman, J. Phys. Chem.,* **65**:1636 (1961).]

to the energy of a quantum of incident monochromatic radiation. Resonant transitions between energy levels of *nuclear* dipoles are the subject of study in nuclear magnetic resonance spectroscopy. The term *electron paramagnetic resonance* was introduced by H. E. Weaver of Varian Associates as a term which would account for contributions from orbital as well as spin angular momentum. This "EPR" designation is still widely used. However, we prefer the term *electron spin resonance* because the phenomenon is approached by way of the spin angular momentum. Even when orbital angular momentum is present, a "spin" hamiltonian is employed; here one uses an "effective" spin which combines contributions from orbital and spin angular momenta.† Having chosen the "ESR" designation, we respectfully bow in the direction of the Clarendon Laboratory, where the term *paramagnetic resonance* was first employed.

Electron spin resonance (ESR) spectra may convey a remarkable wealth of significant chemical information. A brief summary of structural or kinetic information derivable from Figs. 1-3 to 1-5 will foreshadow the diversity of the applications of the method. Each of these spectra will be considered at a later point.

Figure 1-3 represents ESR absorption by a species formed by γ irradiation of NH_4ClO_4. The number of lines, their spacing, and their relative intensities unequivocally indicate a molecule with one nucleus with spin $I = 1$ and three equivalent nuclei with $I = \frac{1}{2}$. There is little doubt that the species is the NH_3^+ radical (see Sec. 4-6).

Figure 1-4 shows ESR spectra of species formed by γ irradiation of a single crystal of XeF_4 (see Sec. 4-6); again the number, spacing, and in-

---

† Exceptions are represented by some systems observed in the gaseous phase. Some molecules have a net *orbital* angular momentum but zero *spin* angular momentum. For example, the $^1\Delta$ state of O_2 has two units of orbital angular momentum about the internuclear axis but zero spin angular momentum. Such molecules may exhibit electron resonance in the gas phase; it would be inappropriate to refer to this phenomenon as electron *spin* resonance.

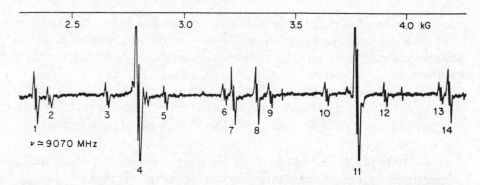

Fig. 1-4 First-derivative ESR spectrum of XeF trapped in a single crystal of XeF$_4$. Numbered lines will be discussed in Sec. 4-6. [*Spectrum kindly supplied by Dr. J. R. Morton; see J. R. Morton and W. E. Falconer, J. Chem. Phys.,* **39**:427 (1963).]

tensity of lines provide positive identification of one xenon and one fluorine atom, i.e., the XeF molecule. Here the positive identification of xenon comes from observation of lines arising from several of its isotopes.

Figure 1-5 shows the ESR spectrum of the CH$_3$ĊHOH radical produced as a transient species in the uv photolysis of a solution of H$_2$O$_2$ in ethanol (see Prob. 4-13*e*). This is an excellent example of the use of ESR

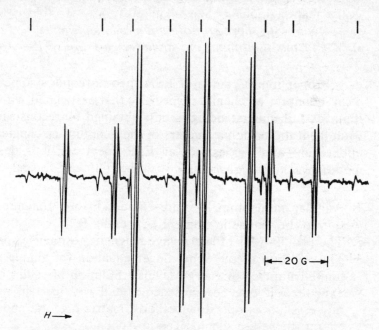

Fig. 1-5 First-derivative ESR spectrum of the CH$_3$ĊHOH radical produced by continuous ultraviolet photolysis of a mixture of H$_2$O$_2$ and CH$_3$CH$_2$OH. The weak lines which are indicated above the spectrum belong to the radical ĊH$_2$CH$_2$OH. [*Taken from R. Livingston and H. Zeldes, J. Chem. Phys.,* **44**:1245 (1966).]

spectra in the identification of radical intermediates in chemical reactions.

ESR spectra are by no means always readily interpretable; however, their complexity may provide important information about the structure or environment of a species.

Electron spin resonance spectroscopy is a technique applicable only to systems with net electron spin angular momentum; nevertheless, a re-respectable number of systems fulfill this condition. These include:

1. *Free radicals in the solid, liquid, or gaseous states.* A free radical is regarded as a molecule containing one unpaired electron.
2. *Some point defects (localized crystal imperfections) in solids.* Best known in this class is the *F* center, an electron trapped at a negative ion vacancy. Deficiency of an electron (a "positive hole") may give rise to a paramagnetic entity also. (See Sec. 8-5.)
3. *Biradicals.* These are molecules containing two unpaired electrons sufficiently remote from one another that interactions between them are very weak. The molecule behaves as two slightly interacting free radicals. (See Sec. 10-11.)
4. *Systems in the triplet state (two unpaired electrons).* Some of these molecules have a triplet ground state; others require excitation, either thermal or optical. (See Chap. 10.)
5. *Systems with three or more unpaired electrons.*
6. *Most transition-metal ions and rare-earth ions.* (See Chaps. 11 and 12.)

Proper interpretation of ESR spectra requires some understanding of basic quantum mechanics, especially that associated with angular momentum. A full understanding is best obtained by reconstruction of the spectrum from the basic parameters of the quantum-mechanical treatment. To understand and reconstruct an ESR spectrum, it is desirable to have a modest working acquaintance with the following topics:

1. Angular momentum, its representation by quantum numbers, and its relation to the magnetic moment (Appendix B).
2. The peculiarities of microwave magnetic resonance spectrometers.
3. The solution of the Schrödinger equation for simple systems having a limited number of energy levels. Solution of such problems is greatly expedited if one becomes acquainted with certain manipulative techniques such as operator methods, matrix algebra, and matrix diagonalization; these are summarized in Appendix A.

The elementary aspects of these topics will be treated where needed in the text or in appendices. *Even the reader who has had no previous training in quantum mechanics should be able to acquire some under-*

standing of the fundamentals of electron spin resonance. We shall undertake the development of the necessary background in a step-by-step fashion. Beyond this fundamental background, there are certain special areas of electron spin resonance which require particular background material:

1. Whereas the properties of many systems are independent of orientation in a magnetic field, i.e., they are *isotropic,* nevertheless there are other *anisotropic* systems for which the magnitude of the observable properties depends strongly on orientation. The description of systems showing anisotropic behavior usually requires that six independent parameters be specified. It is convenient to order these parameters in a symmetric 3×3 array known as a *tensor.* Simple examples of tensors are given in Appendix A, and numerous other examples will be encountered in the text.
2. Consideration of transition-metal ions requires a detailed knowledge of the splitting of orbital and spin levels by crystal fields of various symmetries.
3. For the interpretation of ESR spectra of organic π-electron free radicals, it is helpful to have some guidance in determining the distribution of the unpaired electron within the π-electron system. A useful procedure for getting first-order estimates of electron distributions in many hydrocarbon radicals is a molecular orbital approach due to Hückel (HMO approach, see Chap. 5). Without the benefit of simple computations, the interpretation of spectral parameters may be very difficult. Unfortunately, the HMO method may lead to erroneous results for certain classes of compounds. More refined theoretical interpretations are then required.
4. Time-dependent phenomena such as chemical or electron exchange, or molecular motions (such as internal rotation or reorientation by discrete jumps) can affect ESR spectra. An analysis of these effects leads to information about specific kinetic processes. However, even in the absence of these phenomena, there are time-dependent effects concerned with the lifetime of a spin state. These various phenomena are described in Chap. 9.

1-2 ENERGY OF MAGNETIC DIPOLES IN A MAGNETIC FIELD

There are many ESR spectra which are so simple and unambiguous in their interpretation that even a rudimentary understanding of the basic theory will permit an appreciation of the gross features. The first three chapters provide the groundwork for the interpretation of simple spectra.

For those who are not expert in the quantum mechanics of angular momentum, it is necessary to begin with a study of a set of noninteracting magnetic dipoles in a fixed magnetic field. These may represent either

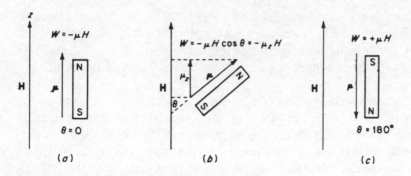

Fig. 1-6 Energy of a classical magnetic dipole in a magnetic field as a function of the angle θ between the magnetic field and the axis of the dipole. (a) $\theta = 0$ (configuration of minimum energy). (b) Arbitrary value of θ. (c) $\theta = 180°$ (maximum energy).

nuclear or idealized electronic dipoles. The magnetic dipole moment μ is defined by the expression

$$W = -\boldsymbol{\mu} \cdot \mathbf{H} = -\mu H \cos (\boldsymbol{\mu},\mathbf{H})\dagger \tag{1-1}$$

Here W is the energy of a magnetic dipole of moment μ in a field $\mathbf{H}$, and $(\boldsymbol{\mu},\mathbf{H})$ represents the angle between μ and $\mathbf{H}$.

The reader who is unfamiliar with the scalar product is referred to Sec. A-4. If the magnitude of μ is positive, then the energy is a minimum if the angle $(\boldsymbol{\mu},\mathbf{H})$ is zero. For a given value of $\mathbf{H}$, there is a maximum energy, $W_{max} = +\mu H$, which occurs when $(\boldsymbol{\mu},\mathbf{H})$ is equal to π, i.e., the dipole is antiparallel to the direction of $\mathbf{H}$ (Fig. 1-6). The units of μ are those of energy divided by magnetic field (ergs per gauss).

If the interaction energy $-\boldsymbol{\mu} \cdot \mathbf{H}$ in fields of a few thousand gauss and at 300 K were large compared with kT, practically all dipoles would be aligned along the direction of $\mathbf{H}$ (corresponding to the case of minimum energy). This would give a macroscopic magnetic moment per unit volume $\mathcal{M}$ (also called the magnetization) which would be approximately equal to $N\mu$, where N is the number of dipoles per unit volume. However, since $\mu H/kT << 1$ except at very low temperatures and at very high field strengths, $\mathcal{M}$ will ordinarily be several orders of magnitude less than $N\mu$, even for the relatively strong electronic magnetic dipoles.

The magnetization is obtained by measuring the force on a sample in an inhomogeneous magnetic field. $\mathcal{M}$ is related to the applied field by a proportionality factor χ, the susceptibility, which for a magnetically dilute sample is given in the simplest case by the expression

† When the directional aspects of vector quantities are being considered, we shall designate them with boldface type. When only magnitudes are involved, we shall employ italic type.

$$\chi = \frac{\mathscr{M}}{H} = \frac{N\mu^2}{3kT} \,\dagger \qquad\qquad (1\text{-}2)$$

The simplest case corresponds to a system with one unpaired electron and zero orbital angular momentum. The experimental determination of χ yields only the product $N\mu^2$; to obtain μ^2 one must determine N from other data. ESR measurements allow N and μ to be determined independently.

1-3 QUANTIZATION OF ANGULAR MOMENTUM

Although the classical description would allow the angle $(\mathbf{\mu},\mathbf{H})$ to assume any value, electrons and nuclei obey *quantum* rather than *classical mechanics*. A useful analogy to the angular-momentum properties of electrons and nuclei is provided by the behavior of a particle of mass m restricted to motion in a ring.‡ The particle has a linear momentum $\mathbf{p} = m\mathbf{v}$ if its instantaneous velocity is $\mathbf{v}$. As a quantum-mechanical particle, it has an associated deBroglie wavelength $\lambda = h/p$. The square of the amplitude of this wave at any point around the ring is a measure of the probability of finding the particle at that point. For this probability to be time-independent, the wave function must be single-valued. That is, the wave must not interfere destructively with itself in propagating about the ring. This requires the circumference to be an integral number M of deBroglie wavelengths. That is,

$$2\pi r = M\lambda = M\,\frac{h}{p}$$

Hence

$$pr = p_\phi = M\,\frac{h}{2\pi}$$

The quantity p_ϕ is the magnitude of the angular momentum of the particle; thus p_ϕ is required to be an integral multiple of $h/2\pi\ (\equiv \hbar)$, with M taking the values 0, 1, 2, 3, . . . , etc. The fact that p_ϕ refers to a fixed axis— which will be designated as the z axis—may be emphasized by using the symbol p_{ϕ_z} to refer to the angular momentum about the z axis.

This very simple model finds a direct application in the case of the angular momentum of electrons about the axis of a diatomic molecule.

† The units of $\mathscr{M}$ are ergs per gauss per cubic centimeter which is equivalent to the units of gauss. This can be seen by substituting $g^{\frac12}\,cm^{-\frac12}\,s^{-1}$ as the units of gauss.

$$\frac{\mathrm{Ergs}}{\mathrm{G\ cm^3}} = \frac{\mathrm{g\ cm^2\ s^{-2}}}{g^{\frac12}\,cm^{-\frac12}\,s^{-1}\,cm^3} = g^{\frac12}\,cm^{-\frac12}\,s^{-1}$$

Hence χ is a dimensionless quantity.

‡ J. W. Linnett, "Wave Mechanics and Valency," chap. III, Methuen & Co., Ltd., London, 1960.

When $M = 0$, *the electrons are said to be in a σ orbital, whereas when*
$M = 1$, *they are said to be in a π orbital.* For most stable free radicals,
the unpaired electron resides in a π orbital.

The important conclusion resulting from the examination of the particle in a ring is the *quantization of angular momentum.* However, the
"derivation" using the deBroglie relation is valid only when, as here, the
potential energy is constant. Another deficiency of the deBroglie derivation is that it leaves one uncertain as to the direction of $\mathbf{p}_\phi$.

For atoms, electronic motion is not restricted to a plane; the corresponding model for orbital angular momentum is the motion of a particle
on the surface of a sphere. This model[†] predicts the quantization of the
total orbital angular momentum, as well as of its component along any direction fixed in space. The allowed values of the orbital angular momentum
for a *single* electron will be of the form $[l(l + 1)]^{\frac{1}{2}}\hbar$, where l take the values
$0, 1, 2, 3, \ldots$, etc. The successive values of the orbital angular-momen-
tum quantum number l of a single electron are represented for atomic spectra
by the spectroscopic symbols $s, p, d, f, \ldots$, etc. The corresponding designations for the *total* orbital angular-momentum quantum number $L = 0, 1,$
$2, 3, \ldots$ are $S, P, D, F, \ldots$, etc.

Whereas the orbital angular momentum of a *single* electron is
$[l(l + 1)]^{\frac{1}{2}}\hbar$, the allowed values along any direction *fixed in space* are $M_l\hbar$,
just as for the particle in the ring. The allowed values of M_l are the inte-
gers $-l, -l + 1, \ldots, l - 1, l$. There are then $2l + 1$ allowed values of M_l.

The solutions to the problem of the particle on the surface of a
sphere are *spherical harmonics;* these functions (which are characterized by l and M_l) describe the angular dependence of $s, p, d, f \ldots$
wave functions.[‡]

The allowed values of the magnitude of the electron spin angular momentum are given by $[S(S + 1)]^{\frac{1}{2}}\hbar$, where by convention S now represents
the spin quantum number. (The significance of S will usually be clear from
the context.) The components of the spin angular momentum vector in a
space-fixed direction are restricted to the values $M_S\hbar$. The value of S which
is to be substituted in this expression is $\frac{1}{2}$ for a single electron; for concise-
ness one says that the spin is $\frac{1}{2}$. For systems of two or more unpaired electrons the maximum value of the spin quantum number M_S is $1, \frac{3}{2}, 2$, etc.
S is equal to the maximum value of M_S. The allowed values of M_S range
in unit increments from $-S$ up to $+S$, giving $2S + 1$ components along an
arbitrary direction. The spin angular momentum vectors and their pro-

[†] J. W. Linnett, "Wave Mechanics and Valency," chap. III, Methuen & Co., Ltd., London,
1960.

[‡] L. Pauling and E. B. Wilson, "Introduction to Quantum Mechanics," p. 133, McGraw-Hill
Book Company, New York, 1935.

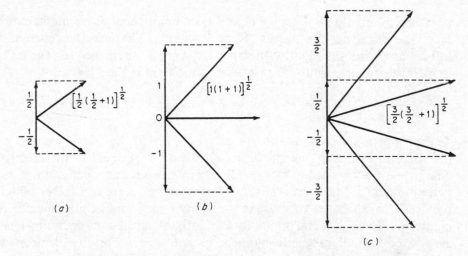

Fig. 1-7 Allowed values of the total spin angular momentum $[S(S + 1)]^{\frac{1}{2}}$ and of the component M_S (in units of $\hbar$) in a fixed direction for (a) $S = \frac{1}{2}$, (b) $S = 1$, and (c) $S = \frac{3}{2}$.

jected components for $S = \frac{1}{2}$, $S = 1$, and $S = \frac{3}{2}$ are represented in Fig. 1-7. The case $S = \frac{1}{2}$ is certainly of most interest for free radicals, whereas $S = 1$ is descriptive of triplet states (Chap. 10). For paramagnetic ions, especially those of the transition-metal and rare-earth elements, states with $S > \frac{1}{2}$ are very common.

Nuclear spin angular momentum is also quantized; the nuclear spin quantum number is designated by I, which may be half-integral or integral.

The orbital and the spin angular momenta have been considered separately; it is important to know the extent to which these are coupled. Hopefully, to a first approximation the two may be considered independently, later introducing a small correction to account for the so-called "spin-orbit" interaction. For free radicals which have essentially zero orbital angular momentum, the spin-orbit interaction is usually very small; hence, for most purposes attention may be focused wholly upon the spin angular momentum. However, spin-orbit interaction must necessarily be included in a discussion of the ESR behavior of transition-metal and rare-earth ions; reference will be made to this interaction in Chaps. 11 and 12.

1-4 RELATION BETWEEN MAGNETIC MOMENTS AND ANGULAR MOMENTA

The quantization of angular momentum has been considered; however, it is the value of the magnetic dipole moment which is of principal interest. Magnetic moment and angular momentum are proportional, both in classical and in quantum mechanics. An analog of an orbital magnetic dipole is a particle of mass m and charge q (expressed in electrostatic units), rotating with velocity $\mathbf{v}$ in a circle of radius r in the xy plane. Associated with a circulating current i is a magnetic field equivalent to that produced by a point magnetic dipole. Such a dipole has a component of magnetic moment $\mu_z = iA$ normal to the plane. Here $A = \pi r^2$. The effective current i is $(q/c)(v/2\pi r)$; the division by the speed of light c is required for conversion to electromagnetic units. The magnetic moment is then given by

$$\mu_z = \frac{q}{c}\frac{v\pi r^2}{2\pi r} = \frac{q}{2mc}\,mvr = \frac{q}{2mc}\,p_{\phi_z} = \gamma p_{\phi_z} = \gamma M_l \hbar \tag{1-3}$$

The proportionality constant $\gamma = q/2mc$ is called the *magnetogyric ratio* (or sometimes the gyromagnetic ratio). If q is set equal to the electronic charge $-e$, the classical factor $-e/2mc$ converts angular momentum to magnetic moment. More generally, $\gamma = -ge/2mc$, where g is a factor which is required for all cases other than those involving pure orbital angular momentum. Each integral multiple $\hbar$ of orbital angular momentum has an associated orbital magnetic moment of $eh/4\pi mc$. For the electron, this ratio of constants is represented by β and is called the Bohr magneton ($\beta = 9.2741 \times 10^{-21}$ erg gauss^{-1}). (See Table A, inside front cover.) However, for electron spin (which has no classical counterpart), the value of g is very close to 2 (for a free electron $g_e = 2.00232$). The component of electron spin magnetic moment μ_z along the direction of the magnetic field $\mathbf{H}$ is

$$\mu_z = \gamma M_S \hbar = -g\beta M_S \tag{1-4}$$

The negative sign arises because of the negative charge of the electron.

The quantization of spin angular momentum in a specified direction leads to the quantization of energy levels of a system of magnetic dipoles in a magnetic field. Application of the expression $W = -\mu_z H$ to a "spin-only" system and substitution of $-g\beta M_S$ for μ_z gives a set of energies

$$W = g\beta H M_S \tag{1-5}$$

The possible values of M_S are $+\frac{1}{2}$ and $-\frac{1}{2}$. Hence the two possible values of W are $\pm\frac{1}{2}g\beta H$. These are sometimes referred to as the Zeeman energies. The *separation* $g\beta H$ between the Zeeman levels increases linearly with the magnetic field, as is shown in Fig. 1-8.

1-5 INTERACTION OF MAGNETIC DIPOLES WITH ELECTROMAGNETIC RADIATION

Transitions between the two Zeeman levels can be induced by an electromagnetic field of the appropriate frequency ν, if the photon energy $h\nu$ matches the energy-level separation ΔW. Then

$$\Delta W = h\nu = g\beta H_r \tag{1-6}$$

ν is expressed in hertz (Hz),† and H_r‡ is the magnetic field at which the

---

† By international convention, 1 cps (cycle per second) = 1 Hz (hertz).
‡ The notation for particular magnetic fields is explained in the table of symbols following Appendix D.

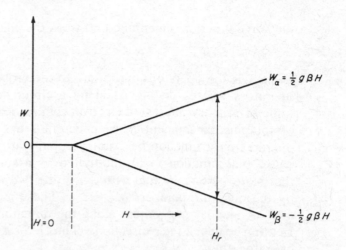

Fig. 1-8 Energy-level scheme for the simplest system showing ESR absorption. W_α and W_β represent the energies of the $M_S = +\frac{1}{2}$ and the $M_S = -\frac{1}{2}$ states, respectively.

resonance condition, Eq. (1-6), is met. Equation (1-6) implies that M_S of Eq. (1-5) increases by one unit, i.e., $\Delta M_S = +1$. The upward and downward transitions indicated in Fig. 1-8 correspond to $\Delta M_S = +1$ and $\Delta M_S = -1$, respectively.

In practice, the remarkably simple proportionality between the irradiation frequency and the resonant field strength is maintained from tens of gauss to tens of kilogauss. For a free electron $g = 2.00232$; hence, at a frequency of 30 MHz (megahertz) the resonant field is 10.7 G; at 30 GHz (gigahertz)(1 GHz = 10^9 Hz) $H_r = 10{,}700$ G. A host of free radicals or transition-metal ions do have $g \approx 2$, but there are also systems which show marked deviations from this value.

The first ESR experiment was carried out in 1944, when Zavoisky[†] detected a peak in the paramagnetic absorption from $CuCl_2 \cdot 2H_2O$. He found a resonant field value of 47.6 G for a frequency of 133 MHz. In this case $g \approx 2$.

It is instructive to use operator methods to rederive the resonance equation [Eq. (1-6)] for a system with $S = \frac{1}{2}$. This enables one to acquire an acquaintance with a technique which is applicable to more complicated systems. (The algebraic manipulation of operators is briefly described in Sec. A-2. General properties of spin operators are given in Appendix B. Their application to the hydrogen atom and the $\cdot RH_2$ radical are given in Appendix C.)

For a system having discrete energy levels described by well-defined quantum numbers, it is always possible to write an eigenvalue equation. That is, if λ_i represents an eigenvalue of a state for which

† E. Zavoisky, *J. Phys. U.S.S.R.*, **9**:211, 245 (1945).

the wave function (eigenfunction) is ψ_i, the eigenvalue equation is

$$\hat{\Lambda}\psi_i = \lambda_i\psi_i \tag{1-7}$$

Here $\hat{\Lambda}$ is the operator appropriate to the property under study. (All operators will be distinguished by a circumflex in this book.) The topic of primary interest in electron spin resonance is the quantization of spin angular momentum. Hence one seeks a *spin operator*, which operates on a function describing a spin state. This operator should cause that function to be multiplied by a constant characteristic of that state. For a system with $S = \frac{1}{2}$, the two states are characterized by the quantum numbers $M_S = \pm\frac{1}{2}$. These measure the components $M_S\hbar$ of angular momentum along the direction of the magnetic field. Let this be called the z direction. Thus if $\hat{S}\hbar$ is the angular momentum operator, then

$$\hat{S}_z\psi_i = M_S\psi_i \tag{1-8}$$

or

$$\hat{S}_z\psi(M_S = +\tfrac{1}{2}) = +\tfrac{1}{2}\psi(M_S = +\tfrac{1}{2})$$

and   (1-9)

$$\hat{S}_z\psi(M_S = -\tfrac{1}{2}) = -\tfrac{1}{2}\psi(M_S = -\tfrac{1}{2})$$

The symbolism for the representation of an eigenfunction can readily be simplified. Since the functions are distinguished by their quantum numbers, one may enclose these numbers in a distinctive way to represent the function. Dirac suggested the notation $|n\rangle$ for an eigenfunction. (A function represented in such a way is called a "ket.") (See Sec. A-5d.) Then Eqs. (1-9) are rewritten as

$$\hat{S}_z|+\tfrac{1}{2}\rangle = +\tfrac{1}{2}|+\tfrac{1}{2}\rangle$$

and   (1-10)

$$\hat{S}_z|-\tfrac{1}{2}\rangle = -\tfrac{1}{2}|-\tfrac{1}{2}\rangle$$

A still simpler notation uses the symbols α and β for states with $M_S = +\frac{1}{2}$ and $-\frac{1}{2}$, respectively, that is,

$$|+\tfrac{1}{2}\rangle \equiv |\alpha\rangle \equiv \psi_\alpha$$
$$|-\tfrac{1}{2}\rangle \equiv |\beta\rangle \equiv \psi_\beta$$

The energies W_i of systems for which M_S is a precise measure of the components of spin angular momentum are obtained from the expression

$$\hat{\mathscr{H}}\psi_i = W_i\psi_i \tag{1-11}$$

Here $\hat{\mathscr{H}}$ is the operator for the energy; it is called the spin hamiltonian operator. The importance of Eq. (1-11) taken together with Eqs. (1-9) is that *the same ψ_i is an eigenfunction of the z component of spin* *angular momentum and of the energy.*† Hence

$$\hat{\mathcal{H}}|\alpha\rangle = W_\alpha|\alpha\rangle$$

and   (1-12)

$$\hat{\mathcal{H}}|\beta\rangle = W_\beta|\beta\rangle$$

Since $W = -\boldsymbol{\mu} \cdot \mathbf{H} = -\mu_z H$, one requires a relation between the magnetic moment and the spin angular momentum. It has been shown in Eq. (1-4) that $\mu_z = \gamma M_S \hbar$. Thus one may expect that the magnetic moment *operator* $\hat{\mu}_z$ should be proportional to the spin *operator* $\hat{S}_z$. That is,

$$\hat{\mu}_z = \gamma \hat{S}_z \hbar = -g\beta \hat{S}_z \tag{1-13}$$

Equations (1-1) and (1-13) may be combined to give the spin hamiltonian operator $\hat{\mathcal{H}}$ as

$$\hat{\mathcal{H}} = g\beta H \hat{S}_z \tag{1-14}$$

Then

$$\hat{\mathcal{H}}|\alpha\rangle = g\beta H \hat{S}_z|\alpha\rangle$$

$$= \frac{g\beta H}{2} |\alpha\rangle \tag{1-15a}$$

and

$$\hat{\mathcal{H}}|\beta\rangle = \frac{-g\beta H}{2} |\beta\rangle \tag{1-15b}$$

One may infer from Eqs. (1-15) that

$$W_\alpha = \frac{g\beta H}{2} \tag{1-16a}$$

and

$$W_\beta = \frac{-g\beta H}{2} \tag{1-16b}$$

Thus $\Delta W = W_\alpha - W_\beta = g\beta H_r = h\nu$ for a transition between states $|\alpha\rangle$ and $|\beta\rangle$ as noted in Eq. (1-6).

A more general procedure for determining the energy W_i in Eqs. (1-16) involves multiplication of both sides of Eq. (1-11) from the left by ψ_i^* ‡:

$$\psi_i^* \hat{\mathcal{H}} \psi_i = \psi_i^* W_i \psi_i$$

$$= W_i \psi_i^* \psi_i \qquad \text{since } W_i \text{ is a constant} \tag{1-17}$$

---

† This is a general property of linear operators $\hat{A}$ and $\hat{B}$ which commute (see Sec. A-2), that is, $\hat{A}\hat{B} - \hat{B}\hat{A} = 0$.

‡ ψ_i^* is the complex conjugate of ψ_i. (See Sec. A-1.)

Multiplication of both sides by $d\tau$ (where τ represents one or more variables of integration) and integration over the full range of the variables τ gives

$$\int_\tau \psi_i^* \hat{\mathcal{H}} \psi_i \, d\tau = W_i \int_\tau \psi_i^* \psi_i \, d\tau \qquad (1\text{-}18)$$

Hence

$$W_i = \frac{\displaystyle\int_\tau \psi_i^* \hat{\mathcal{H}} \psi_i \, d\tau}{\displaystyle\int_\tau \psi_i^* \psi_i \, d\tau} \qquad (1\text{-}19)$$

If the functions ψ_i are normalized, i.e., if they satisfy the condition

$$\int_\tau \psi_i^* \psi_i \, d\tau = 1 \qquad (1\text{-}20)$$

then

$$W_i = \int_\tau \psi_i^* \hat{\mathcal{H}} \psi_i \, d\tau \qquad (1\text{-}21)$$

It is appropriate to rewrite Eqs. (1-17) to (1-21) in the Dirac notation used in Eqs. (1-12) and (1-15). The symbol appropriate to multiplication from the left by ψ_i^* is $\langle \psi_i |$. (Dirac called this function a "bra.") When $\langle \psi_i |$ is combined with $| \psi_i \rangle$, integration over the full range of all variables is implied. Thus the combination $\langle \psi_i | \psi_i \rangle$, i.e., *bra[c]ket*, suggests the origin of the notation. Then Eq. (1-18) becomes

$$\langle \psi_i | \hat{\mathcal{H}} | \psi_i \rangle = W_i \langle \psi_i | \psi_i \rangle \qquad (1\text{-}22)$$

For normalized functions

$$\langle \psi_i | \psi_i \rangle = 1$$

and hence

$$W_i = \langle \psi_i | \hat{\mathcal{H}} | \psi_i \rangle \qquad (1\text{-}23)$$

For states $|\alpha\rangle$ and $|\beta\rangle$ of our simple problem, one writes

$$W_\alpha = \langle \alpha | \hat{\mathcal{H}} | \alpha \rangle = \tfrac{1}{2} g \beta H \qquad (1\text{-}24a)$$

and

$$W_\beta = \langle \beta | \hat{\mathcal{H}} | \beta \rangle = -\tfrac{1}{2} g \beta H \qquad (1\text{-}24b)$$

At this point, the reader who is unfamiliar with wave functions and their manipulation is urged to turn to Sec. A-2b, where the problem of the particle in a ring is considered both in terms of the angular-momentum operator and the energy operator.

The transitions between the Zeeman levels involve a change in the orientation of the electron magnetic moment. Hence, transitions can occur only if the electromagnetic radiation brings about such a reorientation. Assume that the electromagnetic radiation is polarized such that the oscillating magnetic field is oriented parallel to the static magnetic field. Then the effect of the radiation will be to cause an oscillation in the *energies* of the Zeeman levels, through Eq. (1-5). However, no *reorientation* of the electron magnetic moment will occur. In this case no transitions are possible. To make transitions possible, the electromagnetic radiation must be polarized such that the oscillating magnetic field has a component *perpendicular* to the static magnetic field. (The justification of this statement will be given in Sec. C-4.) The requirement of an oscillating perpendicular magnetic field is easily met at microwave frequencies.

From Eq. (1-6) one may infer that there are two approaches to the detection of resonant absorption by a paramagnetic sample. In the first case the separation of the Zeeman levels is fixed by holding the magnetic field constant; the microwave frequency is then varied until a resonant absorption is found. In the second case the microwave frequency is fixed; the magnetic field is then varied. For experimental reasons, the second method is used. *The characteristic aspect of ESR (and of NMR) spectroscopy is the variation of the energy-level separation by variation of the magnetic field.* In other branches of molecular spectroscopy, the molecular energy levels are fixed and the first method must be used.

Everything that has been said about electron spin energy levels and transitions is also applicable to nuclear spin systems. Indeed, six years before any ESR experiment was conducted on a sample in a condensed phase,[†] nuclear spin transitions were being induced in atomic or molecular beams by radio-frequency fields.[‡] It was soon shown[§] that the method was generally applicable for transitions between spin energy levels. The nuclear Zeeman levels are given by an expression analogous to Eq. (1-5), namely, $W = -g_N\beta_N H M_I$; g_N is the nuclear g factor, $\beta_N = eh/4\pi m_N c$, m_N is the proton mass, and M_I is the component of the nuclear spin angular momentum vector in the z direction. In analogy to the electron-spin case, only transitions for which $\Delta M_I = \pm 1$ are allowed; hence

$$\Delta W = h\nu = g_N\beta_N H_r \tag{1-25}$$

The phenomenon for nuclei is commonly referred to as nuclear magnetic resonance.

1-6 CHARACTERISTICS OF THE g FACTOR

The g factor in the simple resonance equation

$$g = \frac{h\nu}{\beta H_r} \tag{1-26}$$

† E. Zavoisky, *J. Phys. U.S.S.R.*, **9**:211, 245 (1945).
‡ I. I. Rabi, S. Millman, P. Kusch, and J. R. Zacharias, *Phys. Rev.*, **55**:526 (1939).
§ J. M. B. Kellogg, I. I. Rabi, N. F. Ramsey, Jr., and J. R. Zacharias, *Phys. Rev.*, **57**:677 (1940).

is independent of field direction only in isotropic systems. For example, an electron in a negative-ion vacancy (F center) in an alkali halide is at the center of a regular octahedron of cations. The g factor is isotropic, as are other properties of a system with local octahedral symmetry.

There are examples of systems for which the g factor is sufficiently distinctive so as to provide a reasonable identification of the paramagnetic species. Consider the spectrum of x-irradiated MgO shown in Fig. 1-9 for a resonant frequency $\nu = 9.41756$ GHz. We seek to establish the origin of the very intense line which obscures the sixth member of the octet arising from $^{59}Co^{++}$, for which $g = 4.2785$. Substitution of the value 1629.06 G for the magnetic field at the center of the intense line gives its g factor as

$$g = \frac{h\nu}{\beta H_r} = \frac{(6.62620 \times 10^{-27} \text{ erg s})(9.41756 \times 10^9 \text{ Hz})}{(9.27410 \times 10^{-21} \text{ erg G}^{-1})(1629.06 \text{ G})} = 4.1304$$

A g factor of this magnitude is unusual, and it gives an important clue as to the ion responsible for the intense line. It is generally observed that isoelectronic ions (i.e., ions which have the same electronic configuration) in environments of similar symmetry have similar g factors. An ion which is isoelectronic with the $3d^7$ Co^{++} ion is Fe^+. Since iron in natural abundance consists mostly (>97.7 percent) of nuclides of zero spin, a single line is expected for Fe^+ (except when the line is narrow, since the nuclear moment of ^{57}Fe is one of the smallest of any nuclide for which $I = \frac{1}{2}$). Con-

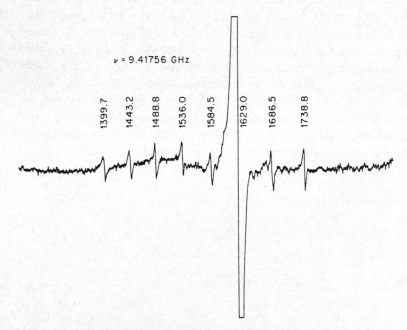

$\nu = 9.41756$ GHz

1399.7 1443.2 1488.8 1536.0 1584.5 1629.0 1686.5 1738.8

Fig. 1-9 First-derivative ESR spectrum of Fe^+ and Co^{++} in MgO at 4.2 K. The Fe^+ spectrum consists of a single intense line while the Co^{++} spectrum is a hyperfine octet due to ^{59}Co with $I = \frac{7}{2}$. (*Spectrum kindly supplied by Mr. Fricis Dravnieks.*)

sidering the large deviation of both the g factors 4.279 and 4.1304 from the free-spin g factor 2.0023, the two magnitudes may be considered similar enough to be due to isoelectronic ions. Hence the intense line is assigned to Fe^+. The disappearance of the Fe^+ line on heating the crystal to 400 K is consistent with expectation for an unstable valence state. It should be remarked that ESR lines for both the Fe^{++} and the Fe^{3+} ions may be observed in these crystals. It is typical of isoelectronic ions in an environment of similar symmetry that their ESR spectra are observable under comparable experimental conditions. Neither for Co^{++} nor for Fe^+ does one observe a resonance line at 77 K; one does find strong absorption for both at 20 K and lower. This similarity is confirmatory evidence for the identification of Fe^+. Inability to see lines at room temperatures or even at 77 K will be shown in Chap. 9 to be due to excessive broadening of lines as a result of their very short relaxation times.

There are other magnetically isotropic systems such as dilute liquid solutions of low viscosity; their isotropic behavior is the result of rapid, random reorientation of solute molecules. However, if these solutions are frozen or even cooled sufficiently, the ESR spectrum may consist of a broad unstructured line. An asymmetric ESR line suggests that the individual molecules responsible for it have anisotropic magnetic properties. At this point, it may be helpful to some readers to cite a more familiar physical property. They may be aware that the magnetic susceptibility of an anisotropic crystal is a function of its orientation in the magnetic field. For example, the absolute value of the susceptibility measured with the field perpendicular to the layer plane in graphite is many times as great as that with the field parallel to the plane. However, to describe the susceptibility quantitatively, it is not necessary to have an indefinitely large number of parameters. Any anisotropic system, however low its symmetry, has three mutually perpendicular directions (principal axes) such that the susceptibility values (principal values) measured along these directions succinctly define the susceptibility properties. (See Sec. A-6.) Analogous statements may be made about the optical properties (e.g., optical absorption behavior or refractive index) of an anisotropic crystal.

The description of the resonance properties of an anisotropic system is closely analogous. The resonant field value is a function of the field orientation relative to crystal (or molecular) axes; the angular dependence is attributed to a variation in the g factor. Hence it is customary to append subscripts on g to specify the field orientation defining it. If principal axes in the molecule are labeled X, Y, and Z,[†] g_{XX} is to be interpreted for our simple case as $h\nu/\beta H_X$, that is, the g factor for $\mathbf{H}$ along the X axis of the molecule.

[†] Henceforth, x, y, and z will be used for laboratory-fixed axes and X, Y, and Z for axes fixed with respect to the paramagnetic species.

It is apparent that a truly isotropic system is one for which

$$g_{XX} = g_{YY} = g_{ZZ}$$

For paramagnetic species in a liquid system of low viscosity the measured, apparently isotropic, g factor is to be regarded as an effective value averaged over all orientations.

Some systems will have threefold, fourfold, or sixfold axes of symmetry. Systems with an n-fold axis of symmetry ($n \geqslant 3$) are described as having *axial* symmetry. For these, X and Y are equivalent. The unique axis is usually designated as Z, and the value of g for $\mathbf{H} \parallel Z$ will be called $g_{\parallel}$. The g factor for $\mathbf{H}$ in the XY plane, that is, $\mathbf{H} \perp Z$, will have the constant value $g_{\perp}$. Simple cases exhibiting axial symmetry will be considered in Chap. 7.

PROBLEMS

1-1. Explain why one might wish to perform an ESR experiment on a free-radical system in preference to determining its magnetic susceptibility.

1-2. What would be the numerical value of the angular momentum of an electron rotating in an orbit of 1 Bohr radius (0.529×10^{-8} cm) with a frequency ν of 10^{13} Hz? Note that the magnitude of the linear velocity v of a rotating particle is given by $v = r\omega$, where $\omega = 2\pi\nu$.

1-3. The separation of two lines (splitting) in a free-radical spectrum is given as 75.0 MHz and $g = 2.0050$. Express the splitting in gauss and in reciprocal centimeters.

1-4. Is it possible to obtain ESR spectra with NMR equipment? Assuming $g = 2.0050$, what magnetic field would be required for $\nu = 60$ MHz?

1-5. A classical rotating magnetic dipole placed in a magnetic field will precess about the magnetic field direction with an angular frequency $\omega = 2\pi\nu$ which is given by $\omega = \gamma H$.

    (*a*) What is the magnetogyric ratio γ for a free electron?

    (*b*) At what frequency will it precess in a field $H = 3,500$ G?

    (*c*) What would be the value of γ for an electron trapped in a negative ion vacancy in KBr? $g = 1.985$.

1-6. Calculate the ratio of resonant frequencies of a free electron and a deuteron in the same magnetic field.

1-7. From the data in Table C (inside back cover) compute the ratio of NMR frequencies at a magnetic field of 10^4 G for hydrogen (^{1}H) and for deuterium (^{2}H) atoms.

2
Basic Instrumentation of Electron Spin Resonance

2-1 A SIMPLE ESR SPECTROMETER

The detection of electron spin resonance absorption in a system containing unpaired electrons requires a spectrometer which includes a static magnetic field.   As with any spectrometer, one requires a source of radiation and some means of detecting absorption by the sample.   The simplest experimental arrangement which fulfills these requirements was indicated in Fig. 1-2b.   From this figure one notes that there are two important differences between an ESR and an optical spectrometer.   First, the microwave source, a klystron, emits monochromatic radiation; hence a dispersing element such as a prism or a grating (i.e., a monochromator) is not required.   Second, an ESR spectrometer operates at a fixed microwave frequency and scans an ESR spectrum by a linear variation of the static magnetic field.   Such operation is possible because the spacing of energy levels can be altered by varying the magnetic field.   It is extremely fortunate that this is true, since it is ordinarily very difficult to achieve high sensitivity if the microwave frequency is varied.   The difficulty arises largely from the fixed-frequency characteristics of the microwave-resonant cavity (Sec. 2-3a).

Absorption lines will be observed in the spectrum when the separa-

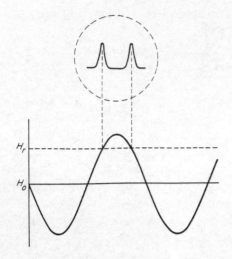

Fig. 2-1 Detection of ESR absorption by large-amplitude field modulation. When $H_0 \approx H_r$, the paramagnetic system will pass through the resonance condition $2\nu_m$ times per second if the field is modulated with sufficient amplitude at the frequency ν_m. The detected signal may be displayed on an oscilloscope as shown above.

tion of two energy levels is equal to the quantum energy $h\nu$ of the incident microwave photons. The absorption of these photons by the sample in Fig. 1-2*b* will be indicated by a change in the detector current.

The direct detection of the absorption signal, as in Fig. 1-2*b*, is possible only for samples containing a high concentration of unpaired electrons; noise components over a wide range of frequencies appear with the signal, making its detection difficult. In the optical spectrometer, the signal-to-noise ratio may be improved by chopping the light beam at a preselected frequency. This permits narrow-band amplification of the detected signal; hence, noise components are limited to those in a narrow band centered at the chopping frequency.

In a magnetic resonance spectrometer, the analog of a light chopper is the field modulator. This device superimposes an alternating component on the static field H_0 such that the magnetic field passes periodically through the resonant field H_r. The useful signal at the detector is an alternating voltage at the modulation frequency. This signal may be amplified in a narrow-band amplifier. If the amplified signal is displayed on an oscilloscope which is swept at the modulation voltage, a trace such as that in Fig. 2-1 is obtained. For such display, the amplitude of the field modulation should be several times the width of a line to be observed.

2-2 CHOICE OF EXPERIMENTAL CONDITIONS

In principle, the resonance condition $h\nu = g\beta H$ is valid for any frequency. ESR absorption has been detected from magnetic fields of a few gauss ($\nu \sim 3$ MHz) to fields as large as 20,000 G ($\nu \sim 60$ GHz).† A sensitive magnetometer used by geologists utilizes ESR absorption to measure the earth's magnetic field which is only $\simeq 0.5$ G. There are several considera-

† If there is an internal magnetic field, as from nuclei, one may detect resonance at zero applied field. [See T. Cole. T. Kushida, and H. C. Heller, *J. Chem. Phys.*, **38**:2915 (1963).]

tions which limit the choice of the radiation frequency. The primary concern is one of sensitivity; this requirement dictates that the frequency be as high as possible, since the sensitivity of an ESR spectrometer increases approximately as ν^2. Three factors place a limit on the microwave frequency employed. One is the size of the sample; at high frequencies (~ 30 to 40 GHz), the microwave-resonant-cavity dimensions are of the order of a few millimeters. Thus, although the sensitivity *per unit volume* is high, the sample volume is limited to about 0.02 cm³. Second, high frequencies require high magnetic fields homogeneous over the sample volume. With conventional magnets, sufficiently homogeneous magnetic fields in excess of 25,000 G are difficult to produce. Superconducting magnets may produce fields of the order of 100,000 G. Third, for aqueous samples, dielectric absorption seriously impairs sensitivity as the frequency increases. These and other factors have resulted in a choice of about 9.5 GHz as the working frequency of most commercial spectrometers. Radiation at this frequency is readily propagated in a so-called X-band waveguide; this waveguide is appropriate to the frequency range of 8.2 to 12.4 GHz. For single crystals and for samples with a low dielectric loss, one may find it profitable to work at about 35 GHz; this frequency falls within the range 33.0 to 50.0 GHz, referred to as the Q band.

2-3 TYPICAL SPECTROMETER ARRANGEMENT

Modern spectrometers are designed to achieve a high sensitivity; numerous instruments now approach a sensitivity close to the theoretical limit. (See Sec. D-1.) A typical spectrometer is blocked out in Fig. 2-2 according to the function of groups of components. The region labeled "source" contains those components which control or measure the frequency and the intensity of the microwave beam. The "cavity system" includes the components which hold the sample and which direct and control the microwave beam to and from the sample. The "detection" and "modulation systems" monitor, amplify, and record the signal. Finally, the "magnet system" provides a stable, linearly variable and homogeneous magnetic field of arbitrary magnitude. The functions of the individual components within each of the blocks will now be considered. Because of its central importance, the resonant cavity will be considered first.

2-3a. The cavity system The heart of an ESR spectrometer is the resonant cavity containing the sample. The reader is doubtless familiar with the phenomenon of acoustic resonance in organ pipes or in nearly enclosed vessels. The latter are referred to as cavity resonators; their properties were described in detail by Helmholtz.† Reflection of sound waves from walls leads to destructive interference at wavelengths which are not sub-

† H. L. F. Helmholtz, "On the Sensations of Tone," transl. by A. J. Ellis, pp. 68, 579, Longmans, Green & Co., London, 1875.

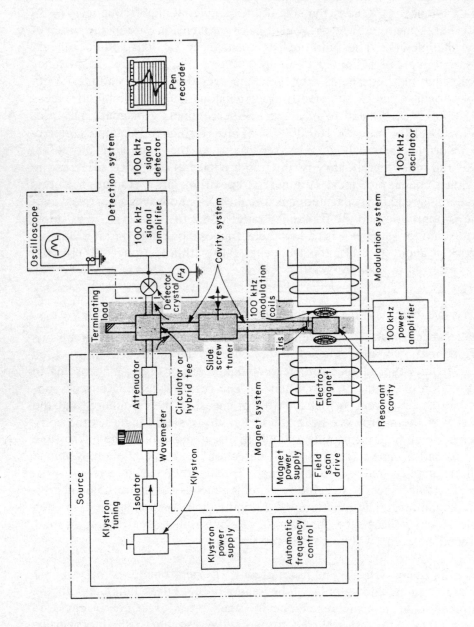

Fig. 2-2 Block diagram of a typical X-band ESR spectrometer employing 100-kHz phase-sensitive detection.

multiples of one dimension of such resonant cavities. The frequency at which one-half of a wavelength corresponds to a cavity dimension is called the *fundamental resonant frequency;* this frequency decreases with increasing dimensions of the cavity. A cavity resonator *may* be excited to produce more than one type of standing wave pattern or "mode." A cylindrical cavity may be made tunable by fitting a piston at one end; this property is utilized in the wavemeter (a tunable resonant cavity) to make it possible to measure wavelength.

The energy density associated with a traveling wave is usually small; however, considerable acoustical energy may be stored in the *standing waves* of a resonant cavity. One can readily observe acoustical resonance because the wavelengths of sound are in the range of a few centimeters to a few meters. For microwaves, the wavelength is also typically of the order of centimeters. Hence the dimensions of a resonant microwave cavity will be conveniently large. As with acoustical resonators, the choice of geometry is arbitrary. Unlike the acoustical case, one has both electric (E_1) and magnetic (H_1) fields to consider. The positions of maxima of E_1 and H_1 fields are different; their relative location depends on the mode. To be useful for electron spin resonance, a cavity mode should: (*a*) permit a high-energy density, (*b*) allow placement of the sample at a maximum of H_1, and (*c*) have H_1 perpendicular to the static field H. Cavities commonly employed for ESR spectrometers are shown in Figs. 2-3*a* and 2-4*a*. The modes of both are referred to (in North America) as transverse electric (TE); subscripts designate the number of half-wavelengths along the several dimensions. For the rectangular-parallelepiped cavity of Fig. 2-3*a*, the mode is TE_{102}, since there are, respectively, one and two half-waves along A and C; the fields do not vary along B. The spatial distribution of electric and of magnetic fields for the TE_{102} cavity may be seen from Figs. 2-3*b* and *c*, respectively. Such a cavity permits the use of large samples of low dielectric constant without drastic reduction in the energy density. It is especially useful for liquid samples, which may extend through the entire height of the cavity.

The cylindrical cavity of Fig. 2-4*a* may have a very high-energy density when operated in the TE_{011} mode shown in Figs. 2-4*b* and 2-4*c*. This energy density is typically higher by a factor of at least 3 compared to that in a TE_{102} cavity under comparable conditions. This cavity is especially useful for observing transitions in gaseous systems since very large (25 mm) diameter sample tubes may be used.

The "sharpness" of response of any resonant system is commonly described by a factor of merit, universally represented by the symbol Q. For example, the response curve of a parallel-tuned circuit in the range of resonance is given in Fig. 2-5. The voltage across the LC combination is a maximum at the resonant frequency ν_r, given by $1/(2\pi\sqrt{LC})$. If one determines the difference $\Delta\nu$ of the frequency values at the half-power points,

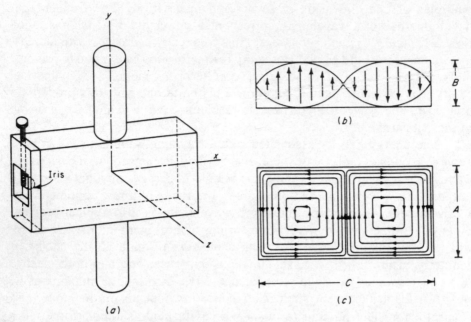

Fig. 2-3 A rectangular parallelepiped TE_{102} microwave cavity. (*a*) Cylindrical extensions above and below the cavity prevent excessive leakage of microwave radiation out of the cavity and act as positioning guides for a sample. The microwave energy is coupled into the cavity through the iris hole at the left. This coupling may be varied by means of the iris screw. (*b*) The electric field contours in the *xz* plane. One half-wavelength in the *x* direction corresponds to the shortest distance between points of equal field intensity but of opposite phase. (*c*) Magnetic field flux in the *xy* plane. *A* is approximately one half-wavelength and *C* is exactly two half-wavelengths. The *B* dimension is not critical but should be less than one half-wavelength.

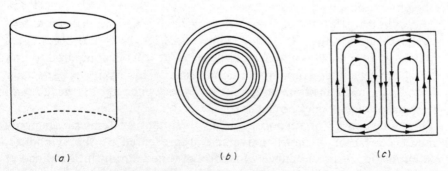

Fig. 2-4 Cylindrical cavity operating in the TE_{011} mode. (*a*) Cavity; the height and diameter of the cylinder govern the resonant frequency. (*b*) Electric field contours. (*c*) Magnetic field contours.

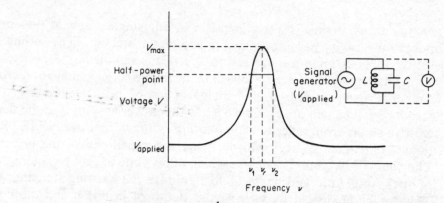

Fig. 2-5 Frequency-response curve of a parallel-tuned circuit to an externally applied oscillating voltage of magnitude $V_{applied}$. $(V_{max} - V_{applied})/\sqrt{2}$ represents the half-power points on the response curve.

i.e., where the voltage is $(V_{max} - V_{applied})/\sqrt{2}$, then Q may be defined as follows:

$$Q = \frac{\nu_r}{\Delta\nu} \tag{2-1}$$

A similar definition is applicable to the response of a resonant cavity as one varies the frequency of the incident microwaves through the region of resonance. The power reflected is a minimum when $\nu = \nu_r$.

An alternative definition, equivalent to the above, is the following:

$$Q = \frac{2\pi \text{ (maximum microwave energy stored in the cavity)}}{\text{energy dissipated per cycle}} \tag{2-2}$$

One infers correctly from this definition that energy storage (and hence Q) generally increases with cavity volume for a fixed frequency. The definition also implies that Q may be increased by decreasing energy losses from currents flowing in the cavity walls or sample. Heavy silver plating is effective in maximizing Q; a final thin gold plate prevents deterioration of the silver. The Q will be lowered if a sample having a high dielectric constant extends into regions of appreciable microwave electric field.

Microwave energy is coupled into and out of the sample cavity by a small hole referred to as an *iris*. (See Fig. 2-3a.) The iris serves a function analogous to that of an impedance-matching transformer in a typical electrical circuit. It is desirable to reduce the fractional reflection of microwave energy from the cavity to some small value. The fractional change in reflected power at the sample resonance is thus enhanced. In any transmission line, the reflection is a minimum if the line is terminated with the characteristic impedance of the line. When this requirement is met, the energy transmission is a maximum. An adjustable screw at the entrance port to the cavity permits optimal impedance matching. The setting of the screw depends upon the size and nature of the sample in the cavity.

Discontinuities in the waveguide, or imperfect matching of microwave elements results in reflection of some fraction of the incident energy; hence standing waves will arise. One desires the power appearing at the detector to arise solely from reflections *originating at the microwave cavity*. Reflections from other sources will decrease the sensitivity of the instrument. The *slide-screw tuner* is a device which allows one to set up standing waves of such amplitude and phase as to minimize the existing standing waves. This is accomplished by varying the depth of insertion and the position along the waveguide of a small metallic probe.

A *circulator* is a nonreciprocal device; i.e., its properties are not identical for waves traveling in the forward and in the reverse directions. It serves to pass the former with little loss, but it strongly attenuates the latter. The circulator is used to direct microwave power to the cavity and to direct the signal reflected from the cavity to the detector. The operation of a four-port circulator is indicated in Fig. 2-6. A terminating load serves to absorb any power which might be reflected from the detector arm.

2-3b. The source The usual ESR spectrometer employs a klystron as the source of microwave radiation. A klystron is a vacuum tube which can produce microwave oscillations centered on a small range of frequency; the output as a function of frequency is referred to as the *klystron mode*. A klystron may operate in several modes; one usually selects that corresponding to the highest power output.

The mode may be displayed on an oscilloscope if one modulates the klystron frequency over the range of the mode. (See the oscilloscope trace

Four-port microwave circulator

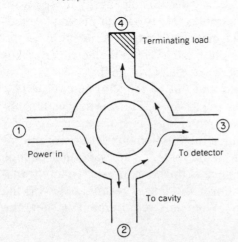

Fig. 2-6 A four-port microwave circulator showing the directions of microwave transmission among the several arms.

in Fig. 2-2.) Owing to the resonant absorption by the cavity, there will be a sharp dip at that region of the mode which corresponds to the cavity resonance frequency. The klystron is tuned so that the dip occurs at the center of the mode.

The frequency of the monochromatic radiation is determined by the voltages applied to the klystron. The klystron can thus be tuned over the range of a mode. Adjustment of a tuning stub on the klystron makes it possible to vary the center frequency of a given mode over a certain range. It is desirable that the klystron frequency be very stable, since the energy density in the resonant cavity is very sensitive to the frequency of the incident radiation. (Frequency variations should be small in comparison with the true linewidth.) Stabilization may be accomplished by an automatic-frequency-control system. The klystron frequency may readily be "locked" to the resonant frequency of the sample cavity.

Serious perturbations of the klystron frequency may occur if there are significant backward reflections of microwave energy from the system fed by the klystron. The *isolator,* like the circulator, is a nonreciprocal device which readily passes microwave energy in the forward direction, while it strongly attenuates any reflections. It thus minimizes variations in the klystron frequency due to backward reflections in the region between the klystron and the circulator.

The *wavemeter* is a cylindrical resonant cavity, the length of which is adjustable to an integral number of half-wavelengths by means of a micrometer. When the resonance frequency of the wavemeter corresponds to the frequency of the incident microwaves, there is a decrease in the power detected by a silicon crystal. The wavemeter is usually calibrated in frequency units (megahertz) instead of wavelength; it can usually be read to ± 1 MHz, but the accuracy may be generally ± 9 MHz. If greater accuracy is required, one may couple a frequency counter to the microwave system, which allows measurement of the frequency to ± 10 kHz or better.

The *attenuator* adjusts the level of the microwave power incident upon the sample. It contains an absorptive element and corresponds to a neutral filter in light-absorption measurements.

2-3c. The magnet system The magnet has thus far been mentioned only as a source of a static magnetic field. This field should be stable and uniform over the sample volume; field variations should be kept within ± 10 mG for organic free radicals in liquid solution.

However, for most inorganic systems, a stability and uniformity of ~ 1 G will usually be adequate. Stability is achieved by energizing the magnet with a highly regulated power supply. Modern magnet energizing systems may utilize a Hall-effect device to detect directly any variation in the magnetic field and to stabilize the field. A scanning system connected to the power supply allows the field to be varied in a linear fashion. The scan should be very reproducible and linear in order to simplify interpretation of the spectra.

Measurement of the magnetic field at the sample may be accomplished by means of a nuclear magnetic resonance (NMR) probe placed beside (or even inside) the microwave cavity. Detection of the NMR signal and measurement of the corresponding resonance frequency with a counter permit a measurement of the magnetic field to about one part in 10^5. Measurement of H at two or three points calibrates the entire spectrum if the field scan is truly linear. If only field differences between lines are required, one may use a double cavity, with a standard (which gives a multiple-line spectrum) in the reference cavity. (See Appendix D.)

2-3d. The modulation and detection systems The principal disadvantage of the spectrometer outlined in Fig. 1-2*b* is the contribution of noise components (principally at low frequencies) to the output signal. The phase-sensitive detection technique employed in the spectrometer of Fig. 2-2 utilizes small-amplitude magnetic field modulation to limit the noise-contributing components to frequencies very close to the modulation frequency. Modulation at the commonly used frequency of 100 kHz is achieved by placing small Helmholtz coils on each side of the cavity along the axis of the static field. Very thin cavity walls must be employed to allow the penetration of the 100-kHz field. Under these conditions the rectified signal at the detector will be amplitude modulated at 100 kHz.

The most commonly used detector is a silicon crystal which acts as a microwave rectifier. If the average incident power is fixed at a level of about 1 mW, the detector current varies as the square root of the microwave power. A crystal detector produces an inherent noise which is inversely proportional to the frequency of the detected signal ("1/*f* noise"). The widespread use of 100 kHz as a modulation frequency is based on the fact that at this frequency the 1/*f* detector noise is less than the noise from other sources.

A possible disadvantage of 100 kHz as a modulation frequency is the presence of modulation sidebands at ± 100 kHz (± 36 mG) from the center of an absorption line. (See Appendix D.) Whereas normally these are not resolved, they prevent resolution of very narrow lines (< 50 mG). Use of very low modulation amplitudes will allow further resolution, but at the expense of sensitivity. If the modulation frequency could be reduced with no increase in the detector noise, improved resolution would be possible without loss of sensitivity. The "backward diode" permits operation at modulation frequencies as low as 10 kHz with the same sensitivity as the silicon crystal at 100 kHz.

Another detector less frequently used is a bolometer, which consists of a resistive wire heated by the microwave electric field. The resulting change in the resistance of the wire can be detected with a Wheatstone bridge. While it is very convenient at low frequencies, its sensitivity is extremely low at modulation frequencies above 1,000 Hz. The bolometer finds additional use in microwave power measurements.

A sensitive detection system which is unexcelled at low microwave power makes use of the superheterodyne principle, i.e., the mixing of a signal with the output of a local oscillator so as to produce an intermediate frequency which is then amplified and detected. In the ESR case, the microwave signal reflected from the sample cavity is mixed with the output of a local-oscillator klystron; the latter generally operates at a frequency 30 MHz above or below that of the signal klystron. The resulting 30 MHz difference frequency produced at the output of the mixing device contains all the information of interest. At such high frequencies the $1/f$ detector noise is negligible. A low-frequency field modulation can be used without loss of sensitivity, since the noise added at the detection frequency of 30 MHz is negligible. Some samples require the use of very low levels of microwave power. The superheterodyne system has unparalleled sensitivity for these cases.

We now resume discussion of the spectrometer shown in Fig. 2-2. After detection, the 100-kHz signal undergoes narrow-band amplification. At this point one can achieve still further reduction of noise by use of a phase-sensitive detector. This reduction in noise is achieved by rejection of all noise components except those in a very narrow band ($\sim\pm1$ Hz), about 100 kHz. The operation of the phase-sensitive detector is readily understood with the aid of Fig. 2-7. The amplified signal is mixed with the output of the modulating 100-kHz oscillator from the secondary winding of the transformer. If the two signals are opposite in phase, the output from the system is a minimum; if exactly in phase, a maximum.

If the amplitude of the 100-kHz field modulation is kept small compared to the linewidth, the amplitude of the detected 100-kHz signal will be approximately proportional to the slope of the absorption curve at the center point of the modulating field. This can be seen from Fig. 2-8. The tangent to the absorption curve at a chosen point is taken to approximate the small portion of the curve traversed by the modulating magnetic field.

Fig. 2-7 Schematic diagram of a phase-sensitive detector. A transformer with a split secondary coil can be used to combine the amplified output of the crystal detector with a fraction of the output of the oscillator driving the modulation coils. The combined signal is rectified, filtered, and recorded. The output signal depends markedly on the relative phase and amplitude of the signal and reference voltages.

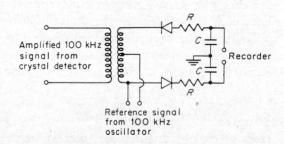

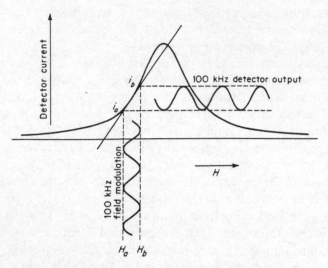

Fig. 2-8 Effect of small-amplitude 100-kHz field modulation on the crystal detector output current. The static magnetic field is modulated between the limits H_a and H_b. The corresponding crystal current varies between the limits i_a and i_b.

As the total field varies between the limits H_a and H_b, the detector current will vary sinusoidally between the limits i_a and i_b. When the slope of the absorption curve is zero (i.e., at field values far from resonance and also *at* the resonant field), the 100-kHz component at the detector will be zero. At the inflection points, where the slope is a maximum, the amplitude of the output signal will also be a maximum. The output polarity of the phase-sensitive detector is governed by the sign of the slope; hence for small modulation amplitudes, the output signal appears approximately as the first derivative of the absorption signal. Use of modulation amplitudes approaching the linewidth leads to a distorted line shape. (See Sec. D-2a.) Usually the output signal is filtered by a selected resistor-capacitor combination. The filter time constant (in seconds) is given by the product RC, where R is the resistance in megohms and C the capacitance in microfarads.

Typical ESR absorption lines are shown in Figs. 2-9a and 2-10a; the corresponding first-derivative presentations are given in Figs. 2-9b and 2-10b. The derivative signal may be increasing or decreasing on the left-hand side, depending on the phase adjustment of the phase detector or the polarity of the recorder connection. *The sense of the recorded signal is totally irrelevant to its interpretation.* (One is the mirror image of the other.) Recorded spectra of either phase are found in the literature and in this book.

2-4 LINE SHAPES AND INTENSITIES

It is often of considerable interest to determine the relative intensity of different lines in a spectrum. The most accurate procedure for determining

Table 2-1 Properties of lorentzian and gaussian lines

	Line shape	
	Lorentzian	Gaussian
Equation for normalized line	$Y = Y_{max} \dfrac{\Gamma^2}{\Gamma^2 + (H - H_r)^2}$	$Y = Y_{max} \exp\left[\dfrac{(-\ln 2)(H - H_r)^2}{\Gamma^2}\right]$
Peak amplitude	$Y_{max} = \dfrac{1}{\pi\Gamma}$	$Y_{max} = \left(\dfrac{\ln 2}{\pi}\right)^{\frac{1}{2}} \dfrac{1}{\Gamma}$
Half-width at half-height	Γ	Γ
Equation for first derivative	$Y' = -Y_{max} \dfrac{2\Gamma^2(H - H_r)}{[\Gamma^2 + (H - H_r)^2]^2}$	$Y' = -Y_{max} \dfrac{2(\ln 2)(H - H_r)}{\Gamma^2} \exp\left[\dfrac{(-\ln 2)(H - H_r)^2}{\Gamma^2}\right]$
Peak-to-peak amplitude	$2Y'_{max} = \dfrac{3\sqrt{3}}{4\pi} \dfrac{1}{\Gamma^2}$	$2Y'_{max} = 2\left(\dfrac{2}{\pi e}\right)^{\frac{1}{2}} \dfrac{\ln 2}{\Gamma^2}$
Peak-to-peak width	$\Delta H_{pp} = \dfrac{2}{\sqrt{3}}\,\Gamma$	$\Delta H_{pp} = \left(\dfrac{2}{\ln 2}\right)^{\frac{1}{2}} \Gamma$
Equation for second derivative	$Y'' = -Y_{max} 2\Gamma^2 \left(\dfrac{\Gamma^2 - 3(H - H_r)^2}{[\Gamma^2 + (H - H_r)^2]^3}\right)$	$Y'' = -Y_{max} \dfrac{2\ln 2}{\Gamma^4} \{\Gamma^2 - 2(\ln 2)(H - H_r)^2\} \exp\left[\dfrac{(-\ln 2)(H - H_r)^2}{\Gamma^2}\right]$
Peak amplitude of positive lobe	$A = Y_{max}\left(\dfrac{1}{2\Gamma^2}\right)$	$A = Y_{max} \dfrac{4e^{-\frac{3}{2}}\ln 2}{\Gamma^2}$
Peak amplitude of negative lobe	$B = -Y_{max}\left(\dfrac{2}{\Gamma^2}\right)$	$B = -Y_{max} \dfrac{2\ln 2}{\Gamma^2}$

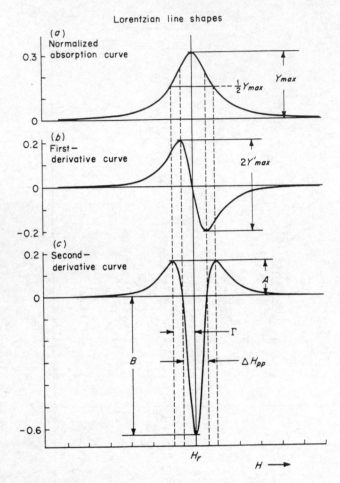

Fig. 2-9 Lorentzian line shapes. (*a*) Absorption spectrum. (*b*) First-derivative spectrum. (*c*) Second-derivative spectrum. For an explanation of the symbols, see Table 2-1.

the intensity of a line is to integrate the full absorption curve. To obtain the intensity from a first-derivative presentation, one may perform two consecutive integrations. This is a tedious procedure; however, if linewidths of two components are equal, then peak-to-peak amplitudes of the derivative lines will be proportional to their intensities. Even when linewidths are different, the approximate relative intensity $\mathscr{I}$ of a line may be obtained from the expression

$$\mathscr{I} \propto Y'_{\max}(\Delta H_{pp})^2$$

where $2Y'_{\max}$ is the peak-to-peak derivative amplitude and ΔH_{pp} is the peak-to-peak width. (See Figs. 2-9*b* and 2-10*b*.)

One will obtain a trace which is approximately the second derivative of the absorption line by tuning the amplifier to twice the modulation fre-

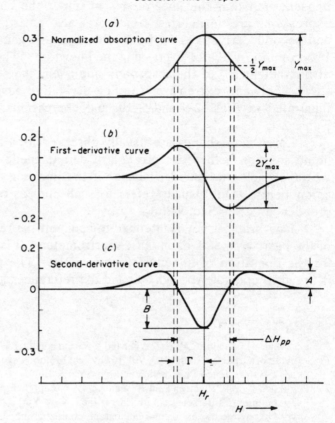

Fig. 2-10 Gaussian line shapes. (*a*) Absorption spectrum. (*b*) First-derivative spectrum. (*c*) Second-derivative spectrum. For an explanation of the symbols, see Table 2-1.

quency (e.g., 200 kHz). It can be seen from Figs. 2-9*c* and 2-10*c* that the shape of the second-derivative presentation is very sensitive to the nature of the absorption line.

The shapes of ESR lines are usually described by comparison with the lorentzian and the gaussian line shapes. (See Figs. 2-9*a* and 2-10*a*, respectively.) Analytical expressions for these lines are as follows:

Lorentzian: $y = \dfrac{a}{1 + bx^2}$ (2-3)

Gaussian: $y = a \exp(-bx^2)$ (2-4)

Detailed expressions in terms of measurable experimental parameters are given in Table 2-1 (see p. 33).

The parameters describing these lines are the maximum amplitude Y_{max} and the full width at half-height 2Γ. These parameters have been chosen such that the integrated area under the curves is unity; i.e., the expressions are normalized. Table 2-1 also gives expressions for the first and second derivatives for both lorentzian and gaussian lines. There are two common ways of expressing the linewidth. The first is the half-width at half-height (Γ) of the absorption line itself; the second is the full width (ΔH_{pp}) between extrema of the first-derivative curve. These widths are illustrated in Figs. 2-9 and 2-10, and expressions for them are given in Table 2-1.

Lorentzian shapes are usually observed for ESR lines of systems in liquid solution if the concentration of paramagnetic centers is low. Lines will approach the gaussian shape† if the line is a superposition of many components. One usually refers to such lines as being inhomogeneously broadened. (See Sec. 9-3a.)

This brief survey is intended to acquaint the reader with the function of the basic elements of an ESR spectrometer. It is inadequate as a guide for the operation of an actual spectrometer. The potential experimenter is urged to consult Appendix D and the references given below.

REFERENCES

1. Fraenkel, G. K.: in A. Weissberger (ed.), Paramagnetic Resonance Absorption, "Technique of Organic Chemistry," vol. 1, Physical Methods, part IV, chap. XLII, pp. 2801–2872, Interscience Publishers, Inc., New York, 1960.
2. Poole, Jr., C. P.: "Electron Spin Resonance—A Comprehensive Treatise on Experimental Techniques," Interscience Publishers, Inc., New York, 1967. This book contains a very extensive discussion and bibliography on experimental methods.
3. Alger, R. S.: "Electron Paramagnetic Resonance: Techniques and Applications," John Wiley & Sons, Inc., New York, 1968. This book contains numerous drawings of experimental equipment and details of construction.

PROBLEMS

2-1. (a) Complete the discussion of analogies between the components of the spectrometers of Figs. 1-1a and b. Indicate where there is a lack of correspondence.

(b) Some aspects of the optical spectrometer are somewhat unusual. The arrangement is a possible one and was chosen to show analogies. Point out the unusual components involved.

2-2. Why is it that when a sample tube filled with water is placed inside a microwave cavity, the Q factor is greatly decreased, whereas when the same tube contains benzene it has little effect on Q?

2-3. Explain the function of field modulation, using simple diagrams to illustrate the details of application of the technique.

2-4. (a) Plot on the same piece of graph paper a lorentzian line and a gaussian line, each with the same value of Y_{max} and Γ.

† Lines of gaussian form give a linear plot on gaussian graph paper (available from Canadian Charts and Supplies, Ltd., Oakville, Montreal).

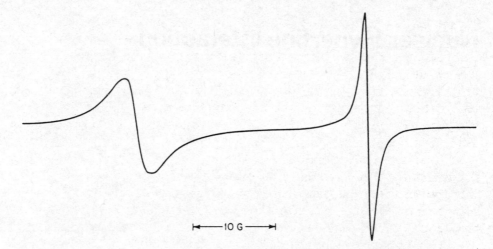

Fig. 2-11 An ESR spectrum showing two lines of different amplitude and width.

(*b*) On a second piece of graph paper plot the first derivative curves for these two lines.

(*c*) On the basis of *a* and *b*, what criteria would you use to determine the line shape of an experimental trace?

2-5. What variables in the ESR spectrometer would you expect to affect the signal-to-noise ratio?

2-6. What methods could be employed to determine the intensity of an ESR line?

2-7. Compute the relative intensities of the two lines shown in Fig. 2-11.

2-8. For a modulation frequency of 15 kHz, compute the positions (in gauss) of the modulation sidebands relative to the central line.

3
Nuclear Hyperfine Interaction

3-1 INTRODUCTION

If the interaction of an electron with a magnetic field were the only effect operative, then all ESR spectra would consist of one line. The only useful information to be garnered from these spectra would be the g factor. The ESR technique would thus provide rather limited information. Fortunately there are other interactions which can produce spectra rich in line components. For the present, the discussion will be restricted to molecules containing one unpaired electron ($S = \frac{1}{2}$), although much of the following will apply equally well to molecules containing more than one unpaired electron ($S \geq 1$). States with $S = \frac{1}{2}$ are referred to as *doublet* states since the multiplicity ($2S + 1$) is equal to 2.

The first additional interaction to be considered is that of the electron spin magnetic dipole with nuclei in its vicinity. It was noted in Chap. 1 that some nuclei possess an intrinsic spin angular momentum. The spin quantum number I of these magnetic nuclei takes on one of the values $\frac{1}{2}, 1, \frac{3}{2}, 2, \ldots$, etc., with a corresponding multiplicity of nuclear spin states given by ($2I + 1$). Analogous to the electron case, there will be a magnetic moment associated with the nuclear spin angular momentum. The spins and magnetic moments of some common nuclei are listed in Table C (inside back cover). Note that $I = 0$ for all nuclei for which the atomic mass *and* the atomic number are *even*. If the atomic number is *odd* and the atomic mass is *even*, I is an integer; if the atomic mass is *odd*, I is a half-integer.

The field H_r in the resonance equation (1-6) represents the magnetic field experienced by the unpaired electron. If there are any local magnetic fields $\mathbf{H}_{local}$, these will add vectorially to the external magnetic field $\mathbf{H}$ to give an effective field $\mathbf{H}_{eff}$. That is

$$\mathbf{H}_{eff} = \mathbf{H} + \mathbf{H}_{local}$$

This chapter will be concerned with local magnetic fields arising from magnetic nuclei.

The interaction of an unpaired electron and a magnetic nucleus is called nuclear hyperfine interaction. The term "hyperfine splitting" was used in atomic spectra to designate the splitting of certain lines as a result of interaction with magnetic nuclei. The hyperfine interaction may be either anisotropic (orientation-dependent) or isotropic (independent of the orientation of $\mathbf{H}$ with respect to a molecular axis).

The simplest system exhibiting nuclear hyperfine interaction is the hydrogen atom. This system will first be considered in a qualitative fashion. The details of the origin of the hyperfine interaction and the calculation of energy levels will be discussed later in this chapter. An ESR spectrum of hydrogen atoms produced in a nonoriented solid by irradiation is given in Fig. 3-1. Instead of a single line characterized by $H_r = h\nu/g\beta$ with $g = 2.00$,

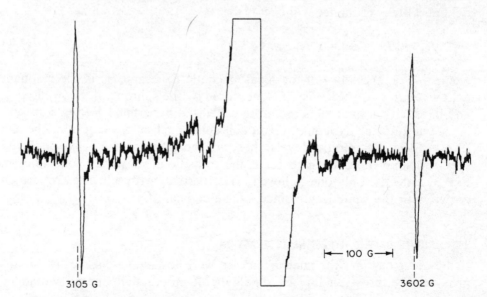

3105 G   100 G   3602 G

Fig. 3-1 ESR spectrum of hydrogen atoms in an x-rayed human tooth at room temperature. $\nu = 9.495$ GHz. [*See T. Cole and A. H. Silver, Nature,* **200**:700 (1963).]

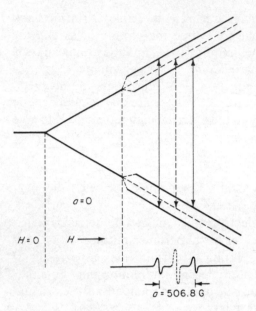

$a = 0$

$H = 0$ $H \longrightarrow$

$a = 506.8\,G$

Fig. 3-2 Energy levels of the hydrogen atom as a function of magnetic field at constant microwave frequency. The dotted transition would be observed if a were zero.

one observes a *pair* of lines. The average position of these lines (3,353.5 G) corresponds to $g \approx 2$.[†] The presence of more than two spin energy levels is implied by the observation of two magnetic resonance lines in a spectrum. Since the proton has a spin $I = \frac{1}{2}$, M_I has the two allowed values: $M_I = \pm \frac{1}{2}$. Hence, at the electron there will be one of two possible local fields contributed by the proton. There will then be two values of the external magnetic field at which resonance will occur, that is

$$H_r = \left(H' \mp \frac{a}{2} \right) = (H' - aM_I) \tag{3-1}$$

where $a/2$ is the value of the local magnetic field, and H' is the resonant field when $a = 0$. Note that a is measured by the splitting of the two hyperfine lines. This interval is therefore called the hyperfine *splitting* constant.

The nuclear hyperfine interaction splits each of the electron Zeeman levels of Fig. 1-8 into two levels; these are shown in Fig. 3-2. The observed spectrum may be accounted for if the transitions shown as solid lines in Fig. 3-2 are the only ones allowed. The vertical arrows in Fig. 3-2 are all drawn with the same length since $h\nu$ is a constant.

3-2 ORIGINS OF THE HYPERFINE INTERACTION

If the electron- and the nuclear dipoles were to behave classically, an approximate expression for the dipole-dipole interaction energy would be

---

[†] This procedure for obtaining the g factor leads to a small error when the separation of hyperfine lines is large, that is, > 10 G. (See Sec. C-6.)

given by†

$$W_{\text{dipolar}} = \frac{(1 - 3 \cos^2 \theta)}{r^3} \mu_{N_z} \mu_{e_z} = H_{\text{local}} \mu_{e_z} \qquad (3\text{-}2)$$

Here, the components of the electron and nuclear dipole moments along a magnetic field are μ_e and μ_N, respectively; the dipoles are separated by the distance r. θ is the angle between the magnetic field direction and a line joining the two dipoles. This classical system is shown in Fig. 3-3. Depending on the value of θ, the local-field contribution H_{local} at the electron can either aid or oppose the external magnetic field. From Fig. 3-3 and Eq. (3-2) it is apparent that H_{local} depends markedly on the instantaneous value of θ.

Since the electron is not localized at one position in space, the interaction energy H_{local} must be averaged over the electron probability distribution function. If all values of θ are equally probable (as for an electron in an s orbital centered on nucleus N), then the average local field is obtained by inserting the value of $\cos^2 \theta$ averaged over a sphere‡:

$$\langle \cos^2 \theta \rangle = \frac{\int_0^{2\pi} \int_0^{\pi} \cos^2 \theta \sin \theta \, d\theta \, d\phi}{\int_0^{2\pi} \int_0^{\pi} \sin \theta \, d\theta \, d\phi} = \frac{1}{3} \qquad (3\text{-}3)$$

† This equation is valid only if the applied field is much greater than the hyperfine field present at the nucleus as a result of the electron-nuclear interaction. (See Sec. 7-6.)

‡ The average value of a quantity $g(q)$ weighted by the probability function $P(q)$ is given by

$$\langle g \rangle = \frac{\int g P(q) \, dq}{\int P(q) \, dq}$$

The integration is taken over the allowed range of q.

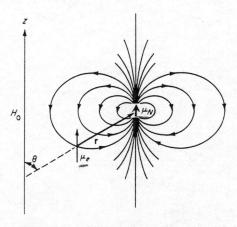

Fig. 3-3 Interaction of dipoles arising from electron (μ_e) and from nuclear spin (μ_N). θ is the angle between the vector **r** and the applied field **H**. The vector μ_N is indicated for the state $M_I = \frac{1}{2}$.

In spherical polar coordinates $\sin \theta \, d\theta \, d\phi$ is the element of surface area on a sphere. Since $\langle \cos^2 \theta \rangle = \frac{1}{3}$, H_{local} in Eq. (3-2) vanishes. Consequently, the classical dipolar interaction *cannot* be the origin of the hyperfine splitting in the hydrogen atom since the electron distribution in a $1s$ orbital is spherically symmetric.

A clue to the origin of the hyperfine interaction in the hydrogen atom is obtained by examining the radial dependence of the hydrogen $1s$ orbital shown in Fig. 3-4. One notes that the $1s$ electron density at the nucleus is nonzero; it is precisely this finite density which gives rise to the *isotropic* hyperfine interaction. It is clear from Fig. 3-4 that *only electrons in s orbitals have a nonzero probability of being at the nucleus; p, d, f*, etc., orbitals all have nodes at the nucleus. Electrons in $2s$, $3s$, . . . , etc.,

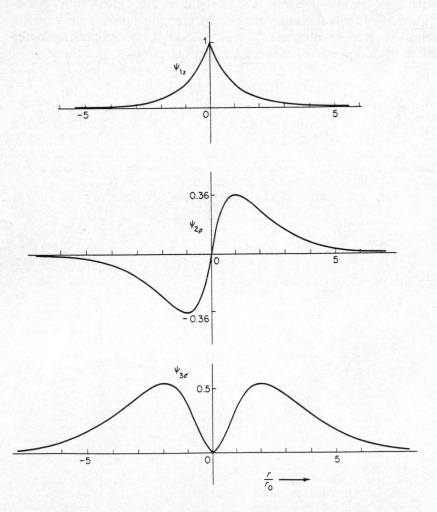

Fig. 3-4 Radial dependence of hydrogenic $1s$, $2p$, and $3d$ wave functions. r_0 is the Bohr radius.

orbitals also have a nonzero electron density at the nucleus and will give rise to isotropic hyperfine interactions. In Table C (inside back cover), column 6, the isotropic hyperfine interactions for valence s electrons of some atoms are seen to attain enormous values.

Fermi[†] has shown that for systems with one electron the isotropic interaction energy is given approximately by

$$W_{\text{iso}} = -\frac{8\pi}{3}\,|\psi(0)|^2\mu_{e_z}\mu_{N_z} \tag{3-4}$$

Here $\psi(0)$ represents the wave function evaluated at the nucleus. For example, the hydrogen $1s$ wave function is given by

$$\psi_{1s} = \left(\frac{1}{\pi r_0^3}\right)^{\frac{1}{2}} \exp\left(\frac{-r}{r_0}\right) \tag{3-5}$$

where r_0 is the radius of the first Bohr orbit (0.0529 nm). At $r = 0$, $|\psi(0)|^2 = 1/\pi r_0^3$. Using this value, one can then calculate a value of W_{iso} with the aid of Eq. (3-4). This calculation is the subject of Prob. 3-3.

3-3 ENERGY LEVELS OF A SYSTEM WITH ONE UNPAIRED ELECTRON AND ONE NUCLEUS WITH $I = \frac{1}{2}$

The hamiltonian operator for the hyperfine interaction may be obtained from Eq. (3-4) by replacing the magnetic moments by their corresponding operators

$$\hat{\mu}_{e_z} = -g\beta\hat{S}_z \tag{3-6a}$$

$$\hat{\mu}_{N_z} = g_N\beta_N\hat{I}_z \tag{3-6b}$$

following the procedure outlined in Sec. 1-5. The operator equation corresponding to Eq. (3-4) is

$$\hat{\mathscr{H}}_{\text{iso}} = \frac{8\pi}{3}\,g\beta g_N\beta_N|\psi(0)|^2\hat{S}_z\hat{I}_z \tag{3-7a}$$

$$= hA_0\hat{S}_z\hat{I}_z \tag{3-7b}$$

A_0 is called the isotropic hyperfine *coupling* constant (measured in hertz), and hA_0 measures the interaction energy between the electron and the nucleus. Strictly speaking, Eqs. (3-7) should be written with the factor $\hat{\mathbf{S}} \cdot \hat{\mathbf{I}}$ in place of $\hat{S}_z\hat{I}_z$; however, it is shown in Sec. C-7 that when the hyperfine interaction hA_0 is small compared to the electron Zeeman interaction $g\beta H$, Eq. (3-7b) is adequate.

The spin hamiltonian operator for the hydrogen atom (and other isotropic systems) is obtained by adding Eqs. (1-14) and (3-7b)

$$\hat{\mathscr{H}} = g\beta H\hat{S}_z + hA_0\hat{S}_z\hat{I}_z \tag{3-8}$$

[†] E. Fermi, *Z. Physik*, **60**:320 (1930).

For completeness, a nuclear Zeeman term $-g_N\beta_N H\hat{I}_z$ should have been added; however, it has been omitted from Eq. (3-8) because it does not affect the *transition* energies. (See Sec. C-1 for the more complete hamiltonian.)

Since the eigenvalues M_S of $\hat{S}_z$ are $\pm\frac{1}{2}$ and those of I_z are $M_I = \pm\frac{1}{2}$, there will be four possible spin states of the hydrogen atom,

$$|\alpha_e,\alpha_n\rangle \quad |\beta_e,\alpha_n\rangle$$
$$|\alpha_e,\beta_n\rangle \quad |\beta_e,\beta_n\rangle$$

These states are labeled as in Sec. 1-5; for example,

$$\hat{S}_z|\alpha_e,\beta_n\rangle = +\tfrac{1}{2}|\alpha_e,\beta_n\rangle \tag{3-9a}$$

$$\hat{I}_z|\alpha_e,\beta_n\rangle = -\tfrac{1}{2}|\alpha_e,\beta_n\rangle \tag{3-9b}$$

The energies of these states are obtained by evaluating expressions analogous to Eqs. (1-24); for example,

$$W_{\alpha_e,\alpha_n} = \langle\alpha_e,\alpha_n|\hat{\mathscr{H}}|\alpha_e,\alpha_n\rangle = \langle\alpha_e,\alpha_n|g\beta H\hat{S}_z + hA_0\hat{S}_z\hat{I}_z|\alpha_e,\alpha_n\rangle$$
$$= \tfrac{1}{2}g\beta H + \tfrac{1}{4}hA_0 \tag{3-10a}$$

Similarly

$$W_{\alpha_e\beta_n} = +\tfrac{1}{2}g\beta H - \tfrac{1}{4}hA_0 \tag{3-10b}$$

$$W_{\beta_e\beta_n} = -\tfrac{1}{2}g\beta H + \tfrac{1}{4}hA_0 \tag{3-10c}$$

$$W_{\beta_e\alpha_n} = -\tfrac{1}{2}g\beta H - \tfrac{1}{4}hA_0 \tag{3-10d}$$

These energy levels are shown in Fig. 3-5a under conditions of constant magnetic field. The situation for constant microwave frequency is depicted in Fig. 3-5b.

From these quantitative expressions for the energy levels of the hydrogen atom in a magnetic field, note the equality of splitting of each of the $M_S = +\frac{1}{2}$ and $M_S = -\frac{1}{2}$ states. Second, note that the ordering of the M_I levels is reversed in the lower set of levels as compared with the upper set. Finally, note that transitions occur only between levels with the same value of M_I. These conclusions justify the assertions regarding the spectrum of the hydrogen atom, which were made in Sec. 3-1.

In Chap. 1 it was indicated that the selection rule for an electron spin transition is $\Delta M_S = \pm 1$. Such transitions correspond to a change in spin angular momentum of $\pm\hbar$. A photon has an intrinsic angular momentum equal to $\hbar$. Hence, if $\Delta M_S = +1$ when a photon is absorbed, then M_I must remain unchanged to conserve the total angular momentum. Thus the selection rules are $\Delta M_S = \pm 1$ and $\Delta M_I = 0$.[†] These selection rules break down due to the mixing of states when the resonant magnetic field ap-

---

[†] The selection rules $\Delta M_S = 0$, $\Delta M_I = \pm 1$ apply in the case of NMR spectroscopy, i.e., when the system is irradiated at the nuclear resonance frequency.

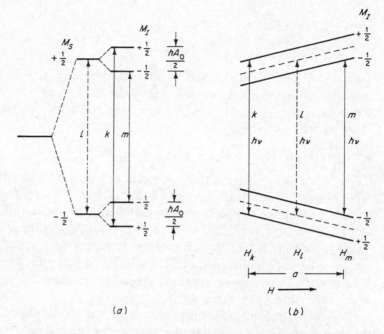

Fig. 3-5 (a) Energy levels of the hydrogen atom at constant magnetic field. Here the dashed line l would be the transition corresponding to $h\nu = g\beta H$ in the absence of hyperfine interaction. The solid lines marked k and m correspond to the allowed transitions with hyperfine coupling operative; $h\nu = g\beta H \pm \frac{1}{2}hA_0$. A_0 (megahertz) is the isotropic hyperfine coupling constant. (b) Energy levels of the hydrogen atom with variable magnetic field. The dashed line l corresponds to the transition in the absence of hyperfine interaction. The solid lines k and m refer to transitions induced by a microwave quantum $h\nu$ of the same energy as for the transition l. Here the resonant field values corresponding to these two transitions are given by $H_k = h\nu/g\beta - a/2$ and $H_m = h\nu/g\beta + a/2$ where a (measured in gauss) is the hyperfine splitting constant at constant frequency and is given by $H_m - H_k$.

proaches zero. For solid-state systems, mixing of states occurs at high microwave frequencies (see Sec. 7-8).

The energies of the two allowed transitions are

$$\Delta W_1 = W_{\alpha_e \alpha_n} - W_{\beta_e \alpha_n} = g\beta H + \tfrac{1}{2}hA_0 \qquad (3\text{-}11a)$$
$$\Delta W_2 = W_{\alpha_e \beta_n} - W_{\beta_e \beta_n} = g\beta H - \tfrac{1}{2}hA_0 \qquad (3\text{-}11b)$$

If there were no hyperfine interaction ($A_0 = 0$), then under conditions of constant H, a single transition would occur at a frequency $\nu_l = h^{-1}g\beta H$. (See transition marked l in Fig. 3-5a.) For a nonzero hyperfine interaction, transitions will occur at the two frequencies:

$$\nu_k = h^{-1}g\beta H + \tfrac{1}{2}A_0 \qquad M_I = +\tfrac{1}{2}$$

and

$$\nu_m = h^{-1}g\beta H - \tfrac{1}{2}A_0 \qquad M_I = -\tfrac{1}{2}$$

(See transitions marked k and m, respectively, in Fig. 3-5a.) Note that the two transitions occur between levels of identical M_I value. When the microwave frequency is held constant at ν_0, the resonance equation is appropriately written

$$h\nu_0 = g\beta(H_r + aM_I) = g\beta H_r + hA_0 M_I \qquad (3\text{-}12)$$

If A_0 were zero, a transition would occur at the resonant magnetic field $H_l = h\nu_0/g\beta$. (See transition marked l in Fig. 3-5b.) With $A_0 \neq 0$, transitions will occur at the two resonant fields: $H_k = h\nu_0/g\beta - a/2$, $M_I = +\tfrac{1}{2}$ and $H_m = h\nu_0/g\beta + a/2$, $M_I = -\tfrac{1}{2}$; here $a = hA_0/g\beta$. (See transitions marked k and m in Fig. 3-5b.) It is important to note that a and A_0 cannot be converted into each other without knowledge of the g factor, since from one paramagnetic species to another, the g factor should always be specified when reporting hyperfine splittings in gauss.

Although Fig. 3-5b corresponds to the usual experimental arrangement, it is easier to visualize the splitting of the energy levels in Fig. 3-5a. If one notes that the *higher* energy transition (that is, ν_k) in Fig. 3-5a corresponds to a transition at the *lower* magnetic field (that is, H_k) in Fig. 3-5b, one can correlate the two diagrams. In subsequent chapters, energy-level schemes such as that in Fig. 3-5a will be utilized.

3-4 THE ENERGY LEVELS OF A SYSTEM WITH $S = \tfrac{1}{2}$ AND $I = 1$

The $^2_1$H (deuterium) atom is a simple example of a system with $S = \tfrac{1}{2}$ and $I = 1$. As in Sec. 3-3, the energy levels are computed using the hamiltonian operator $\hat{\mathscr{H}}$ [Eq. (3-8)]. There are now six spin states which are represented by $|M_S, M_I\rangle$:

$$|\tfrac{1}{2},+1\rangle \qquad |-\tfrac{1}{2},-1\rangle$$
$$|\tfrac{1}{2},\ 0\rangle \qquad |-\tfrac{1}{2},\ 0\rangle$$
$$|\tfrac{1}{2},-1\rangle \qquad |-\tfrac{1}{2},+1\rangle$$

The energies are given by expressions analogous to Eq. (3-10a); these energies are

$$W_{\tfrac{1}{2},1} = \tfrac{1}{2}g\beta H + \tfrac{1}{2}hA_0 \qquad W_{-\tfrac{1}{2},-1} = -\tfrac{1}{2}g\beta H + \tfrac{1}{2}hA_0$$
$$W_{\tfrac{1}{2},0} = \tfrac{1}{2}g\beta H \qquad W_{-\tfrac{1}{2},0} = -\tfrac{1}{2}g\beta H$$
$$W_{\tfrac{1}{2},-1} = \tfrac{1}{2}g\beta H - \tfrac{1}{2}hA_0 \qquad W_{-\tfrac{1}{2},1} = -\tfrac{1}{2}g\beta H - \tfrac{1}{2}hA_0 \qquad (3\text{-}13)$$

By virtue of the selection rules $\Delta M_S = \pm 1$ and $\Delta M_I = 0$, there are three allowed transitions. These are depicted in Fig. 3-6a; a typical derivative spectrum in an increasing magnetic field is shown in Fig. 3-6b. Under con-

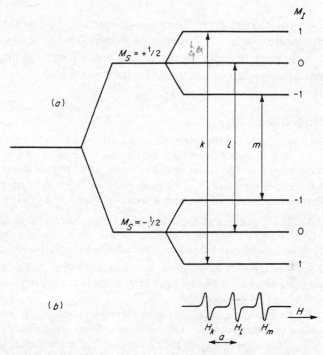

Fig. 3-6 (*a*) Energy levels and allowed transitions for the deuterium atom at constant field. (*b*) Spectrum at constant frequency.

ditions of constant microwave frequency, transitions will occur at the resonant fields:

$$H_k = \frac{h\nu_0}{g\beta} - a \qquad H_l = \frac{h\nu_0}{g\beta} \qquad \text{and} \qquad H_m = \frac{h\nu_0}{g\beta} + a$$

These lines will be of equal intensity, since there is no coincidence of states; i.e., the states are all nondegenerate.

The extension to systems with $S = \frac{1}{2}$ and $I > 1$ is straightforward. For $I = \frac{3}{2}$, four transitions of equal intensity will be observed. In general, for a single nucleus interacting with one unpaired electron, there will be $2I + 1$ lines of equal intensity; adjacent lines are separated by the hyperfine splitting a.

3-5 SUMMARY

In this chapter expressions have been obtained for the energy levels of systems in which a single electron interacts with *one* magnetic nucleus. Expressions have been derived for the transition frequencies in a constant magnetic field and for the resonant field values at a constant microwave frequency.

In most free radicals the unpaired electron interacts with a number of magnetic nuclei. Examples of practical procedures for determining qualitative hyperfine splitting patterns if more than one magnetic nucleus interacts with the unpaired electron are given in Chap. 4.

PROBLEMS

3-1. (a) Carry out the integrations indicated in Eq. (3-3) and verify the result.

(b) Compute $\langle \cos^2 \theta \rangle$ if μ_e and μ_N are confined to a plane containing the magnetic field, for example, $\phi = 0$. Assume that all values of θ are equally probable.

3-2. Given the value of μ_N for hydrogen from Table A (inside front cover) compute the local field at an electron 0.2 nm from a proton when $\theta = 0°$ and when $\theta = 90°$.

3-3. The experimental hyperfine coupling constant A_0 for the hydrogen atom is 1,420.40573 MHz, and $g = 2.002256$. Compare the value of A_0 with that calculated using Eqs. (3-4) and (3-5). (The data are taken from atomic beam experiments.)

3-4. Use the data on magnetic moments in Table C (inside back cover) and the hyperfine coupling constant of the hydrogen atom to check the hyperfine coupling constant of the deuterium atom given in that table.

3-5. Calculate the energy levels of the $\cdot RH_2$ radical at high magnetic fields, using the hamiltonian operator equation (3-8). The spin states should be written as $|M_S, M_I\rangle$ where $M_I = M_{I_1} + M_{I_2}$ (M_{I_1} and M_{I_2} are the z components of the nuclear spins of nuclei 1 and 2). Plot the energy levels as a function of magnetic field; indicate the allowed transitions and their relative intensities.

3-6. Use the hamiltonian operator of Eq. (3-8) to calculate the spin energy levels of the ground state (^2S) of the sodium atom ($S = \frac{1}{2}, I = \frac{3}{2}$). Determine the allowed transitions and evaluate the resonant fields corresponding to the predicted transitions when the microwave frequency is 9.25 GHz.

4

Analysis of Electron
Spin Resonance Spectra of
Systems in the Liquid Phase

4-1 INTRODUCTION

Most free radicals contain several magnetic nuclei; in some molecules these may be grouped into magnetically equivalent sets. Sometimes the nuclei may be equivalent by virtue of the symmetry of the molecule; in other cases the equivalence may be accidental. The hyperfine splittings from radicals having numerous magnetic nuclei may give spectra rich in line components. The analysis of these spectra may be straightforward; more often a successful analysis requires some experience. This chapter presents a large number of experimental spectra ranging from the simple to the complex. The reader is urged to consider each spectrum carefully and to understand its analysis before proceeding to the next. Section 4-7 presents a number of rules which should aid in the analysis of complex spectra.

It was shown in Chap. 3 that the effect of the hyperfine interaction with a single proton ($I = \frac{1}{2}$) is to split the electron energy levels into two, one for each of the $M_S = +\frac{1}{2}$ and $M_S = -\frac{1}{2}$ states. Interaction with a deuteron ($I = 1$) leads to splitting of each electron level into three levels. In general, if the nuclear spin is I, there will be $2I + 1$ energy levels for each value of M_S.

Sets of equivalent nuclei can be treated by considering that they interact as *one* nucleus with a nuclear spin equal to the sum of all the nuclear spins in the equivalent set. Thus, if there are n equivalent nuclei of spin I, the number of levels will be $(2nI + 1)$ for each M_S value. This procedure gives the correct energy levels but does not account for the degeneracy (multiplicity) of some of the levels. However, the degeneracies can be obtained from simple rules which will be given later.

4-2 ENERGY LEVELS OF RADICALS CONTAINING A SINGLE SET OF EQUIVALENT PROTONS

For a system of one unpaired electron interacting with two equivalent protons, it is possible to obtain the appropriate hyperfine energy levels by replacing the two nuclei with one nucleus having $I = 1$. The energy-level sequence will then be the same as that in Fig. 3-6 for the deuterium atom; there the levels are labeled according to the value of $M_I(+1, 0,$ or $-1)$. Alternatively, the energy-level scheme may be obtained by successive splitting of levels as shown in Fig. 4-1a. Interaction with the first nucleus causes the $M_S = +\frac{1}{2}$ and $M_S = -\frac{1}{2}$ levels to be split by $hA_0/2$; interaction with the second nucleus causes each level to be split again by $hA_0/2$, since equivalence implies identity of hyperfine coupling constants. Figure 4-1a demonstrates that there is a coincidence of the intermediate levels ($M_I = 0$) in both the $M_S = +\frac{1}{2}$ and the $M_S = -\frac{1}{2}$ groups. Note that the twofold degeneracy is associated with the two possible permutations of nuclear spins

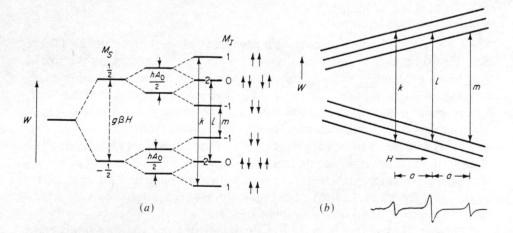

(a)   *(b)*

Fig. 4-1 *(a)* Energy levels and transitions at constant field for two equivalent nuclei with $I = \frac{1}{2}$. k, l, and m are the allowed transitions. l will be twice as intense as k or m, owing to the fact that the l transition occurs between doubly degenerate energy levels. The various possible configurations of the nuclear spins are shown at the right. *(b)* Energy levels and transitions at constant frequency for two equivalent nuclei with $I = \frac{1}{2}$. The three transitions shown correspond respectively to the transitions in *(a)*.

which give a net spin of zero. (See Fig. 4-1a.) The factor of 2 in populations of the $M_I = 0$ states, as compared with the $M_I = +1$ or -1 states, is reflected in the $1:2:1$ relative intensities of the allowed transitions.[†] These transitions are shown at constant field in Fig. 4-1a and at constant frequency in Fig. 4-1b. The selection rules are $\Delta M_S = \pm 1$, $\Delta M_I = 0$, just as for the single-nucleus case. Here I refers to the combined nuclear spin.

For three equivalent nuclei with $I = \frac{1}{2}$, the repetitive splitting procedure leads to four levels for both the $M_S = +\frac{1}{2}$ and $M_S = -\frac{1}{2}$ states (Fig. 4-2). The inner levels are threefold-degenerate, corresponding to the number of nuclear spin states having $M_I = +\frac{1}{2}$ or $-\frac{1}{2}$. Alternatively, the degeneracy

[†] This statement applies as long as $kT \gg hA_0$; experimentally this condition almost always applies.

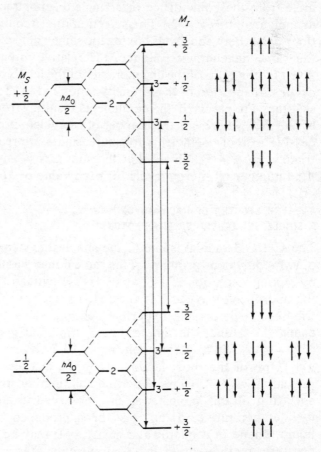

Fig. 4-2 Hyperfine levels and nuclear spin orientations for three equivalent nuclei of spin $\frac{1}{2}$ in a fixed magnetic field. Degeneracies are shown by a numeral at the center of a level. The various possible configurations of the nuclear spins are shown at the right.

																n
								1								0
							1		1							1
						1		2		1						2
					1		3		3		1					3
				1		4		6		4		1				4
			1		5		10		10		5		1			5
		1		6		15		20		15		6		1		6
	1		7		21		35		35		21		7		1	7
1		8		28		56		70		56		28		8		1	8

Fig. 4-3 The binomial triangle representing the coefficients in the expansion of $(1 + x)^n$. This triangle is usually attributed to B. Pascal. Note that the sum across any row is 2^n.

of these levels may be viewed as a result of the fact that the $M_I = \pm\frac{1}{2}$ levels arise from the coincidence of a single level and a doubly degenerate level, as indicated in Fig. 4-2. The selection rules require that the allowed transitions occur between levels having the same value of M_I, and therefore having the same degeneracy. Hence, the relative intensity of observed lines is given by the ratio of the degeneracies of the levels between which transitions occur. Inspection of the intensity ratios $1:1$, $1:2:1$, $1:3:3:1$, etc., reveals that they are precisely the coefficients resulting from the binomial expansion of $(1 + x)^n$, where n is the number of equivalent protons in the set. The successive sets of coefficients for increasing n are readily found from Pascal's triangle (Fig. 4-3). Note that the sum across any row is 2^n, which is the total number of energy levels for each value of M_S.

4-3 ESR SPECTRA OF RADICALS CONTAINING A SINGLE SET OF EQUIVALENT PROTONS

The $\dot{C}H_2OH$ radical is one of the simplest systems which contains one set of two equivalent protons. This radical may be produced in a flow system by mixing a solution of Ti^{3+} and CH_3OH with an H_2O_2 solution just outside the microwave cavity.[†] If the pH is sufficiently low, the OH proton contributes no detectable hyperfine splitting since this proton is rapidly exchanged. Hence, a simple $1:2:1$ triplet arises from the CH_2 protons (Fig. 4-4). The hyperfine splitting is 17.2 G.

One of the most familiar $1:3:3:1$ quartets is that due to the $CH_3\cdot$ radical (Fig. 4-5). The hyperfine splitting constant is 23.0 G. The ESR spectrum of this radical was first observed in an irradiated rigid solid at low temperature.[‡] It has also been produced by electron irradiation of liquid ethane in a microwave cavity.[§] It may be made even at room temperatures by interaction, in a flow system, of the products of the Ti^{3+}—H_2O_2

† W. T. Dixon and R. O. C. Norman, *J. Chem. Soc.*, **1963**:3119; H. Fischer, *Mol. Phys.*, 9:149 (1965).
‡ T. Cole, H. O. Pritchard, N. R. Davidson, and H. M. McConnell, *Mol. Phys.*, 1:406 (1958).
§ R. W. Fessenden and R. H. Schuler, *J. Chem. Phys.*, 39:2147 (1963).

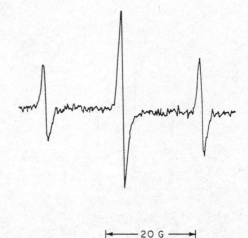

Fig. 4-4 ESR spectrum of the $\dot{C}H_2OH$ radical at pH 1.03. Under these conditions no hyperfine splitting from the OH protons is resolved. (*Spectrum kindly supplied by Dr. H. Fischer.*)

├─── 20 G ───┤

reaction with a molecule such as dimethylsulfoxide $(CH_3)_2SO.$† The methyl radical occupies a central position in the theory of proton hyperfine splitting (Chap. 6), since the unpaired electron resides in a single $2p_z$ orbital.

A frequently studied $1:4:6:4:1$ quintet ESR spectrum of a radical containing four equivalent protons is that of the *p*-benzosemiquinone anion (Fig. 4-6).‡ This radical is readily formed by the autoxidation of hydroquinone in alkaline ethanol solution. The splitting constant, 2.368 G, is among the most accurately measured.§

The cyclopentadienyl radical is an example of a monocyclic hydrocarbon in which all protons are equivalent in solution; in this case the splitting constant is 6.00 G.¶ In the solid state above 120 K one observes a

† F. Dravnieks and J. E. Wertz (unpublished work).
‡ B. Venkataraman and G. K. Fraenkel, *J. Chem. Phys.*, **23**:588 (1955).
§ B. Venkataraman, B. G. Segal, and G. K. Fraenkel, *J. Chem. Phys.*, **30**:1006 (1959).
¶ R. W. Fessenden and S. Ogawa, *J. Am. Chem. Soc.*, **86**:3591 (1964).

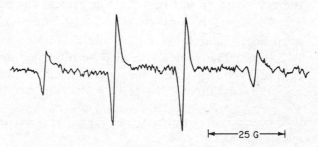

├──25 G──┤

Fig. 4-5 ESR spectrum of the methyl radical $(\dot{C}H_3)$ at 25°C in aqueous solution. (*Spectrum kindly supplied by Mr. Fricis Dravnieks.*)

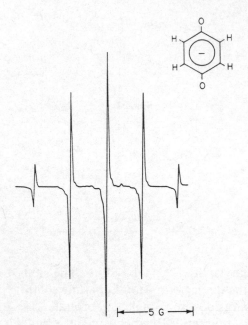

Fig. 4-6 ESR spectrum of the *p*-benzosemiquinone anion showing hyperfine structure from four equivalent protons. The solvent is alkaline ethanol.

|←——5 G ——→|

$1:5:10:10:5:1$ sextet from this radical, produced by irradiation of cyclopentadiene. (See Fig. 4-7.) This spectrum indicates that even at low temperatures the radical is rotating in its own plane. If the magnetic field is not along the fivefold axis of the molecule, the hyperfine interaction will in general be different at each proton. (See Sec. 7-6.) However, rapid in-plane rotation will lead to equivalence of the hyperfine interactions.

The benzene anion (Fig. 4-8) with its $1:6:15:20:15:6:1$ spectrum has also played a central role in the understanding of proton hyperfine splittings. This anion is readily made at 170 K[†]; it is unstable at room temperature. The splitting constant is 3.75 G at 210 K.

The $1:7:21:35:35:21:7:1$ octet of the cycloheptatrienyl (or tropyl) radical is shown in Fig. 4-9. The radical can be made by ultraviolet irradiation of liquid bitropyl (C_7H_7—C_7H_7),[‡] or by electron irradiation of liquid cycloheptatriene.[§] In liquid solution the hyperfine splitting is 3.95 G at ~120 K. The radical can also be observed on irradiation of a solid solution of cycloheptatriene in naphthalene.[¶] Another method of preparation uses the reaction of Ti^{3+}—H_2O_2 with cycloheptatriene in aqueous solution.[††]

† T. R. Tuttle, Jr., and S. I. Weissman, *J. Am. Chem. Soc.*, **80:**5342 (1958).

‡ G. Vincow, M. L. Morrell, W. V. Volland, H. J. Dauben, Jr., and F. R. Hunter, *J. Am. Chem. Soc.*, **87:**3527 (1965).

§ R. W. Fessenden and S. Ogawa, *J. Am. Chem. Soc.*, **86:**3591 (1964).

¶ D. E. Wood and H. M. McConnell, *J. Chem. Phys.*, **37:**1150 (1962).

†† A. Carrington and I. C. P. Smith, *Mol. Phys.*, **7:**99 (1963).

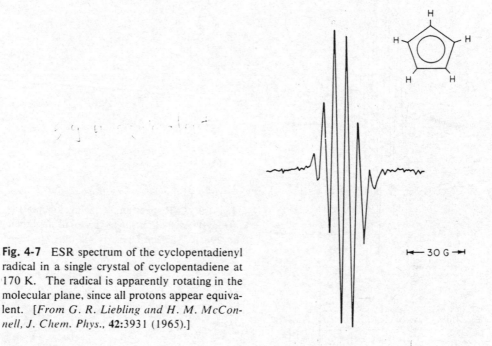

Fig. 4-7 ESR spectrum of the cyclopentadienyl radical in a single crystal of cyclopentadiene at 170 K. The radical is apparently rotating in the molecular plane, since all protons appear equivalent. [*From G. R. Liebling and H. M. McConnell, J. Chem. Phys.*, **42**:3931 (1965).]

←— 30 G —→

Observation of the $1:8:28:56:70:56:28:8:1$ spectrum of the cyclooctatetraene anion (see Fig. 4-10) is significant in that it shows that the eight protons are equivalent. In conjunction with other evidence, this demonstrates that the anion is planar, whereas the neutral molecule is not. The hyperfine splitting is 3.21 G.†

† T. J. Katz and H. L. Strauss, *J. Chem. Phys.*, **32**:1873 (1960).

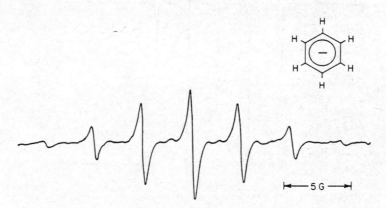

←—5 G—→

Fig. 4-8 ESR spectrum of the benzene anion in a solution of dimethoxyethane and tetrahydrofuran at 170 K. [*Spectrum taken from J. R. Bolton, Mol. Phys.*, **6**:219 (1963).]

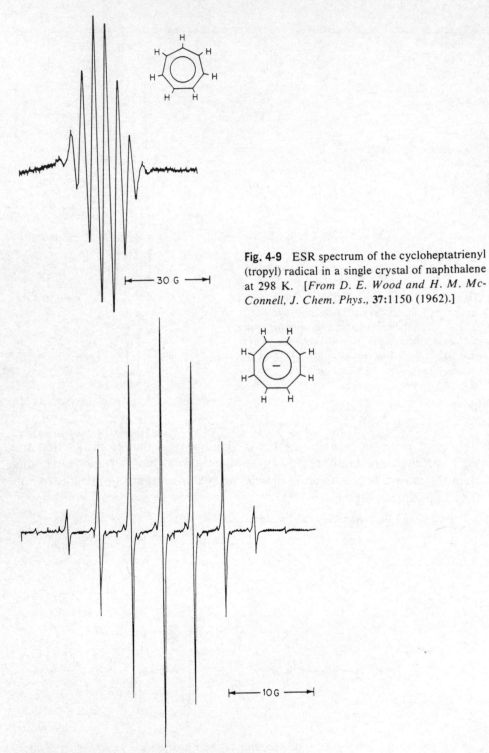

Fig. 4-9 ESR spectrum of the cycloheptatrienyl (tropyl) radical in a single crystal of naphthalene at 298 K. [*From D. E. Wood and H. M. Mc-Connell, J. Chem. Phys.*, **37**:1150 (1962).]

Fig. 4-10 ESR spectrum of the cyclooctatetraene anion. The lines of low intensity on each side of the major lines arise from molecules containing one ^{13}C nucleus. (*Spectrum kindly supplied by Professor G. K. Fraenkel.*)

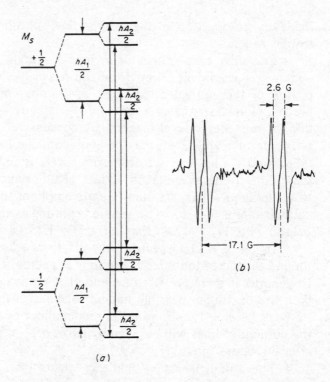

Fig. 4-11 (*a*) Energy level splitting by two nonequivalent nuclei of spin $\frac{1}{2}$ in a constant magnetic field. (*b*) ESR spectrum of the HOĊHCOOH radical as an example of two nonequivalent protons. The larger splitting is due to the CH proton and the smaller to the OH proton.

4-4 ESR SPECTRA OF RADICALS CONTAINING MULTIPLE SETS OF EQUIVALENT PROTONS

The energy levels for hyperfine interaction with a single set of equivalent protons may be obtained by repetitive equal splitting of hyperfine levels of the $M_S = +\frac{1}{2}$ and $M_S = -\frac{1}{2}$ states. Magnetically nonequivalent protons will, in general, have different splitting constants. Consider a radical containing two nonequivalent protons, having hyperfine coupling constants of A_1 and A_2, respectively $(A_1 \gg A_2)$. The energy-level diagram may be constructed by representing the splitting $hA_1/2$ due to the first proton. Next, each of the four resulting levels is split into two levels separated by $hA_2/2$. These energy levels are shown in Fig. 4-11*a*. The allowed transitions are again those for which $\Delta M_S = \pm 1$ and $\Delta M_I = 0$. The spectrum shown in Fig. 4-11*b* is that of the HOĊHCOOH radical. Here $a(\text{COH}) = 2.56$ G and $a(\text{CH}) = 17.13$ G. It is not possible to assign the hyperfine splitting constants solely from an analysis of this spectrum; the assignment may be made by comparison of these hyperfine splittings with those from other similar radicals.

The next case is that of a radical containing three protons, two of which are equivalent. Let A_i be the hyperfine coupling constant for one proton of the equivalent pair and A_j the coupling constant for the single proton. Consider the case for $A_i \gg A_j$. Figure 4-12a shows the construction of the energy-level diagram by successive splitting of energy levels. Crossing of many levels may be avoided if the larger hyperfine coupling is taken first. The final set of energy levels is independent of the order in which the splittings are taken. The $\dot{C}H_2OH$ radical produced by photolysis of a methanol–H_2O_2 solution is an example of this case. The spectrum is shown in Fig. 4-12b. The smaller splitting is apparently that of the OH proton. $a(CH_2) = 17.4$ G and $a(OH) = 1.15$ G.

In the $\dot{C}H_2OH$ spectrum the six lines arise from a doubling by the unique proton of the three transitions expected for two equivalent protons. In general, if there are sets of n and of m equivalent protons in a molecule, then the maximum possible number of lines in the spectrum will be given by $(n + 1)(m + 1)$. For an arbitrary number of sets of equivalent protons the number of lines will be given by $\Pi_i(n_i + 1)$, where Π_i indicates a product over all values of i.

The butadiene anion, made by electrolysis in liquid NH_3,[†] is an example of a molecule with protons equivalent in sets of two and of four. Without guidance from theory, one would not know whether to predict a spectrum of three quintets or one of five triplets. In Chap. 5 it is shown from elementary molecular orbital theory that the latter spectrum is expected. The spectrum in Fig. 4-13a is readily interpreted in terms of five fully resolved groups of $1:2:1$ triplets. Here $a_1 = a_4 = 7.62$ G and $a_2 = a_3 = 2.79$ G. It is necessary to construct the set of energy levels for only one of the M_S spin states, since the two sets of levels are mirror images. When the energy levels are plotted to scale as in Fig. 4-13b, the relative separation of levels corresponds to the separations of lines in the ESR spectrum. (Compare Fig. 4-13b with Fig. 4-13a.) A line is drawn with an amplitude proportional to the degeneracy of the corresponding level. The relative amplitudes then correspond to the predicted relative intensities of the ESR lines. The reconstruction of the spectrum is illustrated in Fig. 4-13b. The positions of lines in this spectrum are a function of the hyperfine *splitting* constants a_i. The kth line in the spectrum will be found at the field H_k given by

$$H_k = H' - \sum_i a_i \tilde{M}_i \qquad (4\text{-}1)$$

† D. H. Levy and R. J. Myers, *J. Chem. Phys.*, **41**:1062 (1964).

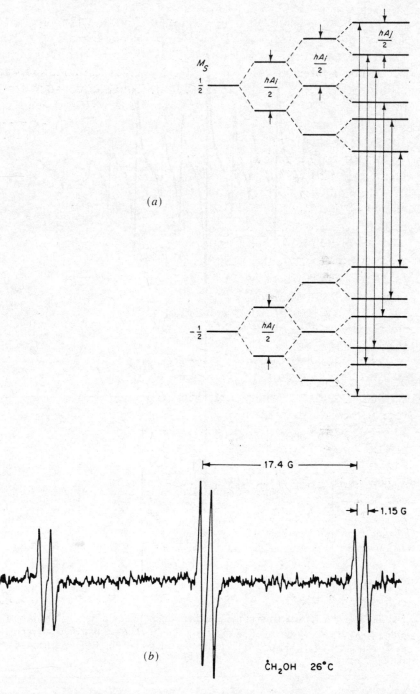

Fig. 4-12 (*a*) Energy-level diagram for the hyperfine interaction of an electron with a set of two equivalent protons and a single proton. (*b*) ESR spectrum of the $\dot{C}H_2OH$ radical in methanol. [*R. Livingston and H. Zeldes, J. Chem. Phys.* **44**:1245 (1966).]

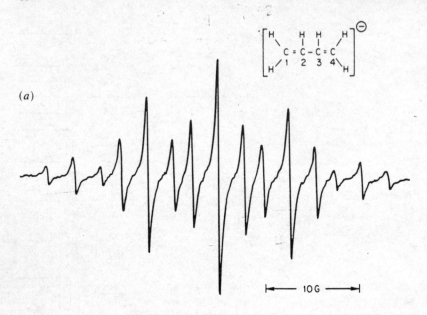

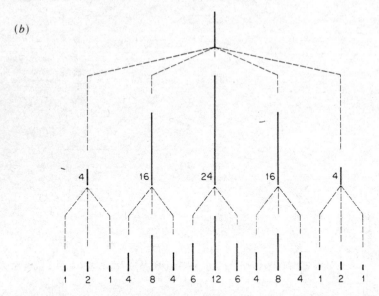

Fig. 4-13 (*a*) ESR spectrum of the butadiene anion in liquid NH_3. [*Spectrum kindly supplied by Professor R. J. Myers. See D. H. Levy and R. J. Myers, J. Chem. Phys.,* **41**:1062 (1964).] (*b*) Reconstruction of the ESR spectrum of butadiene anion.

Here H' is the magnetic field at the center of the spectrum, and $\tilde{M}_i$ is the sum of the M_I values for the protons of the ith set. It is convenient to assume that a_i is negative; as a result, the $\tilde{M}_i$ values will be positive toward the high-field side of the spectrum. It will be shown in Chap. 6 that hyperfine splitting constants may have either positive or negative signs. However, the sign of the hyperfine splitting constant does *not* affect line positions in the spectrum.

The next case to be considered is that of $n = m = 4$, with $a_n \gg a_m$. The predicted *spectrum* is drawn first for the n protons (labeled a_n in Fig. 4-14). The further splitting due to the m protons (labeled a_m) is also shown. The lines are labeled with the appropriate relative intensities. Note that the central line of the final spectrum has an intensity 36 times that of the outermost components. The sum of the relative intensities of all the lines is $2^8 = 256$. (See legend of Fig. 4-3.) This is the number of energy levels for one value of M_S, if eight protons are interacting.

Figure 4-15, showing the biphenylene anion spectrum, is an excellent example of the case $a_n \gg a_m$ ($n = m = 4$). Theoretical considerations, such as those outlined in Chap. 5, are helpful in making a tentative assignment of

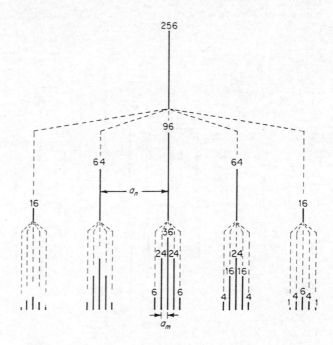

Fig. 4-14 Reconstruction of an ESR spectrum for a radical containing two sets of four equivalent protons with very different hyperfine splitting constants. The numbers on the lines indicate the relative degeneracies.

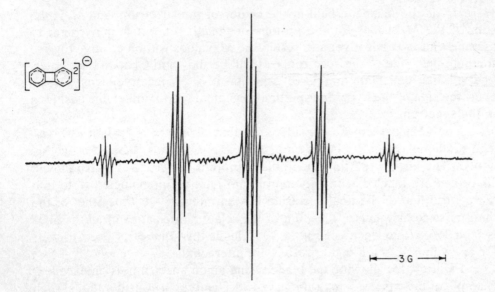

Fig. 4-15 ESR spectrum of the biphenylene anion. Numerous lines of low intensity appearing between the major groups arise from molecules containing one ^{13}C nucleus. (*Spectrum kindly supplied by Dr. J. dos Santos Veiga.*)

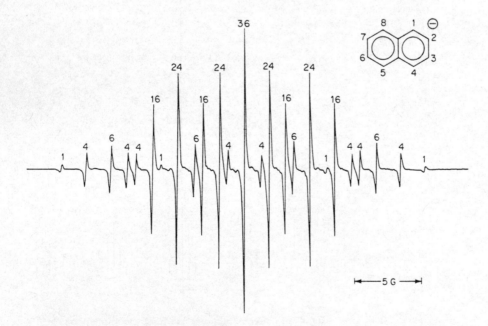

Fig. 4-16 ESR spectrum of the naphthalene anion (K^+ is the counterion) at 298 K. The numbers above each line are the degeneracies of the corresponding nuclear spin states. These numbers correspond approximately to the relative derivative amplitudes.

the splittings. Deuterium substitution at appropriate known sites may permit an experimental verification of the assignment. A single deuteron gives a triplet (instead of a doublet) hyperfine pattern. However, if all other factors are identical, the deuteron splitting is only one-seventh that of the proton. Often the deuterium splitting is too small to be seen.

When $a_n > a_m$, but a_m is sufficiently large, there will be a crossing of line groups. This is true for the naphthalene anion (Fig. 4-16). The splittings are given by $a_1 = 4.90$ G and $a_2 = 1.83$ G. Here the analysis may not be apparent at first sight. The separation of the outermost to the next line is always the smallest hyperfine splitting.

As an aid in the analysis, the degeneracy of the nuclear spin states for each transition is given above the corresponding line. The naphthalene anion is of special historical interest since it was the first radical for which proton hyperfine splitting was observed in solution.[†]

If the difference between a_n and a_m is small, one may fail to see all lines because of overlapping. Whenever large numbers of protons are involved, one must expect at least partial overlapping. If $a_n = ka_m$, where k is an integer or reciprocal of an integer, the spectrum will have fewer than the expected number of lines, and the intensities will not follow a binomial distribution. There are numerous instances in the literature in which erroneous assignments have been made because of such accidental relations. When the *difference* of two splitting constants is exactly or nearly a multiple of another splitting constant, there is a further hazard of misassignment.

In some spectra one finds deviations from the binomial distribution of *amplitudes*. Such deviations are to be expected if the *linewidths* are different. (See Table 2-1 for relations between amplitude and width.) However, the integrated line *intensities* should follow the binomial distribution.

A simple example of a radical with three sets of symmetry-equivalent protons is that of the anthracene anion (Fig. 4-17). The spectrum may be reconstructed by the extension of the procedure for biphenylene; there is an additional triplet splitting arising from the protons at the 9 and 10 positions. $a_1 = 2.73$ G, $a_2 = 1.51$ G, and $a_9 = 5.34$ G[‡]; it will be simplest to consider the splittings in order of decreasing magnitude. A step-by-step analysis of this spectrum is given in Sec. 4-7.

Another radical with three sets of protons is the biphenyl anion (Fig. 4-18). The spectrum consists of a set of nine equally spaced quintets. As there could be $5 \times 5 \times 3 = 75$ lines, it is apparent that there must be some coincidences. The resolved quintet structure indicates that one of the sets of four equivalent protons has a splitting constant (0.39 G) markedly different from the rest. It is not readily apparent how a set of nine lines with the observed intensities arises from a set of two and a set of four protons.

[†] D. Lipkin, D. E. Paul, J. Townsend, and S. I. Weissman, *Science*, 117:534 (1953).

[‡] J. R. Bolton and G. K. Fraenkel, *J. Chem. Phys.*, 40:3307 (1964).

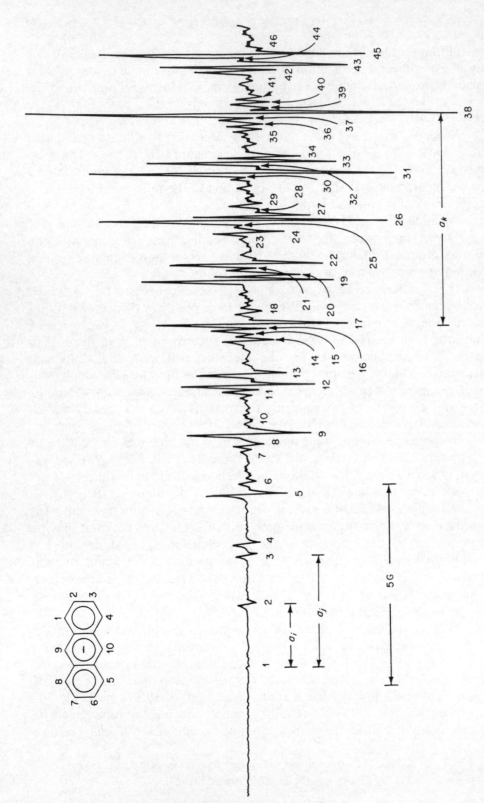

Fig. 4-17 Low-field portion of the ESR spectrum of the anthracene anion. Proton hyperfine lines are numbered; unnumbered lines arise from ^{13}C splittings. The three proton splitting constants are indicated.

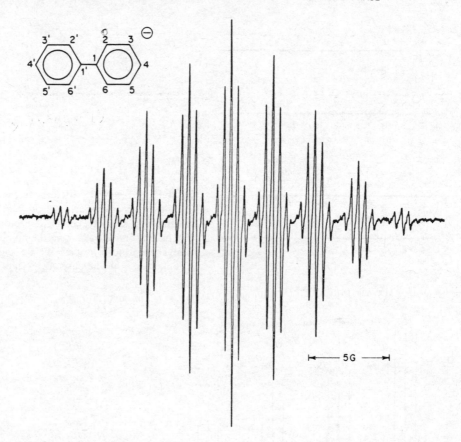

Fig. 4-18 ESR spectrum of the biphenyl anion. K^+ is the counterion.

It is necessary to assume that one splitting is some multiple of the other. As a first guess one may take the splitting from the two equivalent protons to be one-half that from one of the sets of four equivalent protons. The procedure given in Fig. 4-19a will be useful in determining the correct pattern. Eleven sets of lines are predicted for this case; hence, this guess is incorrect. The alternative assumption takes the splitting from a set of four equivalent protons to be one-half that from the two equivalent protons. The resulting splitting pattern shown in Fig. 4-19b correctly predicts nine groups of lines. Using the hyperfine splittings $a_4 = 5.40$ G, $a_2 = 2.70$ G, and $a_3 = 0.39$ G, one obtains a reconstruction which compares favorably with the spectrum in Fig. 4-18. Under conditions of higher resolution (see Fig. 4-20) one finds that the more precise hyperfine splittings are $a_4 = 5.387$ G, $a_2 = 2.675$ G, and $a_3 = 0.394$ G. The expected 75 lines are now resolved. One cannot always be so fortunate as to achieve complete resolution of all possible lines. The assignment of the quintet splittings can readily be made on the basis of elementary theory given in Chap. 5. While this theory would predict a ratio of only 1.8:1 for the two largest splittings, this theoretical prediction is adequate for assignment.

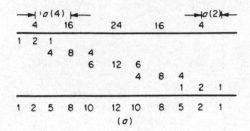

$$\rightarrow|a(4)|\leftarrow \qquad\qquad \rightarrow|a(2)|\leftarrow$$

		4	16	24	16	4				
1	2	1								
		4	8	4						
			6	12	6					
				4	8	4				
					1	2	1			

| 1 | 2 | 5 | 8 | 10 | 12 | 10 | 8 | 5 | 2 | 1 |

(a)

$$\rightarrow|\ a(2)\ |\leftarrow \qquad\qquad \rightarrow|a(4)|\leftarrow$$

	16		32		16			
1	4	6	4	1				
	2	8	12	8	2			
			1	4	6	4	1	

| 1 | 4 | 8 | 12 | 14 | 12 | 8 | 4 | 1 |

(b)

Fig. 4-19 Procedure for determining the relative intensities of lines arising from successive splittings by sets of equivalent protons. (*a*) The case $a(4) = 2a(2)$ where the numbers of equivalent protons in each set are shown in parentheses. (*b*) The case $a(2) = 2a(4)$.

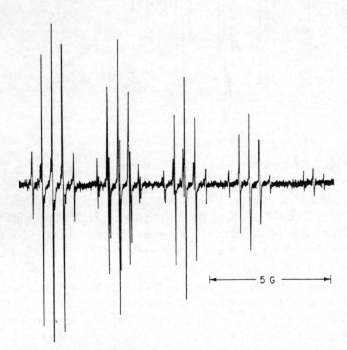

|← 5 G →|

Fig. 4-20 High-field portion of the ESR spectrum of the biphenyl anion at higher resolution than in Fig. 4-18.

4-5 HYPERFINE SPLITTINGS FROM OTHER NUCLEI WITH $I = \frac{1}{2}$

In organic radicals the most common nuclei with $I = \frac{1}{2}$ are ^1H, ^{13}C, ^{19}F, and ^{31}P. Proton hyperfine splittings have already been discussed at length. Hyperfine splitting from ^{19}F or ^{31}P is usually indistinguishable from proton hyperfine splittings. It is an important characteristic of ESR spectra that an analysis yields only the *spin* of the interacting nucleus and the *hyperfine splitting*. Other evidence is required to identify the interacting nucleus. For this reason everything that has been said about the analysis and reconstruction of the spectra involving proton splittings will also apply to ^{19}F and ^{31}P. Sometimes the unusually large magnitude of a ^{31}P splitting will allow one to make an assignment. For ^{19}F, variations in linewidths can sometimes be utilized to make an assignment.[†] ^{13}C splittings are a special case to be considered shortly.

^{19}F hyperfine splittings have been observed in many organic radicals such as perfluoro-*p*-benzosemiquinone.[‡] An especially interesting example is the CF_3 radical[§]; here the ^{19}F hyperfine splitting is 144.75 G.

An example of a radical showing ^{31}P hyperfine splitting is PO_3^{--}. This radical[¶] has a very large isotropic splitting (~ 600 G); this indicates that PO_3^{--} has a pyramidal structure with approximately sp^3 hybridization. If PO_3^{--} were planar, the radical would show a much smaller isotropic splitting. (See Sec. 8-4 for a discussion of the dependence of hyperfine splitting on structure.)

As an example of a molecule showing hyperfine splitting from both ^{31}P and ^{19}F, consider the spectrum attributed to FPO_2^-[††,‡‡] which is shown in Fig. 4-21. The strongest lines in the spectrum are those from this ion. The four equally intense lines can be reconstructed on the basis of two different hyperfine splittings from nuclei with spin $I = \frac{1}{2}$.

The natural abundance of the isotope ^{13}C is 1.1 percent. The more abundant ^{12}C has zero nuclear spin. In some cases, a higher gain in taking an ESR spectrum of an organic radical will reveal satellite lines due to ^{13}C hyperfine splittings. Consider the simple case of a molecule containing one carbon atom (for example, CO_2^-). On the average, 1.1 percent of the CO_2^- molecules will be $^{13}CO_2^-$. For these molecules two lines will arise from the ^{13}C hyperfine splitting. The $^{12}CO_2^-$ spectrum will consist of only one line, since ^{12}C and ^{16}O have zero nuclear spin. The intensity of the $^{13}CO_2^-$ spectrum will be divided between two lines; hence, *each* line of the

† M. Kaplan, J. R. Bolton, and G. K. Fraenkel, *J. Chem. Phys.*, **42:**955 (1965).
‡ D. H. Anderson, P. J. Frank, and H. S. Gutowsky, *J. Chem. Phys.*, **32:**196 (1960).
§ R. W. Fessenden and R. H. Schuler, *J. Chem. Phys.*, **43:**2704 (1965).
¶ A. Horsfield, J. R. Morton, and D. H. Whiffen, *Mol. Phys.*, **4:**475 (1961).
†† J. R. Morton, *Can. J. Phys.*, **41:**706 (1963).
‡‡ R. W. Fessenden, *J. Magnetic Resonance*, **1:**277 (1969).

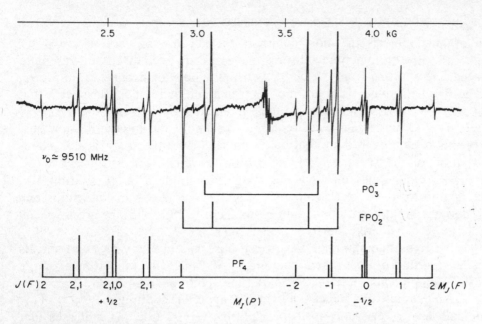

Fig. 4-21 The ESR spectrum of a γ-irradiated single crystal of NH_4PF_6. The three radicals produced have been attributed to PF_4, FPO_2^-, and PO_3^{--}. The splittings of the PF_4 lines are due to second-order interactions. [*Spectrum kindly supplied by Dr. J. R. Morton. See J. R. Morton, Can. J. Phys.*, **41**:706 (1963); *R. W. Fessenden, J. Mag. Res.*, **1**:277 (1969).] Here J(F) refers to the total fluorine nuclear spin quantum number in the coupled representation.

$^{13}CO_2^-$ spectrum will have an intensity of 0.55 percent relative to the intensity of the central line of the $^{12}CO_2^-$ spectrum.

For molecules which contain n equivalent carbon atoms, the intensity of each satellite relative to that of the central component will be $0.0055n$. Consider the spectrum of the benzene negative ion in Fig. 4-22. Here each satellite has an intensity of 3.3 percent of its central ^{12}C component.

The radical arising from 2,5-dihydroxy-p-benzosemiquinone in alkaline solution is an example of a case in which two different ^{13}C hyperfine splittings are observed. The ESR spectrum at high amplification is shown in Fig. 4-23. It consists of a $1:2:1$ triplet from the two ring protons. (These lines are off scale in the center of the spectrum.) The ^{13}C lines are apparent as satellites on the wings. The two ^{13}C splittings are indicated on the spectrum. The relative intensities of the satellites allow the ^{13}C splittings to be assigned. The smaller ^{13}C splitting arises from the carbonyl carbon atoms.

4-6 HYPERFINE SPLITTINGS FROM NUCLEI WITH $I > \frac{1}{2}$

The most commonly encountered examples of nuclei with $I = 1$ are ^{14}N and ^2H (or D). To say that a nucleus has a spin $I = 1$ means that in a mag-

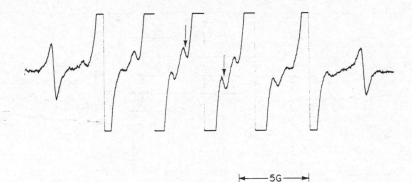

Fig. 4-22 ESR spectrum of the benzene anion. A hyperfine doublet (arrows) about each proton line arises from ^{13}C nuclei. [*Taken from J. R. Bolton, Mol. Phys.. 6:219 (1963).*]

netic field, three orientations will be allowed; these are labeled by the values of $M_I = 0 \pm 1$. (See Chap. 3.) These states are nondegenerate in a magnetic field, in contrast to the case for two equivalent protons. (See Sec. 4-2.) Hence, the spectrum should consist of three equally intense lines. The energy-level diagrams for one and for two equivalent nuclei of spin $I = 1$ are shown in Figs. 3-6 and 4-24, respectively. An example of a single ^{14}N hyperfine splitting is shown in Fig. 4-25, for the di-*t*-butyl nitroxide radical.

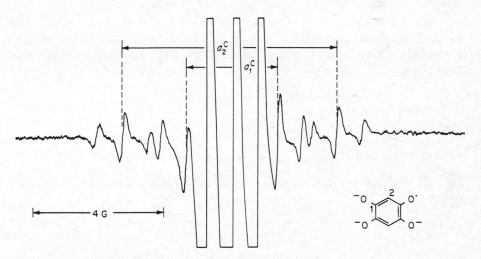

Fig. 4-23 ESR spectrum of the 2,5-dihydroxy-*p*-benzosemiquinone anion. The off-scale triplet in the center arises from molecules having all ^{12}C nuclei. The satellite lines on the wings arise from molecules containing one ^{13}C nucleus. [*Taken from D. C. Reitz, F. Dravnieks, and J. E. Wertz, J. Chem. Phys., 33:1880 (1960).*]

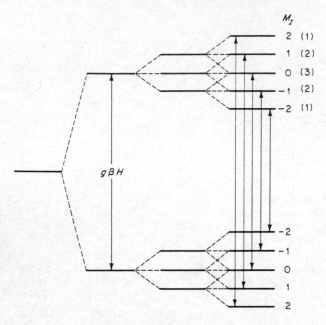

Fig. 4-24 Energy level diagram showing the hyperfine levels for two equivalent nuclei with $I = 1$. Numbers in parentheses indicate the degeneracy of each level.

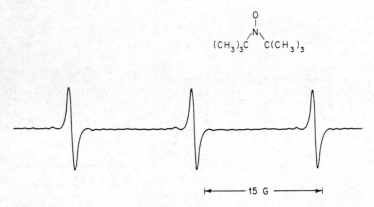

Fig. 4-25 ESR spectrum of the di-t-butyl nitroxide radical in an ethanol solution.

The spectrum of the radical identified as NH_3^+ was given in Fig. 1-3. One notes patterns of $1:1:1$ triplets both for the weaker and the stronger lines. The triplet arises from the single ^{14}N nucleus. Three $1:3:3:1$ quartet patterns are also apparent; these arise from three equivalent protons.

For two equivalent $I = 1$ nuclei, one expects a pattern of five lines with an intensity distribution of $1:2:3:2:1$. (See Fig. 4-24.) The spectrum in Fig. 4-26 exhibits such a pattern.

As mentioned before, analysis of the ESR spectrum does not identify the interacting nuclei. One of the methods of assigning hyperfine splittings is the use of isotopic substitution. The most widely used isotope has been deuterium. In the spectrum a pattern of lines arising from one less proton may hopefully be identified, and hence an assignment made. The two hyperfine splittings in the naphthalene anion were assigned by this procedure.[†]

The most common nuclei with spin $I = \frac{3}{2}$ are ^7Li, ^{23}Na, ^{33}S, ^{35}Cl, ^{37}Cl, ^{39}K, ^{53}Cr, ^{63}Cu, and ^{65}Cu. There are four nuclear spin states ($M_I = +\frac{3}{2}$, $+\frac{1}{2}$, $-\frac{1}{2}$, $-\frac{3}{2}$), and consequently one should observe four hyperfine components of equal intensity. Sometimes ESR spectra of radical anions exhibit small hyperfine splittings from alkali-metal cations. Such splittings indicate the presence of ion pairs in solution. Figure 4-27a shows a spectrum of the pyrazine anion prepared by sodium reduction of pyrazine in dimethoxyethane.[‡] If there were no splitting from ^{23}Na, there would be 25 lines as in Fig. 4-27b. However, due to the ion pairing, each of the lines is split into a quartet by the ^{23}Na interaction.

† T. R. Tuttle, Jr., R. L. Ward, and S. I. Weissman, *J. Chem. Phys.*, **25**:189 (1956).
‡ J. dos Santos Veiga and A. F. Neiva-Correia. *Mol. Phys.*, **9**:395 (1965).

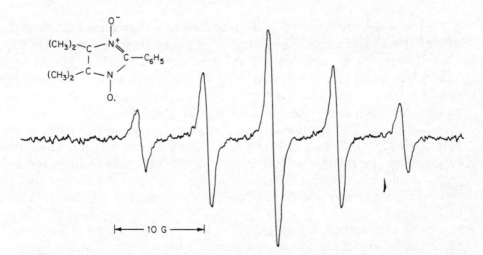

Fig. 4-26 ESR spectrum of a substituted nitronylnitroxide, showing a splitting due to two equivalent nitrogen atoms. [*See J. H. Osiecki and E. F. Ullman, J. Am. Chem. Soc.*, **90**:1078 (1968). Spectrum kindly supplied by Dr. E. F. Ullman.]

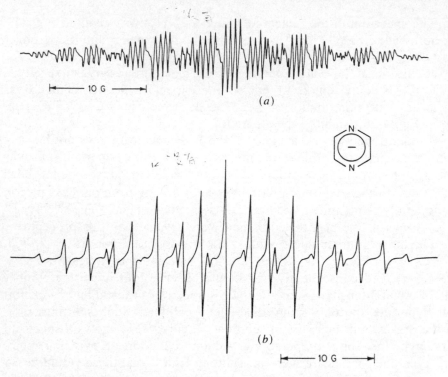

Fig. 4-27 (*a*) ESR spectrum of the pyrazine anion. Na$^+$ is the counterion. [*Spectrum taken from J. dos Santos Veiga and A. F. Neiva-Correia, Mol. Phys.*, **9**:395 (1965).] (*b*) ESR spectrum of the pyrazine anion. K$^+$ is the counterion; no hyperfine splitting from ^{39}K is observed. [*Spectrum taken from A. Carrington and J. dos Santos Veiga, Mol. Phys.*, **5**:21 (1962).]

   The ESR spectrum of XeF produced in a γ-irradiated single crystal of XeF$_4$† was given in Fig. 1-4. It is a simple example of a system in which separate spectra are observed for different isotopic species. Here it is the xenon isotopes ^{129}Xe ($I = \frac{1}{2}$) and ^{131}Xe ($I = \frac{3}{2}$) which give rise to hyperfine splitting; the isotopes of even mass number do not. The relative abundances of ^{129}Xe and ^{131}Xe are 26.4 and 21.2 percent, respectively. The remaining 52.4 percent is distributed among the isotopes of mass number 124, 126, 128, 130, 132, 134, and 136. The XeF radicals containing these isotopes will be referred to collectively as $^{(even)}$XeF. Analysis of the spectrum should begin with a tabulation of expected line patterns and the relative line intensities for different XeF species. Such a tabulation is given in Table 4-1.

   The total intensity of *all* lines for a given species will be proportional to the relative abundance of the Xe isotope in the radical. The Xe hyperfine

† J. R. Morton and W. E. Falconer, *J. Chem. Phys.*, **39**:427 (1963).

Table 4-1

Species	Pattern of lines	Line numbers	Expected relative intensity of lines	Xenon magnetic moments (nuclear magnetons)	Mean xenon hyperfine splitting†	Mean fluorine hyperfine splitting†
(even)XeF	One doublet	4, 11	1.000	—	—	959 G
^{129}XeF	Two doublets	1, 7; 8, 14	0.252	−0.7725	862 G	960 G
^{131}XeF	Two quartets	2, 3, 5, 6; 9, 10, 12, 13	0.101	+0.6868	255 G	960 G

† Mean values of the measured separation in gauss of corresponding line components in the spectrum of XeF in Fig. 1-4.

splittings for the various radical species should be proportional to the magnetogyric ratios (i.e., proportional to μ/I). This ratio for the ^{129}Xe to ^{131}Xe nuclei is 3.374. This compares very well with the ratio 3.380 of the mean measured hyperfine splittings. A small doublet splitting on each line is due to a neighboring fluorine nucleus in the XeF_4 host. Because of the very large magnitude of these hyperfine splittings, the measured values are not constant across the spectrum and do not correspond exactly to $hA/g\beta$, where A is the hyperfine *coupling* constant in megahertz. (See Secs. 4-9, C-7, and C-9.) For ^{129}XeF, the hyperfine coupling constants are $A^{Xe} = 2,368$ MHz and $A^F = 2,637$ MHz; for ^{131}XeF, $A^{Xe} = 701$ MHz and $A^F = 2,653$ MHz; finally for $^{(even)}$XeF, $A^F = 2,649$ MHz. These coupling constants have been determined to within ± 10 MHz. The g factor for all species is 1.9740. All these data have been quoted for **H** parallel to the XeF axis.

4-7 USEFUL RULES FOR THE INTERPRETATION OF SPECTRA

The following are a few important rules which aid in the interpretation of isotropic ESR spectra.

1. The positions of lines in a spectrum are expected to be symmetric about a center point. Asymmetry may be caused by superposition of two spectra, as is shown in Fig. 4-28, due to differences in their g factors. If hyperfine splittings are large, second-order splittings can cause asymmetry of line positions. (See Sec. 4-9.) Variations in spectral linewidths may arise from a slow tumbling rate of the radical. (See Sec. 9-7.) This may also give an appearance of asymmetry.
2. A spectrum having no intense center line indicates the presence of an

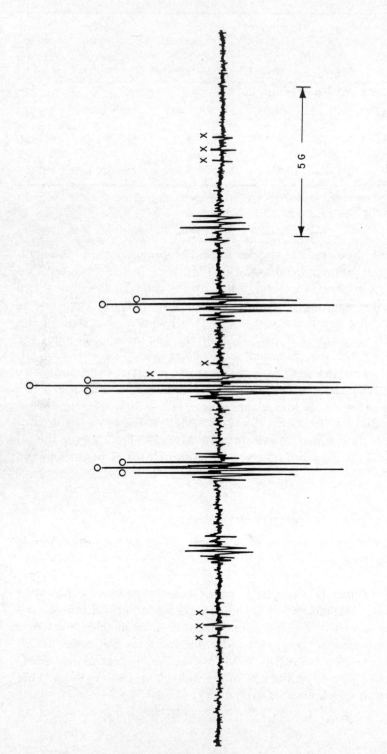

Fig. 4-28 ESR spectrum of a solution containing both the biphenyl and the biphenylene anions. Lines marked with an x arise from the biphenyl anion, while those marked with an o arise from the biphenylene anion. (*Spectrum kindly supplied by Dr. J. dos Santos Veiga.*)

5 G

odd number of equivalent nuclei of half-integral spin. The observation of a center line does not exclude the presence of an odd number of nuclei.

3. For nuclei with $I = \frac{1}{2}$, the sum of the hyperfine splitting constants (absolute values) for *all* nuclei must equal the separation in gauss between the outermost lines, which may be very weak and may therefore be missed. This sum is $\Sigma_i n_i a_i$ where n_i is the number of nuclei with hyperfine splitting a_i.

4. The stick-plot reconstruction, if it is correct, should match the experimental line positions, especially in the wings of the spectrum. If the widths of all lines are equal and there is little overlap, the relative amplitudes should correspond to the degeneracies. If the widths are unequal or the overlap is serious, it may be desirable to undertake a computer simulation of the spectrum. (See 9 below.)

5. The separation of the two outermost lines is always the smallest hyperfine splitting.

6. The total number of energy levels in the system for one value of M_S is given by $\Pi_i (2I_i + 1)^{n_i}$ where n_i is the number of nuclei with spin I_i.

7. The maximum possible number of lines (when second-order splittings are not resolved) is given by $\Pi_i (2n_i I_i + 1)$ where n_i is the number of equivalent nuclei with spin I_i.

8. The positions of all the lines in a spectrum will be given by

$$H_k = H' - \sum_i a_i \tilde{M}_i$$

9. In cases where resolution is poor or lines are too numerous, it may be advantageous to carry out a computer simulation of the spectrum based on assumed hyperfine splittings and linewidths. Such programs have been described by several authors.[†,‡,§] When several hyperfine splittings are present or more than one radical is present, it is imperative to carry out a computer simulation as a test of the analysis.

The anthracene anion spectrum (Fig. 4-17) will now be analyzed as an example of the application of some of these rules. By rule 5 the smallest hyperfine splitting is clearly the separation of lines 1 and 2. For reference, this splitting will be designated a_i. By using a pair of dividers, one may step off equal intervals from the outermost line. These intervals identify a pattern of five lines (lines 1, 2, 4, 6, 10) with amplitudes approximately $1:4:6:4:1$.

Thus the splitting a_i is to be assigned to a set of four equivalent protons. Line 3 is the first line which has not thus far been identified. Using the same divider spacing as before and starting from line 3, one may step

† H. M. Gladney, Ph.D. dissertation, Princeton University, 1963.
‡ E. W. Stone and A. H. Maki, *J. Chem. Phys.*, **38**:1999 (1963).
§ M. Kaplan, Ph.D. thesis, Columbia University, 1965.

off another quintet of lines with relative amplitudes $1:4:6:4:1$ (i.e., lines 3, 5, 9, 13, 18). The interval between corresponding lines of this quintet and the first quintet is the next hyperfine splitting (e.g., the separation between lines 1 and 3). This second hyperfine splitting will be identified as a_j. Starting from the center line of the first quintet (i.e., line 4) and stepping off an interval corresponding to the separation of lines 1 and 3, one may identify a quintet of lines with approximate relative amplitudes of $1:4:6:4:1$ (i.e., lines 4, 9, 17, 27, 39). Thus the second hyperfine splitting a_j must also arise from four equivalent protons.

To find the third and last hyperfine splitting a_k, which must correspond to a set of two equivalent protons (protons 9 and 10), one may invoke rule 3. The separation between lines 1 and 38, i.e., half the total extent of the spectrum, is given by

$$\Delta H(1\text{--}38) = 2a_i + 2a_j + a_k$$

The interval $2a_i$ from line 1 brings one to line 4; moving to the right a distance $2a_j$ from line 4, one arrives at line 17. Hence, the interval between lines 17 and 38 must be a_k, i.e., the hyperfine splitting to be assigned to the two equivalent protons.

At this point the reader is urged to begin the construction of a stick-plot diagram using the methods outlined in Figs. 4-13b and 4-14. It is desirable to start with the largest hyperfine splitting (that is, a_k).

4-8 OTHER PROBLEMS ENCOUNTERED IN THE ESR SPECTRA OF FREE RADICALS

Most of the ESR spectra encountered in this chapter refer to liquid samples in which the radicals are free to reorient rapidly. The reorientation averages out any anisotropy in the g factor and in the hyperfine splittings. Free radicals are often encountered in a rigid matrix. If the host is a single crystal, one may obtain a maximum amount of information from the ESR spectra taken as a function of orientation of the crystal in the magnetic field. If the radicals are randomly oriented, one may be able to extract a significant amount of information about their structure from their spectrum. The analysis of such spectra requires a detailed understanding of the nature of anisotropic interactions. This subject is treated in Chap. 7, and numerous examples are discussed in Chap. 8.

For the spectra given in this chapter, all the lines have amplitudes which are proportional to the intensities of the lines. For other systems, this is frequently not the case, especially if radicals are in media of high viscosity which still allow some reorientation or if the radical can exhibit some degree of internal reorganization. Then the linewidths in a given spectrum will vary, sometimes manyfold. The analysis of such spectra can yield kinetic information. This subject is treated in Chap. 9.

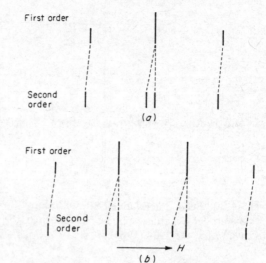

Fig. 4-29 (a) Second-order splitting of a 1:2:1 triplet arising from two equivalent nuclei of spin $\frac{1}{2}$ when splittings become large. (b) Second-order splittings for a 1:3:3:1 quartet for three equivalent nuclei of spin $\frac{1}{2}$.

4-9 SECOND–ORDER SPLITTINGS

The analysis of hyperfine splittings which has been presented is valid only in cases where the hyperfine coupling energy hA_0 is very much smaller than the electron Zeeman energy $g\beta H$. Where hyperfine couplings are *very large* or when small magnetic fields are used, additional splitting of some lines can occur. This additional splitting is usually called "second-order" splitting since the energies of the levels must be calculated to second order, using perturbation theory. A detailed analysis is not presented here (see Secs. C-7 and C-9); however, the behavior for equivalent nuclei with $I = \frac{1}{2}$ will be outlined.†

Consider the case of two equivalent nuclei with $I = \frac{1}{2}$. Note that the central line of the $1:2:1$ triplet arises from transitions between degenerate energy levels. When hA_0 becomes a significant fraction of $g\beta H$, this degeneracy is lifted, and one observes four equally intense lines (see Fig. 4-29a). In fact, all lines except one of the central lines have been shifted downfield from the "first order" positions.

In general, when the nuclear spins **I** of a set of equivalent nuclei are added vectorially to form a total spin vector **J** with components M_J, the second-order field shift of the lines will be given by

$$\Delta H^{(2)} = -\frac{hA_0^2}{2g\beta\nu_0}\left[J(J+1) - M_J^2\right] \tag{4-2}$$

As a further example, consider the case of three equivalent nuclei of spin $I = \frac{1}{2}$. The first- and second-order spectra are sketched in Fig. 4-29b. The spectrum of the CF_3 radical, shown in Fig. 4-30, displays the second-

† For detailed treatment consult R. W. Fessenden. *J. Chem. Phys.*, **37**:747 (1962) and also R. W. Fessenden and R. H. Schuler, *J. Chem. Phys.*, **43**:2704 (1965).

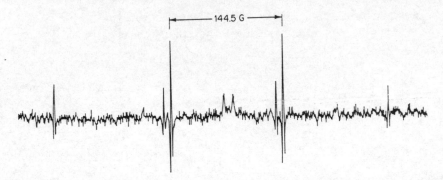

Fig. 4-30 Second-derivative ESR spectrum of the CF_3 radical showing resolved second-order hyperfine splitting. [*From R. W. Fessenden and R. H. Schuler, J. Chem. Phys.*, **43**:2704 (1965).]

order splittings given by Eq. (4-2). The first-order spectrum of the ethyl radical ($CH_3\dot{C}H_2$) is displayed in Fig. 4-31a. Under higher resolution the second-order splittings become apparent as in Fig. 4-31b.

The second-order splittings briefly considered here are observed only for relatively large hyperfine splittings. They are noted here to alert the

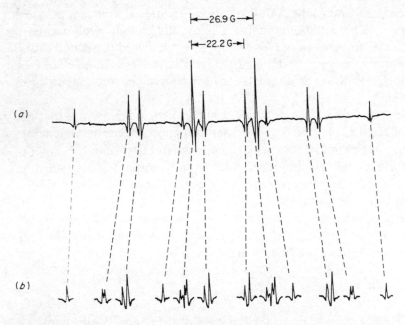

Fig. 4-31 (a) ESR spectrum of the ethyl radical observed by continuous electron bombardment of liquid ethane. The lines are second derivatives of the absorption lines under conditions of low resolution. [*From R. W. Fessenden and R. H. Schuler, J. Chem. Phys.*, **33**:935 (1960).] (b) ESR spectrum of the ethyl radical under conditions of higher resolution. The additional splittings are due to second-order effects. [*Spectra taken from R. W. Fessenden. J. Chem. Phys.*, **37**:747 (1962).]

reader to their occurrence. However, they provide no information which is not already obtainable from the first-order spectrum.

PROBLEMS

4-1. Show that in synthesizing an ESR spectrum the sequence of application of splitting constants does not affect the position or intensity of lines. Demonstrate this for a system with three nonequivalent protons, taking $a_1 = 7.0$ G, $a_2 = 3.0$ G, and $a_3 = 1.0$ G.

4-2. What is the maximum possible number of hyperfine ESR lines to be expected in the first-order spectra of the following radicals in liquid solution (all protons interact)?

CH₃

Toluene anion

Acenaphthene anion
(all four methylene protons
are symmetry-equivalent)

4-3. Use Eq. (4-1) and the hyperfine splittings of the butadiene anion to specify the relative positions of all lines in the spectrum. Use the scale on Fig. 4-13*a* to measure the field value of each line relative to the center; compare with the computed values.

4-4. Construct an energy-level diagram for one of the M_S states of the biphenyl anion, using the splittings given in the text. Compare the results with the spectra shown in Figs. 4-18 and 4-20.

4-5. Interpret the spectrum of the catechol anion (Fig. 4-32). Determine the hyperfine splittings and show successive splittings of one of the M_S states.

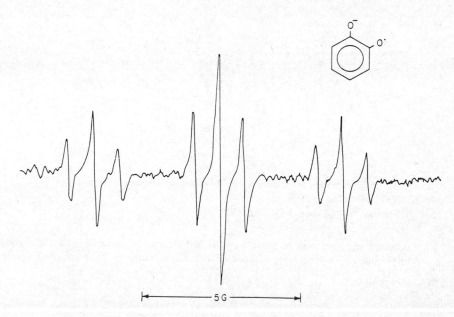

Fig. 4-32   ESR spectrum of catechol anion at pH 7.6. (*Figure kindly supplied by Dr. I. C. P. Smith.*)

4-6. Analyze the spectra (Figs. 4-33 and 4-34) of neutral (*a*) and cationic (*b*) *p*-benzosemiquinones to obtain the hyperfine splittings.

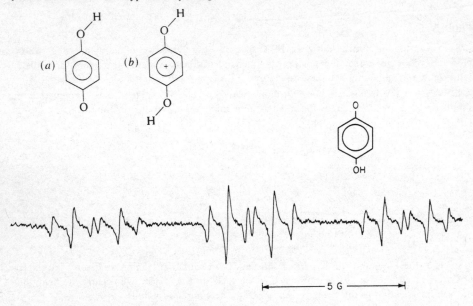

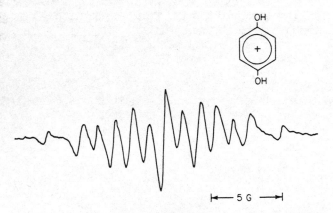

Fig. 4-33 ESR spectrum of the neutral *p*-benzosemiquinone radical OC_6H_4OH. (*Spectrum kindly supplied by Dr. T. Gough.*)

Fig. 4-34 ESR spectrum of the cation of *p*-benzosemiquinone ($HOC_6H_4OH^+$). (*The spectrum is taken from J. R. Bolton and A. Carrington, Proc. Chem. Soc., 1961:*385.)

4-7. Analyze the spectrum of the phenyl trimethylgermane anion (Fig. 4-35) to obtain the hyperfine splittings (only the ring protons interact appreciably with the unpaired electron).

$Ge(CH_3)_3$

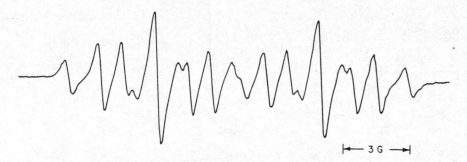

Fig. 4-35 ESR spectrum of the anion of phenyltrimethylgermane $[C_6H_5Ge(CH_3)_3]$.
[*The spectrum is taken from J. A. Bedford. J. R. Bolton, A. Carrington, and R. H. Prince, Trans. Faraday Soc.,* **59**:53 (1963).]

4-8. Complete the assignment of lines in the anthracene anion spectrum (Fig. 4-17), using the splittings given in the text.

4-9. Interpret the ESR spectrum of the pyrene anion (Fig. 4-36). (Hint: see Fig. 4-19.)

4-10. Draw the energy-level diagram for the interaction of one electron, in the presence of a strong magnetic field, with
  (*a*) Three equivalent nuclei of spin $I = 1$
  (*b*) Two equivalent nuclei of spin $I = \frac{3}{2}$

4-11. How many ESR hyperfine lines would you expect in the completely resolved spectrum of the following radicals in liquid solution (all magnetic nuclei interact)?
  (*a*) CF_2H
  (*b*) CH_2D
  (*c*) $(CF_3)_2NO$

(*d*)

(*e*)

4-12. (*a*) When deuterium (^2H) is substituted for hydrogen (^1H) in a free radical, can one predict the value of a^D if a^H is known for the undeuterated radical? Assume that no other changes occur.

  (*b*) Figure 4-37 displays the spectrum of a mixture of CH_2D and CHD_2. Identify the lines belonging to each spectrum and give values for each of the hyperfine splittings. Compute the ratio a^H/a^D.

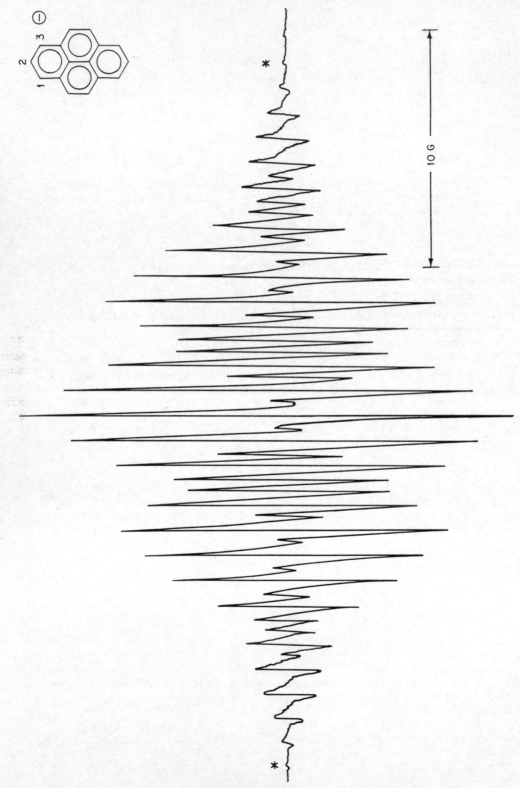

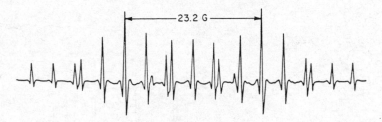

Fig. 4-37 Second-derivative ESR spectrum of a mixture of CHD_2 and CH_2D. [*Spectrum kindly supplied by Dr. R. W. Fessenden. See R. W. Fessenden, J. Phys. Chem.*, **71**:74 (1967).]

4-13. Interpret and draw a stick-plot diagram of the following ESR spectra:

(*a*) $(CF_3)_2NO$ (Fig. 4-38)

(*c*) (Fig. 4-27*a*)

(*b*) (Fig. 4-39)

(*d*) (Fig. 4-40)

(*e*) $CH_3\dot{C}HOH$ and $\dot{C}H_2CH_2OH$ (Fig. 1-5)

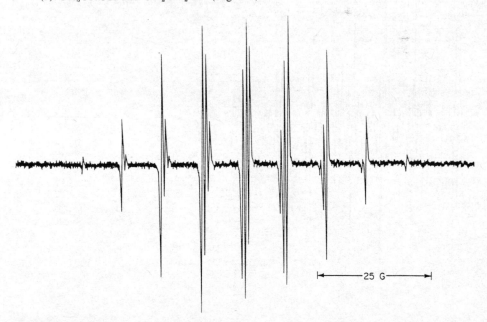

Fig. 4-38 ESR spectrum of $(CF_3)_2NO$. [*From P. J. Scheidler and J. R. Bolton, J. Am. Chem. Soc.*, **88**:371 (1966).]

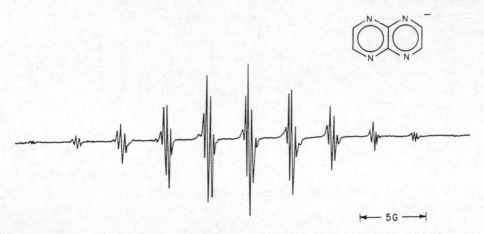

Fig. 4-39 ESR spectrum of tetra-azanaphthalene anion. [*From F. Gerson and W. L. F. Armarego, Helv. Chim. Acta,* **48:**112 (1965).]

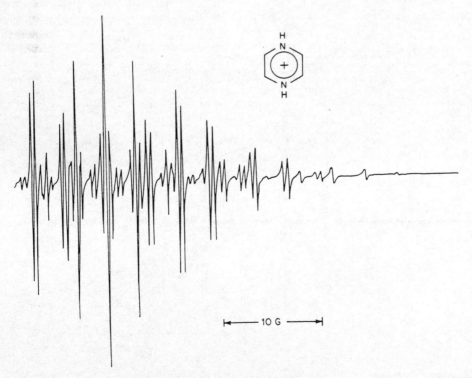

Fig. 4-40 High-field portion of the ESR spectrum of the dihydropyrazine cation in dimethylformamide. (*Spectrum kindly supplied by Dr. George K. Fraenkel.*)

4-14. Figure 4-41 represents the spectrum obtained when a single crystal of KCl, doped with ^{33}S, is γ-irradiated. The crystal has been prepared from a sample enriched in ^{33}S to an extent of 60 percent. The spectrum has been ascribed to S_2^-.

(a) What are the various possible S_2^- species, and what are their relative abundances?

(b) How many lines are to be expected for each of these S_2^- species?

(c) Compute the ^{33}S hyperfine splitting.

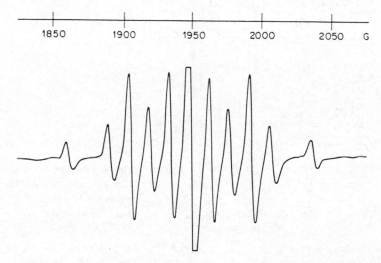

1850   1900   1950   2000   2050   G

Fig. 4-41   ESR spectrum at 9.550 GHz of S_2^- in a γ-irradiated single crystal of KCl doped with ^{33}S (60%). [*Spectrum kindly supplied by Dr. J. R. Morton*). *See L. E. Vannotti and J. R. Morton, Phys. Rev.*, **161**:282 (1967).]

4-15. The spectrum shown in Fig. 4-21 contains some lines which have been assigned to the radical PF_4. The isotropic hyperfine coupling constants and g factor are as follows:

$a^P = 3.769$ MHz

$a^F = 549$ MHz

$g = 1.9985$

The fluorine splitting pattern clearly shows resolved second-order splittings.

(a) If the unshifted line in the low-field fluorine splitting pattern (i.e., the line with $J = 0$ and $M_J = 0$) occurs at 2,501.1 G, predict the positions of the other eight lines in the low-field group.

(b) The unshifted line of the high-field group occurs at 3,848.4 G. The separation of this line and that of the low-field group corresponds to the ^{31}P hyperfine *splitting*. Why is this not quite the same as $hA_0{}^P/g\beta$?

4-16. The spectrum of peroxylamine disulfonate in Fig. 4-42 shows eight satellites. Explain their origin, and relate their intensities to those of the central lines.

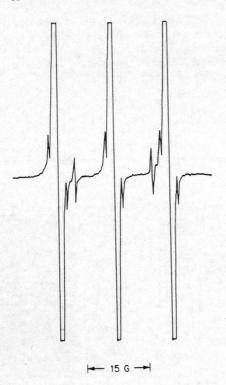

├── 15 G ──┤

Fig. 4-42 The ESR spectrum at high gain of peroxylamine disulfonate [(SO₃)₂NO⁻⁻]. [*Spectrum kindly supplied by Dr. J. J. Windle. See J. J. Windle and A. K. Wiersema, J. Chem. Phys.,* **39:**1139 (1963).]

5

Interpretation of Hyperfine
Splittings in π-type
Organic Radicals

5-1 INTRODUCTION

For a number of the spectra examined in Chap. 4, it was not possible to
assign the observed hyperfine splittings on the basis of the spectrum alone.
For the naphthalene anion (Fig. 4-16), it is not obvious which set of four
equivalent protons should be assigned the larger hyperfine splitting. The
same uncertainty is found for the biphenylene anion (Fig. 4-15). Furthermore, the two quintet splittings in the anthracene anion (Fig. 4-17) were not
assigned. It would be desirable to have some rational basis for the assignment of these hyperfine splittings. In addition, it would be helpful if the
relative magnitudes of hyperfine splittings could be predicted without making
extremely detailed calculations. Fortunately, a very useful yet simple
procedure has long been available; this is the Hückel molecular orbital
method.†

All of the radicals cited above as well as many others examined in
Chap. 4 are conjugated molecules. Benzene is perhaps the classical example of a conjugated molecule. The distinguishing characteristic of all
these compounds is the overlap of $2p_z$ orbitals on adjacent atoms. Such

† E. Hückel, *Z. Physik.*, **70**:204 (1931); **76**:628 (1932); **83**:632 (1933).

overlapping permits the electrons in these $2p_z$ orbitals to be delocalized over the molecular skeleton. One may describe the energy states of these electrons in terms of *molecular orbitals* generated from linear combinations of the atomic $2p_z$ orbitals.

In diatomic molecules, orbitals for which the electron distribution is cylindrically symmetric about the internuclear axis are referred to as σ orbitals. Orbitals for which the electron distribution has a nodal plane containing the internuclear axis are referred to as π orbitals. Each of the $2p_z$ orbitals in benzene has a node in the molecular plane. Hence, the molecular orbitals arising from combinations of $2p_z$ orbitals are referred to as π orbitals. If the unpaired electron in a radical resides in a π-molecular orbital, the radical is called a π-type radical. Most organic free radicals belong to this class. However, there are a limited number of free radicals in which the unpaired electron occupies a σ orbital. This orbital is often principally localized at a position in the molecule where an atom is missing. As an example, one may cite the vinyl radical

Discussion of such σ radicals is deferred until Sec. 8-3. This chapter is devoted exclusively to π-type radicals.

A brief outline of the Hückel molecular orbital (HMO) approach to the calculation of orbital energies and unpaired-electron distributions in π-electron systems will be given here. This outline is presented for the benefit of the reader who is unfamiliar with the application of the method. Detailed molecular orbital calculation procedures and tabulation of the results for many molecules are given in the references at the end of this chapter.

5-2 MOLECULAR ORBITAL ENERGY CALCULATIONS

The assignment of an unpaired electron to a particular molecular orbital of a radical requires knowledge of the number and ordering of the set of orbital energies of the system. Knowledge of the molecular orbital which contains the unpaired electron enables one to obtain the distribution of the unpaired electron over the radical. Since an exact solution of the Schrödinger equation for a polyatomic radical is hopelessly complicated, one is forced to make a number of approximations, some of them drastic.

The HMO approach applies to planar molecules in which the $2s$, $2p_x$, and $2p_y$ orbitals of carbon atoms hybridize to form three equivalent (sp^2) σ bonds which make an angle of $120°$ with one another.

(See Fig. 5-1.)   The $2p_z$ orbital is perpendicular to the plane of the other three.   Overlap of $2p_z$ orbitals on adjacent atoms permits a considerable degree of delocalization of π electrons over the molecular framework established by the σ bonds.   It is explicitly assumed that the interactions between a π electron and electrons in the σ bonds can be neglected for magnetic resonance considerations.   Thus it is possible to construct an energy-level scheme involving only the π electrons.

For a molecular framework of n atoms, it is assumed that the molecular orbital wave function ψ_i can be written as a linear combination of atomic $2p_z$ orbitals.

$$\psi_i = c_{i1}\phi_1 + c_{i2}\phi_2 + \cdots + c_{ij}\phi_j + \cdots + c_{in}\phi_n$$

$$= \sum_{j=1}^{n} c_{ij}\phi_j \quad (5\text{-}1)$$

The energy W_i corresponding to this wave function is obtained from

$$W_i = \frac{\displaystyle\int_\tau \psi_i^* \hat{\mathscr{H}} \psi_i \, d\tau}{\displaystyle\int_\tau \psi_i^* \psi_i \, d\tau} \quad (5\text{-}2)$$

[See Eq. (1-19).]

Here $\hat{\mathscr{H}}$ is a hamiltonian which takes account of the interaction of a single π electron with the potential field arising from the nuclei and the σ electrons. $\hat{\mathscr{H}}$ does not include specific interactions between π electrons.

Being ignorant of the *exact* wave function ψ, one finds it helpful to have some guide in the choice of an *approximate* wave function.   The variational principle states that when the exact wave function is sub-

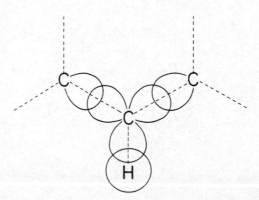

Fig. 5-1   A fragment of a planar conjugated hydrocarbon showing the trigonal array of σ bonds (sp^2 hybrids) about each carbon atom.   One set of σ bonds is shown between carbon and hydrogen atoms.

stituted in Eq. (5-2), a minimum energy, which is the exact energy, is obtained. Hence of several trial wave functions, the most satisfactory one is that which leads to a minimum energy. As an illustration of the variational principle, consider the simplest of molecules, viz., H_2^+, which has only one electron and two protons. A molecular wave function for H_2^+ may be approximated by

$$\psi = c_1\phi_1 + c_2\phi_2 \tag{5-3}$$

ϕ_1 and ϕ_2 are hydrogenlike $1s$ atomic orbitals; c_1 and c_2 are adjustable coefficients. In the following, it is assumed that the wave functions are real, that is, $\psi^* = \psi$. Then

$$W = \frac{\displaystyle\int_\tau (c_1\phi_1 + c_2\phi_2)\hat{\mathcal{H}}(c_1\phi_1 + c_2\phi_2)\ d\tau}{\displaystyle\int_\tau (c_1\phi_1 + c_2\phi_2)^2\ d\tau}$$

$$= \frac{\displaystyle\int_\tau [(c_1\phi_1\hat{\mathcal{H}}c_1\phi_1) + (c_1\phi_1\hat{\mathcal{H}}c_2\phi_2) + (c_2\phi_2\hat{\mathcal{H}}c_1\phi_1) + (c_2\phi_2\hat{\mathcal{H}}c_2\phi_2)]\ d\tau}{\displaystyle\int_\tau (c_1{}^2\phi_1{}^2 + 2c_1c_2\phi_1\phi_2 + c_2{}^2\phi_2{}^2)\ d\tau} \tag{5-4}$$

For this form of the molecular orbital of H_2^+, the variational principle provides a criterion for choosing c_1 and c_2. The best wave function is that for which the values of c_1 and c_2 are such that $\partial W/\partial c_1$ and $\partial W/\partial c_2$ are equal to zero. Subsequent calculations will be simplified by the definition of the quantities H_{ij} and S_{ij}:

$$H_{11} = \int_\tau \phi_1\hat{\mathcal{H}}\phi_1\ d\tau \qquad S_{11} = \int_\tau \phi_1{}^2\ d\tau$$

$$H_{22} = \int_\tau \phi_2\hat{\mathcal{H}}\phi_2\ d\tau \qquad S_{22} = \int_\tau \phi_2{}^2\cdot d\tau$$

$$H_{12} = \int_\tau \phi_1\hat{\mathcal{H}}\phi_2\ d\tau \qquad S_{12} = \int_\tau \phi_1\phi_2\ d\tau$$

$$H_{21} = \int_\tau \phi_2\hat{\mathcal{H}}\phi_1\ d\tau \tag{5-5}$$

Since $\hat{\mathcal{H}}$ is a hermitian operator (see Sec. A-2a), $H_{12} = H_{21}$. H_{11} is called the coulomb integral and H_{12} the resonance integral. W may be written as follows

$$W = \frac{c_1{}^2H_{11} + 2c_1c_2H_{12} + c_2{}^2H_{22}}{c_1{}^2S_{11} + 2c_1c_2S_{12} + c_2{}^2S_{22}} \tag{5-6}$$

Next

$$\frac{\partial W}{\partial c_1} = \frac{(c_1^2 S_{11} + 2c_1 c_2 S_{12} + c_2^2 S_{22})(2c_1 H_{11} + 2c_2 H_{12})}{(c_1^2 S_{11} + 2c_1 c_2 S_{12} + c_2^2 S_{22})^2}$$
$$- \frac{(c_1^2 H_{11} + 2c_1 c_2 H_{12} + c_2^2 H_{22})(2c_1 S_{11} + 2c_2 S_{12})}{(c_1^2 S_{11} + 2c_1 c_2 S_{12} + c_2^2 S_{22})^2} = 0 \quad (5\text{-}7)$$

Simplifying

$$2c_1 H_{11} + 2c_2 H_{12}$$
$$= \frac{(c_1^2 H_{11} + 2c_1 c_2 H_{12} + c_2^2 H_{22})}{(c_1^2 S_{11} + 2c_1 c_2 S_{12} + c_2^2 S_{22})}(2c_1 S_{11} + 2c_2 S_{12}) \quad (5\text{-}8)$$

Since the fraction is just W,

$$c_1 H_{11} + c_2 H_{12} = W(c_1 S_{11} + c_2 S_{12}) \quad (5\text{-}9)$$

or alternatively,

$$c_1(H_{11} - WS_{11}) + c_2(H_{12} - WS_{12}) = 0 \quad (5\text{-}10)$$

The equation resulting from setting $\partial W/\partial c_2 = 0$ is

$$c_1(H_{12} - WS_{12}) + c_2(H_{22} - WS_{22}) = 0 \quad (5\text{-}11)$$

Equations (5-10) and (5-11) are two simultaneous linear equations; these may be solved by setting the determinant of the coefficients equal to zero, recalling that c_1 and c_2 are the variables. Then

$$\begin{vmatrix} H_{11} - WS_{11} & H_{12} - WS_{12} \\ H_{12} - WS_{12} & H_{22} - WS_{22} \end{vmatrix} = 0 \quad (5\text{-}12)$$

Equations (5-10) and (5-11) are referred to as the *secular equations*, and Eq. (5-12) is the *secular determinant*. By symmetry, $H_{11} = H_{22}$ and $S_{11} = S_{22}$; both S_{11} and S_{22} are set equal to unity because normalized atomic orbitals are used. S_{12}, the overlap integral, is arbitrarily set equal to zero. The drastically simplified determinantal equation is then

$$\begin{vmatrix} H_{11} - W & H_{12} \\ H_{12} & H_{11} - W \end{vmatrix} = 0 \quad (5\text{-}13)$$

Each term may be divided by H_{12}. The determinant is further simplified by defining $x = (H_{11} - W)/H_{12}$. Then

$$\begin{vmatrix} x & 1 \\ 1 & x \end{vmatrix} = 0 \qquad x^2 - 1 = 0 \qquad x = \pm 1 \quad (5\text{-}14)$$

The two solutions are

$$W_1 = H_{11} + H_{12} \quad (5\text{-}15)$$
$$W_2 = H_{11} - H_{12} \quad (5\text{-}16)$$

Thus there are two possible orbital energies for H_2^+, one (W_1) of lower energy than that of isolated H atoms and one (W_2) of higher energy. In the ground state† the single electron occupies the orbital corresponding to W_1, since H_{12} is a negative quantity (as is also H_{11}).

Substituting $W_1 = H_{11} + H_{12}$ into the equation $c_1(H_{11} - W) + c_2 H_{12} = 0$, one obtains the ratio $c_1/c_2 = 1$. Alternatively, upon substituting the value W_2, one obtains $c_1/c_2 = -1$. Hence the wave functions corresponding to W_1 and W_2 are

$$\psi_1 = c_1(\phi_1 + \phi_2) \tag{5-17}$$

$$\psi_2 = c_1(\phi_1 - \phi_2) \tag{5-18}$$

The value of c_1 is determined from the normalization condition

$$\int_\tau \psi^2 \, d\tau = 1$$

Substituting Eq. (5-17),

$$\int_\tau c_1{}^2 (\phi_1 + \phi_2)^2 \, d\tau = c_1{}^2 \left(\int_\tau \phi_1{}^2 \, d\tau + 2 \int_\tau \phi_1 \phi_2 \, d\tau + \int_\tau \phi_2{}^2 \, d\tau \right)$$

$$= c_1{}^2 (S_{11} + 2S_{12} + S_{22}) \tag{5-19}$$

Since S_{12} has been set equal to zero and $S_{11} = S_{22} = 1$, $c_1 = 1/\sqrt{2}$, and therefore

$$\psi_1 = \frac{1}{\sqrt{2}} (\phi_1 + \phi_2) \tag{5-20a}$$

$$\psi_2 = \frac{1}{\sqrt{2}} (\phi_1 - \phi_2) \tag{5-20b}$$

ψ_1 will be called the symmetrical or bonding wave function, since the interchange of nuclei 1 and 2 does not change the sign of ψ_1. ψ_2, the antisymmetric or antibonding orbital, has a change of sign, and therefore a node between the two nuclei. It is a very general result that the energy increases with an increasing number of nodes; examples of this property will be given later for benzene.

Next consider the simplest of π-type organic molecules, viz., ethylene. The molecular orbital for a *single* electron moving over the carbon skeleton is represented as a linear combination of the carbon $2p_z$ atomic orbitals, in close analogy to the treatment of H_2^+. In fact, for

---

† One must carefully distinguish among the terms orbital, configuration, and state. An *orbital* is a wave function which represents a solution of a one-electron hamiltonian.

A *configuration* designates an assignment of electrons to a set of orbitals according to the Pauli principle. A configurational wave function consists of an antisymmetrized product of occupied orbitals.

A *state* of a molecule in general may be a linear combination of configurations. In this book we shall usually be dealing with states represented by a single configuration.

the π system, the expressions for the orbital energies and for the molecular wave function are analogous to those for H_2^+. There are now two π electrons; both of these are assigned to the lower energy (bonding) orbital, with opposite spin, according to the Pauli principle. This configuration will be taken to represent the ground state.

For a linear conjugated molecule of n atoms, there will be n molecular orbitals, each a linear combination of n atomic orbitals. The resulting $n \times n$ secular determinant is set equal to zero

$$\begin{vmatrix} H_{11} - WS_{11} & H_{12} - WS_{12} & H_{13} - WS_{13} & \cdots & H_{1n} - WS_{1n} \\ H_{12} - WS_{12} & H_{22} - WS_{22} & H_{23} - WS_{23} & & \cdots \\ H_{13} - WS_{13} & H_{23} - WS_{23} & H_{33} - WS_{33} & & \cdots \\ \cdot & \cdot & \cdot & \cdots & \cdot \\ H_{1n} - WS_{1n} & \cdots & & & H_{nn} - WS_{nn} \end{vmatrix} = 0$$

(5-21)

A number of simplifying assumptions are now made:

1. $S_{ii} = 1$, $S_{ij} = 0$ $(i \neq j)$.
2. All $H_{ij}(j \neq i) = \beta$ if atoms are bonded and zero otherwise. β is called the resonance integral.
3. All $H_{ii} = \alpha$. α is called the coulomb integral.

For the allyl molecule

$$\begin{vmatrix} \alpha - W & \beta & 0 \\ \beta & \alpha - W & \beta \\ 0 & \beta & \alpha - W \end{vmatrix} = 0 \quad (5\text{-}22)$$

All terms are divided by β and the substitution $x = (\alpha - W)/\beta$ is made. Then

$$\begin{vmatrix} x & 1 & 0 \\ 1 & x & 1 \\ 0 & 1 & x \end{vmatrix} = 0$$

(5-23)

Expansion of the determinant gives

$$x^3 - 2x = 0$$
$$x = 0, \pm\sqrt{2}$$

Hence

$$W_1 = \alpha + \sqrt{2}\beta$$
$$W_2 = \alpha$$

and

$$W_3 = \alpha - \sqrt{2}\beta$$

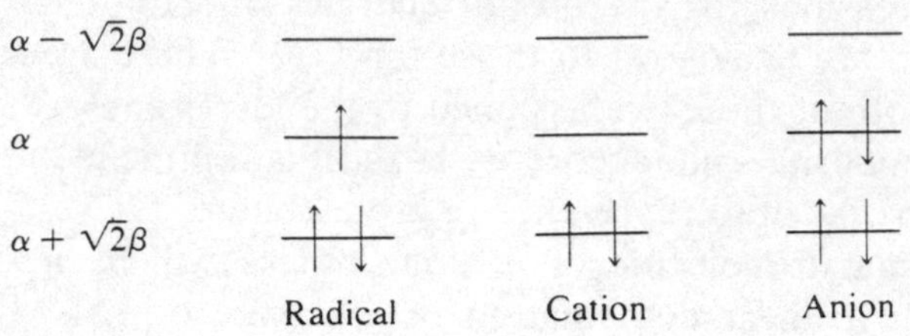

$\alpha - \sqrt{2}\beta$ _____   _____   _____ $\psi_3 = \frac{1}{2}\phi_1 - \frac{1}{\sqrt{2}}\phi_2 + \frac{1}{2}\phi_3$

α $\underline{\uparrow}$ _____ $\underline{\uparrow\downarrow}$ $\psi_2 = \frac{1}{\sqrt{2}}\phi_1 + 0\phi_2 - \frac{1}{\sqrt{2}}\phi_3$

$\alpha + \sqrt{2}\beta$ $\underline{\uparrow\downarrow}$ $\underline{\uparrow\downarrow}$ $\underline{\uparrow\downarrow}$ $\psi_1 = \frac{1}{2}\phi_1 + \frac{1}{\sqrt{2}}\phi_2 + \frac{1}{2}\phi_3$

Radical   Cation   Anion

Fig. 5-2 The orbital energies and the wave functions of the allyl radical, cation, and anion. Here ϕ_i is a $2p_z$ atomic orbital on carbon atom i.

The orbital energies and the ground-state configuration for the allyl radical, cation, and anion are shown in Fig. 5-2.

The set of corresponding molecular orbitals can be obtained from the secular determinant, Eq. (5-23), by writing each line as an equation and substituting one value of x; for example, $-\sqrt{2}$:

$$-\sqrt{2}c_1 + c_2 = 0$$

$$c_1 - \sqrt{2}c_2 + c_3 = 0$$

$$c_2 - \sqrt{2}c_3 = 0 \qquad (5\text{-}24)$$

Solution of Eqs. (5-24) yields only the ratios of coefficients. The coefficients themselves are determined by invoking the normalization condition

$$\int_\tau \psi^2 \, d\tau = \int_\tau (c_1\phi_1 + c_2\phi_2 + c_3\phi_3)(c_1\phi_1 + c_2\phi_2 + c_3\phi_3) \, d\tau = 1$$

$$= c_1{}^2 + c_2{}^2 + c_3{}^2 = 1 \qquad (5\text{-}25)$$

since ϕ_1, ϕ_2, and ϕ_3 are assumed to be orthogonal and normalized. The generalized form of Eq. (5-25)

$$\sum_{i=1}^{n} c_i{}^2 = 1 \qquad (5\text{-}26)$$

will apply to systems with n π centers. The combination of Eqs. (5-25) and (5-24) yields the coefficients in ψ_1 of Fig. 5-2. Substitution of $x = 0$ and $x = +\sqrt{2}$ in turn gives the coefficients in ψ_2 and ψ_3, respectively.

The calculation of the four Hückel molecular orbitals and energies of butadiene is given as a problem at the end of this chapter; the results are quoted in Table 5-1.

The neutral butadiene molecule has four π electrons. Following the rules, these must be assigned to the molecular orbitals of lowest energy (i.e., two to ψ_1 and two to ψ_2, since β is negative) for the ground state.

For other conjugated systems one may proceed in an analogous fashion. The secular determinant for linear conjugated systems will have the values $\alpha - W$ on the diagonal, β one off the diagonal, and zero elsewhere. For cyclic systems there are other nonzero off-diagonal terms. Solution of the resulting $n \times n$ secular determinant, although readily feasible by computers, is laborious by hand. The secular determinant can often be factored

Table 5-1   Molecular orbitals and energies of butadiene

$$\psi_4 = 0.371\phi_1 - 0.600\phi_2 + 0.600\phi_3 - 0.371\phi_4; \quad W_4 = \alpha - \left(\frac{\sqrt{5}+1}{2}\right)\beta$$

$$\psi_3 = 0.600\phi_1 - 0.371\phi_2 - 0.371\phi_3 + 0.600\phi_4; \quad W_3 = \alpha - \left(\frac{\sqrt{5}-1}{2}\right)\beta$$

$$\psi_2 = 0.600\phi_1 + 0.371\phi_2 - 0.371\phi_3 - 0.600\phi_4; \quad W_2 = \alpha + \left(\frac{\sqrt{5}-1}{2}\right)\beta$$

$$\psi_1 = 0.371\phi_1 + 0.600\phi_2 + 0.600\phi_3 + 0.371\phi_4; \quad W_1 = \alpha + \left(\frac{\sqrt{5}+1}{2}\right)\beta$$

if use is made of the symmetry properties of the molecule by straightforward methods of group theory.[†],[‡]

A simple geometric construction (Fig. 5-3) will allow one to obtain the Hückel orbital energies for monocyclic π systems.[§] The appropriate regular polygon is inscribed in a circle of radius 2β such that there is a vertex at the lowest point. Horizontal lines drawn through each vertex will represent to scale the set of orbital energies for a particular molecule. The vertical separations x from the position of the center (taken as the energy α) give the energies $W = \alpha - \beta x$. For each of these systems, some pairs of orbitals are degenerate; the orbital of lowest energy is always nondegenerate. These properties may be seen from Fig. 5-3. Orbitals with energies less than α are called *bonding* orbitals, whereas those energies greater than α are called *antibonding* orbitals. If an orbital has the energy α, it is called a *nonbonding* orbital.

5-3 UNPAIRED–ELECTRON DISTRIBUTIONS

The unpaired electron of a π-type radical is expected to be distributed over the molecular framework. In the benzene anion the time-average probability of finding the unpaired electron in the vicinity of any one carbon atom should be $\frac{1}{6}$, as required by symmetry. For other monocyclic radicals a similiar uniform distribution should be found. The equivalence of each position in a given monocyclic radical is demonstrated by the hyperfine splitting patterns shown in Figs. 4-7 to 4-10.

[†] A. Streitwieser, Jr., "Molecular Orbital Theory," chap. 3. John Wiley & Sons, Inc., New York, 1961.
[‡] F. A. Cotton, "Chemical Applications of Group Theory," chap. 7. Interscience Publishers, a division of John Wiley & Sons, Inc., New York, 1963.
[§] A. A. Frost and B. Musulin, *J. Chem. Phys.*, **21**:572 (1953).

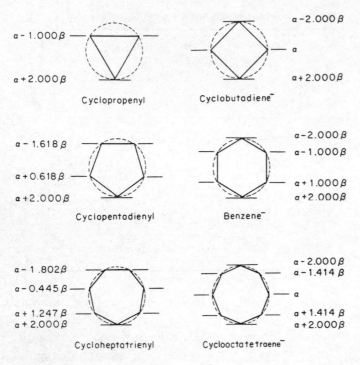

Fig. 5-3 Orbital energies for monocyclic systems. The reference energy is taken as α.

For radicals with lower symmetry (e.g., the butadiene anion) there is no such obvious guide to the unpaired-electron distribution. The HMO approach provides valuable guidance toward determining this distribution, though in some cases its predictions are not fully valid. The information of interest is contained in the expression for the molecular orbital ψ_k occupied by the unpaired electron

$$\psi_k = c_{k1}\phi_1 + c_{k2}\phi_2 + \cdots + c_{kn}\phi_n \qquad (5\text{-}27)$$

If ψ_k is to be normalized, $c_{k1}{}^2 + c_{k2}{}^2 + \cdots + c_{kn}{}^2 = 1$. The square of the coefficient c_{kj} of the atomic orbital ϕ_j is the probability that the electron in the molecular orbital ψ_k is on atom j. Thus $c_{kj}{}^2$ measures the *unpaired-π-electron density* ρ_j on the atom j, that is,

$$\rho_j = c_{kj}{}^2 \qquad (5\text{-}28)$$

As an example, consider the radical anion of 1,3-butadiene.[†] The ESR spectrum is displayed in Fig. 4-13a. It was analyzed in Chap. 4 on the basis of a quintet of lines of relative intensity $1:4:6:4:1$ with a hyperfine splitting of 7.62 G; each line of the quintet is split further into a $1:2:1$ triplet with a hyperfine splitting of 2.79 G. The structure of the molecule dictates that the quintet hyperfine splitting be assigned to the four equivalent protons

[†] D. H. Levy and R. J. Myers, *J. Chem. Phys.*, **41**:1062 (1964).

at 1 and 4 positions; the triplet hyperfine splitting is then assigned to the two equivalent protons at 2 and 3 positions. The considerable difference in the hyperfine splitting constants suggests a highly nonuniform unpaired-electron distribution.

The butadiene anion has five π electrons. Reference to the molecular orbitals of Table 5-1 shows that the unpaired electron must reside in ψ_3. Utilizing Eq. (5-28) the unpaired-electron densities are found to be $\rho_1 = \rho_4 = 0.360$ and $\rho_2 = \rho_3 = 0.140$. The HMO theory predicts that the end carbon atoms should have the higher unpaired-electron densities. These are also the positions at which the larger proton hyperfine splittings are observed. Note that the ratio of the hyperfine splittings, $a_1/a_2 = 2.73$, agrees satisfactorily with the ratio of the unpaired-electron densities, $\rho_1/\rho_2 = 2.61$. This correspondence seems to point to some sort of linear relation between the unpaired-π-electron densities and the proton hyperfine splittings in π-type organic radicals. Indeed, such a relation has been proposed;[†,‡,§] it may be written as

$$a = Q\rho \tag{5-29}$$

where Q is a proportionality constant. The origin of Eq. (5-29) will be considered in Chap. 6; for the present, its validity will be assumed. An examination of Fig. 5-4 shows that for most π-type organic radicals the correlation is reasonably good.

Theoretical estimates of the magnitude of Q place it in the range of -20 to -30 G. The significance of the negative sign will be explained in

† H. M. McConnell, *J. Chem. Phys.*, **24**:764 (1956); H. M. McConnell and D. B. Chesnut, *J. Chem. Phys.*, **28**:107 (1958).
‡ S. I. Weissman, *J. Chem. Phys.*, **25**:890 (1956).
§ R. Bersohn, *J. Chem. Phys.*, **24**:1066 (1956).

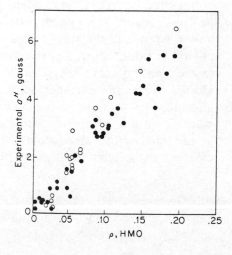

Fig. 5-4 Proton hyperfine splittings vs. HMO unpaired electron densities for a group of aromatic hydrocarbon radical ions. Open circles refer to positive ions and full circles to negative ions. [*Taken from I. C. Lewis and L. S. Singer, J. Chem. Phys.*, **43**:2712 (1965).]

Table 5-2 Hyperfine parameters for monocyclic radicals

Radical	Temperature,[†] K	a, G	Q, G	Reference
C_5H_5	≈ 200	6.00	30.0	‡
$C_6H_6^-$	173	3.75	22.5	§
$C_6H_6^+$	298	4.28	25.7	¶
C_7H_7	298	3.95	27.7	††,‡‡
$C_8H_8^-$	≈ 298	3.21	25.7	§§

† Some of these hyperfine splittings have been found to be temperature dependent.‡,¶,‡‡

‡ R. W. Fessenden and S. Ogawa, *J. Am. Chem. Soc.*, **86**:3591 (1964).

§ J. R. Bolton, *Mol. Phys.*, **6**:219 (1963).

¶ M. K. Carter and G. Vincow. *J. Chem. Phys.*, **47**:292 (1967).

†† A. Carrington and I. C. P. Smith, *Mol. Phys.*, **7**:99 (1963).

‡‡ G. Vincow, M. L. Morrell, W. V. Volland, H. J. Dauben, Jr., and F. R. Hunter, *J. Am. Chem. Soc.*, **87**:3527 (1965).

§§ T. J. Katz and H. L. Strauss, *J. Chem. Phys.*, **32**:1873 (1960).

Chap. 6. However, in certain molecules, it is possible to establish a value semiempirically from the experimental hyperfine splittings. For instance, in the planar cyclic polyene radicals C_5H_5, $C_6H_6^\pm$, C_7H_7, and $C_8H_8^-$, the unpaired-electron density is known from the symmetry of these molecules. Thus an experimental determination of a provides an estimate of Q. Table 5-2 gives the experimental values of a and the corresponding values of Q for these monocyclic radicals.

There is considerable variation in Q. If one compares the values for the two neutral radicals or for the two negatively charged radicals, the variation is much smaller, suggesting that the charge on the radical may have some effect on Q. This effect will be explored in Sec. 6-5.

An understanding of the hyperfine splitting properties of protons in conjugated hydrocarbon radicals is aided by their classification as *alternant* or *nonalternant*. A system is defined as alternant if one may identify *alternate* positions of the skeleton (as with an asterisk) and have no two adjacent positions both "starred" or "unstarred." All linear systems are alternant, as are also those cyclic systems which have no rings with an odd number of atoms. Thus the cyclopentadienyl and cycloheptatrienyl radicals are nonalternant, as is the azulene anion. When there are alternative ways of starring atoms, by convention one adopts that designation which gives the larger number of starred atoms. For example, in the benzyl radical (I), one chooses the second designation below.

(I)   (II)

Table 5-3 Benzyl radical splittings

Protons on carbon atoms	Splitting const., exptl.†,‡	HMO calc.
2, 6	−4.9	−4.0
4	−6.1	−4.0
7	−15.9	(−15.9)
3, 5	1.5	0.0

† W. T. Dixon and R. O. C. Norman, *J. Chem. Soc.*, **1964**:4857.
‡ A. Carrington and I. C. P. Smith, *Mol. Phys.*, **9**:137 (1965).

Odd-alternant hydrocarbon radicals have a very useful property which permits rapid calculation of the unpaired electron densities without actually determining molecular orbital coefficients. Using the allyl radical (II) as a simple example of an odd-alternant hydrocarbon, one notes that for the non-bonding molecular orbital ψ_2 (see Fig. 5-2), unstarred positions have zero coefficients, and the coefficients about an unstarred position sum to zero. As a second example, consider the nonbonding molecular orbital ψ_4 of the benzyl radical (I):

$$\psi_4 = 0\phi_1 - 0.378\phi_2 + 0\phi_3 + 0.378\phi_4 + 0\phi_5 - 0.378\phi_6 + 0.756\phi_7$$

Having starred this odd-alternant radical appropriately, one assigns equal and opposite coefficients about unstarred positions having two neighbors. One begins by assigning the coefficient $-x$ to atom 2, $+x$ to 4, $-x$ to 6 and finally $+2x$ to 7 to cancel the contributions from atoms 2 and 6. The squares of the coefficients must sum to unity; hence $x = 1/\sqrt{7}$. The unpaired-electron density is then $\frac{1}{7}$ at atoms 2, 4, and 6 and $\frac{4}{7}$ at atom 7. The simple procedure employed here for determining unpaired-electron densities saves much effort as compared with the direct HMO calculation. This procedure may also be applied to even-alternant hydrocarbons if nonbonding orbitals are present (e.g., cyclooctatetraene).

The experimental hyperfine splittings for the benzyl radical are given in Table 5-3. Using the splitting for position 7 to fix Q, the hyperfine splittings for positions 2, 4 and 6 are calculated to be -4.0 G. No hyperfine splitting would be expected for protons in positions 3 and 5 because the atomic orbital coefficients are zero. The significance of the small positive hyperfine splitting observed for protons at these positions will be discussed in Chap. 6. Although there are significant deviations from predictions, one can regard the calculated values as being in remarkable agreement with experiment, considering the crudity of the approach.

5-4 THE BENZENE ANION AND ITS DERIVATIVES

The set of orbital energies for benzene is of special interest. In common with the other monocyclic systems pictured in Fig. 5-3, the π-molecular

Table 5-4 Molecular orbitals and energies for benzene

Molecular orbitals	*Orbital energies*
$\psi(b) = \dfrac{1}{\sqrt{6}}\,(\phi_1 - \phi_2 + \phi_3 - \phi_4 + \phi_5 - \phi_6)$	$W(b) = \alpha - 2\beta$
$\psi(e_2) = \dfrac{1}{2}\,(\phi_2 - \phi_3 + \phi_5 - \phi_6)$	$W(e_2) = \alpha - \beta$
$\psi(e_2) = \dfrac{1}{\sqrt{12}}\,(2\phi_1 - \phi_2 - \phi_3 + 2\phi_4 - \phi_5 - \phi_6)$	$W(e_2) = \alpha - \beta$
$\psi(e_1) = \dfrac{1}{2}\,(\phi_2 + \phi_3 - \phi_5 - \phi_6)$	$W(e_1) = \alpha + \beta$
$\psi(e_1) = \dfrac{1}{\sqrt{12}}\,(2\phi_1 + \phi_2 - \phi_3 - 2\phi_4 - \phi_5 + \phi_6)$	$W(e_1) = \alpha + \beta$
$\psi(a) = \dfrac{1}{\sqrt{6}}\,(\phi_1 + \phi_2 + \phi_3 + \phi_4 + \phi_5 + \phi_6)$	$W(a) = \alpha + 2\beta$

orbital of lowest energy in benzene is nondegenerate; some of the higher orbitals form degenerate pairs. (See Table 5-4.) It is customary to use the group-theoretical labeling; here it is sufficient to note that e always refers to degenerate pairs of orbitals, whereas a and b refer to nondegenerate orbitals.

In the benzene anion, the extra electron is in the e_2 set of orbitals, whereas in the cation an electron is missing from the e_1 set of orbitals. The set of six molecular orbitals for benzene is given in Table 5-4 in order of *increasing* energy (bottom to top). The bracketed orbitals are degenerate. Note that in the a orbital there is no change of sign and hence no node. This is the orbital of lowest energy. In increasing order of energy, the e_1 orbitals have one change of sign, whereas the e_2 orbitals have two changes of sign, and hence two nodal planes. The b orbital, that of highest energy, has three sign changes and three nodal planes. It will be shown shortly from hyperfine-splitting data that substituents may cause a removal of the degeneracy of the e_2 molecular orbitals of benzene.

The ESR spectrum of the benzene anion at $-100°$ C in the presence of alkali metal was shown in Fig. 4-8. The spectrum consists of seven lines with intensities characteristic of hyperfine interaction from six equivalent protons. This result is expected from the symmetry of the molecule, but it is instructive to see how it arises from the Hückel molecular orbitals given in Table 5-4. The six π electrons of the neutral benzene go into the six bonding molecular orbitals, but the addition of an extra electron to make the benzene anion creates a new problem. The lowest unoccupied molecular orbital in benzene is doubly degenerate. Hence, the unpaired electron on a time average will occupy equally the two e_2 antibonding molecular orbitals. The coefficients at each of the atoms for these orbitals are given at the right

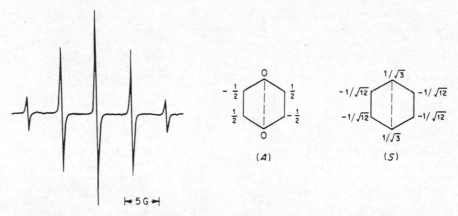

Fig. 5-5 ESR spectrum of the *p*-xylene anion [*from J. R. Bolton and A. Carrington, Mol. Phys.*, 4:497 (1961)], with the atomic orbital coefficients of the antisymmetric *A* and symmetric *S* molecular orbitals of benzene at the right. The symmetry is defined with respect to the perpendicular plane (dotted) passing through the center of the molecule.

of Fig. 5-5. It is evident that *A* is antisymmetric with respect to reflection in a plane passing through carbon atoms 1 and 4; *S* is symmetric with respect to reflection in the same plane. *A* will be termed the "antisymmetric" orbital and *S* the "symmetric" orbital.

The average unpaired-electron density at a given position is obtained by taking one-half the sum of the electron densities (squares of coefficients) at that position for each of the two orbitals. For example, at position 1, $\rho_1 = \frac{1}{2}(0 + \frac{1}{3}) = \frac{1}{6}$, and at position 2, $\rho_2 = \frac{1}{2}(\frac{1}{4} + \frac{1}{12}) = \frac{1}{6}$. These results are expected from the symmetry of the benzene molecule.

The effect of substituents on the ESR spectrum of the benzene anion is best understood by considering the limiting spectra anticipated when the unpaired-electron distribution approximates that of the *A* or the *S* orbitals. The spectrum in Fig. 5-5 is that of the *p*-xylene anion.[†] The splitting from the CH_3 protons is extremely small; this is to be expected if the unpaired electron resides predominantly in the *A* orbital.

Although in the benzene anion the orbitals *A* and *S* are usually equally occupied, the population balance is extremely delicate. The introduction of substituents serves to remove the effective degeneracy, making one orbital more stable than the other. Even the substitution of deuterium for hydrogen alters the relative energies of the *A* and the *S* configurations. The ESR spectrum of the benzene-1-*d* anion[‡] (Fig. 5-6) can be interpreted in terms of the proton hyperfine splittings shown at the right of Fig. 5-6. The observed departure of the unpaired-electron distribution from that of the benzene anion can arise only if *A* is more stable than *S*. Nevertheless, the

† J. R. Bolton and A. Carrington, *Mol. Phys.*, 4:497 (1961).
‡ R. G. Lawler, J. R. Bolton, G. K. Fraenkel, and T. H. Brown, *J. Am. Chem. Soc.*, **86**:520 (1964).

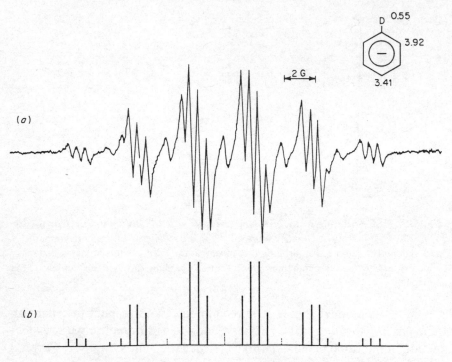

Fig. 5-6 (*a*) ESR spectrum of the benzene-1-*d* anion $(C_6H_5D)^-$. (*b*) Spectrum reconstructed from the splitting constants given at the top right of the spectrum; the dotted lines arise from $C_6H_6^-$ as an impurity. [*Spectrum taken from R. G. Lawler, J. R. Bolton, G. K. Fraenkel, and T. H. Brown, J. Am. Chem. Soc.*, **86**:520 (1964).]

magnitude of the proton splittings indicates that the separation in energy of the *A* and the *S* states is small.

The spectrum of the *p*-xylene anion shows that the introduction of substituents such as CH_3 removes the degeneracy of *A* and *S*.†,‡ The hyperfine splittings which have been observed for the methyl-substituted benzenes are given in Fig. 5-7. The electronic properties of the substituent determine whether the *A* or the *S* orbital will have the lower energy. The methyl group is considered to be electron-releasing in conjugated systems. For the toluene anion, the antisymmetric orbital has a node through the 1 and 4 positions, whereas the symmetric orbital has a large unpaired-electron density ($\frac{1}{3}$) at those positions. Repulsion between the electrons of the methyl group and the large negative charge at position 1 in the *S* orbital causes the latter to be destabilized relative to *A*.

The *Q* value of 22.5 G for the benzene anion may be used to estimate hyperfine splittings. An unpaired-electron density of $\frac{1}{4}$ should give rise to a hyperfine splitting of ≈ 5.6 G. Because of the node through the 1 and 4

† It appears that upon alkyl substitution, there still remains a small contribution to the spin distribution from a configuration of higher energy, namely, the S configuration.‡
‡ E. de Boer and J. P. Colpa, *J. Phys. Chem.*, **71**:21 (1967).

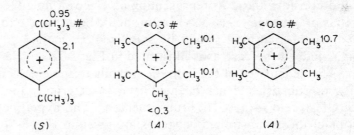

Fig. 5-7 ESR proton hyperfine splittings (in gauss) for substituted benzene anions. (*a*) Methyl-substituted benzene anions. (*b*) Alkyl-substituted benzene anions. (*c*) Silicon and germanium analogs of the *t*-butylbenzene anion. (*d*) Alkyl-substituted benzene cations. *A* and *S* indicate, respectively, that the antisymmetric and the symmetric orbitals lie lowest for these molecules. [(†) *J. R. Bolton and A. Carrington, Mol. Phys.,* **4**:497 (1961). (‡) *J. R. Bolton, J. Chem. Phys.,* **41**:2455 (1964). (§) *J. R. Bolton, A. Carrington, A. Forman, and L. E. Orgel, Mol. Phys.,* **5**:43 (1962). (¶) *J. A. Bedford, J. R. Bolton, A. Carrington, and R. H. Prince, Trans. Faraday Soc.,* **59**:53 (1963). (#) *R. M. Dessau, S. Shih, and E. I. Heiba, J. Am. Chem. Soc.,* **92**:412 (1970).]

positions, one should expect little or no hyperfine splitting from the methyl protons or the proton *para* to the methyl group. The measured hyperfine splittings (Fig. 5-7a) show that the unpaired-electron distribution does approximate that of the A orbital.

The A orbital is also stabilized in the p-xylene anion; in the m-xylene and o-xylene anions, it appears that the S orbital is more stable. In this orbital the methyl groups are at positions of low electron density. To see how well the S orbital distribution is approximated, one may compute the expected hyperfine splittings as before. These are 7.5 and 1.9 G for unpaired-electron densities of $\frac{1}{3}$ and $\frac{1}{12}$, respectively. The agreement with the values given for the m-xylene and o-xylene anions in Fig. 5-7a is very satisfactory.

Successive substitution of methyl groups for the hydrogen atoms of the methyl group in toluene produces an interesting effect in the ESR spectra of the anions. Examination of the hyperfine splittings in Fig. 5-7b indicates that as methyl groups are added, the *para*-proton hyperfine splitting increases and the *ortho* and *meta* splittings decrease.

It is known that the electron-releasing character of alkyl groups in conjugated systems decreases in the order

$$CH_3 > CH_3CH_2 > (CH_3)_2CH > (CH_3)_3C$$

The observed behavior of the hyperfine splittings has been interpreted as a successive decrease in the separation of the A and the S orbitals as the electron-releasing ability of the substituent diminishes.[†,‡]

Some alkyl-substituted benzene cations have been studied.[§] One should consider these as representing an electron hole (+ charge) in the e_1 orbitals of benzene. The alkyl group should be attractive to an electron hole and thus the S orbital should be stabilized for mono- or 1,4-disubstition. This is indeed the case as is illustrated in Fig. 5-7d.

The sensitivity of the benzene orbitals to the electron-releasing properties of substituents has been utilized to determine the behavior of the —Si(CH_3)_3 and —Ge(CH_3)_3 substituents.[¶] The hyperfine splittings of the anions of these substituted benzenes are given in Fig. 5-7c. In contrast to the t-butylbenzene anion (Fig. 5-7b), the S orbital is now the more stable. This observation requires that these substituents be electron-attracting. The d_{xz} orbital of Si or Ge has the correct symmetry to overlap with the π orbitals. This overlap allows delocalization of electrons into the d_{xz} orbital of Si or Ge and thus accounts for the electron-attracting character of these groups.

† J. R. Bolton, A. Carrington, A. Forman and L. E. Orgel, *Mol. Phys.*, **5**:43 (1962).
‡ E. de Boer and J. P. Colpa, *J. Phys. Chem.*, **71**:21 (1967).
§ R. M. Dessau, S. Shih, and E. I. Heiba, *J. Am. Chem. Soc.*, **92**:412 (1970).
¶ J. A. Bedford, J. R. Bolton, A. Carrington and R. H. Prince, *Trans. Faraday Soc.*, **59**:53 (1963).

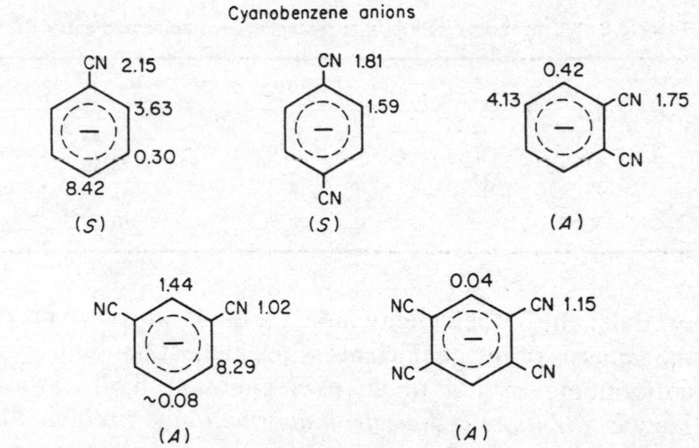

Cyanobenzene anions

CN 2.15
3.63
0.30
8.42
(*S*)

CN 1.81
1.59
CN
(*S*)

0.42
4.13
CN 1.75
CN
(*A*)

1.44
NC
CN 1.02
8.29
~0.08
(*A*)

0.04
NC
CN 1.15
NC
CN
(*A*)

Fig. 5-8 ESR hyperfine splitting constants in cyano-substituted benzenes. [*Taken from P. H. Rieger, I. Bernal, W. H. Rein-muth, and G. K. Fraenkel, J. Am. Chem. Soc.,* **85**:683 (1963).] See legend of Fig. 5-7 for the meaning of *A* and *S*.

Some examples of the effect of an electron-attracting substituent are given in Fig. 5-8 for the cyano-substituted benzene anions. It can be seen that whereas methyl groups stabilize the *A* orbital, cyano groups stabilize the *S* orbital and vice versa.

5-5 THE ANIONS AND CATIONS OF THE POLYACENES

In the HMO approximation, alternant hydrocarbons have orbital energies symmetrically disposed about the central energy α. *Odd*-alternant hydrocarbons have a nonbonding orbital at this energy. Orbitals with energies symmetrically disposed about the energy α involve the same atomic

Table 5-5 Hyperfine splittings in polyacene ions

Molecule	Position	a_+^H, G	a_-^H, G
	9	6.53	5.34
	1	3.06	2.74
	2	1.38	1.51
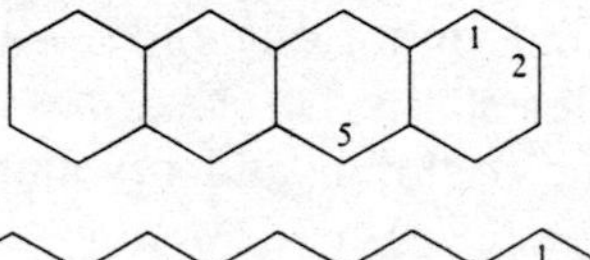	5	5.05	4.23
	1	1.69	1.54
	2	1.03	1.16
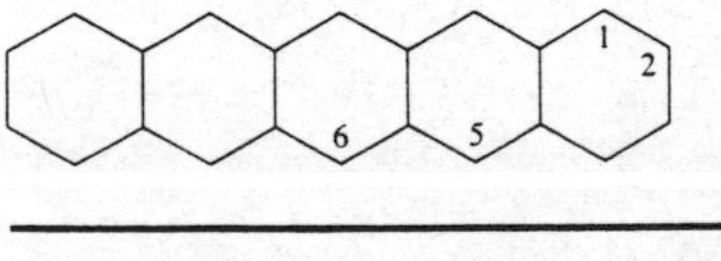	6	5.08	4.26
	5	3.55	3.03
	1	0.98	0.92
	2	0.76	0.87

Table 5-6 ^{13}C hyperfine splittings in the anthracene anion and cation

	Position	$a_+{}^c$, G	$a_-{}^c$, G
	9	8.48	8.76
	11	−4.50	−4.59
	1	—	3.57
	2	±0.37	−0.25

orbitals; their coefficients have the same *absolute* magnitudes. Therefore, the squares of the coefficients of the highest bonding orbital and of the lowest antibonding orbital of an even-alternant hydrocarbon will be identical. Hence, *the unpaired-electron distribution is predicted to be identical in the corresponding cation and anion radicals.* This is one statement of the pairing theorem. This theorem applies to a high degree of approximation.[†,‡]

The ESR spectra of both the anions and cations of some of the polyacenes (anthracene, tetracene, and pentacene) have been studied. The hyperfine splittings for these molecules are listed in Table 5-5.

It is apparent that the hyperfine splittings are similar for protons in corresponding positions in the anion and the cation of a given molecule. These results are in reasonable accord with the pairing theorem. The agreement is even better than is apparent, since Q depends somewhat on the excess charge density. (See Sec. 6-5.) Hyperfine splittings from ^{13}C have been detected in the ESR spectra of the anthracene anion and cation.[§] The ^{13}C splittings are *not* simply proportional to the unpaired-electron density on the same carbon atom. (See Sec. 6-7.) Nevertheless, the ^{13}C splittings should be the same in the anion and cation if the electron distribution is the same. The data are given in Table 5-6.

The similarity of anion and cation hyperfine splittings would appear to confirm the validity of the pairing theorem.

5-6 OTHER ORGANIC RADICALS

Alkyl radicals have thus far not been considered in this chapter. Here the unpaired electron is largely localized on one carbon atom. These radicals are important in the development of the theory of proton hyperfine interactions and will be considered in Sec. 6-4.

Radicals containing N, O, or S form another important class. The HMO method may be applied to these molecules if the heteroatoms are treated as pseudocarbon atoms with appropriate coulomb and resonance integrals. That is,

[†] A. D. McLachlan, *Mol. Phys.*, **2**:271 (1959).
[‡] J. Koultecky, *J. Chem. Phys.*, **44**:3702 (1966).
[§] J. R. Bolton and G. K. Fraenkel, *J. Chem. Phys.*, **40**:3307 (1964).

$$\alpha_X = \alpha_C + h_X \beta_{CC} \tag{5-30a}$$

$$\beta_{CX} = k_{CX} \beta_{CC} \tag{5-30b}$$

Here β_{CC} and α_C are the integrals appropriate for a conjugated hydrocarbon. h_X varies from atom to atom and with the type of bonding. Generally h_X increases with the electronegativity of the atom.[†]

The interpretation of the ESR spectra of inorganic radicals has not been considered in this chapter. Most of these radicals have been observed in the solid state. They present numerous complications which are considered in Sec. 8-4.

5-7 SUMMARY

It is clear that the HMO theory gives valuable assistance in the interpretation of proton hyperfine splittings for π-electron radicals. Indeed, where serious discrepancies have been found for even-alternant radicals, careful examination has sometimes shown that the experiment—rather than the theory—was in error. This is not to say that the HMO theory can predict the exact value of hyperfine splittings, but it does give an indication of the approximate relative magnitudes. One of the most serious shortcomings of the HMO theory is that it predicts zero unpaired-electron densities at positions where small proton hyperfine splittings are observed (e.g., the 3 and 5 positions in the benzyl radical). An understanding of the origin of these splittings requires a different approach to be described in Sec. 6-3.

It is useful at this point to summarize some rules relating to the determination of the energy levels and wave functions in π-electron systems.

Orbital energies of π-electron systems

1. A conjugated system consisting of n atoms, each contributing one $2p_z$ orbital, will have a set of n π-electron orbital energies.
2. To satisfy the Pauli exclusion principle, one puts two electrons with opposed spins in each orbital starting with the lowest, until all π electrons have been allocated. This corresponds to the ground-state configuration.
3. Orbitals with energy below the reference energy α (see Fig. 5-3) are called bonding orbitals, while those of energy above α are called antibonding orbitals. If an orbital has an energy α, it is called a nonbonding orbital.
4. For a hydrocarbon anion, the number of electrons in the π system is $n + 1$; for a cation, the number is $n - 1$.
5. If the number of electrons is odd, the ESR properties are determined primarily by the wave function of the orbital containing the odd electron.

[†] Representative values of h_X and k_{CX} are given in A. Streitwieser, Jr., "Molecular Orbital Theory for Organic Chemists," chap. 5. John Wiley & Sons. Inc., New York, 1961.

6. If there are two electrons to be assigned to a set of two degenerate orbitals, the electrons are distributed one to each orbital, with the electron spins parallel (one of Hund's rules).

Wave functions of π-electron systems

1. A system with n $2p_z$ atomic orbitals gives rise to n wave functions (molecular orbitals), each consisting of a linear combination of the atomic orbitals.
2. The molecular orbital of lowest energy will have all atomic orbital coefficients positive, whereas that of highest energy will have coefficients alternately positive and negative. For orbitals of intermediate energy, the coefficients of some terms in the corresponding molecular orbital may be zero.
3. If the sign of the product of coefficients of adjacent atoms is positive, the wave function is said to be bonding between the two atoms. If the sign is negative, the wave function is antibonding and has a node between the two atoms. In general, the greater the number of nodes in a wave function, the higher the energy of the corresponding orbital.
4. If the ith molecular orbital contains a single electron, the unpaired-electron density on atom j is obtained by squaring the appropriate atomic orbital coefficient in the ith molecular orbital; that is $\rho_j = c_{ij}^2$.

REFERENCES—HMO METHOD

Method and applications

1. Streitwieser, Jr., A.: "Molecular Orbital Theory," John Wiley & Sons, Inc., New York, 1961. Chapters 2 and 3 describe in detail the procedures for calculations of orbital energies and wave functions of hydrocarbons. Chapter 4 describes refinements of the method and chap. 5 deals with applications to molecules having hetero (N, O, S, or halogen) atoms.
2. Cotton, F. A.: "Chemical Application of Group Theory," Interscience Publishers, a division of John Wiley & Sons, Inc., New York 1963. The treatment of monocyclic systems in Chap. 7 is of special interest.
3. Murrell, J. N., S. F. A. Kettle, and J. M. Tedder: "Valence Theory," John Wiley & Sons, Inc., New York, 1965. Chapter 15 deals with the π-electron theory of organic molecules. Section 15-8, "A critique of Hückel theory," gives some insight into the successes of the HMO approach.
4. Salem, L.: "The Molecular Orbital Theory of Conjugated Systems," chap. 1, W. A. Benjamin, Inc., New York, 1966.
5. Dewar, M. J. S.: "The Molecular Orbital Theory of Organic Chemistry," chap. 5, McGraw-Hill Book Company, New York, 1969.

Tabulations of HMO orbitals

1. Coulson, C. A. and A. Streitwieser: "Dictionary of π-Electron Calculations," W. H. Freeman and Company, San Francisco, 1965.
2. Heilbronner, E. and P. A. Straub: "Hückel Molecular Orbitals," Springer-Verlag New York Inc., New York, 1966.

PROBLEMS

5-1. Set up the secular equation for the cyclopropenyl (C_3H_3) radical and solve for the orbital energies. Draw an orbital energy diagram and show the distribution of electrons among the orbitals.

5-2. Set up the secular determinant for the 1,3-butadiene anion and solve for the energies. Substitute the energies into the secular equations and determine the coefficients in the four molecular orbitals. (See Table 5-1.)

5-3. The naphthalene molecule has the symmetry D_{2h}; this implies that perpendicular to what may be called the principal (or Z) axis, there are two other twofold axes. The Z axis is taken to be perpendicular to the plane of the molecule. The Y axis passes through atoms 9 and 10, while the X axis is midway between atoms 2 and 3, or 6 and 7. Fill out the table below, showing the results of performing the operations identity (E), rotation through 180° about $Z(C_2^z)$, $Y(C_2^Y)$, and $X(C_2^X)$. At the bottom, give the number of atom positions unchanged by each symmetry operation.

Atom	E	C_2^Z	C_2^Y	C_2^X
1	1	5	8	4
2				
3				
4				
5				
6				
7				
8				
9				
10	—	—	—	—
Number of positions unchanged				

The objective of this exercise is to point out that atoms which are related by the symmetry operations of the molecule *must* enter molecular orbitals in the same way (excluding differences in sign). That is, if in a specific molecular orbital the coefficient c_1 is nonzero, c_4, c_5, and c_8 *must also* be nonzero, and no symmetry operation can move an atom outside this subgroup (or class, as it is called). This is illustrated in the data of Prob. 5-4. Atoms 2, 3, 6, 7 also form a symmetry class, as do atoms 9 and 10. When dealing with a molecule which would give a large (here 10 by 10) secular determinant, the problem may be greatly simplified by writing separate secular determinants for each class. Group theory provides a detailed procedure for this simplification; it starts with the table which has just been constructed. (See references for this chapter.)

5-4. The following are the lowest-lying Hückel molecular orbitals for naphthalene, in order of increasing energy:

$$\psi_n = c_1\phi_1 + c_2\phi_2 + c_3\phi_3 + \cdots + c_{10}\phi_{10}$$

(a) Without doing any calculations, sketch approximately the set of HMO energies for naphthalene, and show the orbital occupation by electrons.

(b) Compare ψ_5 with ψ_6 and ψ_4 with ψ_7. What identities may be written for corresponding c_i values of the related pairs of molecular orbitals?

	c_1	c_2	c_3	c_4	c_5	c_6	c_7	c_8	c_9	c_{10}
ψ_1	0.301	0.231	0.231	0.301	0.301	0.231	0.231	0.301	0.461	0.461
ψ_2	0.263	0.425	0.425	0.263	−0.263	−0.425	−0.425	−0.263	0	0
ψ_3	0.400	0.174	−0.174	−0.400	−0.400	−0.174	0.174	0.400	0.347	−0.347
ψ_4	0	0.408	0.408	0	0	0.408	0.408	0	−0.408	−0.408
ψ_5	0.425	0.263	−0.263	−0.425	0.425	0.263	−0.263	−0.425	0	0
ψ_6	0.425	−0.263	−0.263	0.425	−0.425	0.263	0.263	−0.425	0	0
ψ_7	0	−0.408	+0.408	0	0	0.408	−0.408	0	0.408	−0.408

(c) What is the significance of a zero value of c_i?

(d) Sketch location of nodal planes for all of these orbitals.

5-5. The hyperfine splittings for the naphthalene anion are 4.95 and 1.87 G. Based on the molecular orbitals of naphthalene (see Prob. 5-4), how should these hyperfine splittings be assigned? How does the ratio of hyperfine splittings compare with the ratio of the squares of the atomic orbital coefficients for the molecular orbital containing the odd electron?

5-6. (a) Using the orbital energies for cyclooctatetrene, given in Fig. 5-3, assign electrons to orbitals of the anion.

(b) Obtain the coefficients of the highest occupied molecular orbitals. The procedure is the same as that used for the benzyl radical (Sec. 5-3). The two possible assignments of starred positions correspond to the degenerate pair of orbitals.

(c) For the methyl cyclooctatetrene anion, predict which of the two highest occupied orbitals will contain the unpaired electron. Compare this prediction with the experimental results given below[†]

5-7. The following are the HMO energies and molecular orbitals for biphenylene:

		Absolute values of the coefficients for atoms:		
Energy		9(10,11,12)	1(4,5,8)	2(3,6,7)
$\alpha + 2.532\beta$	ψ_1	0.422	0.225	0.147
$\alpha + 1.802\beta$	ψ_2	0.164	0.296	0.368
$\alpha + 1.347\beta$	ψ_3	0.225	0.147	0.422
$\alpha + 1.247\beta$	ψ_4	0.296	0.368	0.164
$\alpha + 0.879\beta$	ψ_5	0.147	0.422	0.225
$\alpha + 0.445\beta$	ψ_6	0.368	0.164	0.296
$\alpha - 0.445\beta$	ψ_7	0.368	0.164	0.296
$\alpha - 0.879\beta$	ψ_8	0.147	0.422	0.225

† A. Carrington and P. F. Todd, *Mol. Phys.*, 7:533 (1964).

Using $Q = 27.0$ G, predict the hyperfine splittings to be expected for the biphenylene anion. Compare these with the experimental values of $a_2 = 0.21$ G and $a_3 = 2.86$ G for the anion.

5-8. Using the method outlined for the benzyl radical show that the coefficients in the non-bonding molecular orbital for the perinaphthenyl radical are

$$\psi_{\text{NB}} = \frac{1}{\sqrt{6}}\,(\phi_1 - \phi_3 + \phi_4 - \phi_6 + \phi_7 - \phi_9)$$

Indicate the characteristics of the expected ESR spectrum. Compare with the observed proton hyperfine splittings given in Prob. 6-6. An explanation for the discrepancies is given in Sec. 6-3.

6

Mechanism of Hyperfine Splittings in Conjugated Systems

6-1 ORIGIN OF PROTON HYPERFINE SPLITTINGS

In Chap. 5 it was assumed that in planar conjugated radicals the proton hyperfine splittings are proportional to the unpaired π-electron density on the carbon atom adjacent to the proton. That is

$$a_i = Q\rho_i \qquad (6-1)$$

Isotropic proton hyperfine splittings were shown in Chap. 3 to arise when there is a net unpaired-electron density at the proton. In π radicals, the unpaired electron resides in a π-molecular orbital constructed from a linear combination of $2p_z$ carbon atomic orbitals. However, each $2p_z$ orbital has a node in the plane of the molecule. Since this plane also contains the protons, there should be no unpaired-electron density at the proton and hence no hyperfine splitting. In spite of this node, the numerous spectra in Chap. 4 demonstrate that isotropic proton hyperfine splittings *do* occur in π radicals.

The concept of unpaired-electron density must be reexamined in order to resolve this paradox. It was assumed that when an electron is added to a conjugated molecule to make a negative ion, the other electrons in the molecule would be unaffected. The density of electron spin (or spin density) would then be just the density of the added unpaired electron. However, the other electrons in the molecule *are* affected slightly by the addition of the extra electron. As a result, in some regions of the molecule the "paired" electrons become slightly unpaired. (This is one of several effects which go under the name of "electron correlation.") Thus the *spin density* will not be equal to the *unpaired-electron density*. This is why we have been careful to use the latter term up to now in the one-electron HMO treatment. Since isotropic hyperfine splittings arise from a net *spin* density at the nucleus, one requires a precise definition of this term.

Spin density is a function of many electrons and may be defined by the expression

$$\rho_i = P_i(\alpha) - P_i(\beta) \tag{6-2}$$

where ρ_i is the spin density in the region i of the molecule and $P_i(\alpha)$ and $P_i(\beta)$ are the total probability densities of electrons with α and β spin, respectively, in the region i. $P_i(\alpha)$ and $P_i(\beta)$ are obtained by adding the densities of all electrons with α spin and β spin, respectively.†

Consider a C—H fragment of a conjugated system. If spin α is assigned to the one electron in the $2p_z$ orbital on the carbon atom, there are two possibilities for assigning the spins in the C—H σ bond; these are shown in Fig. 6-1.

Here it is assumed that the carbon atom has a $2p_z$ orbital perpendicular to the C—H bond; the $2p_x$ and $2p_y$ orbitals, plus the $2s$ orbital of the carbon atom, form trigonal sp^2 hybrids. The hydrogen atom bonds with one of these sp^2 hybrids.

If there were no electron in the $2p_z$ orbital, the electron configurations a and b of Fig. 6-1 would be equally probable; hence the *spin* density at the proton would be zero. However, when the $2p_z$ electron is present, config-

† The densities referred to here are "normalized." Unpaired-electron densities and spin densities are normalized to unity over the whole molecule; *electron* densities are normalized to the total number of electrons.

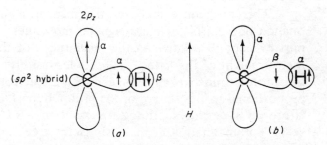

Fig. 6-1 Possible configurations of electron spin in the σ orbital bonding the hydrogen atom of a C—H fragment, for α spin in the $2p_z$ orbital of carbon. (*a*) Spins parallel in the σ bonding orbital and the $2p_z$ orbital of carbon. (*b*) Corresponding spins antiparallel.

urations a and b are no longer equally probable. It has been demonstrated from atomic spectroscopy that when two different orbitals on the same atom are singly occupied by electrons, the more stable arrangement for the ground state is that with the electron spins parallel (one of Hund's rules). Thus configuration a, in which the two electrons on the carbon atom have parallel spins, will be more stable and hence more probable than b, for which the spins are antiparallel. That is, there will be a net *negative* spin density (i.e., excess of β spin over α spin) at the proton. Note also that there will be a *positive* spin density at the carbon nucleus. If there is *one* unpaired electron in the $2p_z$ orbital, the negative spin density at the proton will produce a negative proton hyperfine splitting, the magnitude of which is denoted by Q. This effect is often called spin polarization. However, in a conjugated radical the unpaired-electron density ρ_i at a given carbon atom will usually be less than unity. The hyperfine splitting a_i is then obtained from Eq. (6-1). The negative sign of a_i has been confirmed experimentally, as is demonstrated in Sec. 6-2. The above discussion indicates qualitatively why Q should be negative in Eq. (6-1). Detailed calculations utilizing both the molecular orbital and valence bond approaches have confirmed both the form of Eq. (6-1) and the negative sign of Q.[†]

6-2 SIGN OF THE HYPERFINE SPLITTING CONSTANT

Equation (6-1) indicates that the spin density at the proton of a C—H fragment should be *negative* if the spin density in the adjacent carbon $2p_z$ orbital is *positive*. Thus proton hyperfine splittings in conjugated radicals will be expected to be *negative*. The first confirmation of the negative sign of Q in Eq. (6-1) was obtained by an analysis of the splittings in the malonic acid radical.[‡] Since the argument involves a consideration of the anisotropic hyperfine coupling, the details will be deferred until Sec. 8-2b. Instead, a verification of the signs by a proton magnetic resonance method will be examined.

This procedure involves the measurement of proton magnetic resonance line shifts for paramagnetic molecules. The NMR lines must be narrow enough relative to the magnitudes of the line shifts to permit the measurement of the latter. These lines are broadened by the relaxation of the proton spins in the presence of the electron spin. The broadening is proportional to the square of the proton hyperfine splitting. (See Sec. 9-5.) If proton hyperfine splittings are less than ~ 6 G, it may be possible to observe paramagnetic chemical shifts for free radicals in solution at room temperature.[§,¶]

The NMR spectrum of the biphenyl anion at room temperature is

[†] H. M. McConnell and D. B. Chesnut. *J. Chem. Phys.*, **28**:107 (1958).

[‡] T. Cole, C. Heller, and H. M. McConnell, *Proc. Natl. Acad. Sci., U.S.*, **45**:525 (1959).

[§] E. de Boer and C. MacLean. *Mol. Phys.*, **9**:191 (1965).

[¶] K. H. Hausser. H. Brunner, and J. C. Jochims, *Mol. Phys.*, **10**:253 (1966).

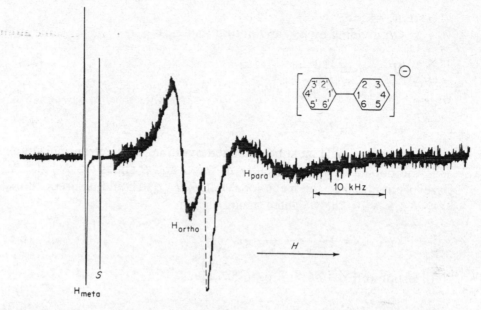

Fig. 6-2 Proton NMR spectrum at 60 MHz of a $1M$ solution of the biphenyl anion in diglyme [CH_3—O—(CH_2CH_2—O)$_2$—CH_3] at room temperature. The concentration of neutral biphenyl is negligible. The line labeled S arises from the solvent. All peaks have been measured with different r.f. power, gain and modulation. [*Taken from G. W. Canters and E. de Boer, Mol. Phys.*, **13**:395 (1967).]

shown in Fig. 6-2. The chemical shifts in this spectrum are huge compared to those found for protons in diamagnetic molecules. The large chemical shifts are due to local magnetic fields generated by the hyperfine interaction.

There will be a negative (downfield) chemical shift if a_i is positive and vice versa. The magnitude of the chemical shift is given by

$$\Delta H = H_i - H' = -\frac{a_i}{2} \frac{|\gamma_e|}{\gamma_p} \frac{g\beta H'}{2kT} \tag{6-3}$$

where H_i is the resonance field for the shifted line and H' is the field corresponding to the unshifted proton resonance line.†,‡

The relation (6-3) can be derived as follows. For a proton with splitting a_i, the hyperfine contribution to the proton resonance transition energy is $\pm g\beta a_i/2 = \pm\hbar|\gamma_e|a_i/2$. Here the plus sign applies when M_S is positive. The total proton transition energy is then

$$h\nu_p = \hbar\omega_p = \hbar\gamma_p H_i \pm \hbar|\gamma_e|\frac{a_i}{2} \tag{6-4}$$

† H. M. McConnell and C. H. Holm, *J. Chem. Phys.*, **27**:314 (1957).
‡ D. R. Eaton and W. D. Phillips, Nuclear Magnetic Resonance of Paramagnetic Molecules, in J. S. Waugh (ed.), "Advances in Magnetic Resonance," vol. 1, Academic Press Inc., New York, 1965.

Here $\omega_p = 2\pi\nu_p$.

On dividing by $\hbar\gamma_p$ and noting that $\omega_p = \gamma_p H$

$$H' = H_i \pm \frac{|\gamma_e|}{\gamma_p} \frac{a_i}{2} \tag{6-5a}$$

or

$$H_i = H' \mp H_{\text{local}} \tag{6-5b}$$

Now the local field must be averaged over the two electron spin states. If the populations of molecules in the α and β states are N_α and N_β respectively, then the mean local field $\langle H_{\text{local}} \rangle$ should be proportional to $N_\alpha - N_\beta$. The weighted mean is then

$$\langle H_{\text{local}} \rangle = \frac{N_\alpha - N_\beta}{N_\alpha + N_\beta} \frac{|\gamma_e|}{\gamma_P} \frac{a_i}{2} \tag{6-6}$$

Insertion of Eq. (6-6) in Eq. (6-5a) gives

$$H_i - H' = \frac{a_i}{2} \frac{|\gamma_e|}{\gamma_p} \frac{N_\alpha - N_\beta}{N_\alpha + N_\beta} \tag{6-7}$$

For a system having a set of states with energies $W_1, W_2, \ldots, W_n$, the relative population in the state i is given by the Boltzmann expression

$$N_i = \frac{Ne^{-W_i/kT}}{e^{-W_1/kT} + e^{-W_2/kT} + \cdots + e^{-W_n/kT}} \tag{6-8}$$

Here N is the total population of states, that is, $\Sigma_i N_i$; in this case, $N_\alpha + N_\beta = N$.

Since the energy W_α of the $M_S = +\frac{1}{2}$ state is $\frac{1}{2}g\beta H'$ and that of W_β is $-\frac{1}{2}g\beta H'$

$$\frac{N_\alpha - N_\beta}{N_\alpha + N_\beta} = \frac{N}{N} \frac{[\exp(-g\beta H'/2kT) - \exp(-(-)g\beta H'/2kT)]}{[\exp(-g\beta H'/2kT) + \exp(g\beta H'/2kT)]} \tag{6-9}$$

For small values of x in the quantity e^x, $e^x \approx 1 + x$. Hence

$$\frac{N_\alpha - N_\beta}{N_\alpha + N_\beta} \approx \frac{1 - g\beta H'/2kT - (1 + g\beta H'/2kT)}{1 - g\beta H'/2kT + (1 + g\beta H'/2kT)} = \frac{-g\beta H'}{2kT} \tag{6-10}$$

Substitution of Eq. (6-10) into (6-7) gives the desired result,

$$H_i - H' = -\frac{a_i}{2} \frac{|\gamma_e|}{\gamma_p} \frac{g\beta H'}{2kT} \tag{6-11}$$

Referring to Fig. 6-2, one notes that there are two lines shifted to the high-field side of the resonance position for diamagnetic molecules. These correspond to a *negative* value of a_i for two sets of protons in the radical. This result is expected from Eq. (6-1) with a negative value of Q. However, one line is shifted downfield; it must therefore correspond to a *positive* a_i for one set of protons. This result may be explained in terms of a negative spin density, which will be discussed in Sec. 6-3.

If no more than one proton is attached to each carbon atom of a conjugated radical and if all π-spin densities are positive, the extent of the ESR spectrum cannot exceed the value of Q. The spectral extent for a number of conjugated radicals is given in Table 6-1. For most of the radicals, the spectral extent does not exceed about 27 G. However, the extent of the perinaphthenyl radical spectrum is considerably in excess of this value.

A large spectral extent can be understood if negative π-spin densities occur. The normalization condition for spin density requires that the *algebraic* sum of all spin densities be unity. If some spin densities are negative, then others must be correspondingly more positive. Consequently, the sum of the *absolute values* of the spin densities can be greater than unity. Since the spectral extent depends only on the absolute magnitude of the hyperfine splittings, negative spin densities will result in an unusually large spectral extent.

In the biphenyl anion spectrum in Fig. 6-2, the down-field-shifted line must be assigned to positions at which the π-spin density is *negative*. One would not have inferred this fact from the spectral extent; however, there are appreciable spin densities at positions which have no protons. The magnitude of the shift for the low-field line indicates that this line arises from protons having the *smallest* magnitude of hyperfine splitting. From the solution ESR spectrum, the smallest splitting arises from a set of four equivalent protons. These can be either the protons at positions 2, 6, 2', 6' or 3, 5, 3', 5' (see Fig. 6-2). Molecular orbital studies indicate that the latter assignment should be made.[†]

[†] A. D. McLachlan, *Mol. Phys.*, **3**:233 (1960).

Table 6-1

Compound	Spectral extent, Gauss
Benzene⁻	22.5
Biphenyl⁻	22.9
Naphthalene⁻	27.3
Anthracene⁻	27.7
Tetracene⁻	27.7
Tropyl	27.4
Pyrene⁻	29.2
Perinaphthenyl	43.1

6-3 EXTENSION OF THE MOLECULAR ORBITAL
THEORY TO INCLUDE ELECTRON CORRELATION

The experimental results outlined in Sec. 6-2 suggest that at some carbon positions in a conjugated radical the sign of the π-spin density may be *negative*. To account for negative π-spin densities, the simple molecular orbital theory which has been used to this point must be extended. The HMO theory provides only the *density of the unpaired electron* and considers that all other electrons are completely paired. However, the presence of the unpaired electron does cause a slight "unpairing" of the other π electrons by the same mechanism which accounts for the presence of an isotropic hyperfine splitting in conjugated radicals (Sec. 6-1). The effect is to make the spatial distributions different for the two electrons in a given orbital. Thus computation of a spin density using Eq. (6-2) requires that *all* π electrons be considered.

There are several ways of introducing this electron correlation into the molecular orbital theory.† One of the methods is to assign a different spatial orbital to every electron. Consider the Hückel molecular orbitals for the allyl radical as given in Fig. 5-2 and below,

$$\psi_3 = \frac{1}{2}\,\phi_1 - \frac{1}{\sqrt{2}}\,\phi_2 + \frac{1}{2}\,\phi_3$$

$$\psi_2 = \frac{1}{\sqrt{2}}\,\phi_1 \qquad\qquad - \frac{1}{\sqrt{2}}\,\phi_3$$

$$\psi_1 = \frac{1}{2}\,\phi_1 + \frac{1}{\sqrt{2}}\,\phi_2 + \frac{1}{2}\,\phi_3 \tag{6-12}$$

The configurational wave function corresponding to the ground state may be written as

$$\Phi_0 = \frac{1}{\sqrt{3!}}\,\|\psi_1(1)\alpha(1)\psi_1(2)\beta(2)\psi_2(3)\alpha(3)\| \tag{6-13}$$

Equation (6-13) is a determinantal wave function (see Sec. A-3). For this wave function, the spin densities are given by the squares of the coefficients of ψ_2 of Eq. (6-12). The Hückel molecular orbital theory ignores interactions among the several π electrons. In this case there are two π electrons in ψ_1 and one in ψ_2. Detailed calculations indicate that the interaction between the α electron in ψ_2 and the α electron in ψ_1 is *less* than that with the β electron in ψ_1. This effect is analogous to that which leads to one of Hund's rules for atoms. That is, electrons in different spatial orbitals will have a lower energy if the spins are par-

---

† See, for instance, L. Salem, "The Molecular Orbital Theory of Conjugated Systems," chap. 2, W. A. Benjamin, Inc., New York, 1966.

allel. Thus Φ_0 can be improved if one assigns different *spatial* functions to the electrons in ψ_1. This can be accomplished by admixing different amounts of ψ_3,[†] that is,

$$\psi_1' = \psi_1 + \epsilon\psi_3 \tag{6-14a}$$

$$\psi_1'' = \psi_1 - \epsilon\psi_3 \tag{6-14b}$$

On substitution of the Hückel functions into Eqs. (6-14) one obtains

$$\psi_1' = \frac{1}{2}(1+\epsilon)\phi_1 + \frac{1}{\sqrt{2}}(1-\epsilon)\phi_2 + \frac{1}{2}(1+\epsilon)\phi_3 \tag{6-15a}$$

$$\psi_1'' = \frac{1}{2}(1-\epsilon)\phi_1 + \frac{1}{\sqrt{2}}(1+\epsilon)\phi_2 + \frac{1}{2}(1-\epsilon)\phi_3 \tag{6-15b}$$

The coefficients of ϕ_1 and ϕ_3 are larger in ψ_1' than in ψ_1'', and hence ψ_1' will have a lower energy due to the greater electron delocalization. Since the ψ_1 orbital with α spin has the lower energy, ψ_1' must contain an electron with α spin, whereas ψ_1'' contains the electron with β spin. Using Eq. (6-2), the spin densities may be calculated by summing the appropriate squares of coefficients. Here c_{ki} represents the coefficient of the ith atom in the kth molecular orbital.

$$\begin{aligned}
\rho_1 &= |c_{11}'|^2 + |c_{21}|^2 - |c_{11}''|^2 = \rho_3 \\
&= \tfrac{1}{4}(1 + 2\epsilon + \epsilon^2) + \tfrac{1}{2} - \tfrac{1}{4}(1 - 2\epsilon + \epsilon^2) \\
&= \tfrac{1}{2} + \epsilon \tag{6-16a} \\
\rho_2 &= |c_{12}'|^2 + |c_{22}|^2 - |c_{12}''|^2 \\
&= \tfrac{1}{2}(1 - 2\epsilon + \epsilon^2) - \tfrac{1}{2}(1 + 2\epsilon + \epsilon^2) \\
&= -2\epsilon \tag{6-16b}
\end{aligned}$$

Since ϵ is positive, the spin density on the central carbon atom is predicted to be negative. In Sec. 8-2c an experimental verification of this negative spin density will be given. The determination of ϵ for the allyl radical is given as a problem at the end of this chapter.

Usually ϵ is less than 0.1; hence, a perturbation calculation can give good results. McLachlan[‡] has developed such a method using Hückel orbitals as the unperturbed functions. If there are n carbon atoms in the conjugated system, this approach leads to the following expression for the spin density at carbon atom t

$$\rho_t = c_{mt}{}^2 + \lambda \sum_{r=1}^{n} \pi_{rt} c_{mr}{}^2 \tag{6-17}$$

† L. Salem, "The Molecular Orbital Theory of Conjugated Systems," p. 264, W. A. Benjamin, Inc., New York, 1966.
‡ A. D. McLachlan, *Mol. Phys.*, 3:233 (1960); see also L. Salem, "The Molecular Orbital Theory of Conjugated Systems," chap. 5, W. A. Benjamin, Inc., New York, 1966.

Here c_{mt} is the coefficient of atom t in the mth molecular orbital which contains the unpaired electron. λ is a parameter which may be varied to provide a best fit to the spectral extent. It is usually given a value between 1.0 and 1.2. π_{rt} is the dimensionless mutual atom-atom polarizability defined by

$$\pi_{rt} = -4\beta \overset{\text{bonding}}{\underset{j}{\sum}} \overset{\substack{\text{anti-}\\\text{bonding}}}{\underset{k}{\sum}} \frac{(c_{jr}c_{kr}^*)(c_{jt}^*c_{kt})}{W_k - W_j} \tag{6-18}$$

The c's are the Hückel coefficients for atoms r and t in the molecular orbitals j and k. W_k and W_j are the Hückel energies of the k and j levels. The summations in Eq. (6-18) do not include nonbonding levels since their effect cancels out in the summations.

As an example, the spin densities in the allyl radical will be calculated from the Hückel molecular orbitals and the energies in Fig. 5-2, taking $\lambda = 1.1$:

$$\rho_1 = c_{21}^2 + 1.1(\pi_{11}c_{21}^2 + \pi_{31}c_{23}^2)$$

$$\rho_2 = 0 + 1.1(2\pi_{21}c_{21}^2)$$

$$\pi_{11} = -4\beta \frac{\left(\frac{1}{2}\right)\left(\frac{1}{2}\right)\left(\frac{1}{2}\right)\left(\frac{1}{2}\right)}{(\alpha - \sqrt{2}\beta) - (\alpha + \sqrt{2}\beta)} = \frac{1}{8\sqrt{2}} = \pi_{13}$$

$$\pi_{21} = -4\beta \frac{(1/\sqrt{2})\left(\frac{1}{2}\right)(-1/\sqrt{2})\left(\frac{1}{2}\right)}{(\alpha - \sqrt{2}\beta) - (\alpha + \sqrt{2}\beta)} = -\frac{1}{4\sqrt{2}}$$

$$\therefore \quad \rho_1 = 0.500 + 0.097 = 0.597$$

$$\rho_2 = 0.000 - 0.194 = -0.194$$

These results may be compared directly with the experimental hyperfine splittings[†]

Using appropriate values of Q (see Sec. 6-4) one calculates the spin densities to be $\rho_1 = 0.589$ and $\rho_2 = -0.155$.

[†] R. W. Fessenden and R. H. Schuler, *J. Chem. Phys.*, **39**:2147 (1963).

[‡] It is unusual to have two different hyperfine splittings for two hydrogen atoms on the same carbon atom. This implies that Q is not the same for the two hydrogen atoms. An explanation for this effect has been proposed.[§]

[§] A. Hinchliffe and N. M. Atherton, *Mol. Phys.*, **13**:89 (1967).

The availability of high-speed computers has permitted the use of more sophisticated molecular orbital theories which include all valence electrons and which also allow for electron correlation. An example of this type of method is the INDO (intermediate neglect of differential overlap) approach. This method has been of considerable value in interpreting ESR data for a wide range of radicals.[†]

6-4 ALKYL RADICALS—A STUDY OF Q VALUES

Perhaps the reader has now become suspicious that Q of Eq. (6-1) may vary in unpredictable ways. However, the neutral radicals—in particular, the alkyl radicals—do show an orderly variation in Q. The alkyl radicals are distinct from the conjugated radicals in that the unpaired electron is localized primarily on a particular carbon atom. This will be referred to as the α atom, whereas its neighbors in a linear chain are successively referred to as β, γ, δ, . . . , carbon atoms, respectively. The splittings for four selected alkyl radicals are given in Table 6-2. Starting with the methyl radical, one notes that successive substitution of CH_3 groups for H atoms causes a small decrease of the α-proton splittings a_α. The splittings of the β protons decrease rather more rapidly with increasing methyl substitution. Except for the CH_3 radical, the assignment of a Q value must be preceded by an assignment of the spin density at the central carbon atom (i.e., the α-carbon atom). A satisfactory expression for ρ_α appears to be[‡],[§]

$$\rho_\alpha = (1 - 0.081)^m = 0.919^m \tag{6-19}$$

where m is the number of methyl groups attached to the α-carbon atom. The values of Q_α and Q_β may then be assigned immediately. Note that this spin-density distribution gives an essentially constant value of Q_β in Table 6-2.

[†] J. A. Pople, D. L. Beveridge, and P. A. Dobosh, *J. Chem. Phys.*, **47**:2026 (1967); *J. Am. Chem. Soc.*, **90**:4201 (1968).
[‡] D. B. Chestnut, *J. Chem. Phys.*, **29**:43 (1958).
[§] D. Lazdins and M. Karplus, *J. Chem. Phys.*, **44**:1600 (1966).

Table 6-2[†]   Hyperfine parameters for alkyl radicals

Radical	ρ	$a_\alpha{}^H$, G	Q_α, G	$a_\beta{}^H$, G	Q_β, G
$\dot{C}H_3$	1.000	23.04	23.04	—	—
$CH_3\dot{C}H_2$	0.919	22.38	24.35	26.87	29.25
$(CH_3)_2\dot{C}H$	0.844	22.11	26.20	24.68	29.25
$(CH_3)_3\dot{C}$	0.776	—	—	22.72	29.30

[†] R. W. Fessenden and R. H. Schuler, *J. Chem. Phys.*, **39**:2147 (1963).

Table 6-3 Variation of Q_α with substituent in radicals of the type $CH_3\text{—}\dot{C}H\text{—}X^\dagger$

X	a_α, G	a_β, G	Q_α, G
CH_3	21.11	24.68	26.2
H	22.38	26.87	24.4
$COCH_2CH_3$	18.45	22.59	23.9
COOH	20.18	24.98	23.7
OH	15.04	22.61	19.5
$O\text{—}CH_2CH_3$	13.96	22.28	18.3

† H. Fischer, *Z. Naturforsch.*, **20A**:428 (1965).

Another series of radicals which has been studied may be represented by $CH_3\text{—}\dot{C}H\text{—}X$. The proton splittings for various substituents X are given in Table 6-3. The value of Q_β for $\dot{C}\text{—}CH_3$ protons is taken as 29.25 G, and the spin densities on the central carbon atom are computed from the methyl-proton splittings. Hence, the values of Q_α may be obtained from the C—H splitting constants. These Q_α values are given in Table 6-3. At the present time the reason for the variation in Q_α is not clear. It has been suggested that the variation may be related to the inductive effect of the substituent since the Q_α values are closely correlated with proton-resonance chemical shifts of α protons in the corresponding diamagnetic compounds.†

Although the Q_α values in Table 6-3 exhibit considerable variation, those for protons in closely related compounds appear to be relatively constant, at least for neutral radicals. The allyl radical provides a test of the transferability of the Q_α values from one molecule to another. For the 1 and 3 positions which have two protons, the Q_α value of 24.4 G from the ethyl radical should be used. For the central 2 position with one proton, the Q_α value of 26.2 G from the isopropyl radical should be used. Using the average of the two CH_2 splittings of 13.93 and 14.83 G and with Q_α equal to 24.4 G, $\rho_1 = 0.589$. With $a_2 = 4.06$ G and $Q_\alpha = 26.2$ G, $\rho_2 = -0.155$. The spin densities sum to 1.023, which is very close to the expected value of 1.000. Agreement for this and other neutral hydrocarbon radicals gives some confirmation of the general applicability of the Q_α values in Table 6-2. In such radicals, a Q_α value of about 27 G seems to give best agreement with experiment at carbon atoms having one proton.

6-5 THE EFFECT OF EXCESS CHARGE ON THE PARAMETER Q

Although Eq. (6-1) holds rather well, some systematic deviations have been observed. As an example, consider the proton hyperfine splittings listed in Table 5-5. If the pairing theorem holds (see Sec. 5-5), the π-spin densities

† H. Fischer, *Z. Naturforsch.*, **20A**:428 (1965).

for corresponding positions in the cation and anion of a given alternant hydrocarbon should be identical. The fact that the cation splittings are generally larger than those of the corresponding anion (especially for large hyperfine splittings) suggests that the Q value of Eq. (6-1) may depend somewhat on the excess charge at the carbon atom. (The excess charge ϵ_i is defined as $\epsilon_i = 1 - q_i$, where q_i is the total π-electron density at carbon atom i.) The following extension of Eq. (6-1) has been proposed to account for this charge effect[†]:

$$a_i = [Q(0) + K\epsilon_i]\rho_i \qquad (6\text{-}20)$$

Here $Q(0)$ is the parameter in Eq. (6-1) appropriate for neutral radicals, and K is a constant associated with the excess charge. Correlations of experimental values with theoretical calculations[‡] have demonstrated the applicability of Eq. (6-20). The values of the constants which correlate best with the experimental data are: $Q(0) = -27$ G and $K = -12$ G. The approximation $|\epsilon_i| = \rho_i$ holds well for alternant conjugated hydrocarbons. This allows one to calculate "experimental" spin densities from observed hyperfine splittings. For instance, in the 9 position of anthracene, $a_9 = 6.53$ G in the cation and $a_9 = 5.34$ G in the anion. Using the excess-charge relation, Eq. (6-20), one calculates $\rho_9 = 0.224$ and 0.215 for the cation and anion, respectively. Although the excess-charge correction is a useful extension, it should be noted that for many applications the simple relation, Eq. (6-1), gives a good approximation. Even in extreme cases, the excess-charge correction changes Q by only ± 15 percent from the value for neutral radicals.

There have been other attempts to explain the trend of results in Table 5-5. The following relation has been suggested[§]

$$\rho_i = Q_1 c_{mi}^2 + Q_2 \sum_j c_{mi} c_{mj} \qquad (6\text{-}21)$$

where the c's are Hückel coefficients and j refers to an atom bonded to atom i. In many cases this relation gives just as good a correlation as Eq. (6-20).[¶] However, for the tropyl radical and its corresponding dinegative ion, Eq. (6-21) would predict identical hyperfine splittings. The observed values[††] are 3.92 G for the neutral radical and 3.52 G for the dinegative ion, in good agreement with the predictions of Eq. (6-20).

[†] J. P. Colpa and J. R. Bolton, *Mol. Phys.*, 6:273 (1963); J. R. Bolton, *J. Chem. Phys.*, 43:309 (1965).
[‡] T. C. Sayetta and J. D. Memory, *J. Chem. Phys.*, 40:2748 (1964); J. R. Bolton, *J. Phys. Chem.*, 71:3099, 3702 (1967).
[§] G. Giacometti, P. L. Nordio, and M. V. Pavan, *Theoret. Chim. Acta (Berlin)*, 1:404 (1963).
[¶] L. C. Snyder and T. Amos, *J. Chem. Phys.*, 42:3670 (1965).
[††] N. L. Bauld and M. S. Brown, *J. Am. Chem. Soc.*, 89:5417 (1967).

6-6 METHYL-PROTON HYPERFINE SPLITTINGS—HYPERCONJUGATION

Examination of Fig. 5-7 reveals that splittings from some methyl protons exceed those of some ring protons. Hence, there must be some mechanism which effectively couples the methyl protons to the π system.

An effective coupling mechanism is that of *hyperconjugation*, which provides a direct link of the methyl hydrogen atoms with electrons in the π system. It is well known that the interaction of two fragments of a molecule is enhanced if there is a correspondence in the symmetry properties of their wave functions. A single $2p_z$ orbital or a π orbital is antisymmetric with respect to the plane of the molecule, i.e., it changes sign upon reflection in the plane. The atomic orbitals of the three hydrogen atoms may be combined to give a molecular orbital with the same symmetry as a π orbital. Such a combination is

$$\psi = c_1\phi_1 - c_2(\phi_2 + \phi_3) \tag{6-22}$$

This symmetry is shown schematically in Fig. 6-3. ψ can be considered as a *pseudo* π orbital. Hence, it may be regarded as part of the π system. The surprisingly large magnitudes of the methyl-proton hyperfine splittings may be attributed to this *direct* coupling of the protons into the π system. Because the methyl protons form a part of the π system, the spin density at the protons will be *positive*. It is to be recalled that the hyperfine splitting a_i of proton i is proportional to the *square* of the wave function at the proton, that is, $|\psi(0)|^2$. Hence the splittings of protons H_a, H_b, and H_c of Fig. 6-3 all have the same sign.

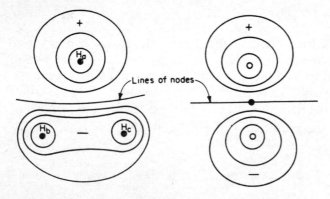

Fig. 6-3 Schematic representation of a three-hydrogen-atom molecular orbital of the same symmetry as the π orbitals in a conjugated radical. (*Figure kindly supplied by Prof. C. A. Coulson: See C. A. Coulson, "Valence," p. 362. Oxford University Press, London, 1961.*)

That the spin density at

$$\dot{C}-\overset{\overset{\displaystyle H}{|}}{\underset{\underset{\displaystyle H}{|}}{C}}-H$$

protons is opposite to that of

$$\overset{\displaystyle \diagdown}{\underset{\displaystyle \diagup}{\dot{C}}}-H$$

protons was established by observing an opposite shift of the two types of protons in a nuclear magnetic resonance experiment.[†,‡,§]

6-7 HYPERFINE SPLITTING BY NUCLEI OTHER THAN PROTONS

When isotropic proton hyperfine splittings were considered in Sec. 6-1, it was necessary to consider only the interaction of the π-spin density with the σ electrons in one bond (i.e., the C—H bond). However, in the case of nuclei which form part of the framework of a conjugated molecule, the interactions with several bonds must be considered. The hyperfine splittings by ^{13}C will be considered first, but the model should be generally applicable to other nuclei, such as ^{14}N, ^{17}O, ^{19}F, ^{33}S, etc.[¶] This model is essentially a generalization of the treatment given in Sec. 6-1 for the C—H fragment. In that analysis, it was shown that a *positive* spin density is induced at the carbon nucleus by the same mechanism which produces a *negative* spin density at the proton of a C—H fragment. It has been observed that experimental ^{13}C hyperfine splittings are not simply proportional to the π-spin density on the same carbon atom. Thus, it is necessary to consider contributions from the π-spin densities on neighboring carbon atoms. Figure 6-4 illustrates the several interactions which are characterized by various Q parameters. The notation used is as follows: The superscript designates the atom giving rise to the hyperfine splitting; the subscript next to the Q designates the atom on which the π-spin density is contributing to the spin polarization; the two subscripts together indicate the bond which is being polarized. S^C is a parameter which characterizes the polarization of the carbon $1s$ electrons by the local π-spin density.

By analogy with the C—H fragment, Q_{CH}^C and $Q_{CC'}^C$ are expected to be *positive*, whereas $Q_{C'C}^C$ and Q_{CH}^H should be *negative*. A consideration of the combined contributions leads to the following relation[¶]:

† Although the hyperconjugative coupling does account for the presence of methyl-proton hyperfine splittings, this effect alone leads to an incorrect ordering of the orbital energies of the toluene anion. Addition of a small *inductive* effect does lead to the correct ordering.§

‡ A. Forman, J. N. Murrell, and L. E. Orgel, *J. Chem. Phys.*, **31**:1129 (1959).

§ D. Lazdins and M. Karplus, *J. Am. Chem. Soc.*, **87**:920 (1965).

¶ M. Karplus and G. K. Fraenkel, *J. Chem. Phys.*, **35**:1312 (1961).

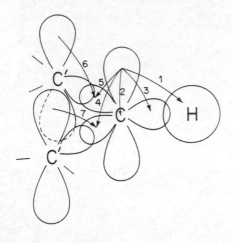

Fig. 6-4 Spin polarization contributions to the ^{13}C and to the proton hyperfine splittings in a

$$C' \atop C' \Big\rangle C{-}H$$ fragment. The numbered interac-

tions are (1) Q_{CH}^H; (2) S^C; (3) Q_{CH}^C; (4,5) $Q_{CC'}^C$; (6,7) $Q_{C'C}^C$.

$$a_i^C = \left(S^C + \sum_{j=1}^{3} Q_{CX_j}^C\right)\rho_i + \sum_{j=1}^{3} Q_{X_jC}^C \rho_j \tag{6-23}$$

where the atoms X_j are bonded to the carbon atom i. Quantitative calculations of the spin-polarization constants in Eq. (6-23) yield the following results[†]:

$$S^C = -12.7 \text{ G} \qquad Q_{CH}^C = +19.5 \text{ G} \qquad Q_{CC'}^C = +14.4 \text{ G} \qquad Q_{C'C}^C = -13.9 \text{ G}$$

Inserting these values into Eq. (6-23), one obtains

$$a_i^C = 35.6\rho_i - 13.9 \sum_j \rho_j \tag{6-24a}$$

$$= 30.5\rho_i - 13.9 \sum_j \rho_j \tag{6-24b}$$

where Eq. (6-24a) applies to a $CC_2'H$ fragment and Eq. (6-24b) to a CC_3' fragment.

The simplest example of a system exhibiting a ^{13}C hyperfine splitting is $^{13}CH_3$. Here from Eq. (6-23) one calculates $a^C = 45.8$ G for a planar radical. The experimental value is 38.5 G.[‡] The agreement is considered satisfactory and provides further evidence that CH_3 is planar.

Sometimes an independent estimate of the spin densities can be obtained from proton hyperfine splittings with the aid of Eq. (6-1). This has been done in the case of the anthracene cation and anion.[§] The spin densities and ^{13}C hyperfine splittings were calculated using $Q_{CH}^H(0) = -27.0$ G and the normalization condition, $\Sigma_i\rho_i = 1$; these are listed in Table 6-4. The agreement is very satisfactory, considering that the ^{13}C parameters were calculated from an approximate theory. Similar comparisons for other

† M. Karplus and G. K. Fraenkel, *J. Chem. Phys.*, **35**:1312 (1961).
‡ R. W. Fessenden and R. H. Schuler, *J. Chem. Phys.*, **43**:2704 (1965).
§ J. R. Bolton and G. K. Fraenkel, *J. Chem. Phys.*, **40**:3307 (1964).

Table 6-4   Calculated and experimental ^{13}C splittings and spin densities
in the anthracene cation and anion

| Position | ^{13}C splittings, a_i^C, G | | | Experimental spin densities[†] |
	Neg. ion	Pos. ion	Calc.	
9	8.76	8.48	8.42	0.220
11	−4.59	−4.50	−4.90	−0.021
1	3.57	—	3.37	0.107
2	−0.25	±0.37	−0.33	0.054

† Calculated from proton hyperfine splittings using $Q_{CH}^H(0) =$ −27.0 G.

radicals show that Eq. (6-23) is widely applicable for ^{13}C splittings in aromatic hydrocarbons.

In nitrogen heterocyclic aromatic molecules ^{14}N substitutes for carbon atoms; hence, one might expect that Eq. (6-23) would also apply to ^{14}N. This is probably correct; however, experience has shown that the effect of π-spin densities on neighboring atoms is small. This implies that the factor $Q_{C'N}^N$ must be small; certain estimates place it in the range −4 to +4 G.[†] In view of the small contribution from neighbors, many workers have used a simpler equation similar to Eq. (6-1) for ^{14}N hyperfine splittings

$$a_i^N = Q_{N(C_2H)}^N \rho_N \qquad (6\text{-}25a)$$

$$a_i^N = Q_{N(C_2P)}^N \rho_N \qquad (6\text{-}25b)$$

Here the first equation applies for protonated nitrogen sites and the second for nonprotonated sites; P stands for the lone pair of electrons on N. $Q_{N(C_2H)}^N$ has been variously placed in the range +27 to +30 G and $Q_{N(C_2P)}^N$ in the range +23 to +26 G. The major contribution to the nitrogen hyperfine splitting comes from spin density on the nitrogen atom itself. Thus one expects a positive hyperfine splitting; this has been confirmed experimentally.[‡]

^{17}O hyperfine splittings have been detected in enriched samples of some semiquinones and ketyls. An equation of the type exemplified by Eq. (6-23) appears to hold. The values $Q_{OC}^O = -44.5$ G[§] and $Q_{CO}^O = -14.3$ G appear satisfactory for the p-benzosemiquinone anion and cation[¶],[††]; however, there is some dispute about these values.[‡‡]

† R. L. Ward, *J. Am. Chem. Soc.*, **84**:332 (1962); J. C. M. Henning and C. deWaard, *Phys. Letters*, **3**:139 (1962); D. H. Geske and G. R. Padmanabhan, *J. Am. Chem. Soc.*, **87**:1651 (1965); J. C. M. Henning, *J. Chem. Phys.*, **44**:2139 (1966); C. L. Talcott and R. J. Myers, *Mol. Phys.*, **12**:549 (1967).
‡ J. H. Freed and G. K. Fraenkel, *J. Chem. Phys.*, **40**:1815 (1964).
§ The term S^0 is included in Q_{OC}^O.
¶ W. M. Gulick, Jr., and D. H. Geske, *J. Am. Chem. Soc.*, **88**:4119 (1966).
†† P. D. Sullivan, J. R. Bolton, and W. E. Geiger, Jr., *J. Am. Chem. Soc.*, **92**:4176 (1970).
‡‡ M. Broze, Z. Luz, and B. L. Silver, *J. Chem. Phys.*, **46**:4891 (1967).

It might be expected that since fluorine substitutes for hydrogen in aromatic molecules, an equation such as Eq. (6-1) would also hold for fluorine hyperfine splittings. That is, if ρ_C is *positive*, one expects that a^F would be *negative*. However, it has been shown conclusively that fluorine hyperfine splittings are *positive* in such molecules.[†] The nonbonding p electrons on the fluorine apparently participate in partial double bonding with the conjugated system to which the fluorine atom is attached. That is, some of the electron density in fluorine p orbitals is delocalized into the π system of the molecule. This electron transfer results in a net positive π-spin density on the fluorine atom. One should thus consider ^{19}F as being analogous to ^{13}C rather than to ^1H. An expression similar to Eq. (6-23) should be used for calculating ^{19}F splittings. One expects that the local contribution to a^F (i.e., from π-spin density on F) will predominate; this would result in a *positive* fluorine hyperfine splitting.

Hyperfine splittings have been measured for ^{33}S in a number of ring systems such as thianthrene. For these, it is possible to use an equation similar to Eq. (6-1):

$$a^S = Q^S_{S(C_2P)}\rho_S \tag{6-26}$$

where $Q^S_{S(C_2P)} = \sim 33$ G.[‡]

BIBLIOGRAPHY

Kaiser, E. T. and L. Kevan (eds.), "Radical Ions," Interscience Publishers, a division of John Wiley & Sons, Inc., New York, 1968. Chapters 1, 4, 5, 6 deal respectively with spin densities, radical cations, orbital degeneracy in substituted benzenes, and anion radicals.

Memory, J. D., "Quantum Theory of Magnetic Resonance Parameters," chaps. 7 and 8, McGraw-Hill Book Company, New York, 1968. Relations between hyperfine couplings and spin densities are treated in terms of valence-bond and molecular orbital theories.

PROBLEMS

6-1. The statement is often made that the value of Q determines the total extent of a spectrum. For benzene the extent is about 22.5 G; for CH_3, about 69 G; and for perinaphthenyl (formula in Prob. 5-8), about 43 G. Comment on the reasons for such divergent values of the spectral extent.

6-2. The allyl radical shows splittings of 4.06 G for the proton on the central carbon atom and 13.93 and 14.83 G for the protons on the end carbon atoms. Assume the average of the two splittings for the CH_2 protons and calculate ϵ of Eqs. (6-16) from the ratio of the hyperfine splittings.

6-3. The NMR spectra of ethylbenzene at 56.4 MHz is shown in Fig. 6-5 in the presence of various concentrations of the corresponding anion. From the shifts apparent in Fig. 6-5b,

† D. R. Eaton, A. D. Josey, W. D. Phillips, and R. E. Benson, *Mol. Phys.*, **5**:407 (1962).
‡ P. D. Sullivan, *J. Am. Chem. Soc.*, **90**:3618 (1968).

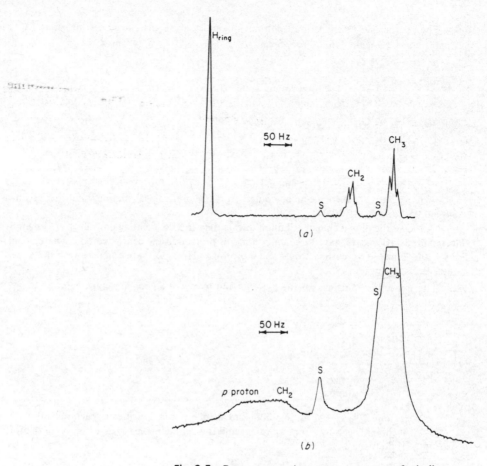

Fig. 6-5 Proton magnetic resonance spectra of ethylbenzene at $-75°$ C in deuterated tetrahydrofuran at 56.4 MHz and at the following concentrations: (a) 1.93M ethylbenzene; (b) 1.93M ethylbenzene plus $4.5 \times 10^{-2}M$ ethylbenzene anion. Peaks marked S are due to an impurity. [*Taken from E. de Boer and J. P. Colpa. J. Phys. Chem.,* **71**:21 (1967).]

confirm that the hyperfine splittings for the CH_2 and the *para* protons of the group are $+0.80$ and -0.87 G, respectively. In this system, electron exchange is rapid so that all ethylbenzene molecules participate; the shifts are proportional to the fraction of the reduced form.

6-4. The hyperfine splittings for the 1,3-butadiene negative ion are -7.62 and -2.79 G.

(a) What is the average value of Q?

(b) Can you explain why Q is so low? (Usually $Q \approx -25$ to -30 G.)

(c) Compute the spin densities with the average value of Q from part a; then use Eq. (6-20) to reevaluate the spin densities. (Use $Q(0) = -27$ G and $K = -12$ G, and assume $|\epsilon_i| = \rho_i$.)

6-5. Compute "experimental" spin densities for the cation and anion of pentacene. The hyperfine splittings are given in Table 5-5. [Use $Q(0) = -27$ G and $K = -12$ G, and assume $|\epsilon_i| = \rho_i$.] Compare these with the Hückel unpaired-electron densities: $\rho_1 = 0.0353$, $\rho_2 = 0.0250$, $\rho_5 = 0.1060$, and $\rho_6 = 0.1412$.

6-6. The proton and ^{13}C hyperfine splittings (including the signs) have been measured for the perinaphthenyl radical.† (See Prob. 5-8.)

$$a_1{}^H = -6.270 \text{ G}$$

$$a_2{}^H = +1.833 \text{ G}$$

$$a_1{}^C = +9.79 \text{ G}$$

$$a_2{}^C = -7.92 \text{ G}$$

$$a_{10}{}^C = -7.92 \text{ G}$$

$$a_{13}{}^C = +3.32 \text{ G}$$

(a) Assume $Q_{CH}^H = -27$ G, and calculate ρ_1 and ρ_2.

(b) ρ_{10} and ρ_{13} have been computed from theoretical calculations and are given as $\rho_{10} = -0.054$ and $\rho_{13} = +0.044$.

(c) Use the above spin distribution to calculate the ^{13}C splitting constants. (Remember that positions 10 and 13 have three carbon atoms bonded to the central carbon, whereas positions 1 and 2 have two carbon atoms and a proton.)   How do these compare with the experimental ^{13}C splittings?

6-7. The hyperfine splittings for the pyridine and pyrazine anions are given below‡:

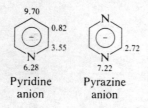

Pyridine anion   Pyrazine anion

(a) Assume $Q_{CH}^H = -27$ G, and compute all the π-spin densities in each radical.

(b) Fit the nitrogen hyperfine splittings and the spin densities to an equation of the form of Eq. (6-23), that is,

$$a^N = Q_{(C_2P)}^N \rho_N + Q_{C'N}^N \sum_j \rho_j$$

where the atoms j are bonded to the nitrogen atom.

† S. H. Glarum and J. H. Marshall, *J. Chem. Phys.*, **44**:2884 (1966).
‡ C. L. Talcott and R. J. Myers, *Mol. Phys.*, **12**:549 (1967).

7

Anisotropic Interactions in Oriented Systems with $S = \frac{1}{2}$

7-1 INTRODUCTION

The preceding six chapters have dealt with the ESR spectra of systems with $S = \frac{1}{2}$ in dilute solutions of low viscosity. For these, one may refer to the g factor or the hyperfine coupling A_0 as scalar constants of the system.

In contrast, for solid-state systems *even the qualitative aspects of an ESR spectrum may be markedly dependent upon the orientation of a crystalline sample in a magnetic field.* Some of the important classes of systems which show such anisotropy include:

1. *Free radicals oriented in solids.* Such radicals are most often generated by irradiation. They are considered in detail in Chap. 8.
2. *Transition-metal ions in single crystals.* For most of these, $S > \frac{1}{2}$, and the ESR spectrum shows additional complexities described in Chaps. 11 and 12.
3. *Paramagnetic point defects in single crystals.* These include electrons or holes at suitable trapping sites. A few examples are given in Sec. 8-5.

7-2 A SIMPLE EXAMPLE OF ANISOTROPY OF g

Before undertaking a general discussion of the anisotropy of g, it is instructive to examine one of the simplest of isotropic systems, namely, a cubic

Mg^{2+} O^{2-} Mg^{2+}

O^- ☐ O^{2-}

Mg^{2+} O^{2-} Mg^{2+} **Fig. 7-1**   Model of the V_1 center in MgO.

crystal in which there is octahedral symmetry about any normal lattice site. For such a crystal, g is strictly a scalar *constant;* the hamiltonian has the form

$$\hat{\mathscr{H}} = g\beta(H_x\hat{S}_x + H_y\hat{S}_y + H_z\hat{S}_z) \tag{7-1}$$

The symmetry may be reduced from octahedral to tetragonal by applying an external stress along one of the three [100]-type directions.† Alternatively, one may introduce an imperfection along one of these axes. The V_1 center in MgO (rock-salt structure) is an example of such a "defect center."‡,§,†† If an electron is lost from an oxygen atom adjacent to a magnesium-ion vacancy, there will be a small displacement of the resulting O^- ion away from the vacancy. The geometry of this defect center is shown in Fig. 7-1. This distortion leaves a fourfold axis of symmetry. It is generally customary to label the unique axis as the Z direction. In terms of the Miller notation, the axis is then $\langle 001 \rangle$.

If **H** is parallel to the $\langle 001 \rangle$ axis and $\nu = 9.0650$ GHz, a line is observed at 3,233.1 G. When the crystal is rotated so that **H** remains in the YZ or (100) plane, the line shifts from 3,233.1 to 3,177.1 G as the field direction changes from [001] to [010]. The variation in line position with orientation is shown in Fig. 7-2. Then

$$g_{\parallel} = \frac{h\nu}{\beta H_{\parallel}} = \frac{6.6262 \times 10^{-27} \times 9.0650 \times 10^9}{9.2741 \times 10^{-21} \times 3233.1} = 2.0033 \tag{7-2a}$$

$$g_{\perp} = \frac{h\nu}{\beta H_{\perp}} \qquad\qquad\qquad = 2.0386 \tag{7-2b}$$

† Miller indices enclosed in parentheses refer to planes, e.g., (001); in square brackets to directions, e.g., [011]; and in angular brackets to axes, e.g., $\langle 111 \rangle$.

‡ There is an unfortunate confusion in the usage of the term "V_1 center." Seitz proposed that the V_1 optical band in the alkali halides corresponds to a positive hole trapped adjacent to a cation vacancy.¶ This model corresponds to the defect observed in the alkaline-earth oxides (Fig. 7-1); the ESR behavior is given in Fig. 7-2 for MgO. However, the defect which is found to be correlated with the V_1 optical band in KCl has also been called a V_1 center. It is a Cl_4^{3-} ion (*H* center) with a $\langle 110 \rangle$-type axis with a Na^+ ion as a neighbor of one of the inner pair of chlorine atoms.††

§ J. E. Wertz, P. Auzins, J. H. E. Griffiths and J. W. Orton, *Discussions Faraday Soc.*, **28:**136 (1959); W. C. O'Mara, Ph.D. thesis, University of Minnesota, 1969.

¶ F. Seitz, *Rev. Mod. Phys.*, **26:**7 (1954).

†† C. J. Delbecq, E. Hutchinson, D. Schoemaker, E. L. Yasaitis, and P. H. Yuster, *Phys. Rev.*, **187:**1103 (1969), F. W. Patten and F. J. Keller, *Phys. Rev.*, **187:**1120 (1969).

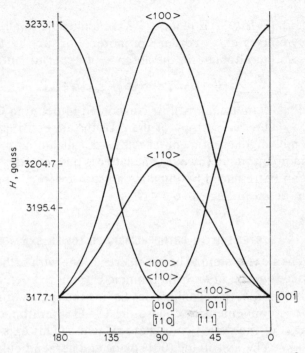

Fig. 7-2 Angular dependence of the ESR spectrum of the V_1 center in MgO. Resonant field values at extrema or at positions of coincidence are given at the left of the figure. The axes of rotation are shown for each line. The direction of the magnetic field at extrema or at coincidence of lines is given at the bottom. When two field directions are given, the upper corresponds to rotation about the $\langle 100 \rangle$ and the lower to the $\langle 110 \rangle$ axis.

Here $g_\parallel$ and $g_\perp$ are the g factors appropriate to the orientations $H_\parallel$ and $H_\perp$ of the field when it is, respectively, parallel and perpendicular to the symmetry axis (that is, Z).

Since these defects are in a cubic crystal, the directions [001], [010], and [100] are all equivalent. Hence there is an equal probability that an O^- ion-vacancy axis will lie along one of these three directions. Defects along each direction give rise to one ESR line; the position of each line varies with the orientation of the field relative to its axis. Thus when $\mathbf{H} \parallel$ [001], the lines from defects with $\langle 100 \rangle$ and $\langle 010 \rangle$ axes coincide at the $g_\perp$ position, as in Fig. 7-2. The coincidences for $\mathbf{H} \parallel$ [011] and for $\mathbf{H} \parallel$ [111] are also shown in Fig. 7-2.

The shape of the curve in Fig. 7-2 is found to be well represented by an effective g factor (g_{eff}) given by

$$g_{\text{eff}}^2 = g_\parallel^2 \cos^2 \theta + g_\perp^2 \sin^2 \theta \tag{7-3}$$

where θ is the angle between $\mathbf{H}$ and the symmetry axis of the defect. It will be shown later that Eq. (7-3) is a special case of a more general expression

[Eq. (7-16b)].   Equation (7-3) is applicable to all $S = \frac{1}{2}$ systems possessing a symmetry axis of order 3 or more.   For such systems of axial symmetry, the spin hamiltonian (in the absence of hyperfine interaction) is

$$\hat{\mathscr{H}} = \beta[g_\perp(H_X\hat{S}_x + H_Y\hat{S}_y) + g_{\parallel}H_Z\hat{S}_z] \tag{7-4}$$

This hamiltonian will be considered in detail in Chap. 11.

The anisotropy of the g factor arises from a coupling of the electron spin angular momentum with a small amount of induced orbital angular momentum.   However, the latter is not considered explicitly since its effect can be replaced by imputing anisotropy to g.   This approach is considered in detail in Sec. 11-6.

7-3   SYSTEMS WITH ORTHORHOMBIC OR LOWER SYMMETRY

The next system to be considered is one with orthorhombic symmetry.   This need *not* be in an orthorhombic crystal; the defect center shown in Fig. 7-3 is found in the alkali halides having the rock-salt structure.   (See Sec. 8-5b.) It is convenient to choose the O—O interatomic direction in O_2^- as the Z axis of the defect coordinate system.   (This axis is $\langle 110 \rangle$ in the cubic crystal.)   The axis in the (001) plane and perpendicular to Z will be called the X axis.   The Y axis of the right-handed coordinate system will be directed out of the plane.   The X and the Y axes are not equivalent, and hence the defect does not have axial symmetry.   The spin hamiltonian (assuming no hyperfine interaction) is the following:

$$\hat{\mathscr{H}} = \beta(g_{XX}H_X\hat{S}_x + g_{YY}H_Y\hat{S}_y + g_{ZZ}H_Z\hat{S}_z)\dagger \tag{7-5}$$

If it were possible to orient all defects along the Z axis, the spectrum would consist of only a single line.   If the line position is ascertained for the field, respectively, along the X, Y, and Z directions, the positions are described

† The significance of the double subscripts is explained in Secs. 7-4 and A-6.

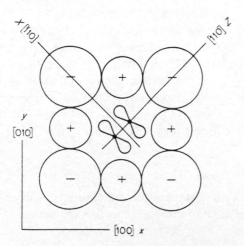

Fig. 7-3   Orientation of the O_2^- ion in an alkali halide crystal of the rock salt structure.

by g_{XX}, g_{YY}, and g_{ZZ}. The effective value of g for an arbitrary orientation is then given by

$$g_{\text{eff}}^2 = g_{XX}^2 \cos^2 \theta_{HX} + g_{YY}^2 \cos^2 \theta_{HY} + g_{ZZ}^2 \cos^2 \theta_{HZ} \qquad (7\text{-}6a)$$

$$= g_{XX}^2 l_X^2 + g_{YY}^2 l_Y^2 + g_{ZZ}^2 l_Z^2 \qquad (7\text{-}6b)$$

Here θ_{HX}, θ_{HY}, and θ_{HZ} are, respectively, the angles between the field $\mathbf{H}$ and the X, Y, and Z axes. It is convenient to represent the cosines of these angles by the symbols l_X, l_Y, and l_Z, respectively, as in Eq. (7-6b). These are referred to as the "direction cosines."† Note that Eq. (7-6b) is equivalent to the product $\mathbf{l} \cdot {}^d\mathbf{g}^2 \cdot \mathbf{l}$, where $\mathbf{l}$ is a vector with components l_X, l_Y, and l_Z and ${}^d\mathbf{g}^2$ is a diagonal tensor with principal values g_{XX}^2, g_{YY}^2, and g_{ZZ}^2. (See Secs. A-4, A-5 and A-6 for a discussion of matrices, vectors, and tensors.) That is,

$$g_{\text{eff}}^2 = \begin{bmatrix} l_X & l_Y & l_Z \end{bmatrix} \begin{bmatrix} g_{XX}^2 & 0 & 0 \\ 0 & g_{YY}^2 & 0 \\ 0 & 0 & g_{ZZ}^2 \end{bmatrix} \begin{bmatrix} l_X \\ l_Y \\ l_Z \end{bmatrix} \qquad (7\text{-}7)$$

It might be suspected that the simple form of Eq. (7-6) or of the tensor in Eq. (7-7) is the consequence of foreknowledge of the principal axes of the defect system, as well as the direct determination of the g components by measurement of the resonant field when $\mathbf{H}$ is parallel to a principal axis. This is indeed the case. More generally, when one measures g factors in ignorance of the principal axes, the off-diagonal elements of the g tensor are nonzero.‡ Indeed, it would have been very logical to have measured line positions as a function of rotation about the $\langle 100 \rangle$-type axes of the cubic crystal. The technique of arriving at the values in the diagonal $\mathbf{g}^2$ tensor from such measurements will be discussed in the next section.

7-4 EXPERIMENTAL DETERMINATION OF THE g TENSOR IN ORIENTED SOLIDS

In recognition of the fact that g is, in general, a tensor, the spin hamiltonian of Eq. (7-5) may be rewritten as

$$\hat{\mathscr{H}} = \beta \hat{\mathbf{S}} \cdot \mathbf{g} \cdot \mathbf{H} \qquad (7\text{-}8)$$

If values of g are measured with respect to an arbitrary set x, y, z of orthogonal crystal fixed axes,

---

† The three direction cosines are related by the trigonometric identity $l_X^2 + l_Y^2 + l_Z^2 = 1$, so that two variable parameters are sufficient to specify direction, as with spherical polar coordinates. However, the more symmetrical expressions from the use of direction cosines make this the preferred system.

‡ Strictly, $\mathbf{g}$ is not a tensor from the mathematical standpoint, whereas, $\mathbf{g}^2$ is a true tensor. However, from an operational standpoint one may refer to $\mathbf{g}$ as a "tensor." For a discussion of the subtleties of this problem the reader is referred to A. Abragam and B. Bleaney, "Electron Paramagnetic Resonance of Transition Ions," pp. 651ff., Clarendon Press, Oxford, England, 1970.

$$\mathscr{H} = \frac{\beta[\hat{S}_x \quad \hat{S}_y \quad \hat{S}_z]}{} \begin{bmatrix} g_{xx} & g_{xy} & g_{xz} \\ g_{yx} & g_{yy} & g_{yz} \\ g_{zx} & g_{zy} & g_{zz} \end{bmatrix} \begin{bmatrix} H_x \\ H_y \\ H_z \end{bmatrix} \qquad (7\text{-}9)$$

One may interpret the double subscripts as follows. A g component g_{yx} may be considered as the contribution to g along the y axis when the magnetic field is applied along x. That such contributions are to be expected may be seen from Fig. 7-3. Since the x axis is not orthogonal to the X or Z axes of the paramagnetic center, a field H_x has components along both the X and Z directions. There will thus be components of magnetization along the X and Z axes.† Thus even at this special orientation there will be the additional component g_{yx}. However, since $z(\equiv Y)$ is a principal axis, the components g_{xz}, g_{zx}, g_{yz}, and g_{zy} will vanish for the present case.

The product $\mathbf{g} \cdot \mathbf{H}$ may be regarded as a transformation of the actual field $\mathbf{H}$ to an effective field $\mathbf{H}_{\text{eff}} = \mathbf{g} \cdot \mathbf{H} = g_{\text{eff}}\mathbf{H}$. The spin angular momentum will be quantized along $\mathbf{H}_{\text{eff}}$ such that the two electron energy levels are

$$W_1 = +\tfrac{1}{2}\beta g_{\text{eff}}H$$

and

$$W_2 = -\tfrac{1}{2}\beta g_{\text{eff}}H \qquad (7\text{-}10)$$

The energy-level separation is thus

$$\Delta W = \beta g_{\text{eff}}H \qquad (7\text{-}11)$$

g_{eff} for an arbitrary field orientation is unknown until the tensor $\mathbf{g}$ has been established. Since energy is a scalar, the factor multiplying β in Eq. (7-11) must also be a scalar. The quantity $\mathbf{g} \cdot \mathbf{H}$ is a vector. The square of the magnitude of a vector is obtained by taking the scalar product of the vector with itself. (See Sec. A-4.) Hence‡

$$(\Delta W)^2 = \beta^2 g_{\text{eff}}^2 H^2 = \beta^2(\mathbf{H} \cdot \mathbf{g}) \cdot (\mathbf{g} \cdot \mathbf{H}) \qquad (7\text{-}12)$$

$$= \beta^2\mathbf{H} \cdot \mathbf{g}^2 \cdot \mathbf{H} \qquad (7\text{-}13)$$

where $\mathbf{g}^2$ is the matrix product of $\mathbf{g}$ with its transpose. (See Sec. A-6.)§

† Consider the following crude analogy: Assume that a small cube with highly polished surfaces is at the origin of a cartesian coordinate system, with axes perpendicular to the cube faces. If a small pencil of light is directed at the cube, exactly along one of the axes, it will be reflected only along that axis. For an arbitrary orientation of the cube there will be components of reflected light along each of the three axes.

‡ J. E. Geusic and L. C. Brown, *Phys. Rev.*, **112**:64 (1958).

§ By its definition, the tensor $\mathbf{g}^2$ will always be symmetric, that is, $(\mathbf{g}^2)_{xy} = (\mathbf{g}^2)_{yx}$. For paramagnetic systems of monoclinic or of triclinic symmetry, it is possible to have inherently asymmetric g tensors. However, for systems of spin $S = \tfrac{1}{2}$, one cannot experimentally establish the asymmetry from measurements of resonant field values. If $S \geq 1$ and there is a splitting of levels in zero magnetic field [see Secs. 10-2, 10-3, and 11-6], the components of the asymmetric tensor may be determined. [See F. K. Kneubühl, *Phys. kondens. Materie*, **1**:410 (1963).]

H on the left is written in terms of its components as

$$H[l_x \quad l_y \quad l_z]$$

while H on the right is denoted by

$$H \begin{bmatrix} l_x \\ l_y \\ l_z \end{bmatrix}$$

Hence

$$g_{\text{eff}}^2 H^2 = H^2 [l_x \quad l_y \quad l_z] \begin{bmatrix} g_{xx} & g_{xy} & g_{xz} \\ g_{yx} & g_{yy} & g_{yz} \\ g_{zx} & g_{zy} & g_{zz} \end{bmatrix} \begin{bmatrix} g_{xx} & g_{yx} & g_{zx} \\ g_{xy} & g_{yy} & g_{zy} \\ g_{xz} & g_{yz} & g_{zz} \end{bmatrix} \begin{bmatrix} l_x \\ l_y \\ l_z \end{bmatrix} \quad (7\text{-}14)$$

or

$$g_{\text{eff}}^2 = [l_x \quad l_y \quad l_z] \begin{bmatrix} (\boldsymbol{g}^2)_{xx} & (\boldsymbol{g}^2)_{xy} & (\boldsymbol{g}^2)_{xz} \\ (\boldsymbol{g}^2)_{yx} & (\boldsymbol{g}^2)_{yy} & (\boldsymbol{g}^2)_{yz} \\ (\boldsymbol{g}^2)_{zx} & (\boldsymbol{g}^2)_{zy} & (\boldsymbol{g}^2)_{zz} \end{bmatrix} \begin{bmatrix} l_x \\ l_y \\ l_z \end{bmatrix} \quad (7\text{-}15)$$

The elements of the tensor in Eq. (7-15) represent the components of the $\boldsymbol{g}^2$ tensor. These need not be more explicitly written out. The $(\boldsymbol{g}^2)_{ij}$ elements will be determined from experiment by successive rotation of the crystal or the field in the xz, yz, and xy planes. For the xz plane, if θ is the angle between H and the z axis, $l_z = \cos\theta$, $l_y = 0$, and $l_x = \sin\theta$. Then

$$g_{\text{eff}}^2 = [\sin\theta \quad 0 \quad \cos\theta] \begin{bmatrix} (\boldsymbol{g}^2)_{xx} & (\boldsymbol{g}^2)_{xy} & (\boldsymbol{g}^2)_{xz} \\ (\boldsymbol{g}^2)_{yx} & (\boldsymbol{g}^2)_{yy} & (\boldsymbol{g}^2)_{yz} \\ (\boldsymbol{g}^2)_{zx} & (\boldsymbol{g}^2)_{zy} & (\boldsymbol{g}^2)_{zz} \end{bmatrix} \begin{bmatrix} \sin\theta \\ 0 \\ \cos\theta \end{bmatrix} \quad (7\text{-}16a)$$

and

$$g_{\text{eff}}^2 = (\boldsymbol{g}^2)_{xx} \sin^2\theta + 2(\boldsymbol{g}^2)_{xz} \sin\theta \cos\theta + (\boldsymbol{g}^2)_{zz} \cos^2\theta \quad (7\text{-}16b)$$

Similarly, for rotation in the yz plane

$$[l_x \quad l_y \quad l_z] = [0 \quad \sin\theta \quad \cos\theta] \quad (7\text{-}17)$$

and

$$g_{\text{eff}}^2 = (\boldsymbol{g}^2)_{yy} \sin^2\theta + 2(\boldsymbol{g}^2)_{yz} \sin\theta \cos\theta + (\boldsymbol{g}^2)_{zz} \cos^2\theta \quad (7\text{-}18)$$

Likewise, for rotation in the xy plane

$$g_{\text{eff}}^2 = (\boldsymbol{g}^2)_{xx} \cos^2\theta + 2(\boldsymbol{g}^2)_{xy} \sin\theta \cos\theta + (\boldsymbol{g}^2)_{yy} \sin^2\theta \quad (7\text{-}19)$$

It is apparent that in each plane only three measurements are necessary. For the xz plane, measurements with $\theta = 0$ and $90°$ give the values $(\boldsymbol{g}^2)_{xx}$ and $(\boldsymbol{g}^2)_{zz}$ respectively. The value of $(\boldsymbol{g}^2)_{xz}$ can be determined with the best precision at $\theta = 45°$ and at $\theta = 135°$.

Following the evaluation of the six independent components of the $\boldsymbol{g}^2$ tensor, it is necessary to transform it to a diagonal form. This is accomplished by finding a matrix $\mathscr{L}$ such that

$$
\underbrace{\begin{bmatrix} l_{Xx} & l_{Xy} & l_{Xz} \\ l_{Yx} & l_{Yy} & l_{Yz} \\ l_{Zx} & l_{Zy} & l_{Zz} \end{bmatrix}}_{\mathscr{L}} \underbrace{\begin{bmatrix} (\boldsymbol{g}^2)_{xx} & (\boldsymbol{g}^2)_{xy} & (\boldsymbol{g}^2)_{xz} \\ (\boldsymbol{g}^2)_{yx} & (\boldsymbol{g}^2)_{yy} & (\boldsymbol{g}^2)_{yz} \\ (\boldsymbol{g}^2)_{zx} & (\boldsymbol{g}^2)_{zy} & (\boldsymbol{g}^2)_{zz} \end{bmatrix}}_{\boldsymbol{g}^2} \underbrace{\begin{bmatrix} l_{Xx} & l_{Yx} & l_{Zx} \\ l_{Xy} & l_{Yy} & l_{Zy} \\ l_{Xz} & l_{Yz} & l_{Zz} \end{bmatrix}}_{\mathscr{L}^\dagger}
$$

$$
= \underbrace{\begin{bmatrix} (\boldsymbol{g}^2)_{XX} & 0 & 0 \\ 0 & (\boldsymbol{g}^2)_{YY} & 0 \\ 0 & 0 & (\boldsymbol{g}^2)_{ZZ} \end{bmatrix}}_{{}^d\boldsymbol{g}^2} \quad (7\text{-}20)
$$

The l components are again the direction cosines connecting the molecular axes XYZ of the paramagnetic defect with the laboratory axes xyz.[†] $\mathscr{L}^\dagger$ is the transpose of $\mathscr{L}$. $\mathscr{L}$ is unitary, so $\mathscr{L}^\dagger = \mathscr{L}^{-1}$. The procedure for finding the $\mathscr{L}$ matrix which will diagonalize $\boldsymbol{g}^2$ is given in Sec. A-5e. Once the principal values of $\boldsymbol{g}^2$ are found, one takes the square root of each diagonal element to obtain the principal g components.[‡] Problem 7-6 gives an opportunity to establish the principal values of a g tensor. An actual computation will be carried out in detail for the closely analogous operation of determining the principal values and direction cosines of the hyperfine coupling tensor (Sec. 7-7).

7-5 ANISOTROPY OF THE HYPERFINE COUPLING

In many oriented systems there may be an anisotropy in the hyperfine splittings as well as in g. This hyperfine anisotropy may be so great that the qualitative appearance of the spectrum may be drastically changed by rotation of a single crystal through a relatively small angle.

A very simple example of a strongly anisotropic hyperfine interaction is that of the V_{OH} center shown in Fig. 7-4. This center in MgO consists of a linear defect $O^- \square\, HO^-$ in which a cation vacancy ($\square$) separates a para-

---

[†] Strictly, X, Y and Z are the principal axes of the tensor $\boldsymbol{g}$. If the molecule has axes of symmetry, they must coincide with X, Y, or Z; if there are planes of symmetry, they must be perpendicular to X, Y, or Z. For molecules of low symmetry, the axes may be in any direction but are necessarily orthogonal to each other. The principal directions, and hence the $\mathscr{L}$ matrix, are the same for $\boldsymbol{g}$ and for $\boldsymbol{g}^2$. One of the principal directions corresponds to a maximum value of g_{eff} and another to a minimum.

[‡] Taking this square root involves an uncertainty of sign, as a 3×3 tensor has eight square roots. If the tensor $\boldsymbol{g}$ arises from small departures from the free-spin g factor 2.0023, then on chemical grounds all the square roots may reasonably be taken as positive. One uses the sign convention that g for a free electron spin is treated as a positive quantity; the true negative magnetogyric ratio of the electron is allowed for by writing the spin hamiltonian $\mathscr{H} = +g\beta\hat{\mathbf{S}} \cdot \mathbf{H}$. By contrast for nuclei one must write $\mathscr{H} = -g_N\beta_N\hat{\mathbf{I}} \cdot \mathbf{H}$, since g_N takes the actual sign of the magnetogyric ratio of the nucleus in question; its sign may be either positive or negative. For transition-metal ions, where g departs greatly from the free-spin value, the correct square root must usually be chosen after a consideration of the wave function. If the resonance experiment can be done with circularly polarized microwaves, the sign of the product $g_{xx}g_{yy}g_{zz}$ can be determined experimentally.

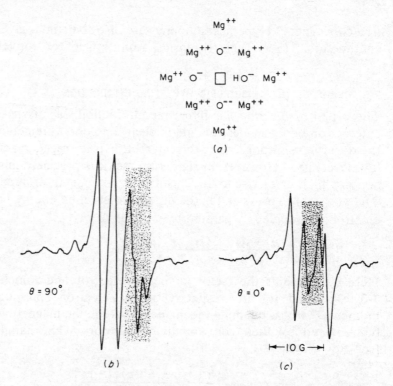

Fig. 7-4 ESR spectra of the V_{OH} center in MgO. These spectra show almost purely anisotropic hyperfine splitting. Lines arising from related defects have been masked. (*a*) Structure of the defect. (*b*) Line components for which **H** is perpendicular to the tetragonal symmetry axis. (*c*) Line components for **H** parallel to the symmetry axis.

magnetic O^- ion and the hydrogen atom of an OH^- impurity ion. The O^-—H separation is about 0.32 nm. If the crystal is rotated in a (100) plane and if the angle between the defect axis and the field direction is θ, the hydrogen hyperfine coupling *for this case* is given by an expression of the form

$$A(\theta) = A_0 + B(3\cos^2\theta - 1) \qquad\qquad (7\text{-}21a)$$

Specifically

$$A(\theta) = [0.0440 + 2.375(3\cos^2\theta - 1)] \qquad \text{MHz} \qquad (7\text{-}21b)$$

The separation of the doublet (Fig. 7-4) ranges from 1.72 G for $\theta = 0°$, becoming zero when $\cos^2\theta = \frac{1}{3}(1 - 0.044/2.375)$, to -0.83 G for $\theta = 90°$. For this system the hyperfine splitting is almost purely anisotropic. In most cases, the isotropic contribution A_0 will be of the same order of magnitude as B. Then Eq. (7-21*a*) is not applicable, and more complicated expressions are required. (See Secs. 7-6 and 7-7.)

For oriented solids, it is customary to express hyperfine couplings in megahertz. The reason is that a frequency times h is a true energy unit, whereas the relation between gauss and energy depends on g_{eff} for the direction concerned. Hyperfine couplings are also sometimes given in reciprocal centimeters. (See Table B, inside front cover, for conversion factors.)

7-6 ORIGIN OF THE ANISOTROPIC HYPERFINE INTERACTION

The origin of the *isotropic* hyperfine interaction was discussed in Chap. 3. Interaction between electron and nuclear dipoles was rejected as a source of the observed splittings, since this interaction is averaged to zero in a liquid of low viscosity. However, in rigid systems, it is precisely this dipolar interaction which gives rise to the anisotropic component of hyperfine coupling. The classical expression for the dipolar interaction energy between a fixed electron and nucleus separated by a distance r is[†]

$$W_{\text{dipolar}} = \frac{\boldsymbol{\mu}_e \cdot \boldsymbol{\mu}_N}{r^3} - \frac{3(\boldsymbol{\mu}_e \cdot \mathbf{r})(\boldsymbol{\mu}_N \cdot \mathbf{r})}{r^5} \tag{7-22}$$

Here $\mathbf{r}$ represents the vector joining an electron and a nucleus. (See Fig. 3-3.) $\boldsymbol{\mu}_e$ and $\boldsymbol{\mu}_N$ are, respectively, the electron and nuclear magnetic moments. For a quantum-mechanical system, the magnetic moments must be replaced by their corresponding operators. The hamiltonian is thus [see Eqs. (3-6)]

$$\hat{\mathscr{H}}_{\text{dipolar}} = -g\beta g_N \beta_N \left[\frac{\hat{\mathbf{S}} \cdot \hat{\mathbf{I}}}{r^3} - \frac{3(\hat{\mathbf{S}} \cdot \mathbf{r})(\hat{\mathbf{I}} \cdot \mathbf{r})}{r^5} \right] \tag{7-23}$$

That $\hat{\mathscr{H}}_{\text{dipolar}}$ involves a tensor interaction can be seen by expanding the vectors in Eq. (7-23),

$$\hat{\mathscr{H}}_{\text{dipolar}} = -g\beta g_N \beta_N \left[\left\langle \frac{r^2 - 3x^2}{r^5} \right\rangle \hat{S}_x \hat{I}_x + \left\langle \frac{r^2 - 3y^2}{r^5} \right\rangle \hat{S}_y \hat{I}_y \right.$$
$$+ \left\langle \frac{r^2 - 3z^2}{r^5} \right\rangle \hat{S}_z \hat{I}_z - \left\langle \frac{3xy}{r^5} \right\rangle (\hat{S}_x \hat{I}_y + \hat{S}_y \hat{I}_x)$$
$$\left. - \left\langle \frac{3xz}{r^5} \right\rangle (\hat{S}_x \hat{I}_z + \hat{S}_z \hat{I}_x) - \left\langle \frac{3yz}{r^5} \right\rangle (\hat{S}_y \hat{I}_z + \hat{S}_z \hat{I}_y) \right] \tag{7-24a}$$

$$= \underbrace{[\hat{S}_x, \hat{S}_y, \hat{S}_z]}_{\hat{\mathbf{S}}} \cdot (-g\beta g_N \beta_N)$$

$$\underbrace{\begin{bmatrix} \left\langle \dfrac{r^2 - 3x^2}{r^5} \right\rangle & -\left\langle \dfrac{3xy}{r^5} \right\rangle & -\left\langle \dfrac{3xz}{r^5} \right\rangle \\[2mm] -\left\langle \dfrac{3xy}{r^5} \right\rangle & \left\langle \dfrac{r^2 - 3y^2}{r^5} \right\rangle & -\left\langle \dfrac{3yz}{r^5} \right\rangle \\[2mm] -\left\langle \dfrac{3xz}{r^5} \right\rangle & -\left\langle \dfrac{3yz}{r^5} \right\rangle & \left\langle \dfrac{r^2 - 3z^2}{r^5} \right\rangle \end{bmatrix}}_{\mathbf{T}} \cdot \underbrace{\begin{bmatrix} \hat{I}_x \\[2mm] \hat{I}_y \\[2mm] \hat{I}_z \end{bmatrix}}_{\hat{\mathbf{I}}} \tag{7-24b}$$

$$= h\hat{\mathbf{S}} \cdot \mathbf{T} \cdot \hat{\mathbf{I}} \tag{7-24c}$$

[†] W. Cheston, "Elementary Theory of Electric and Magnetic Fields," p. 151, John Wiley & Sons, Inc., New York, 1964.

The angular brackets imply an average taken over the electronic wave function, as the electron is not fixed in space.

The full hamiltonian requires the addition of the isotropic hyperfine term $(hA_0\hat{\mathbf{S}} \cdot \hat{\mathbf{I}})$, the electron Zeeman term $(\beta\hat{\mathbf{S}} \cdot \boldsymbol{g} \cdot \mathbf{H})$, and the nuclear Zeeman term $(-g_N\beta_N\mathbf{H} \cdot \hat{\mathbf{I}})$, thus†

$$\hat{\mathcal{H}} = \beta\hat{\mathbf{S}} \cdot \boldsymbol{g} \cdot \mathbf{H} + h\hat{\mathbf{S}} \cdot \boldsymbol{A} \cdot \hat{\mathbf{I}} - g_N\beta_N\mathbf{H} \cdot \hat{\mathbf{I}} \tag{7-25a}$$

where

$$\boldsymbol{A} = A_0\boldsymbol{1} + \boldsymbol{T} \tag{7-25b}$$

A_0 is the isotropic hyperfine coupling, and $\boldsymbol{1}$ is the unit tensor.

Two approximations are usually employed in the application of Eqs. (7-25):

1. The g factor is assumed to be isotropic for the purposes of examining the hyperfine interaction.‡
2. The electron Zeeman term is assumed to be the dominant energy term. This allows one to quantize $\mathbf{S}$ along $\mathbf{H}$.

Equation (7-25a) may then be written as

$$\hat{\mathcal{H}} = g\beta H M_S - g_N\beta_N\mathbf{H}_{\text{eff}} \cdot \hat{\mathbf{I}} \tag{7-26a}$$

Here

$$\mathbf{H}_{\text{eff}} = \mathbf{H} + \mathbf{H}_{\text{hf}} \tag{7-26b}$$

$$\mathbf{H}_{\text{hf}} = \frac{-h}{g_N\beta_N}\hat{\mathbf{S}} \cdot \boldsymbol{A} = \frac{-hM_S}{g_N\beta_N}\boldsymbol{1} \cdot \boldsymbol{A} \tag{7-26c}$$

where $\boldsymbol{1}$ is a unit vector in the direction of $\mathbf{H}$. $\mathbf{H}_{\text{hf}}$ is the magnetic field generated at the nucleus due to the electron-nuclear hyperfine interaction. Clearly, $\mathbf{I}$ must be quantized along the resultant of $\mathbf{H}_{\text{hf}}$ and $\mathbf{H}$, namely, $\mathbf{H}_{\text{eff}}$. (See Fig. 7-5.) Three cases will be considered:

Case a. $|\mathbf{H}| \gg |\mathbf{H}_{\text{hf}}|$. $\mathbf{I}$ may then be taken effectively to be quantized along $\mathbf{H}$ (see Fig. 7-5a); Eq. (7-25a) may then be written

$$\hat{\mathcal{H}} = g\beta H M_S + hM_S M_I\boldsymbol{1} \cdot \boldsymbol{A} \cdot \boldsymbol{1} - g_N\beta_N H M_I \tag{7-27}$$

Since both $\mathbf{S}$ and $\mathbf{I}$ are quantized along $\mathbf{H}$ (taken as the z axis), one may

---

† Additional terms are required if $I \geq 1$ or $S \geq 1$. In the former case, there is an interaction of the quadrupole moment of the nucleus with the gradient of the electric field (see Prob. C-3); in the latter case, there may be energy-level splittings even in the absence of an applied field. These splittings are considered in Chap. 10.

‡ If the g anisotropy is large (i.e., comparable to the hyperfine anisotropy), then complications arise. A full treatment of this problem is found in A. Abragam and B. Bleaney "Electron Paramagnetic Resonance of Transition Ions," pp. 167ff., Clarendon Press, Oxford, England, 1970.

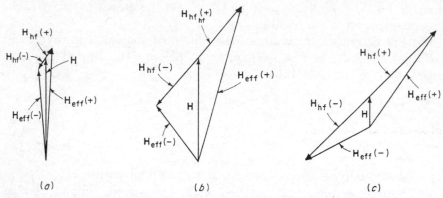

Fig. 7-5 Vector addition of the external field **H** and of the hyperfine field $\mathbf{H}_{hf}$ for $I = \frac{1}{2}$. The plus and minus signs refer to $M_I = +\frac{1}{2}$ and $M_I = -\frac{1}{2}$ respectively. (a) $|\mathbf{H}| >> |\mathbf{H}_{hf}|$; (b) $|\mathbf{H}| \approx |\mathbf{H}_{hf}|$; (c) $|\mathbf{H}| << |\mathbf{H}_{hf}|$.

neglect the x and y components of $\hat{\mathbf{S}}$ and $\hat{\mathbf{I}}$ in Eq. (7-24a). With the substitution $z = r \cos \theta$ for the special case of a p orbital centered on the interacting nucleus, Eq. (7-24a) then reduces to

$$\hat{\mathscr{H}}_{\text{dipolar}} = g\beta g_N\beta_N \left\langle \frac{3 \cos^2 \theta - 1}{r^3} \right\rangle M_S M_I \tag{7-28a}$$

$$= g\beta g_N\beta_N \left\langle \frac{3 \cos^2 \alpha - 1}{2r^3} \right\rangle (3 \cos^2 \theta - 1) M_S M_I \tag{7-28b}$$

$$= hB(3 \cos^2 \theta - 1) M_S M_I \tag{7-28c}$$

θ in Eq. (7-28a) is the angle between **r** and **H**. (See Fig. 3-3.) In Eq. (7-28b) θ is redefined as the angle between **H** and the axis of the p orbital. α is then the angle between **r** and the axis of the p orbital. The angular brackets indicate an average over the electronic wave function.[†] If the electron is interacting with a nucleus not at the center of the p orbital, Eq. (7-28c) still holds, but an appropriate average must be taken over the electronic wave function. General expressions for the interaction of a nucleus with an electron in a p orbital centered at a distance d from the nucleus have been developed.[‡]

The V_{OH} center, considered in Sec. 7-5, is a good example of case a. There $|\mathbf{H}_{hf}| = 538$ G for $\theta = 0$ as compared to $|\mathbf{H}| \sim 3{,}200$ G.

Case b. $|\mathbf{H}| \approx |\mathbf{H}_{hf}|$. In this case the axis of quantization for **I** is not the same for $M_S = +\frac{1}{2}$ or $-\frac{1}{2}$. (See Fig. 7-5b.) This results in the appearance of satellite lines. This case is considered in detail in Sec. 7-8.

Case c. $|\mathbf{H}| << |\mathbf{H}_{hf}|$. This case is the one most commonly encoun-

---

[†] The form of the term inside the angular brackets arises from the fact that if the electron is located at point r on the axis of the p orbital ($\alpha = 0°$), then the result must be $+r^{-3}$.
[‡] H. M. McConnell and J. Strathdee, *Mol. Phys.*. **2**:129 (1959).

tered† and thus will be analyzed in some detail.‡ Complications arise here because in general $\mathbf{H}_{hf}(\approx \mathbf{H}_{eff})$ and $\mathbf{H}$ are not in the same direction. (See Fig. 7-5c.) Thus $\mathbf{S}$ and $\mathbf{I}$ are quantized along different directions (along $\mathbf{H}$ for $\mathbf{S}$ and along $\mathbf{H}_{hf}$ for $\mathbf{I}$). For purposes of illustration, the case of axial symmetry and $I = \frac{1}{2}$ will be considered. Since the electron Zeeman energy is assumed to be the dominant energy term, one chooses axes such that $\mathbf{S}$ is diagonal. This requires that the direction of $\mathbf{H}$ be the z axis. Since $\mathbf{S}$ is quantized along z, terms in $\hat{S}_x$ and $\hat{S}_y$ in Eq. (7-24a) may be neglected. The hyperfine field from the *isotropic* hyperfine interaction will also be oriented along z, as it is a scalar interaction. The contribution to $\mathbf{H}_{hf}$ from the *anisotropic* hyperfine interaction may be resolved into two components, which are respectively, parallel and perpendicular to z. On substitution of $r \cos \theta$ for z and $r \sin \theta$ for x or y in Eq. (7-24a), one may derive (following the treatment for case a), the hamiltonian for an electron in a p orbital centered on the interacting nucleus,

$$\hat{\mathcal{H}} = g\beta H M_S + h M_S \{[A_0 + B(3 \cos^2 \theta - 1)]\hat{I}_z + 3B \cos \theta \sin \theta \, \hat{I}_x\} \tag{7-29a}$$

$$= g\beta H M_S - g_N \beta_N [H_{\parallel} \hat{I}_z + H_{\perp} \hat{I}_x] \tag{7-29b}$$

where

$$H_{\parallel} = \frac{-h M_S}{g_N \beta_N} [A_0 + B(3 \cos^2 \theta - 1)] \tag{7-29c}$$

$$H_{\perp} = \frac{-h M_S}{g_N \beta_N} 3B \sin \theta \cos \theta \tag{7-29d}$$

B and θ are defined in Eqs. (7-28), and A_0 is the isotropic hyperfine coupling.

If $|\alpha_n\rangle$ and $|\beta_n\rangle$ are the wave functions for $\mathbf{I}$ quantized along z, then the hamiltonian matrix is§

$$\mathcal{H} = \begin{array}{c} \\ \langle \alpha_n| \\ \langle \beta_n| \end{array} \begin{array}{cc} |\alpha_n\rangle & |\beta_n\rangle \\ \left[\begin{array}{cc} g\beta H M_S - \dfrac{g_N \beta_N H_{\parallel}}{2} & \dfrac{-g_N \beta_N H_{\perp}}{2} \\[2ex] \dfrac{-g_N \beta_N H_{\perp}}{2} & g\beta H M_S + \dfrac{g_N \beta_N H_{\parallel}}{2} \end{array} \right] \end{array} \tag{7-30}$$

The energies for this system are then

$$W = g\beta H M_S \mp \frac{g_N \beta_N}{2} [H_{\parallel}^2 + H_{\perp}^2]^{\frac{1}{2}} \tag{7-31a}$$

---

† $|\mathbf{H}_{hf}|$ can be very large; e.g., if the proton hyperfine coupling is ~ 100 MHz (a typical value), $|\mathbf{H}_{hf}| = 11,700$ G. $|\mathbf{H}_{hf}|$ is the hyperfine field at the *nucleus* and must be distinguished from the hyperfine field at the electron. The latter would be only 18 G in this case, i.e., half the line separation in the spectrum.

‡ This problem was first considered by H. Zeldes, G. T. Trammell, R. Livingston, and R. W. Holmberg, *J. Chem. Phys.*, **32**:618 (1960). See also S. M. Blinder, *J. Chem. Phys.*, **33**:748 (1960).

§ One must diagonalize the hamiltonian matrix because $\mathbf{I}$ is *not* quantized along z!

$$W = g\beta H M_S \mp \frac{g_N \beta_N}{2} |\mathbf{H}_{hf}| \tag{7-31b}$$

$$= g\beta H M_S \pm \frac{hM_S}{2} \{[A_0 + B(3\cos^2\theta - 1)]^2 + 9B^2 \sin^2\theta \cos^2\theta\}^{\frac{1}{2}} \tag{7-31c}$$

$$= g\beta H M_S \pm \frac{hM_S}{2} [(A_0 - B)^2 + 3B(2A_0 + B)\cos^2\theta]^{\frac{1}{2}} \tag{7-31d}$$

Remarkably, the simple general form of Eq. (7-31d) appears not to have been widely used, although it was published in 1960.[†]

At constant microwave frequency transitions will occur at the resonant fields

$$H_r = H' \pm \frac{a}{2} \tag{7-32a}$$

where

$$a = \frac{h}{g\beta} [(A_0 - B)^2 + 3B(2A_0 + B)\cos^2\theta]^{\frac{1}{2}} \tag{7-32b}$$

It is of interest to consider two limiting cases:

1. $A_0 \approx 0$. a is then given by[‡]

$$a = \frac{hB}{g\beta}(1 + 3\cos^2\theta)^{\frac{1}{2}} \tag{7-33a}$$

2. $A_0 \gg B$. The square root in Eq. (7-31c) may then be expanded to give

$$a \cong \frac{h}{g\beta}[A_0 + B(3\cos^2\theta - 1)] \tag{7-33b}$$

For intermediate cases, the general relation, Eq. (7-32b), must be used. It would appear at first sight that a does not average to a_0 for a molecule tumbling in a liquid. However, one must realize that it is the hyperfine magnetic field at the nucleus which is averaged and not the energy. It is clear that $H_\parallel$ averages to a_0, whereas $H_\perp$ averages to zero, as required. The energy for the tumbling system *may not* be obtained by averaging Eq. (7-31d) over all orientations.

7-7 DETERMINATION OF THE ELEMENTS OF THE HYPERFINE TENSOR

We begin by considering case c of the previous section, since it occurs most frequently. As in Eqs. (7-26), the hyperfine interaction is considered in terms of the hyperfine field $\mathbf{H}_{hf}$ at the nucleus. From Eq. (7-31b) it is ap-

† S. M. Blinder, *J. Chem. Phys.*, **33**:748 (1960).
‡ H. Zeldes, G. T. Trammell, R. Livingston, and R. W. Holmberg, *J. Chem. Phys.*, **32**:618 (1960).

parent that the hyperfine energy is proportional to $|\mathbf{H}_{hf}|$. It now proves convenient to associate the total hyperfine energy ΔW given by

$$\Delta W = M_S^{-1} g_N \beta_N |\mathbf{H}_{hf}| \qquad (7\text{-}34a)$$

with the vector $h\mathbf{l} \cdot \mathbf{A}$ from Eq. (7-26c).

With reference to the allowed transitions k and m of Fig. 3-5a, which could occur for quanta ν_k and ν_m, ΔW is equal to $h(\nu_k - \nu_m)$. The procedure for evaluating the elements of the hyperfine tensor is then exactly analogous to the treatment of the g tensor in Sec. 7-4, since $\mathbf{l} \cdot \mathbf{A}$ is a vector and the energy W is a scalar. In the present case

$$\left(\frac{\Delta W}{h}\right)^2 = (\mathbf{l} \cdot \mathbf{A})^2 = (\mathbf{l} \cdot \mathbf{A}) \cdot (\mathbf{A} \cdot \mathbf{l}) = \mathbf{l} \cdot (\mathbf{A} \cdot \mathbf{A}^\dagger) \cdot \mathbf{l}$$

$$= \mathbf{l} \cdot \mathbf{A}^2 \cdot \mathbf{l} \qquad (7\text{-}34b)$$

$$\left(\frac{\Delta W}{h}\right)^2 = [l_x, l_y, l_z] \begin{bmatrix} (\mathbf{A}^2)_{xx} & (\mathbf{A}^2)_{xy} & (\mathbf{A}^2)_{xz} \\ (\mathbf{A}^2)_{yx} & (\mathbf{A}^2)_{yy} & (\mathbf{A}^2)_{yz} \\ (\mathbf{A}^2)_{zx} & (\mathbf{A}^2)_{zy} & (\mathbf{A}^2)_{zz} \end{bmatrix} \begin{bmatrix} l_x \\ l_y \\ l_z \end{bmatrix} \qquad (7\text{-}35)$$

The task at hand is thus the evaluation of the elements of the tensor $\mathbf{A}^2$. $\mathbf{A}^2$ will be symmetric, and thus there will be only six independent components.[†]

If a crystal is mounted for rotation first about the x axis with the magnetic field in the yz plane, the angle between the direction of the applied field and the z axis of a crystal will be taken as θ. For all directions of the field in the yz plane, $\mathbf{l} = [0, \sin\theta, \cos\theta]$. Then

$$\left(\frac{\Delta W}{h}\right)^2 = [0, \sin\theta, \cos\theta] \begin{bmatrix} (\mathbf{A}^2)_{xx} & (\mathbf{A}^2)_{xy} & (\mathbf{A}^2)_{xz} \\ (\mathbf{A}^2)_{yx} & (\mathbf{A}^2)_{yy} & (\mathbf{A}^2)_{yz} \\ (\mathbf{A}^2)_{zx} & (\mathbf{A}^2)_{zy} & (\mathbf{A}^2)_{zz} \end{bmatrix} \begin{bmatrix} 0 \\ \sin\theta \\ \cos\theta \end{bmatrix}$$

$$= [0, \sin\theta, \cos\theta] \begin{bmatrix} (\mathbf{A}^2)_{xy}\sin\theta + (\mathbf{A}^2)_{xz}\cos\theta \\ (\mathbf{A}^2)_{yy}\sin\theta + (\mathbf{A}^2)_{yz}\cos\theta \\ (\mathbf{A}^2)_{zy}\sin\theta + (\mathbf{A}^2)_{zz}\cos\theta \end{bmatrix}$$

$$\left(\frac{\Delta W}{h}\right)^2 = A_{\text{eff}}^2 = (\mathbf{A}^2)_{yy}\sin^2\theta + 2(\mathbf{A}^2)_{yz}\sin\theta\cos\theta + (\mathbf{A}^2)_{zz}\cos^2\theta \qquad (7\text{-}36a)$$

For a similar rotation about the y axis, the vector $\mathbf{l}$ will be $[\sin\theta, 0, \cos\theta]$. Hence in this case

$$A_{\text{eff}}^2 = (\mathbf{A}^2)_{xx}\sin^2\theta + 2(\mathbf{A}^2)_{xz}\sin\theta\cos\theta + (\mathbf{A}^2)_{zz}\cos^2\theta \qquad (7\text{-}36b)$$

For rotation about the z axis, $\mathbf{l} = [\cos\theta, \sin\theta, 0]$ and

[†] As for g^2, $\mathbf{A}^2$ is a true tensor; however, $\mathbf{A}$ is not a true tensor for the same reasons that g is not. $\mathbf{A}$ need not be symmetric; indeed there are some known examples for which $\mathbf{A}$ is not. See J. M. Baker, E. R. Davis, and J. P. Hurrell, *Proc. Roy. Soc. (London)*, A308:403 (1960); F. K. Kneubühl, *Phys. kondens. Materie*, 1:410 (1963); H. M. McConnell, *Proc. Natl. Acad. Sci. (U.S.)*, 44:766 (1958).

$$A_{\text{eff}}^2 = (\mathbf{A}^2)_{xx} \cos^2 \theta + 2(\mathbf{A}^2)_{xy} \sin \theta \cos \theta + (\mathbf{A}^2)_{yy} \sin^2 \theta \qquad (7\text{-}36c)$$

If the axes x and z correspond to principal axes of a system of axial symmetry, then Eqs. (7-36a) or (7-36b) become

$$A_{\text{eff}}^2 = A_{\parallel}^2 \cos^2 \theta + A_{\perp}^2 \sin^2 \theta \qquad (7\text{-}37)$$

Note that A_{eff} is obtained from the experimental hyperfine splitting a by

$$A_{\text{eff}} = \frac{g\beta a}{h} \qquad (7\text{-}38)$$

These expressions will now be applied to the hyperfine coupling data for the $^-$OOC—ĊF—CF$_2$—COO$^-$ radical obtained by irradiation of sodium perfluorosuccinate.† The crystal structure is monoclinic, with $a = 1.14$ nm, $b = 1.10$ nm, $c = 1.03$ nm, and $\beta = 106°$. Here β is the angle between the c and the a axes. An a^*bc axis system is chosen with a^* perpendicular to

† M. T. Rogers and D. H. Whiffen, *J. Chem. Phys.*, **40**:2662 (1964).

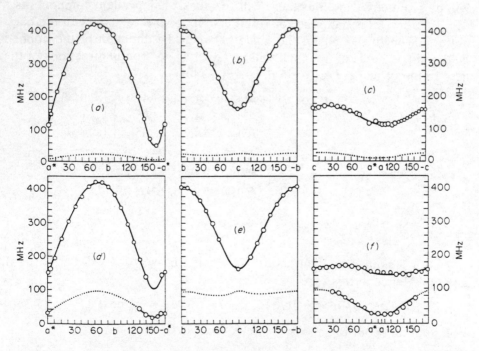

Fig. 7-6 Angular dependence of the hyperfine coupling (megahertz) in the $^-$OOC-CF$_2$ĊFCOO$^-$ radical. The uncertainty of data represented by large circles is greater than that for the small circles. Curves are drawn for the upper signs of the direction-cosine matrix of Table 7-2. Dotted lines correspond to spectral lines with relative intensity less than 20 percent of the total absorption intensity. (a), (b), (c) Microwave frequency 9.000 GHz. **H** is in the a^*b, bc, and a^*c planes in (a), (b), and (c), respectively. (d), (e), (f) Spectra analogous to (a), (b), and (c), but for a frequency of 35.000 GHz. [*Taken from M. T. Rogers and D. H. Whiffen, J. Chem. Phys.*, **40**:2662 (1964).]

the bc plane. In Fig. 7-6 the hyperfine coupling from the α-fluorine atom is plotted as the magnetic field explores the $a*b$, bc, and $a*c$ planes, respectively.

The elements of the $\boldsymbol{A}^2$ tensor may be interpolated from these plots, using values at special angles. (See Table 7-1.) For better precision, a least-squares fit should be made (using plots of A^2_{eff} versus θ) to the experimental data.

The $\boldsymbol{A}^2$ tensor is then

$$\begin{bmatrix} 1.60 & \pm4.78 & 0.64 \\ \pm4.78 & 16.36 & \mp0.16 \\ 0.64 & \mp0.16 & 2.71 \end{bmatrix} \times 10^4 (MHz)^2 \tag{7-39}$$

The elements of $\boldsymbol{A}^2$ have been obtained from Table 7-1 by averaging repeated measurements. The ambiguity in sign of the elements of this tensor arises from the fact that there are two possible radical sites per unit cell.

This matrix may now be diagonalized by subtracting λ from each diagonal element and setting the resulting determinant equal to zero. (See Sec. A-5e.) Expansion of the determinant yields the following cubic equation

$$\lambda^3 - 20.67\lambda^2 + 51.56\lambda - 1.30 = 0 \tag{7-40}$$

The roots of this equation are 17.77, 2.89, and 0.011; hence

$$^d\boldsymbol{A}^2 = \begin{bmatrix} 17.77 & 0 & 0 \\ 0 & 2.89 & 0 \\ 0 & 0 & 0.011 \end{bmatrix} \times 10^4 \quad (MHz)^2 \tag{7-41}$$

Table 7-1   Hyperfine data from Fig. 7-6

Plane	θ. deg	A^2, (MHz)^2	Tensor elements
$a*b$	0	1.61×10^4	$= (\boldsymbol{A}^2)_{a'a'}$
	90	16.24×10^4	$= (\boldsymbol{A}^2)_{bb}$
	45	13.84×10^4	Difference $= 2(\boldsymbol{A}^2)_{a'b}$
	135	4.29×10^4	
bc	0	16.48×10^4	$= (\boldsymbol{A}^2)_{bb}$
	90	2.72×10^4	$= (\boldsymbol{A}^2)_{cc}$
	45	9.67×10^4	Difference $= 2(\boldsymbol{A}^2)_{bc}$
	135	9.99×10^4	
$ca*$	0	2.69×10^4	$= (\boldsymbol{A}^2)_{cc}$
	90	1.59×10^4	$= (\boldsymbol{A}^2)_{a'a'}$
	45	2.69×10^4	Difference $= 2(\boldsymbol{A}^2)_{a'c}$
	135	1.42×10^4	

Table 7-2†

Principal values, MHz	Direction cosines relative to a*bc axes		
422	0.282	±0.958	0.001
170	0.223	∓0.069	0.972
11	0.933	∓0.278	−0.235

† The upper and lower signs refer consistently to the two sets of radical sites related to each other by the twofold b axis, plus translation and possibly inversion.

or

$$^d\mathbf{A} = \begin{bmatrix} 422 & 0 & 0 \\ 0 & 170 & 0 \\ 0 & 0 & 11 \end{bmatrix} \quad \text{MHz} \tag{7-42}$$

The elements of the direction-cosine matrix are obtained by substituting the three values of λ into the equations

$$(1.60 - \lambda_i)l_{i1} \pm 4.78l_{i2} + 0.64l_{i3} = 0 \tag{7-43a}$$

$$\pm 4.78l_{i1} + (16.36 - \lambda_i)l_{i2} \mp 0.16l_{i3} = 0 \tag{7-43b}$$

$$0.64l_{i1} \mp 0.16l_{i2} + (2.71 - \lambda_i)l_{i3} = 0 \tag{7-43c}$$

The set of coefficients (l_{i1}, l_{i2}, l_{i3}) together are called the eigenvector corresponding to the eigenvalue λ_i. The results are collected in Table 7-2, where each row of the direction-cosine matrix is an eigenvector. The corresponding eigenvalues are at the left of the table. The smallest principal value is not precisely determined in this analysis, and there remain ambiguities of signs. Other orientations are required to obtain a more accurate value.

In the present case, small corrections are required to account for the nuclear Zeeman term. (See Sec. 7-8.) The corrected values are listed in Table 7-3.

The above example exhibits the phenomenon of site splitting; two independent hyperfine patterns are observed because there are two possible sites for a paramagnetic center in a unit cell. In analyzing the ESR spectra,

Table 7-3

Principal values, MHz	Direction cosines relative to a*bc axes		
421	+0.267	±0.964	+0.011
165	+0.208	∓0.068	+0.976
11	+0.941	∓0.258	−0.219

one must be careful to separate the contributions from each site. It is only for crystals which belong to the triclinic system that site splitting cannot occur.

As a further example of site splitting, consider the following tensor for the C—H proton of the H—C(OH)CO$_2^-$ radical in H$_2$C(OH)CO$_2$Li. The contributions from the two distinguishable sites are indicated by the $+$ and $-$ signs.[†]

$$A = \begin{bmatrix} -51 & \mp 13 & +14 \\ \mp 13 & -77 & \pm 6 \\ +14 & \pm 6 & -35 \end{bmatrix} \rightarrow \underset{^dA}{\begin{bmatrix} -85 & 0 & 0 \\ 0 & -51 & 0 \\ 0 & 0 & -27 \end{bmatrix}}$$

$$= \underset{A_0 \cdot 1}{-54 \cdot 1} + \underset{^dB}{\begin{bmatrix} -31 & 0 & 0 \\ 0 & +3 & 0 \\ 0 & 0 & +27 \end{bmatrix}} \quad (7\text{-}44)$$

Here the components of A (megahertz) have been measured in terms of axes x and y parallel to the face of a platelet, and z is perpendicular to x and y. By methods just outlined, A may be diagonalized. In the process, one obtains the matrix $\mathscr{L}$ of the direction cosines

$$\mathscr{L} = \begin{bmatrix} 0.43 & \pm 0.88 & -0.23 \\ 0.74 & \mp 0.48 & -0.46 \\ 0.51 & \mp 0.03 & 0.86 \end{bmatrix} \quad (7\text{-}45)$$

Note that the qualitative appearance (see Fig. 7-6) of the plots of hyperfine splittings vs. orientation indicates the relative importance of off-diagonal elements. If an off-diagonal element, e.g., A_{xy}, is relatively small, then the plot of A in the xy plane will be symmetric about $\theta = 90°$. If, however, A_{xy} is relatively large, there will be a marked displacement of the curve. Figure 7-6a is a good example of the latter case, whereas Fig. 7-6b represents the former case.

When there is site splitting, then there may be two overlapping spectra which lead to equal and opposite off-diagonal elements if the cartesian crystal-axis system is appropriately chosen with respect to the crystal symmetry. Equation (7-44) shows an example. Each site will have only one appropriate sign for each off-diagonal element, but care is required to extract the correct pairing. Thus in Eq. (7-44) the value -13 is to be associated with the $+6$ and $+13$ with the -6 and not vice versa. This cannot be determined experimentally from the original three rotations, but can be settled from other crystal positions, especially that with the field in the directions $(3^{-\frac{1}{2}}, 3^{-\frac{1}{2}}, \pm 3^{-\frac{1}{2}})$.

An even more troublesome difficulty is to obtain the correct signs in extracting the square root of A^2, e.g., in passing from Eqs. (7-41) to (7-42).

[†] D. Pooley and D. H. Whiffen, *Trans. Faraday Soc.*, **57**:1445 (1961).

This sign determination is possible only if the nuclear Zeeman term, the final term in (7-27), is significant. This matter is discussed in Sec. 7-8 and accounts for the difference between the 9-GHz separations in Figs. 7-6a, b, and c, for which the nuclear Zeeman term is negligible, and the 35-GHz separations of Figs. 7-6d, e, and f, for which the full theory must be used.

Many ESR spectra of paramagnetic centers in single crystals will exhibit both g and hyperfine anisotropy. In many cases, it is sufficient to measure g from the center of the spectrum and apply the analysis of Sec. 7-4 to obtain the elements of $^d\mathbf{g}$ and the direction cosine matrix. Then Eqs. (7-36) may be used to analyze the hyperfine anisotropy, provided that g in Eq. (7-38) is given by[†]

$$g_{\text{eff}} = [\mathbf{l} \cdot \mathbf{g}^2 \cdot \mathbf{l}]^{\frac{1}{2}} \tag{7-46}$$

7-8   CORRECTIONS TO HYPERFINE–TENSOR ELEMENTS

We shall now proceed to examine case b of Sec. 7-6, i.e., the case $|\mathbf{H}| \approx |\mathbf{H}_{\text{hf}}|$.[‡] Referring to Fig. 7-5b, it is apparent that one must consider the total resultant field $\mathbf{H}_{\text{eff}}$ at the nucleus. I will be quantized along $\mathbf{H}_{\text{eff}}(+)$ for $M_S = +\frac{1}{2}$ and $\mathbf{H}_{\text{eff}}(-)$ for $M_S = -\frac{1}{2}$. The energies [computed by use of Eq. (7-26a) and the treatment for case c] are

$$W_{\alpha_e \alpha'_n} = \tfrac{1}{2} g\beta H - \tfrac{1}{2} g_N \beta_N |\mathbf{H}_{\text{eff}}(+)| \tag{7-47a}$$

$$W_{\alpha_e \beta'_n} = \tfrac{1}{2} g\beta H + \tfrac{1}{2} g_N \beta_N |\mathbf{H}_{\text{eff}}(+)| \tag{7-47b}$$

$$W_{\beta_e \beta''_n} = -\tfrac{1}{2} g\beta H + \tfrac{1}{2} g_N \beta_N |\mathbf{H}_{\text{eff}}(-)| \tag{7-47c}$$

and

$$W_{\beta_e \alpha''_n} = -\tfrac{1}{2} g\beta H - \tfrac{1}{2} g_N \beta_N |\mathbf{H}_{\text{eff}}(-)| \tag{7-47d}$$

The nuclear spin eigenfunctions are not the same for $M_S = +\frac{1}{2}$ and $-\frac{1}{2}$, since the axis of quantization is different in the two cases. By expressing $|\alpha'_n\rangle$ and $|\beta'_n\rangle$ as linear combinations of $|\alpha''_n\rangle$ and $|\beta''_n\rangle$ and by solving the resulting secular determinant, the relation between the nuclear spin states can be shown to be (see Sec. A-5e)

$$|\alpha'_n\rangle = \cos\frac{\theta}{2}\,|\alpha''_n\rangle - \sin\frac{\theta}{2}\,|\beta''_n\rangle \tag{7-48a}$$

$$|\beta'_n\rangle = \sin\frac{\theta}{2}\,|\alpha''_n\rangle + \cos\frac{\theta}{2}\,|\beta''_n\rangle \tag{7-48b}$$

where θ is the angle between $\mathbf{H}_{\text{eff}}(+)$ and $\mathbf{H}_{\text{eff}}(-)$.

The energy levels are given in Fig. 7-7. The four possible transition energies are

[†] See H. Zeldes and R. Livingston, *J. Chem Phys.*, **35**:563 (1961), for an example of this case.
[‡] G. T. Trammell, H. Zeldes, and R. Livingston, *Phys. Rev.*, **110**:630 (1958); J. A. Weil and J. H. Anderson, *J. Chem. Phys.*, **35**:1410 (1961).

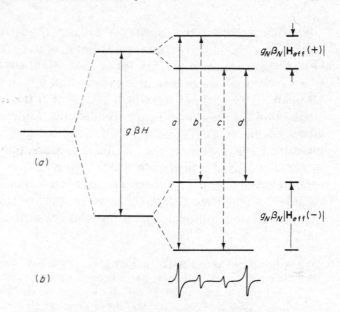

Fig. 7-7 (a) Energy levels at constant field for a system with $S = \frac{1}{2}$ and $I = \frac{1}{2}$ when $|\mathbf{H}|$ is comparable with $|\mathbf{H}_{hf}|$. (See Fig. 7-5b.) a and d are the normally allowed transitions: b and c are usually of much lower intensity. (b) Observed transitions at constant frequency.

$$\Delta W_a = W_{\alpha_c \beta'_n} - W_{\beta_c \alpha''_n} = g\beta H + \tfrac{1}{2}g_N\beta_N\{|\mathbf{H}_{\text{eff}}(+)| + |\mathbf{H}_{\text{eff}}(-)|\} \quad (7\text{-}49a)$$

$$\Delta W_b = W_{\alpha_c \beta'_n} - W_{\beta_c \beta''_n} = g\beta H + \tfrac{1}{2}g_N\beta_N\{|\mathbf{H}_{\text{eff}}(+)| - |\mathbf{H}_{\text{eff}}(-)|\} \quad (7\text{-}49b)$$

$$\Delta W_c = W_{\alpha_c \alpha'_n} - W_{\beta_c \alpha''_n} = g\beta H - \tfrac{1}{2}g_N\beta_N\{|\mathbf{H}_{\text{eff}}(+)| - |\mathbf{H}_{\text{eff}}(-)|\} \quad (7\text{-}49c)$$

$$\Delta W_d = W_{\alpha_c \alpha'_n} - W_{\beta_c \beta''_n} = g\beta H - \tfrac{1}{2}g_N\beta_N\{|\mathbf{H}_{\text{eff}}(+)| + |\mathbf{H}_{\text{eff}}(-)|\} \quad (7\text{-}49d)$$

Since the intensities of the lines are proportional to (see Sec. C-4)

$$|\langle M'_S, M'_I | \hat{S}_x | M_S, M_I \rangle|^2$$

the relative intensities of the lines will be given by

$$\mathscr{I}_a = \mathscr{I}_d = \sin^2 \frac{\theta}{2}$$

$$\mathscr{I}_b = \mathscr{I}_c = \cos^2 \frac{\theta}{2}$$

Since the above treatment is general for cases a, b, and c of Sec. 7-6, it is instructive to examine the results for the two limiting cases a and c.

Case a. Since $|\mathbf{H}| >> |\mathbf{H}_{hf}|$, $\theta \approx 0°$ and $|\alpha'_n\rangle \approx |\alpha''_n\rangle$. Then transitions b and c of Fig. 7-7b will be strong, whereas a and d will be weak. Nevertheless, the separation of the lines b and c is still the hyperfine splitting, since $\Delta W_b - \Delta W_c \approx 2g_N\beta_N|\mathbf{H}_{hf}|$ [cf. Eq. (7-34)].

Case c. Here $|\mathbf{H}| << |\mathbf{H}_{hf}|$, $\theta \approx 180°$, and $|\alpha'_n\rangle \approx |\beta''_n\rangle$. Then transitions a and d will be the strong ones. Again the separation of the lines is the hyperfine splitting, that is, $\Delta W_a - \Delta W_d \approx 2g_N\beta_N|\mathbf{H}_{hf}|$.

Case b. Here all four transitions will be of comparable intensity. Failure to recognize this has led to misassignments of hyperfine splittings.†

The analysis of a complex spectrum, which may contain "forbidden" transitions (e.g., lines for which $\Delta M_I = \pm 1$, $\Delta M_S = \pm 1$), is often aided by using two different microwave frequencies. Figure 7-8 illustrates the spectrum of the $^-OOC\text{-}\dot{C}F\text{-}CF_2\text{-}COO^-$ radical at 9 and 35 GHz. The latter spectrum clearly shows the "forbidden" transitions.

The use of a high microwave frequency has the added advantage that the relative signs of hyperfine-tensor elements may be determined. In Table 7-4 (see page 154) the observed hyperfine splittings are compared with the calculated splittings for various choices of the relative signs of the three

---

† It is instructive to note that $\Delta W_a > \Delta W_d$; hence in case *c*, the strong lines can never coincide, and the relative signs of the principal components of the hyperfine tensor cannot be determined. If $\mathbf{H}_{hf}$ is perpendicular to **H**, then $\Delta W_b = \Delta W_c$, and the two lines coalesce in the center of the spectrum. This case can occur only if two principal elements of the hyperfine tensor are of opposite sign, and it can best be seen experimentally if case *b* applies.

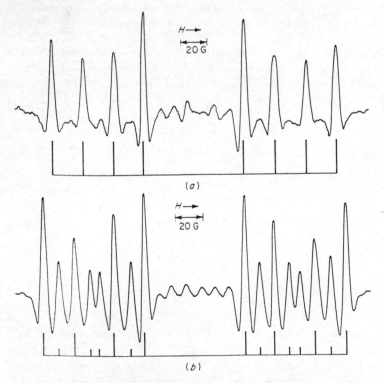

Fig. 7-8 (*a*) Second-derivative spectrum of the perfluorosuccinate ion for **H** ∥ \b\ at 9.000 GHz. (*b*) Similar spectrum at 35.000 GHz showing the greatly increased intensity of the forbidden transitions. [*Spectra taken from*: M. T. Rogers and D. H. Whiffen, *J. Chem. Phys.*, **40**:2662 (1964)].

principal values given in Table 7-3. The measurements were made at 35 GHz. The splittings are computed in the following manner. The hyperfine splitting in the [100] direction will be used as an example.

For $M_S = +\frac{1}{2}$

$$-\frac{1}{2}\mathbf{I} \cdot \mathbf{A} = \begin{array}{c} -\frac{1}{2}[1,0,0] \\ \\ \end{array} \begin{bmatrix} 46.9 & \pm 103.4 & 31.0 \\ \pm 103.4 & 392.7 & \mp 16.0 \\ 31.0 & \mp 16.0 & 157.7 \end{bmatrix} \quad \text{MHz}$$

$$= [-23.5, \mp 51.7, -15.5] \quad \text{MHz} \tag{7-50}$$

$\mathbf{A}$ is obtained from $^d\mathbf{A}$ by the transformation $\mathbf{A} = \mathscr{L}^{\cdot} {}^d\mathbf{A}\mathscr{L}$. $\mathbf{H}$ is oriented along the [100] direction; hence

$$\frac{g_N\beta_N}{h} \mathbf{H}_{\text{eff}}(+) = [(49.5 - 23.5), \mp 51.7, -15.5] \quad \text{MHz} \tag{7-51}$$

since

$$\frac{g_N\beta_N}{h} |\mathbf{H}| = 49.5 \text{ MHz}$$

The magnitude of the vector in (7-51) is then

$$\frac{g_N\beta_N}{h} |\mathbf{H}_{\text{eff}}(+)| = 60 \text{ MHz} \tag{7-52a}$$

Similarly

$$\frac{g_N\beta_N}{h} |\mathbf{H}_{\text{eff}}(-)| = 91 \text{ MHz} \tag{7-52b}$$

The two hyperfine splittings are

$$\frac{\Delta W_a - \Delta W_d}{h} = 151 \text{ MHz} \tag{7-53a}$$

$$\frac{\Delta W_b - \Delta W_c}{h} = 31 \text{ MHz} \tag{7-53b}$$

which agree, as they should, with the observed points for $\mathbf{H} \parallel a^*$ in Figs. 7-6d and f. Since $\theta = 101°$, the relative intensities are 0.59 and 0.41, respectively, for the a, d and b, c transitions.

The other entries in Table 7-4 are calculated in a similar manner. It is clear that the only sign choice which gives good agreement with experiment is that for which all principal values have the same sign. A positive sign is chosen, since the maximum hyperfine coupling is expected to be positive for an unpaired electron in a $2p_z$ orbital on the α-fluorine atom. (See Sec. 8-2b.)

In some systems one may observe additional weak lines not accounted for by considering "forbidden" transitions such as those of Fig. 7-7. An example is the hydrogen atom trapped in irradiated frozen acids such as H_2SO_4; in the ESR spectrum, weak pairs of lines are separated from the

Table 7-4 Observed and calculated splittings (MHz) for α fluorine at 35.000 GHz[†]

Direction cosine of field	Observed	Sign choice			
		a	b	c	d
(1,0,0)	153	151 (0.59)	154 (0.58)	153 (0.58)	154 (0.58)
	29	31 (0.41)	18 (0.42)	21 (0.42)	85 (0.42)
(0,1,0)	407	407 (1.00)	407 (1.00)	407 (1.00)	407 (0.99)
	—	96 (0.00)	96 (0.00)	96 (0.00)	96 (0.01)
(0,0,1)	162	163 (0.96)	164 (0.95)	164 (0.95)	163 (0.96)
	—	97 (0.04)	96 (0.05)	96 (0.05)	97 (0.04)
(cos 30, 0, cos 60)	170	169 (0.75)	171 (0.73)	177 (0.70)	176 (0.70)
	65	61 (0.25)	54 (0.27)	25 (0.30)	32 (0.30)
(cos 50, 0, −cos 40)	148	149 (0.63)	152 (0.62)	156 (0.61)	154 (0.61)
	48	54 (0.37)	44 (0.38)	28 (0.39)	37 (0.39)
(−cos 20, cos 70, 0)	—	110 (0.18)	112 (0.20)	112 (0.20)	111 (0.19)
	17	19 (0.82)	1 (0.80)	46 (0.80)	14 (0.81)
(0, cos 60, cos 30)	252	252 (0.95)	253 (0.94)	266 (0.86)	266 (0.86)
	—	84 (0.05)	83 (0.06)	22 (0.14)	30 (0.14)

	Sign choices	a	+421	+165	+11
		b	+421	+165	−11
		c	+421	−165	+11
		d	+421	−165	−11

(Relative intensities are given in parentheses)

[†] Taken from M. T. Rogers and D. H. Whiffen, *J. Chem. Phys.*, **40**:2662 (1964).

allowed lines by the nuclear resonance frequency of the proton at the field H_r used for the ESR experiment.[†,‡] The weak lines arise from matrix protons which undergo a "spin flip" when the electron spins of nearby trapped hydrogen atoms are reoriented. The coupling is dipolar and the intensity of the weak lines varies approximately as H_r^{-2}.

7-9 LINE SHAPES IN NONORIENTED SYSTEMS

In this section one reaches a middle ground between the effectively isotropic systems of the first six chapters and the highly oriented solids dealt with in the earlier part of this chapter. In powders (and in some other solids as well) one has short-range order, but the principal axes of the paramagnetic system may assume *all* possible angles relative to the direction of the magnetic field. Even in the absence of hyperfine splitting, one expects to have the ESR spectrum spread over the entire field range ΔH determined by the principal g components of the system. Fortunately, however, the lines

[†] H. Zeldes and R. Livingston, *Phys. Rev.*, **96**:1702 (1954).

[‡] G. T. Trammell, H. Zeldes, and R. Livingston, *Phys. Rev.*, **110**:630 (1958).

are not uniformly distributed over ΔH. Otherwise, if ΔH were of the order of hundreds of gauss it would be difficult to detect any absorption.

7-9a. Line shapes for systems with axial symmetry   The first model con--sidered will be that of a system with $S = \frac{1}{2}$ and $I = 0$ and possessing tetragonal symmetry such that for a single crystal one would get ESR lines at positions similar to those given in Fig. 7-2. Upon grinding such a crystal to a fine powder, one expects that all orientations of the tetragonal axis will be equally probable. Hence there are some crystallites in resonance at all fields H_r between $H_\perp$ (the field corresponding to $g_\perp$) and $H_\parallel$ (the field corresponding to $g_\parallel$). H_r is given by

$$H_r = \frac{h\nu}{g_{eff}\beta} = \frac{h\nu}{\beta} [g_\parallel^2 \cos^2 \theta + g_\perp^2 \sin^2 \theta]^{-\frac{1}{2}} \tag{7-54}$$

where θ is the angle between a given symmetry axis and the magnetic field direction.

Since all orientations are equally probable, it is desirable to have a measure of orientation which reflects this. It is convenient to use the concept of a solid angle subtended by a bounded area $\mathscr{A}$ on the surface of a sphere of radius r. The given solid angle Ω is defined to be the ratio of a surface area $\mathscr{A}$ to the total surface area of the sphere.

$$\Omega = \frac{\mathscr{A}}{4\pi r^2} \tag{7-55}$$

For a small powder sample at the center of a sphere, one may translate the statement that all orientations of the tetragonal axis are equally probable into the statement that the number of axes contained in unit solid angle is equal for all regions of the sphere. If the sphere is fixed in a magnetic field, the orientation of axes will be measured by their angle θ relative to the field H_r.

Consider a circular element of area for which the z axis is the field direction (Fig. 7-9). The area of the element is $2\pi(r \sin \theta)r\, d\theta$. Hence the solid angle $d\Omega$ it subtends is given by

$$d\Omega = \frac{2\pi r^2 \sin \theta\, d\theta}{4\pi r^2} = \frac{1}{2} \sin \theta\, d\theta \tag{7-56}$$

Then the solid angle subtended by crystallites with symmetry axes lying between θ and $\theta + d\theta$ measures the probability $P(H)\, dH$ of a system being at a resonant field between H_r and $H_r + dH_r$. Then

$$P(H)\, dH \propto \sin \theta\, d\theta \tag{7-57}$$

or

$$P(H) \propto \frac{\sin \theta}{dH/d\theta} \tag{7-58}$$

It is worthwhile to understand the significance of the numerator and the

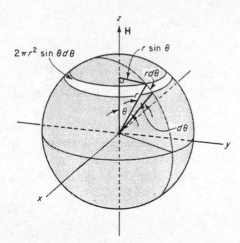

Fig. 7-9 Element of area on the surface of a sphere. (*After G. M. Barrow, "Physical Chemistry," 2d ed., p. 803, McGraw-Hill Book Company, New York, 1966.*)

denominator in Eq. (7-58). The proportionality of $P(H)$ (and therefore of line intensity) to $\sin \theta$ reflects the very large number of systems with axes nearly perpendicular to the field direction, i.e., systems with axes approximately in the equatorial plane about the field direction. By contrast, there will be very few systems with an axis aligned close to the field direction. The value of $P(H)$ is large if $dH/d\theta$ is small. This implies that one has the greatest hope of seeing a line at field values near turning points; $H_\perp$ and $H_\parallel$ represent field extrema and therefore are turning points. Upon taking the derivative $dH/d\theta$ in Eq. (7-54), (7-58) becomes

$$P(H) \propto \frac{\beta}{h\nu} \frac{(g_\parallel^2 \cos^2 \theta + g_\perp^2 \sin^2 \theta)^{\frac{3}{2}}}{(g_\parallel^2 - g_\perp^2) \cos \theta} \tag{7-59}$$

Equation (7-54) may be used to simplify this to

$$P(H) \propto \left(\frac{h\nu}{\beta}\right)^2 \frac{1}{H_r^3 (g_\parallel^2 - g_\perp^2) \cos \theta} \tag{7-60}$$

For $\theta = 0$, $P(H)$ is finite. Since $h\nu/\beta = g H_\parallel$,

$$P(H) \propto \frac{1}{H_\parallel} = \text{constant} \tag{7-61}$$

Owing to the $\cos \theta$ term in the denominator of Eq. (7-60), $|P(H)|$ rises monotonically to infinity as $\theta \to \pi/2$. This behavior is shown in Fig. 7-10a. If various amounts of broadening are added, as shown by the several lines, the absorption line will have the form shown in Fig. 7-10b. Figure 7-10c shows the first-derivative spectrum obtained from a nonoriented system with tetragonal symmetry.

In the case of an orthorhombic system in powder form, there are three turning points. The shape of the absorption line and of its derivative are given in Fig. 7-11.†

† F. K. Kneubühl, *J. Chem. Phys.*, **33**:1074 (1960); J. A. Weil and H. G. Hecht, *J. Chem. Phys.*, **38**:281 (1963).

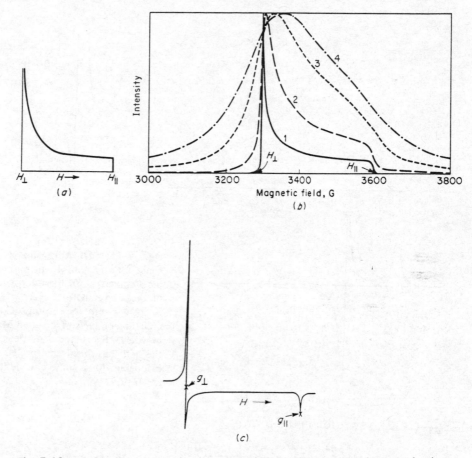

Fig. 7-10 (*a*) Idealized absorption line shape for a randomly oriented system having an axis of symmetry and no hyperfine interaction ($g_{\perp} > g_{\parallel}$). (*b*) Computed line shapes for randomly oriented systems having axial symmetry. The component lorentzian lines are given widths of 1, 10, 50, and 100 G, respectively. [*Taken from J. A. Ibers and J. D. Swalen, Phys. Rev.,* **127**:1914 (1962).] (*c*) ESR powder spectrum for a system of tetragonal symmetry, with $g_{\perp} > g_{\parallel}$. (The V_1 center in MgO; see Sec. 8-5*d*.)

For systems of orthorhombic symmetry, we shall call the Z axis that which has the g component most widely separated from the other two; g_{YY} will be the intermediate g component.

7-9b. Hyperfine line shapes for an isotropic g factor $S = \frac{1}{2}$ and one nucleus with $I = \frac{1}{2}$ The calculation of the expected line shape for hyperfine splitting in a powder is considered for the case of an isotropic g factor, $S = \frac{1}{2}$ and $I = \frac{1}{2}$. Here the dipolar interaction of the unpaired electron with a proton is to be considered for all possible orientations of the electron-nucleus vector **r** of Fig. 3-3. The angle θ between this vector and the applied field can vary from 0 to π. The positions of the allowed hyperfine lines are given by

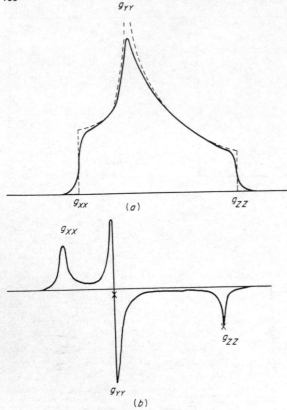

(a)

(b)

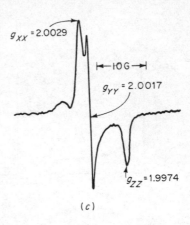

(c)

Fig. 7-11 (a) Absorption line shape for a system with orthorhombic symmetry. (b) First derivative of the curve in (a). Here $g_{XX} > g_{YY} > g_{ZZ}$. (c) Spectrum of the CO_2^- ion on the surface of an MgO powder. The extraneous peak at the left has been interpreted as belonging to a different center. [J. H. Lunsford and J. P. Jayne, J. Phys. Chem., **69**:2182 (1965).]

Eqs. (7-32), rewritten in the form

$$\cos \theta = \pm \left[\frac{4g^2\beta^2(\Delta H)^2 - h^2(A_0 - B)^2}{h^2 3B(2A_0 + B)} \right]^{\frac{1}{2}} \qquad (7\text{-}62)$$

where $\Delta H = H_r - H'$. Differentiating, one obtains

$$\frac{\sin \theta}{d(\Delta H)/d\theta} = \pm \frac{4g^2\beta^2 \, \Delta H}{h^2 3B(2A_0 + B) \, \cos \theta} \qquad (7\text{-}63)$$

$$= \pm \frac{2g\beta}{3hB(2A_0 + B) \, \cos \theta}$$
$$\times [(A_0 - B)^2 + 3B(2A_0 + B) \cos^2 \theta]^{\frac{1}{2}} \propto P(H) \quad (7\text{-}64)$$

From Eq. (7-58), it is apparent that the product in Eq. (7-64) will be proportional to the absorption intensity $P(H)$. The sign is taken such that $P(H)$ is always positive. There will be two separate envelopes for $P(H)$, since $M_I = \pm\frac{1}{2}$.

It is instructive† to rearrange Eq. (7-64) in terms of $\eta = B/A_0$.

$$P(H) \propto \frac{[(1 - \eta)^2 + 3\eta(2 + \eta) \cos^2 \theta]^{\frac{1}{2}}}{\eta(2 + \eta) \cos \theta} \qquad (7\text{-}65)$$

† A relation similar to Eq. (7-65) was developed by S. M. Blinder, J. Chem. Phys., **33**:748 (1960).

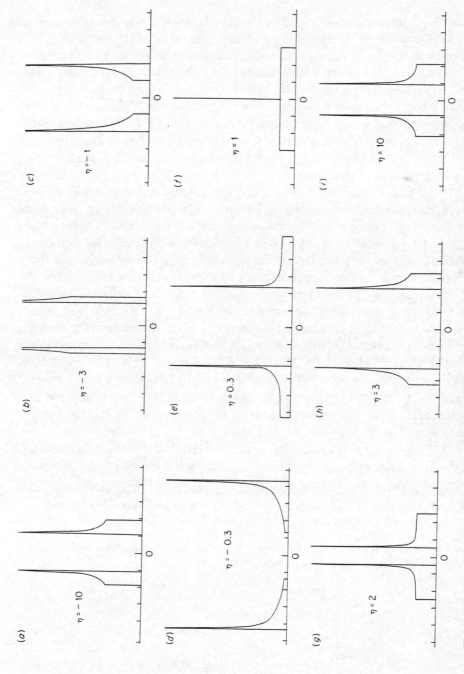

Fig. 7-12 Hyperfine line shapes for a randomly oriented paramagnetic system with $S = \frac{1}{2}$, $I = \frac{1}{2}$ and having an isotropic g factor, for nine selected values of $\eta = B/A_0$. These are plots of $P(H)$ vs. ΔH in arbitrary units, using Eq. (7-65).

159

Figure 7-12 illustrates plots of $P(H)$ versus ΔH for a number of values of η. The shape of the $P(H)$ envelope is very sensitive to the value of η. In particular at $\eta = -2$, the hyperfine splitting would be found to be *independent of orientation!* Such a case could easily be mistaken for a purely isotropic hyperfine interaction. The only way to tell would be to examine (if possible) the system in a liquid of low viscosity where the true isotropic hyperfine splitting would be obtained. Also, at $\eta = +1$, $P(H)$ is independent of ΔH except at $\Delta H = 0$, where a singularity exists. Note that this case (that is, $\eta = +1$) is the only one for which the experimental hyperfine splitting can approach zero.

In all cases (except $\eta = +1$), $P(H)$ has a finite value at $\theta = 0°$ and increases monotonically to infinity at $\theta = 90°$. The curves in Fig. 7-12 have been drawn assuming a delta-function line shape. For a line of finite width, a broadening similar to that given in Fig. 7-10b is found. The derivative line shape will be very similar to that shown in Fig. 7-10c except that there is a repetition for each hyperfine component. Figure 7-13 illustrates the ESR spectrum of the FCO radical which is randomly oriented in a CO matrix at 4.2 K. Although the symmetry is not quite axial, it will be assumed to be axial for purposes of illustration. The separation of the outermost lines is given by $(h/g\beta)|A_0 + 2B| \cong 514$ G; however the separation of the inner lines is given by $(h/g\beta)|A_0 - B| \cong 246$ G. From this one may deduce either that $A_0 \cong \pm 940$ MHz, $B \cong \pm 250$ MHz, *or* that $A_0 \cong \pm 20$ MHz, $B \cong \pm 710$ MHz. The former assignment is the correct one but one requires additional information (such as results for similar radicals) to resolve the ambiguity.[†]

If the symmetry is lower than axial, if more than one nucleus interacts, or if the g anisotropy is significant, it is necessary to utilize other relations.[‡],[§]

[†] F. J. Adrian, E. L. Cochran, and V. A. Bowers, *J. Chem. Phys.*, **43**:462 (1965).

[‡] The case of orthorhombic symmetry and an isotropic g factor is considered by S. M. Blinder, *J. Chem. Phys.*, **33**:748 (1960); E. L. Cochran, F. J. Adrian, and V. A. Bowers, *J. Chem. Phys.*, **34**:1161 (1961).

[§] The case of axial symmetry and comparable hyperfine and g anisotropy is considered by R. Neiman and D. Kivelson, *J. Chem. Phys.*, **35**:156 (1961).

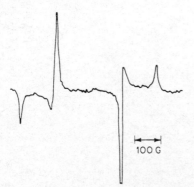

100 G

Fig. 7-13 ESR spectrum of FCO in a CO matrix at 4.2 K. The microwave frequency is 9,123.97 MHz. For this radical, g is essentially isotropic. The hyperfine interaction is not quite axial, as is seen by the partial splitting of the second line from the left. [*Taken from F. J. Adrian, E. L. Cochran, and V. A. Bowers, J. Chem. Phys.*, **43**:462 (1965).]

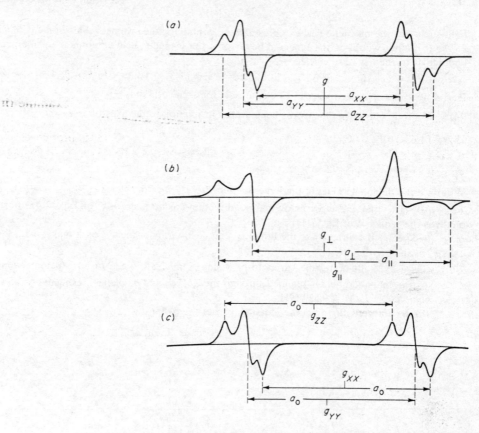

Fig. 7-14 Examples of first-derivative powder spectra of radicals exhibiting hyperfine splitting from one nucleus with $I = \frac{1}{2}$. (a) Isotropic g factors; $a_{ZZ} > a_{YY} > a_{XX}$. (b) Axial symmetry: $g_\parallel < g_\perp$; $a_\parallel > a_\perp$. (c) Isotropic hyperfine splitting; $g_{XX} < g_{YY} < g_{ZZ}$. (*After P. W. Atkins and M. C. R. Symons, "The Structure of Inorganic Radicals," p. 270, Elsevier Publishing Company, Amsterdam, 1967.*)

It *may* be possible in simple cases to determine some or all of the components of **g** and **A**. However, the reader is warned that there are strong possibilities for misassignments. Figure 7-14 illustrates the idealized derivative line shapes for some simple cases. The problems associated with small hyperfine splittings and satellite lines (such as those discussed in Sec. 7-8) can be very considerable.†

PROBLEMS

7-1. Assume that a crystal contains a paramagnetic defect of axial symmetry with $S = \frac{1}{2}$ and a nucleus with $I = \frac{1}{2}$.

(a) What would be the effect on the ESR spectrum of rotating the crystal in the magnetic

† R. Lefebvre and J. Maruani, *J. Chem. Phys.*, **42**:1480 (1965).

. field such that the magnetic field explores a plane containing the symmetry axis of the defect?

(b) What would be the expected behavior if the magnetic field explores the plane perpendicular to the axis of the defect?

7-2. If $g_{XX} = g_{YY}$ show that Eq. (7-6b) reduces to Eq. (7-3). (Hint: $l_x^2 + l_y^2 + l_z^2 = 1$.)

7-3. Show that the mean value of $(3 \cos^2 \alpha - 1)$ for a $2p_z$ orbital is $-\frac{2}{5}$.

7-4. The following hyperfine tensor is measured in relation to the axes of a crystal:

$$\begin{bmatrix} 15 & 0 & 0 \\ 0 & 11 & -1.73 \\ 0 & -1.73 & 9 \end{bmatrix}$$

What are the diagonal values of the hyperfine tensor?

7-5. (a) Show that the eigenvector matrix (direction-cosine matrix) $\mathscr{L}$ of Eq. (7-45) diagonalizes the tensor $\mathbf{A}$ of Eq. (7-44).

(b) Specify the direction of the principal axis Z in terms of the angles it makes with the axes x, y, and z.

7-6. Figure 7-15 illustrates the variation of a single-line ESR spectrum of a paramagnetic defect as the magnetic field is scanned through the ab, ac, and bc planes, respectively, of an *orthorhombic crystal*. $\nu = 9.520$ GHz.

(a) By interpolating from the plots, construct the $\mathbf{g}^2$ tensor.

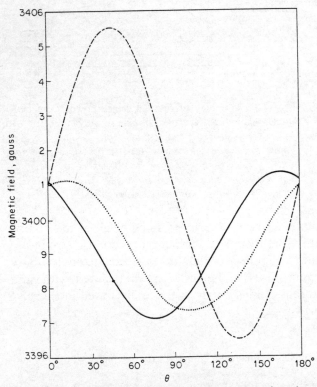

Fig. 7-15 Variation of the resonance field H_r as a function of rotation in the ab (--------), ac (...........) and bc (———) planes of an orthorhombic crystal. Angles are measured with respect to the a axis for the (ab) and (ac) planes, and with respect to the c axis for the (bc) planes.

(*b*) Diagonalize this tensor and hence obtain the principal values of **g** (which are all positive) and the direction-cosine matrix.

7-7. Show that Eq. (7-19) can be written in the equivalent form

$$g^2_{\text{eff}} = \alpha + \beta \cos 2\theta + \gamma \sin 2\theta$$

where

$$g_{XX}{}^2 = \alpha + \beta$$
$$g_{YY}{}^2 = \alpha - \beta$$

and—

$$g_{XY}{}^2 = \gamma$$

7-8. The di-*t*-butyl nitroxide radical may be introduced as a substitutional impurity in 2.2.4.4-tetramethylcyclobutane-1.3-dione. In terms of the *abc'* axis system of the monoclinic crystal, the elements of the A^2 and g^2 tensors are as follows:

$A_{xx}{}^2$	54.86 G^2	$g_{xx}{}^2$	4.03081
$A_{xy}{}^2$	26.66	$g_{xy}{}^2$	-0.00057
$A_{xz}{}^2$	11.10	$g_{xz}{}^2$	$+0.00501$
$A_{yy}{}^2$	1,018.9	$g_{yy}{}^2$	4.00955
$A_{yz}{}^2$	-70.42	$g_{yz}{}^2$	$+0.00092$
$A_{zz}{}^2$	43.86	$g_{zz}{}^2$	4.02834

(*a*) Show that in the principal-axis system

$$A_{XX} = 7.85 \text{ G} \qquad g_{XX} = 2.0087$$
$$A_{YY} = 5.60 \text{ G} \qquad g_{YY} = 2.0061$$
$$A_{ZZ} = 32.01 \text{ G} \qquad g_{ZZ} = 2.0024$$

(*b*) Find the direction cosines of the **A** and the **g** tensors.

(*c*) The crystal structure indicates that the

plane is in the *ac'* plane of the crystal. The N—O bond makes an angle of about 34° with the *a* axis. One usually assumes that the largest principal value of the **A** tensor corresponds to the direction of the nitrogen $2p_z$ orbital; use the direction cosines derived in part *b* to verify the validity of this assumption. [Data taken from W. R. Knolle, Ph.D. dissertation, University of Minnesota, 1970. This system was first studied by O. H. Griffith, D. W. Cornell, and H. M. McConnell, *J. Chem. Phys.*, **43**:2909 (1965).]

7-9. Show that the hyperfine splitting expression

$$\Delta H = \{(a_0 - b)^2 + 3b(2a_0 + b) \cos^2 \theta\}^{\frac{1}{2}}$$

(which is valid for the usual case in which the hyperfine field is much larger than the externally applied field) reduces to

$$\Delta H = b(1 + 3 \cos^2 \theta)^{\frac{1}{2}}$$

if the isotropic splitting $a_0 = 0$. Show also, by a binomial expansion of the square root, that if b^2 can be neglected in comparison with a^2, the general expression reduces to the familiar form

$$\Delta H = a_0 + b(3 \cos^2 \theta - 1)$$

8
Interpretation of the ESR Spectra of Systems in the Solid State

Whereas the ESR spectra of oriented solids may be much more complicated than for liquids, they also contain additional useful information. One may hope to extract details of intra- and intermolecular interactions, molecular configuration, site symmetry, as well as the nature and the location of neighboring atoms. Further, one observes in rigid solids many paramagnetic species which are too reactive or too unstable to be observed in liquid solution.

In this chapter we present numerous examples of information obtained from the analysis of g and of hyperfine tensors for organic free radicals, inorganic free radicals, and point defects in solids. For some examples the ESR spectra are given; for others, it is assumed that values of principal components have already been obtained by the methods of Chap. 7.

The ESR behavior of solutions of high viscosity will be considered in Chap. 9. Chapter 8, with few exceptions, is confined to systems with $S = \frac{1}{2}$. Systems with $S \geq 1$ are considered in Chaps. 10, 11, and 12.

8-1 GENERATION OF FREE RADICALS IN SOLIDS

Free radicals produced by various types of radiation form the largest group of oriented radicals which have been studied. Ultraviolet light, x-rays, or

γ-rays have been widely used for irradiation. Sometimes several different paramagnetic centers are formed, but by annealing and other special treatment, this number can usually be reduced to one or two distinct species. For example, γ irradiation of malonic acid produces not only the free radical HOOC—$\dot{C}$H—COOH† but also the $\dot{C}H_2COOH$ radical.‡ Bombardment of organic single crystals with hydrogen atoms produces some interesting free radicals, especially where the H atoms can add to double bonds.§ In fact, it can be said that almost any kind of high-energy irradiation of organic single crystals will produce free radicals.

The anisotropic properties of stable radicals may be studied if a suitable isomorphic diamagnetic host is available to minimize spin-spin interactions. Perhaps the first radical to be studied in this way was the peroxylamine disulfonate ion $[NO(SO_3)_2{}^{--}]$ in a host crystal of potassium hydroxylamine disulfonate $[K_2NOH(SO_3)_2]$.¶

Many highly reactive radicals have been trapped at low temperatures by the matrix-isolation technique.†† The host is usually an inert substance such as nitrogen, argon, or methane. Even though the radicals are randomly oriented, much information may sometimes be obtained.

8-2 π-TYPE ORGANIC RADICALS

8-2a. Identification The first step in the identification of a radical species is usually a determination of the principal values of the g and of the hyperfine tensors. These may be evaluated by the procedures of Chap. 7. Since most organic free radicals in solution have a nearly isotropic g tensor, with principal values very close to the free-spin value, the g tensor is not always helpful in identification. It is the principal values of the hyperfine tensors which prove to be the most useful properties for the identification of a radical in a solid.

Fortunately, most organic radicals trapped in a single crystal occupy a limited number of sites, usually related by the symmetry operations of the host crystal. For a given radical the number of sites is usually determined by the number of molecules per unit cell. For this reason, a successful analysis of the ESR spectra of such oriented radicals requires a detailed knowledge of the crystal structure of the host. When the magnetic field is oriented parallel or perpendicular to one of the crystallographic axes, some

---

† H. M. McConnell, C. Heller, T. Cole, and R. W. Fessenden, *J. Am. Chem. Soc.*, **82**:766 (1960).
‡ A. Horsfield, J. R. Morton, and D. H. Whiffen, *Mol. Phys.*, **4**:327 (1961).
§ H. C. Heller, S. Schlick, and T. Cole, *J. Phys. Chem.*, **71**:97 (1967).
¶ S. I. Weissman and D. Banfill, *J. Am. Chem. Soc.*, **75**:2534 (1953).
†† A. M. Bass and H. P. Broida, "Formation and Trapping of Free Radicals," Academic Press Inc., New York, 1960.

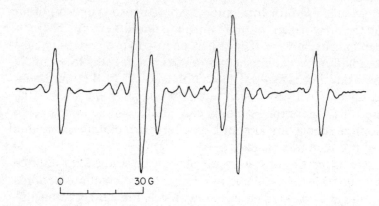

0   30 G

Fig. 8-1 ESR spectrum of an x-irradiated single crystal of β-succinic acid. The applied field is perpendicular to the crystal b axis and makes an angle of 100° with the a axis. [*Taken from C. Heller and H. M. McConnell, J. Chem. Phys.*, **32**:1535 (1960).]

or all of the radical sites may become equivalent; a much simpler spectrum then results. The ESR spectrum in Fig. 8-1 corresponds to a special orientation of the radical produced by the x irradiation of β-succinic acid[†] HOOC—CH$_2$—CH$_2$—COOH. When the magnetic field is perpendicular to the b axis of the crystal and makes an angle of 100° to the a axis, the corresponding principal axes of the radicals in the two sites make equal angles with the magnetic field. The main lines of the spectrum are easily interpreted in terms of a 1:1 doublet, each being split into a slightly smaller 1:2:1 triplet. A reasonable candidate is the radical produced by the removal of an aliphatic hydrogen atom to give

HOOC—CH$_2$—ĊH—COOH

For the orientation chosen, the doublet splitting arises from one of the β-hydrogen atoms, while the 1:2:1 triplet arises from the accidental equivalence of the coupling constants from the other β-hydrogen and the α-hydrogen atoms.

Once the radical has been tentatively identified, ESR spectra are run for many different orientations of the crystal with respect to the magnetic field. Analysis of these spectra, using the techniques outlined in Chap. 7, yields the principal values and the direction cosines of the g tensor and hyperfine tensors. The principal-axis system of the g tensor is determined by the radical site. The principal axes of the g and of the hyperfine tensors need not coincide. For example, the principal axes of the hyperfine tensors for the β-hydrogen atoms of the succinic acid radical do not coincide with each other nor with the principal axes of the g tensor.

In principle, selective deuteration *should* aid in the identification of

† C. Heller and H. M. McConnell, *J. Chem. Phys.*, **32**:1535 (1960).

organic radicals produced by ionizing radiation. However, in some cases, exchange of deuterons or protons with the host occurs after irradiation.[†],[‡],[§] In such cases deuteration may result in misidentification of the radicals.[¶]

If two radicals are present, their saturation behavior under conditions of high microwave power may be different; under favorable conditions, one spectrum may be smeared out, leaving the other unobscured.

A more generally applicable technique involves taking spectra at a higher frequency, for example, 35 GHz. There is then a greater separation of lines from the two different radicals at the higher frequency, owing to differences in g components. However, hyperfine separations in a given radical will be unaffected (in first order). Forbidden lines (i.e., lines for which $\Delta M_I = \pm 1$) of very low intensity at 9.5 GHz may appear much stronger at 35 GHz. (See Fig. 7-8.)

8-2b. Aliphatic radicals Numerous studies of the ESR spectra of single crystals of irradiated aliphatic organic compounds have been published.[††] Such efforts have largely been concentrated on materials which readily form single crystals of known structure. Favorite samples include the saturated dicarboxylic acids, their salts, and some simple amino acids. The most common type of radical detected at room temperature has a π electron centered on a trigonally bonded carbon atom; the radical is usually produced by the ejection of a hydrogen atom from the molecule. In such a radical, the unpaired electron is localized predominantly in a $2p_z$ orbital of the trigonal carbon atom; the principal axis of the p orbital is perpendicular to the radical plane. A model for such a radical is shown in Fig. 8-2. Since

† I. Miyagawa and K. Itoh, *J. Chem. Phys.*, **43**:2915 (1965).
‡ R. F. Weiner and W. S. Koski, *J. Am. Chem. Soc.*, **85**:873 (1963).
§ J. R. Morton, *J. Am. Chem. Soc.*, **86**:2325 (1964).
¶ D. K. Ghosh and D. H. Whiffen, *Mol. Phys.*, **2**:285 (1959).
†† For a review, see J. R. Morton, *Chem. Rev.*, **64**:453 (1964).

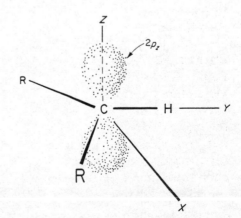

Fig. 8-2 Principal axes of an $\overset{R}{\underset{R}{>}}$C—H radical.

[*After H. M. McConnell, C. Heller, T. Cole, and R. W. Fessenden, J. Am. Chem. Soc.,* **82**:766 (1960).]

^{12}C has zero nuclear spin, only proton hyperfine couplings are usually observed; in certain cases ^{13}C hyperfine couplings have also been observed.

In Chap. 6 it was shown that there should be a negative isotropic hyperfine coupling for an α proton. From Table 6-2 it is noted that $Q_\alpha = 23.04$ G for the CH_3 radical; an average value of Q for a variety of carbon compounds has been taken as 27.0 G. If $g = g_e$, these two values correspond to an isotropic coupling of 64 and 76 MHz, respectively. There will also be an anisotropic hyperfine interaction due to the dipole-dipole interaction between the electron and the proton. A qualitative picture of the anisotropic hyperfine interaction for the proton of a C—H fragment is given in Fig. 8-3.

With **H** along the C—H bond (Fig. 8-3a) the p orbital is largely in a region where $(3 \cos^2 \theta - 1) > 0.†$ [The factor $(1 - 3 \cos^2 \theta)$ in Eq. (3-2) has been multiplied by -1 to take account of the negative magnetic moment of the electron.] Thus the anisotropic hyperfine coupling with **H** in this orientation should be large and positive. In Fig. 8-3b, **H** is perpendicular both to the C—H bond and to the p orbital. Now the p orbital is almost entirely in the region where $(3 \cos^2 \theta - 1) < 0$; hence a large negative contribution to the anisotropic hyperfine coupling is expected. Finally, in Fig. 8-3c, **H** is perpendicular to the C—H bond but parallel to the p orbital;

† Strictly speaking, the angular dependence of the hyperfine splitting will be given by Eq. (7-32b); however, the arguments given here do lead to qualitatively correct conclusions regarding the relative magnitudes of the principal components of the hyperfine tensor.

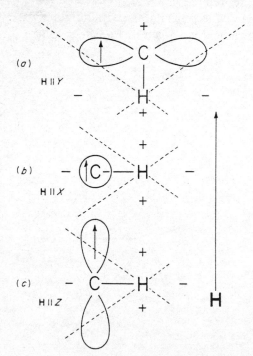

(a)

H ∥ Y

(b)

H ∥ X

(c)

H ∥ Z

H

Fig. 8-3 The sign of the anisotropic hyperfine interaction of α protons in oriented radicals. The dotted lines represent cones which form the nodes of $(3 \cos^2 \theta - 1)$. The X, Y and Z directions are those of Fig. 8-2. (a) **H** parallel to the C—H bond (Y axis). (b) **H** perpendicular to the C—H bond and to the $2p_z$ orbital. (c) **H** parallel to the $2p_z$ orbital. [*After J. R. Morton. Chem. Rev,* **64**:453 (1964).]

the latter extends almost equally over positive and negative regions of $(3 \cos^2 \theta - 1)$. Thus a small anisotropic contribution is expected. If the X, Y, and Z axes are defined as in Figs. 8-2 and 8-3, calculations show that the anisotropic hyperfine tensor should be[†]

$$T^H = \begin{bmatrix} -38 & 0 & 0 \\ 0 & +43 & 0 \\ 0 & 0 & -5 \end{bmatrix} \quad \text{MHz} \qquad (8\text{-}1)$$

Experiments on the malonic acid radical have the especial importance that they first established the sign of the isotropic proton hyperfine coupling for a C—H proton. The following proton hyperfine tensor was established[‡]:

$$A^H = \begin{bmatrix} \pm 91 & 0 & 0 \\ 0 & \pm 29 & 0 \\ 0 & 0 & \pm 58 \end{bmatrix} \quad \text{MHz} \qquad (8\text{-}2)$$

Hence, the isotropic hyperfine coupling A_0^H (one-third the trace of A^H) is ± 59 MHz. The two choices for T^H are then

$$T^H = \begin{bmatrix} -32 & 0 & 0 \\ 0 & +30 & 0 \\ 0 & 0 & +1 \end{bmatrix} \quad \text{or} \quad \begin{bmatrix} +32 & 0 & 0 \\ 0 & -30 & 0 \\ 0 & 0 & -1 \end{bmatrix} \qquad (8\text{-}3)$$

The left-hand tensor in (8-3) corresponds to a negative value of A_0^H and the right-hand tensor to a positive value. Comparison of the tensor (8-3) with the theoretical tensor (8-1) indicates that A_0^H must be *negative*.

Comparison of the magnitudes of the elements of T_0^H in (8-3) with the theoretical tensor (8-1) permits one to make a rough estimate of the π-spin density ρ_C. From the largest components of T^H in Eqs. (8-1) and (8-3), one concludes that $\rho_C \approx 0.75$. This compares favorably with the value of $\rho_C \approx 0.79$ from the isotropic hyperfine coupling, assuming a Q_α value of -76 MHz. (See Sec. 6-4.) Studies of a wide variety of radicals have shown that the above model holds very well. In fact, when such a hyperfine tensor is obtained, it suggests strongly that an α proton is the interacting nucleus.

In the case of β-proton hyperfine couplings, the situation is quite different. These couplings are invariably found to be nearly isotropic. Since β protons are much farther away from the radical carbon atom than α protons, and since the dipole-dipole interaction energy falls off as r^{-3}, it is not surprising that little anisotropy is found for β-hyperfine couplings. It is surprising that the magnitude of the isotropic β-proton hyperfine coupling is about the same as that for α protons. This fact has been rationalized in terms of a hyperconjugation mechanism in which there is a small electron transfer from the β-hydrogen atoms to the p orbital on the radical carbon,

† H. M. McConnell and J. Strathdee, *Mol. Phys.*, 2:129 (1959).
‡ H. M. McConnell, C. Heller, T. Cole, and R. W. Fessenden, *J. Am. Chem. Soc.*, 82:766 (1960).

leading to spin density in the $1s$ orbitals of the β-hydrogen atoms. (See Sec. 6-6.) This interaction will be most effective when the CH bond of the alkyl group is perpendicular to the radical plane and least effective when it is in the radical plane. The following relation has been proposed for β-hydrogen hyperfine couplings[†,‡]:

$$A_\beta = A_1 + A_2 \cos^2 \theta \tag{8-4}$$

Here θ, the dihedral angle, is defined as in Fig. 8-4. The value of A_2 is approximately $+140$ MHz, and A_1 is usually small (0 to 10 MHz).[§]

If the β-hydrogen atoms are freely rotating (as in a CH_3 group), a value of $A_1 + A_2/2$ would be expected for the β-hydrogen hyperfine coupling, since the average value of $\cos^2 \theta$ is $\frac{1}{2}$. In the $CH_3\dot{C}HCOOH$ radical,[§] an isotropic β-proton hyperfine coupling of $+70$ MHz is observed at 300 K with all three β protons equivalent. This spectrum is shown in Fig. 8-5a. However, when the crystal is cooled to 77 K, a marked change in the spectrum takes place, as can be seen from Fig. 8-5b. Analysis shows that now the three β protons are not equivalent; they have hyperfine couplings of $+120$, $+76$, and $+14$ MHz, respectively. The obvious conclusion is that the methyl group has stopped rotating. The magnitudes of the hyperfine couplings indicate an orientation such that $\theta_1 = 18°$, $\theta_2 = 138°$, and $\theta_3 = 258°$.

Hyperfine couplings from ^{13}C or ^{14}N have been observed in a number

† C. Heller and H. M. McConnell, *J. Chem. Phys.*, **32**:1535 (1960).

‡ A. D. McLachlan, *Mol. Phys.*, **1**:233 (1958); P. G. Lykos, *J. Chem. Phys.*, **32**:625 (1960).

§ A. Horsfield, J. R. Morton, and D. H. Whiffen, *Mol. Phys.*, **4**:425 (1961); **5**:115 (1962).

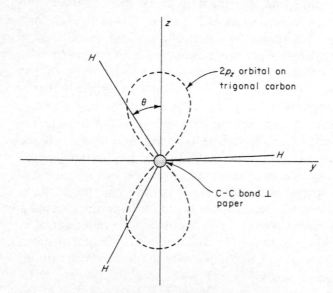

Fig. 8-4 Orientation of the methyl hydrogen atoms in an $R\dot{C}H(CH_3)$ radical.

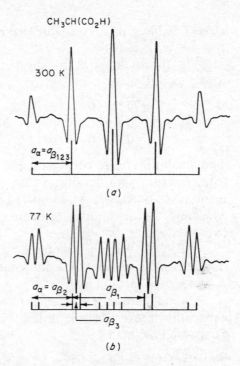

Fig. 8-5 Second derivative of the ESR absorption spectrum of the $CH_3\dot{C}H(CO_2H)$ radical. (*a*) Spectrum taken at 300 K. (*b*) Spectrum taken at 77 K. [*From A. Horsfield, J. R. Morton, and D. H. Whiffen, Mol. Phys.*, 4:425 (1961).]

of irradiated organic single crystals. Usually the interacting nucleus (^{13}C or ^{14}N) is that of the atom with the $2p_z$ orbital containing the unpaired electron. In this case the anisotropic hyperfine tensor will have approximately axial symmetry. If an axial tensor is observed, comparisons with calculated tensor elements can easily be made. (See Table C.) For ^{13}C and ^{14}N the anisotropic hyperfine tensors for one electron in a $2p_z$ orbital are predicted from SCF wave functions to be

$$T^C = \begin{bmatrix} -91 & 0 & 0 \\ 0 & -91 & 0 \\ 0 & 0 & +182 \end{bmatrix} \quad \text{MHz} \tag{8-5}$$

$$T^N = \begin{bmatrix} -48 & 0 & 0 \\ 0 & -48 & 0 \\ 0 & 0 & +96 \end{bmatrix} \quad \text{MHz} \tag{8-6}$$

Here the principal-axis system is the same as that in Fig. 8-2. The maximum value corresponds to the magnetic field along the direction of the p orbital. Axial symmetry arises because of the cylindrical symmetry of a $2p_z$ orbital. That is, the anisotropic hyperfine tensor will be of the form

$$T = \begin{bmatrix} -B & 0 & 0 \\ 0 & -B & 0 \\ 0 & 0 & +2B \end{bmatrix} \tag{8-7}$$

Values of B are tabulated in the fifth column of Table C (inside back cover). They are computed from SCF wave functions by evaluating $B = \frac{2}{5}h^{-1}g\beta g_N\beta_N\langle r^{-3}\rangle$.

In the malonic acid radical, the ^{13}C anisotropic hyperfine tensor is found to be[†]

$$\mathbf{T}^C = \begin{bmatrix} -70 & 0 & 0 \\ 0 & -50 & 0 \\ 0 & 0 & +120 \end{bmatrix} \text{ MHz} \tag{8-8}$$

The small difference between the X and the Y components probably arises from nonzero spin densities in the carboxyl groups. The $+120$ MHz element of (8-8) should be compared with the $+182$ MHz element in (8-5). The ratio 0.66 is another estimate of ρ_C, the π-spin density on the central carbon atom.

$N(SO_3)_2^{--}$ is an example of a radical for which the nitrogen hyperfine tensor shows nearly axial symmetry. The principal values of $\mathbf{T}^N$ are $+70, -35, -35$.[‡]

Some fluorinated radicals have been studied in single crystals. An especially interesting one is that obtained from the disodium salt of perfluorosuccinic acid.[§]

$$NaOOC—CF_2—CF_2—COONa$$

The radical formed appears to be

$$^-OOC—CF_2—\dot{C}F—COO^-$$

Hyperfine couplings were observed for both α- and β-fluorine nuclei. Anisotropic hyperfine tensors are given in Fig. 8-6, along with the approximate directions of the principal axes of the tensors. The α-fluorine tensor was computed in Sec. 7-7.

In contrast to a typical α-proton anisotropic hyperfine tensor, the α-fluorine tensor has its maximum value along the Z axis (i.e., perpendicular to the radical plane). This type of tensor is characteristic of a nucleus interacting primarily with electron-spin density in a p orbital on the same atom. For unit p-orbital spin density on a fluorine atom, the anisotropic tensor is expected to be (see Table C)

$$\mathbf{T}_\alpha^F = \begin{bmatrix} -1,515 & 0 & 0 \\ 0 & -1,515 & 0 \\ 0 & 0 & +3,030 \end{bmatrix} \text{ MHz} \tag{8-9}$$

with the maximum value in the direction of the p orbital. From the numerical magnitude of the Z component of $\mathbf{T}_\alpha^F$, one may deduce that $\rho_{F_\alpha} \approx 0.09$.

[†] H. M. McConnell and R. W. Fessenden. *J. Chem. Phys.*, **31**:1688 (1959); T. Cole and C. Heller. *J. Chem. Phys.*, **34**:1085 (1961).
[‡] A. Horsfield, J. R. Morton, J. R. Rowlands. and D. H. Whiffen, *Mol. Phys.*, **5**:241 (1962).
[§] M. T. Rogers and D. H. Whiffen, *J. Chem. Phys.*, **40**:2662 (1964).

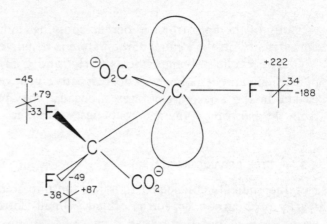

Fig. 8-6 The ^-O_2C—CF_2—$\dot{C}F$—$CO_2{}^-$ radical generated in a single crystal of sodium perfluorosuccinate. The several axis systems represent the orientation of the principal axes of the corresponding fluorine hyperfine tensors. The numbers are the principal values (in megahertz) of the anisotropic fluorine hyperfine tensors. [*Data taken from M. T. Rogers and D. H. Whiffen, J. Chem. Phys.,* **40**:2662 (1964).]

This result may be interpreted as evidence for a partial donation of electron density from the $2p_z$ orbital of fluorine to the $2p_z$ orbital of the carbon atom.

In contrast to β-proton hyperfine interactions, which are almost isotropic, the β-fluorine interaction is very anisotropic. The observed large anisotropic interaction can arise only if there is a net spin density in a p orbital on the fluorine atom. Spin density in an s orbital would produce only an isotropic hyperfine interaction. The orientation of the principal axes of the β-fluorine hyperfine tensors strongly suggests that the interaction which leads to spin density in the β-fluorine p orbitals arises from a direct overlap of these orbitals with the carbon $2p_z$ orbital. There is some evidence from NMR and ESR work in solution that such p-π interactions are important.†

8-2c. Radicals from unsaturated organic compounds Some radicals have been produced in single crystals of unsaturated compounds. One of the most interesting is that produced by x irradiation of glutaconic acid‡

$$HOOC—CH{=}CH—CH_2—COOH$$

This product is the substituted allyl radical

$$HOOC—CH{=\!=\!=}CH{=\!=\!=}CH—COOH$$

† P. Scheidler and J. R. Bolton, *J. Am. Chem. Soc.,* **88**:371 (1966); W. A. Sheppard, *J. Am. Chem. Soc.,* **87**:2410 (1965).
‡ C. Heller and T. Cole, *J. Chem. Phys.,* **37**:243 (1962).

A study of the anisotropic hyperfine couplings indicates that the π-electron spin density on the central carbon atom is negative, as predicted by theory.

The cyclic polyene radicals C_5H_5† and C_7H_7,‡ produced from cyclopentadiene and cycloheptatriene, respectively, have been studied in a single-crystal host. These are of interest because they provide an estimate of Q_α from the isotropic hyperfine couplings. (See Sec. 5-3.)

8-3 σ–TYPE ORGANIC RADICALS

For the radicals considered thus far, the unpaired electron is located primarily in a carbon $2p_z$ (or π) orbital. Small isotropic hyperfine couplings are observed, but these arise primarily from the indirect mechanism described in Chap. 6. The nuclei are usually located at or near the nodal plane of the $2p_z$ orbital.

There are a number of known radicals which exhibit proton hyperfine couplings of the order of 150 to 400 MHz. These couplings are far too large to be explained by the indirect mechanism, and one is forced to conclude that the orbital of the unpaired electron has considerable s character. In these radicals, the unpaired electron may crudely be regarded as "occupying" a position normally occupied by an atom. That is, the unpaired electron is primarily located in the σ orbital which would normally form a σ bond with another atom such as hydrogen. Most σ orbitals have a considerable s-orbital component. Hence, large isotropic hyperfine couplings are characteristic of σ radicals, and the sign of the hyperfine coupling should be positive because of the direct interaction of the proton with the s component of the σ orbital. In the ethynyl radical $\cdot C\equiv C—H$ the unpaired electron primarily occupies the orbital normally directed to a hydrogen atom in the acetylene molecule. Likewise, in the vinyl radical the unpaired electron is primarily in an orbital which would attach a hydrogen atom in the ethylene molecule. Yet another example is the formyl radical

$$H\diagdown \overset{\cdot}{C}{=}O$$

derived from formaldehyde

$$H\diagdown \diagup C{=}O \atop H$$

The bond angle in the formyl radical appears to be 120°.§ A closely related radical is

$$F\diagdown \overset{\cdot}{C}{=}O$$

† G. R. Liebling and H. M. McConnell, *J. Chem. Phys.*, **42**:3931 (1965).
‡ D. E. Wood and H. M. McConnell, *J. Chem. Phys.*, **37**:1150 (1962).
§ F. J. Adrian, E. L. Cochran, and V. A. Bowers, *J. Chem. Phys.*, **36**:1661 (1962).

In each of these cases, the sign of the hyperfine coupling is believed to be positive. The positive coupling doubtless arises from a considerable s component of the C—H or C—F bonds. For instance, in the HCO radical, the spin density in the $1s$ orbital of hydrogen is approximately 0.27, since the proton hyperfine coupling constant is 384 MHz.[†] This is the largest known proton coupling, except for that of the hydrogen atom. It is a general property of σ radicals that the direct (positive) s contribution to the hyperfine coupling of an attached atom will greatly exceed the indirect (negative) contribution discussed in Chap. 6.

The magnitude of isotropic ^{13}C hyperfine splittings provides an additional indication of whether or not there is a significant s-orbital contribution. For example, in the planar $\dot{C}H_3$ radical $a^C = 38.5$ G (see Sec. 6-7); whereas, in the $\dot{C}F_3$ radical $a^C = 271.6$ G.[‡] The large increase in the ^{13}C hyperfine splitting can be explained only in terms of a large pyramidal distortion in $\dot{C}F_3$; this distortion leads to an s character in the unpaired-electron orbital.

It is interesting to compare the couplings in the formyl and the vinyl radicals. For the latter, the couplings are 43.7 MHz for H_1, 95.2 MHz for H_2, and 190 MHz for H_3.[§] (Section 9-6b treats the changes in apparent couplings when this radical is observed in the liquid state.) Even the largest value is considerably less than that for HCO. The difference is largely due to the large variation of coupling constant with bond angle. From the value of 43.7 MHz for H_1 in the vinyl radical, the bond angle is estimated to be 140° to 150°.

Hyperfine couplings in σ-type radicals may exhibit a large anisotropy. For example, in FCO, the principal hyperfine tensor components are 1,437.5, 708.2, and 662.0 MHz.[†]

It is now possible to estimate the spin distribution in σ radicals by using a molecular orbital theory (which includes all valence-shell atomic orbitals) such as the INDO method referred to in Sec. 6-3.

8-4 INORGANIC RADICALS

The assignment and interpretation of the ESR spectra of inorganic radicals has been a very active field of investigation. It is not possible to give a complete coverage; however, we shall attempt to outline the major features with some examples.

8-4a. Identification of radical species As in the case of organic radicals, the values of principal components of hyperfine tensors can provide the major clues in the identification of species resulting from the irradiation of

---

[†] F. J. Adrian, E. L. Cochran, and V. A. Bowers, *J. Chem. Phys.*, **43**:462 (1965).
[‡] R. W. Fessenden and R. H. Schuler, *J. Chem. Phys.*, **43**:2704 (1965).
[§] E. L. Cochran, F. J. Adrian, and V. A. Bowers, *J. Chem. Phys.*, **40**:213 (1964).

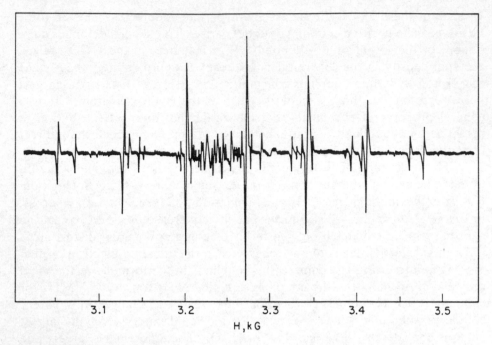

H, kG

Fig. 8-7 ESR spectrum of the V_K center in KCl at 77 K with the magnetic field parallel to the [100] direction in the (100) plane. The microwave frequency is 9.263 GHz. [*Taken from T. G. Castner and W. Känzig, J. Phys. Chem. Solids,* 3:178 (1957).]

inorganic materials. For example, x irradiation of LiF at 77 K produces (among others) a species which exhibits a 1:2:1 triplet ESR spectrum for H ∥ [100]. Such a pattern implies hyperfine interaction with two nuclei of spin $\frac{1}{2}$. The principal values of the g tensor are $g_{XX} = 2.0234$, $g_{YY} = 2.0227$, and $g_{ZZ} = 2.0031$, indicative of nearly axial symmetry. The hyperfine splitting shows axial behavior, with $a_\| = 887$ G and $a_\perp = 59$ G.† The species responsible is doubtless the F_2^- ion (V_K center). If the experiment is done with KCl, the spectra (Fig. 8-7) from the ^{35}Cl—$^{35}Cl^-$, the ^{35}Cl—$^{37}Cl^-$, and the ^{37}Cl—$^{37}Cl^-$ molecules provide redundant and incontrovertible identification of the center. Interpretation of Fig. 8-7 is left as a problem for the reader (Prob. 8-11).

In other cases the appearance of hyperfine structures is not sufficient to provide a positive identification. For example, γ-irradiated KNO_3 exhibits the ESR spectrum shown in Fig. 8-8. There are at least three radical species, each of which contains a nitrogen atom, as evidenced by the triplet hyperfine splittings. However, the assignment to specific radicals requires further information. The reasonable possibilities can be listed as NO_2, NO_2^{--}, NO_3, and NO_3^{--}. The experimental results for the hyperfine and g tensors are listed in Table 8-1.

† T. G. Castner and W. Känzig, *J. Phys. Chem. Solids,* **3**:178 (1957).

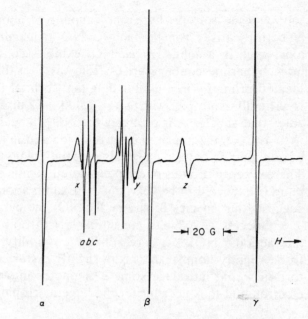

Fig. 8-8 Spectra of radicals obtained on γ irradiation of KNO$_3$. Species A (lines α, β, and γ) has been assigned as the NO$_2$ radical. Species B (lines a, b, and c) has been assigned as the NO$_3$ radical. Species C (lines x, y, and z) has been assigned as the NO$_3{}^{--}$ radical. [*Taken from R. Livingston and H. Zeldes, J. Chem. Phys.*, **41**:4011 (1964).]

The identification requires a knowledge of the theoretical predictions of the structure and orbital sequence in each radical; in addition, one requires information from studies of these radicals in other host matrices. In various hosts, NO$_2$ exhibits a ^{14}N hyperfine coupling with little anisot-

Table 8-1 Hyperfine and g tensors for radical species found in γ-irradiated KNO$_3$

Species	g components	^{14}N hyperfine components, MHz
A	$g_{\parallel} = 2.006$†	$A_{\parallel} = 176$†
	$g_{\perp} = 1.996$	$A_{\perp} = 139$
B	$g_{\parallel} = 2.0031$‡	$A_{\parallel} = 12.08$‡
	$g_{\perp} = 2.0232$	$A_{\perp} = 9.80$
C	$g_{\parallel} = 2.0015$†	$A_{\parallel} = 177.6$†
	$g_{\perp} = 2.0057$	$A_{\perp} = 89.0$

† H. Zeldes, in W. Low (ed.), "Paramagnetic Resonance," vol. 2, p. 764, Academic Press Inc., New York, 1963.
‡ R. Livingston and H. Zeldes, *J. Chem. Phys.*, **41**:4011 (1964).

r.opy and an isotropic hyperfine coupling of about 150 MHz.[†] The small anisotropy arises from the fact that NO_2 is usually rotating about its twofold axis, even in a solid. *Fixed* NO_2 exhibits considerable anisotropy. The large hyperfine coupling arises from the fact that the unpaired electron is located primarily in a nonbonding (sp^3) orbital on nitrogen. The g tensor is virtually isotropic, with $g_{iso} \simeq 2.000$. Comparison with Table 8-1 indicates that species A is probably the NO_2 radical.

In NO_3 the unpaired electron is located in an orbital composed largely of nonbonding oxygen p orbitals perpendicular to the plane of the molecule. The oxygen p orbitals overlap end to end; thus the nitrogen hyperfine coupling is expected to be very small. Examination of the results in Table 8-1 suggests that species B may be the NO_3 radical.

Species C exhibits considerable isotropic and anisotropic hyperfine interaction. NO_3^{--} is a reasonable possibility, since this ion is expected to be slightly nonplanar,[‡] i.e., a slightly distorted π-type radical. The distortion would introduce some s character into the orbital of the unpaired electron and thus account for the large isotropic hyperfine coupling.

8-4b. Structural information When a radical species has been identified, the g and hyperfine tensors can provide considerable information about the detailed geometric and electronic structure of the radical. The NO_2 radical (observed in $NaNO_2$[§]) is an excellent example. From Table C one notes that a single electron in a $2s$ orbital on nitrogen would give rise to an isotropic hyperfine coupling of 1,540 MHz. From the observed value of $A_0^N = 151$ MHz, the spin density in the nitrogen $2s$ orbital is computed to be $\rho_s = \frac{151}{1540} = 0.10$. Similarly, from the maximum value in the anisotropic hyperfine tensor, the spin density in the nitrogen $2p_x$ orbital is computed to be $\rho_p = \frac{12}{48} = 0.25$. Hence the $2p/2s$ ratio is 2.5. A simple consideration of orbital hybridization suggests that the bond angle is between 130 and 140°. This is in good agreement with gas-phase vibrational analysis[¶] and microwave results (134°).[††] Presumably, the reason that the spin densities for the nitrogen $2p$ and $2s$ orbitals do not add up to unity is that there is some spin density in $2p$ orbitals on the oxygen atoms.

When isotropic hyperfine couplings are small, as for species B in Table 8-1, one must beware of interpreting these in terms of a percentage s character in the orbital of the unpaired electron. The indirect mechanism leading to isotropic hyperfine coupling (see Chap. 6) may give the major contri-

---

† P. W. Atkins and M. C. R. Symons. *J. Chem. Soc.*, **1962**:4794.

‡ A. D. Walsh, *J. Chem. Soc.*, **1953**:2296.

§ H. Zeldes and R. Livingston. *J. Chem. Phys.*, **35**:563 (1961).

¶ G. E. Moore. *J. Opt. Soc. Am.*, **43**:1045 (1953).

†† G. R. Bird. *J. Chem. Phys.*, **25**:1040 (1956).

bution.   Generally, if $\rho_s < 0.05$ as computed above, then an interpretation in terms of a bond angle is dubious.

It is interesting to compare the ESR results for an isoelectronic series of radicals.   Table 8-2 contains the data for the ClO_3, SO_3^-, and PO_3^{--} radicals, as well as for the NO_2 and CO_2^- radicals.   It is clear that as the atomic number of the central atom decreases, the tetratomic radicals are becoming more pyramidal (as evidenced by the decreased ρ_p/ρ_s ratio); the triatomic radicals are becoming more bent.

8-5   POINT DEFECTS IN SOLIDS

A *point* defect is defined as any localized imperfection in a crystal, in contrast with a *line* defect such as a dislocation.   Some of the major classes of point defects are:

1. Vacancies
2. Impurity atoms or ions in substitutional or interstitial sites
3. Trapped-electron centers
4. Trapped-hole centers
5. "Broken bonds"

Many point defects are paramagnetic, and in favorable cases, a wealth of detailed information concerning the identification and structure of a defect may be obtained from an analysis of its ESR spectrum.   Vacancies are themselves not paramagnetic, but their existence may permit the formation of some paramagnetic centers.

8-5a.   Generation of point defects   Substitutional defects are often present in host crystals, even those of the highest available purity.   This is especially true of high-melting solids such as oxides.   In the alkali halides, one diamagnetic defect which escaped attention for some time is the OH^- ion; this may be present (especially in NaCl) because of a hydrolysis reaction with water in the starting material.   Another is O_2^- (Fig. 7-3), which is incorporated into alkali-halide crystals.   Impurities other than those incidentally present may be deliberately added to the melt (or the solution) from which a crystal is made.

Substitutional diamagnetic impurities may be changed to a paramagnetic valence state (or if initially paramagnetic to a different paramagnetic valence state) by ionizing radiation (x-rays, γ-rays, or ultraviolet radiation). Such radiation may actually be responsible for generating vacancies in considerable number.   This process occurs in the alkali halides, in which large numbers of anion vacancies are created.   These vacancies may trap one or two electrons freed during the irradiation.   Alternatively, irradiation of

Table 8-2 Comparison of the ESR data for some isoelectronic radicals

Radical	Matrix	g tensor				Hyperfine tensor, MHz				Spin densities				Ref.
		g_{XX}	g_{YY}	g_{ZZ}	g_{iso}	T_{XX}	T_{YY}	T_{ZZ}	A_0	ρ_s	ρ_p	$\dfrac{\rho_p}{\rho_s}$	$\rho_s + \rho_p$	
ClO_3	$KClO_4$	2.0132	2.0132	2.0066	2.0110	−40.5	−40.5	81	342	0.076	0.34	4.5	0.42	[†]
SO_3^-	$K_2CH_2(SO_3)_2$				2.0036	−35	−35	70	358	0.13	0.49	4.5	0.62	[‡]
PO_3^{--}	$Na_2HPO_3 \cdot 5H_2O$	2.001	2.001	1.999	2.000	−148	−148	297	1,660	0.16	0.53	3.3	0.69	[§]
NO_2	$NaNO_2$	2.0057	2.0015	1.9910	1.9994	−22.3	37.0	−14.8	153	0.099	0.44	4.4	0.54	[¶]
CO_2^-	$NaHCO_2$	2.0032	2.0014	1.9975	2.0007	−32.0	78.0	−46.0	468	0.15	0.50	3.3	0.65	[††]

[†] P. W. Atkins, J. A. Brivati, N. Keen, M. C. R. Symons, and P. A. Trevalion, J. Chem. Soc., **1962**:4785.
[‡] G. W. Chantry, A. Horsfield, J. R. Morton, and D. H. Whiffen, Mol. Phys., **5**:233 (1962).
[§] A. Horsfield, J. R. Morton, and D. H. Whiffen, Mol. Phys., **4**:475 (1961).
[¶] H. Zeldes and R. Livingston, J. Chem. Phys., **35**:563 (1961).
[††] D. W. Ovenall and D. H. Whiffen, Mol. Phys., **4**:135 (1961).

solids may cause electrons to be released from sites with reasonably low electron affinities. The resulting hole may be localized at the same site by lattice relaxation (self-trapped hole such as in F_2^-), or it may wander until it is trapped by an impurity ion or by an anion vacancy. If the ESR spectrometer is arranged for data recording *during* irradiation by electrons or any form of ionizing radiation, one may detect species of short life.

In many solids, x or γ irradiation does not cause displacement of atoms from lattice sites. For these substances, irradiation with high-energy proton beams or with neutrons not only creates various types of vacancies but also serves to populate some with electrons.

8-5b. Substitutional or interstitial impurities If a substitutional or interstitial impurity is paramagnetic, its ESR spectrum can usually be detected. However, it may still be detected even if it is not paramagnetic, provided it is adjacent to a paramagnetic center and has a nonzero nuclear spin.

The simplest impurity defects are atoms, which may be generated by irradiation with ultraviolet light, x-rays, or γ-rays. The host matrix must be rigid enough to prevent rapid diffusion which would cause the atoms to recombine. For example, hydrogen atoms may be produced and trapped in acids such as H_2SO_4, H_3PO_4, or $HClO_4$ at 20 K,[†] but at somewhat higher temperatures the ESR spectrum disappears rapidly. However, hydrogen atoms trapped in CaF_2[‡] or $CaSO_4 \cdot \frac{1}{2}H_2O$[§] may be kept for years at room temperature. Regardless of the host, a large doublet splitting is always observed with a hyperfine coupling constant close to that of the free-atom value of 1,420 MHz; actual values vary from 1,391 to 1,460 MHz. Hyperfine couplings in excess of the free-atom value (e.g., H atoms from HI, H_2O, or NH_3 irradiated in rare-gas matrices[¶]) are attributed to a "compression" of the wave function of the electron by overlap interactions with the host. Smaller couplings presumably arise from a "spreading" of the wave function by van der Waals interactions.[††]

The trapped hydrogen atom in CaF_2 is an excellent example of a center which may be described in detail by its ESR spectrum.[‡] The center is generated by a two-stage treatment:

1. Heat a CaF_2 single crystal with H_2 in the presence of a metal such as aluminum; this leads to the formation of hydride ions (H^-) at fluoride-ion sites.[‡‡]

[†] R. Livingston, H. Zeldes, and E. H. Taylor, *Discussions Faraday Soc.*, **19**:166 (1955).
[‡] J. L. Hall and R. T. Schumacher, *Phys. Rev.*, **127**:1892 (1962).
[§] H. Kon, *J. Chem. Phys.*, **41**:573 (1964).
[¶] S. N. Foner, E. L. Cochran, V. A. Bowers, and C. K. Jen, *J. Chem. Phys.*, **32**:963 (1960); *Phys. Rev.*, **104**:846 (1956).
[††] F. J. Adrian, *J. Chem. Phys.*, **32**:972 (1960).
[‡‡] R. G. Bessent, W. Hayes, and J. W. Hodby, *Proc. Roy. Soc. (London)*, **A297**:376 (1967).

2. X irradiation causes the loss of an electron from H⁻, followed by ejection of the hydrogen atom to an interstitial site some distance away from the vacancy. The final environment of the hydrogen atom is unambiguously indicated by the spectrum shown in Fig. 8-9a. The large hydrogen-atom doublet is further split into a nine-line pattern by eight equivalent fluorine nuclei. This pattern indicates that the hydrogen atom lies at the center of a cube of fluoride ions. This simple spectrum is obtained if the field is oriented along a [100] direction of the cubic crystal so that all hydrogen-fluorine axes make the same angle with the magnetic field. The small lines between the major lines arise from "forbidden" transitions among the energy levels of the hydrogen atom. (See Sec. 7-8.) Interpretation of the spectrum in Fig. 8-9b for **H** ‖ [110] is left as Prob. 8-7.

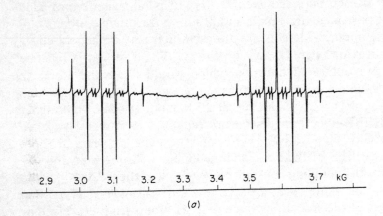

(a)

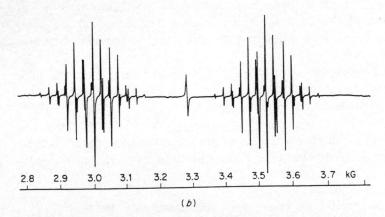

(b)

Fig. 8-9 ESR spectrum of interstitial hydrogen atoms in x-irradiated CaF₂. (a) **H** ‖ [100] in the (001) plane. ' The weak lines are "forbidden" transitions analogous to transitions b and c of Fig. 7-7a. (b) The hydrogen atom in CaF₂ for **H** ‖ [110] in the (001) plane. (The central line is due to DPPH used as a g marker.) [*Taken from J. L. Hall and R. T. Schumacher, Phys. Rev.,* **127**:1892 (1962).]

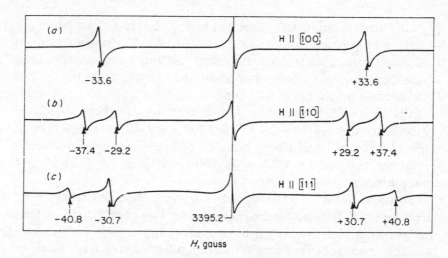

Fig. 8-10 ESR spectrum of nitrogen atoms in diamond at 295 K. (*a*) **H** ‖ [100]. (*b*) **H** ‖ [110]. (*c*) **H** ‖ [111]. [*Taken from W. V. Smith, P. P. Sorokin, I. L. Gelles, and G. J. Lasher, Phys. Rev.,* **115**:1546 (1959).]

The U_2 center in the alkali halides consists of a hydrogen atom simultaneously surrounded by a tetrahedron of halogen ions and a tetrahedron of alkali ions. The ESR spectrum consists of two sets of thirteen lines separated by about 500 G.† For **H** ‖ [100], the halide ("superhyperfine") couplings are 45 MHz in NaCl, 25 MHz in KCl, and 133 MHz in KBr‡; the alkali cation splittings are not resolved.

Nitrogen atoms have been studied in a variety of hosts.§ One of the most interesting of these is the type 1 diamond for which spectra are given in Fig. 8-10.¶ The interpretation of these spectra is given as Prob. 8-8.

There are a wide variety of substitutional paramagnetic impurity defects. A few of these have been considered in Sec. 8-4. Phosphorus, arsenic, antimony, and bismuth in silicon act as electron donors, since they bring one more electron per atom than the host. Boron, aluminum, gallium, and indium serve as electron acceptors. The donors have been extensively studied both by ESR and by ENDOR techniques (Chap. 13). The donor nucleus gives rise to hyperfine splitting. The hydrogenic wave function of the donor electron can be mapped in detail by observation of ^{29}Si splittings in successive shells of neighbors.†† In fact, this was the first major problem attacked by the ENDOR technique.

† C. J. Delbecq, B. Smaller, and P. H. Yuster, *Phys. Rev.*, **104**:599 (1956).
‡ F. Kerkhoff, W. Martienssen, and W. Sander, *Z. Physik*, **173**:184 (1963).
§ T. Cole and H. M. McConnell, *J. Chem. Phys.*, **29**:451 (1958); C. K. Jen, S. N. Foner, E. I. Cochran, and V. A. Bowers, *Phys. Rev.*, **112**:1169 (1958); D. W. Wylie, A. J. Shuskus, C. G. Young, O. R. Gillian, and P. W. Levy, *Phys. Rev.*, **125**:451 (1962).
¶ W. V. Smith, P. P. Sorokin, I. L. Gelles, and G. J. Lasher, *Phys. Rev.*, **115**:1546 (1959).
†† G. Feher, *Phys. Rev.*, **114**:1219 (1959).

The O_2^- ion in KCl, depicted in Fig. 7-3, shows a 13-line spectrum of relative intensities $1:4:10:20:31:40:44:40 \cdots$ when the magnetic field is perpendicular to the molecular axis; an analogous result is found for the rubidium halides.† These hyperfine lines arising from the four equivalent ^{39}K nearest neighbors in KCl confirm the orientation of the $p\pi$ orbital shown in Fig. 7-3. This conclusion derived from an interpretation of the principal g components $g[1\bar{1}0] = 2.4359$, $g[110] = 1.9512$, and $g[001] = 1.9551$. A similar orientation of the $p\pi$ orbital is found in the rubidium halides. However, for the sodium halides the $p\pi$ orbital is found to lie along the [100] direction.

Some substitutional impurities may be detected from an ESR spectrum even if they are diamagnetic. The V_{OH} center cited in Sec. 7-5 is a simple example. Another is the F_A center in the alkali halides. In KCl, if a sodium atom substitutes for potassium adjacent to a trapped electron in an anion vacancy, the linewidth is increased from 46 to 71 G. However, for Li in KCl, the ENDOR technique serves to verify in detail the identity and location of the adjacent nucleus. The effect of alkaline-earth cations in alkali halides can also be detected by the ENDOR technique. Aluminum in quartz was one of the very early examples of a nucleus detected as a neighbor to a paramagnetic center.‡

Numerous transition-metal or rare-earth ions as substitutional impurities in a large number of hosts have been the subjects of a formidable array of investigations. The ions Cr^{3+}, Mn^{++}, and Fe^{3+} may easily be detected in a large number of single crystals. Since these systems involve numerous complications not yet considered, their treatment even at the simplest level is deferred to Chaps. 11 and 12.

8-5c. Trapped-electron centers

A single electron trapped at an anion vacancy is usually referred to as an *F center*. The term "*F center*" represents a partial translation of the original *Farbzentrum*, a name assigned to centers which in the alkali halides are associated with characteristic visible absorption bands.

The *F* centers in alkali halides have been the most intensively studied. Figure 8-11 shows the ESR spectrum of the *F* center in NaH, which may

† W. Känzig and M. H. Cohen, *Phys. Rev. Letters*, **3**:509 (1959); W. Känzig, *J. Phys. Chem. Solids*, **23**:479 (1962).

‡ J. H. E. Griffiths, J. Owen, and I. M. Ward, in Reports Conference on Defects in Crystalline Solids, p. 81, Bristol, 1954, The Physical Society, London, 1955.

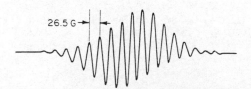

26.5 G

Fig. 8-11 ESR spectrum of *F* centers in NaH, showing a well resolved 19-line spectrum at 77 K. The microwave frequency was 9.1539 GHz. [*Taken from W. T. Doyle and W. L. Williams, Phys. Rev. Letters*, **6**:537 (1961).]

be regarded as a pseudo alkali halide.[†] The 19 lines with intensity ratios $1:6:21:56:120:216:336:456:546:580:546 \cdots$ arise from hyperfine interaction with the six equivalent nearest-neighbor ^{23}Na ions ($I = \frac{3}{2}$). For other alkali halide F centers (except for LiF, NaF, RbCl, and CsCl) the ESR spectra have such a large number of overlapping hyperfine components that only their envelope is observed. The ENDOR technique (see Chap. 13) permits the resolution of hyperfine components from nuclei out to the eighth shell of neighbors in LiF.[‡] Hyperfine-coupling data from nuclei distributed some distance from the F center provide a detailed description of the spatial distribution of the wave function of the trapped electron. Problem 8-9 considers the number of possible lines in NaF, in which essentially all of the cation or anion hyperfine couplings arise from single nuclides. The abundances of ^{39}K, ^{41}K, ^{35}Cl, and ^{37}Cl in KCl are such as to increase greatly the complexity of its F-center spectrum.

An electron trapped at an anion vacancy in an alkaline-earth oxide such as MgO (rock salt structure) may also be called an F center. Owing to the divalence of the ions, the trapped electron finds itself in a much deeper potential well than in the alkali halides. Hence, the wave function of the electron (as measured by hyperfine splitting of neighbors) is far more localized than in the latter. Problem 8-10 concerns itself with the special features of this spectrum. F centers have been observed in all the other oxides, sulfides, and selenides of the alkaline-earth metals, as well as in such salts as NaN_3,[§] CaF_2, SrF_2, BaF_2,[¶] BeO and ZnO.[††]

Clustering of anion vacancies in the alkali halides gives rise to F-aggregate centers. The M center consists of two anion-vacancy-plus-electron centers along an [011] or equivalent direction. Interaction of the two electrons gives rise to a diamagnetic ($S = 0$) ground state plus an excited triplet state ($S = 1$). (See Sec. 10-5.) The R center consists of three adjacent anion vacancies in a (111) plane; each vacancy holds one electron. The ground state is an orbitally degenerate doublet ($S = \frac{1}{2}$), and observation of the ESR spectrum requires low temperatures. There is also an excited quartet state ($S = \frac{3}{2}$) which can be detected by its ESR spectrum during optical irradiation. The R center is discussed further in Sec. 10-12.

8-5d. Trapped-hole centers

Trapped-hole centers are electron-deficient centers formed by the removal of electrons. Removal of an electron from

† W. T. Doyle and W. L. Williams, *Phys. Rev. Letters*, **6**:537 (1961).

‡ W. C. Holton and H. Blum, *Phys. Rev.*, **125**:89 (1962).

§ G. J. King, F. F. Carlson, B. S. Miller, and R. C. McMillan, *J. Chem. Phys.*, **34**:1499 (1961); **35**:1441 (1961).

¶ J. Arends, *Phys. Status Solidi*, **7**:805 (1965); W. Hayes, in "Paramagnetic Resonance," p. 271, C. K. Coogan, N. S. Ham, S. N. Stuart, J. R. Pilbrow, and G. V. H. Wilson (eds.), Plenum Press, New York, 1970.

†† R. C. DuVarney, A. K. Garrison, and R. H. Thorland, *Phys. Rev.*, **188**:657 (1969); J. M. Smith and W. E. Vehse, *Phys. Letters*, **31A**:147 (1970).

an anion leaves behind a net positive charge ("positive hole," or simply "hole").† In an otherwise perfect lattice which does not relax so as to trap the hole, the latter will migrate freely until it is trapped at an impurity atom of variable valence or at a site of net negative charge such as a cation vacancy.

A hole trapped at the latter site has been called a V_1 center. Although this defect structure was postulated for the alkali halides, no centers of such geometry have yet been found in them.‡ However, V_1 centers are a characteristic defect in irradiated MgO or CaO crystals. This defect is cited as an example of tetragonal symmetry in Sec. 7-2.

In the alkali halides, the loss of an electron from a halogen ion leads to a lattice relaxation such that the halogen atom (X) associates with one neighboring X^- anion along a [110]-type direction. The defect should then be considered to be an X_2^- molecule. This V_K center is shown in Fig. 8-7 and was referred to as an inorganic radical in Sec. 8-4. An X_2^- molecule may also occupy a *single* anion site, in which case it will interact strongly with its nearest anion neighbors along its <110>-type axis; this so-called H center may be regarded as a linear X_4^{3-} ion. The outer atoms contribute a secondary splitting about one-tenth that of the primary splitting; the spin density on the outer pair is 0.04 to 0.10.§

In this chapter it has been possible to present only a brief survey of a rapidly expanding research area. We have tried to give examples of the geometric and electronic structural information which may be obtained from a study of paramagnetic species in the solid state. For a more detailed coverage of other radicals and defects, the reader is urged to consult the following references.

REFERENCES

Organic radicals

Morton, J. R.: *Chem. Rev.*, **64**:453 (1964).

Inorganic radicals

Atkins, P. W. and M. C. R. Symons: "The Structure of Inorganic Radicals," Elsevier Publishing Company, Amsterdam, 1967.

Point defects

Alkali halides

Seidel, H., and H. C. Wolf: *Phys. Status Solidi*, **11**:3 (1965) (in German); J. J. Markham, F-Centers in Alkali Halides, chaps. 6 to 8, suppl. 8 of "Solid State Physics," F. Seitz and D. Turnbull (eds.), Academic Press Inc., New York, 1966.

† Note that it is essential to refer to the site of a *missing atom* or *ion* as a *vacancy*.
‡ The geometry of the defect corresponding to "V_1" optical band in alkali halides is totally different. (See the second footnote on page 132.)
§ W. Känzig and T. O. Woodruff, *J. Phys. Chem. Solids*, **9**:70 (1958).

Oxides

Henderson, B. and J. E. Wertz: *Advan. Phys.*, **17**:749 (1968).

Semiconductors

Ludwig, G. W. and H. H. Woodbury: in F. Seitz and D. Turnbull (eds.), "Solid State Physics," vol. 13, Academic Press Inc., New York, 1962.
Feher, G.: in W. Low (ed.), "Paramagnetic Resonance," p. 715, Academic Press Inc., New York, 1963.

PROBLEMS

8-1. In the radical

$$HOOC—CH_2—CH—COOH,$$

the two isotropic β-proton hyperfine couplings are found to be $+100$ and $+80$ MHz. Assume $A_1 = 0$ and $A_2 = 121$ MHz, and calculate the dihedral angle of each β proton relative to an axis passing through the β-carbon atom and perpendicular to the radical plane. (See Fig. 8-4.)

8-2. The following hyperfine coupling tensors have been measured for the $^-O_3S—\dot{C}H—SO_3^-$ radical

	X	Y	Z	
A_α^H	-95	-28	-57	MHz
A^C	$+62$	$+56$	$+260$	MHz

where the axis system is the same as in Fig. 8-2.[†]
(*a*) Compute ρ_C from the isotropic and anisotropic proton hyperfine couplings.
(*b*) Compute ρ_C from the ^{13}C anisotropic hyperfine coupling.

8-3. The radical

has the following principal values of the hyperfine tensors[‡]

	X	Y	Z	
A_α^H	-96	-31	-63	MHz
A^F	-11	-45	$+530$	MHz

where the last value corresponds to the p-orbital direction in each case.
(*a*) Compute the isotropic hyperfine couplings and the anisotropic tensors.
(*b*) Compute ρ_C and ρ_F from the anisotropic tensors.
(*c*) How would you explain the significant spin density on fluorine in terms of the organic chemist's view of the interaction of a halogen with a conjugated system?

† A. Horsfield, J. R. Morton, J. R. Rowlands, and D. H. Whiffen, *Mol. Phys.*, **5**:241 (1962).
‡ R. J. Cook, J. R. Rowlands, and D. H. Whiffen, *Mol. Phys.*, **7**:31 (1963).

8-4. γ irradiation of glycine ($NH_3^+CH_2COO^-$) enriched in ^{13}C yields spectra from which the following ^{13}C hyperfine tensor may be obtained relative to the a^*bc axes of the monoclinic crystal. (a^* is a direction perpendicular to the bc crystal plane.) The radical formed is $NH_3^+\dot{C}HCO_2^-$.[†]

$$\mathbf{A}^C = \begin{bmatrix} 55.3 & \pm 38.2 & 33.2 \\ \pm 38.2 & 225.7 & \pm 48.7 \\ 33.2 & \pm 48.7 & 99.2 \end{bmatrix} \quad \text{MHz}$$

The direction-cosine matrix relative to the principal axes of this tensor is

$$\begin{bmatrix} 0.902 & \mp 0.080 & -0.424 \\ 0.360 & \mp 0.404 & 0.841 \\ 0.239 & \pm 0.911 & 0.335 \end{bmatrix}$$

(a) Diagonalize the ^{13}C hyperfine tensor.

(b) Determine the angle that the axis of the carbon $2p_z$ orbital makes relative to the a^*bc axes.

8-5. The following table gives the principal values of the hyperfine tensors for ^{129}Xe and ^{19}F in XeF and KrF. (The XeF spectrum is given in Fig. 1-4.)[‡]

Radical	$A_\parallel^{Xe}$, MHz	$A_\perp^{Xe}$, MHz	$A_\parallel^F$, MHz	$A_\perp^F$, MHz
^{129}XeF	2,368	1,224	2,637	526
KrF	—	—	3,531	759

(a) Using the data from Table C, compute the s- and p-spin densities. Comment on the nature of the orbital of the unpaired electron.

(b) Compare the fluorine p-spin density for KrF and XeF. Interpret the difference in terms of the relative electronegativity of Kr and Xe. (Note: The molecular orbital containing the unpaired electron is antibonding.)

8-6. γ Irradiation of single crystals of $NH_4H_2PO_2$ yields a radical with the following principal values of hyperfine and g tensors.[§]

	Principal values	Direction cosines		
$\mathbf{A}^I$, MHz	1698	± 0.54	0.84	0.00
	1228	∓ 0.84	0.54	0.00
	1228	0.00	0.00	1.00
$\mathbf{A}^{II}$, MHz	238	∓ 0.50	0.87	0.00
	227	± 0.87	0.50	0.00
	224	0.00	0.00	1.00
g	2.0019	± 0.54	0.84	0.00
	2.0035	∓ 0.84	0.54	0.00
	2.0037	0.00	0.00	1.00

Each of the hyperfine tensors corresponds to the coupling of a single nucleus with $I = \frac{1}{2}$. The crystal is orthorhombic, and the direction cosines are specified relative to the abc axes of the crystal.

[†] J. R. Morton, *J. Am. Chem. Soc.*, **86**:2325 (1964).
[‡] J. R. Morton and W. E. Falconer, *J. Chem. Phys.*, **39**:427 (1963).
[§] J. R. Morton, *Mol. Phys.*, **5**:217 (1962).

(a) On the basis of the magnitude of the hyperfine couplings and the degree of anisotropy, assign specific nuclei to the hyperfine tensors.

(b) Determine the orientation of the symmetry axis of the radical in the abc axis system. Consider the g tensor to be axial.

(c) Compare the direction-cosine matrices for the hyperfine and g tensors. What does this comparison require concerning the location of the two interacting nuclei relative to the principal axes of the g tensor?

(d) Comment on the probable identity and geometric configuration of the radical. Verify your answer by consulting J. R. Morton, *Mol. Phys.*, **5**:217 (1962).

8-7. (a) In Fig. 8-9a there are a number of weak forbidden lines between the strong lines. Use the concepts of Sec. 7-8 and Fig. 7-7 to explain the existence of these lines.

(b) Given the hyperfine couplings $A_\parallel^F = 173.8$ MHz and $A_\perp^F = 69.0$ MHz, make an assignment of the lines in Fig. 8-9b.

(c) Compute the relative intensities of the forbidden lines, and compare with experiment.

8-8. The nitrogen spectrum in type 1 diamond, shown in Fig. 8-10 for several orientations of the magnetic field, is a beautifully simple example of symmetry-equivalent hyperfine axes in a crystal. Here $g = 2.0024 \pm 0.0005$.

(a) In a crystal of the diamond structure (tetrahedral array of atoms), which sets of directions make an equal angle with the field **H** when it is along a [100] direction? (Regard the tetrahedron as inscribed in a cube for the sake of seeing the axis system more clearly. Figure 11-5 shows such an inscribed tetrahedron. It will be helpful to construct a model of this system.) What are then the principal axes of the nitrogen hyperfine interaction in diamond?

(b) Show that the axes selected in part a are consistent with the number and the relative intensity of the hyperfine components in the three spectra of Fig. 8-10.

(c) Obtain an appropriate relation for the angular dependence of the hyperfine splittings of Fig. 8-10. (See Sec. 7-7.) Check the applicability of this relation and obtain the isotropic and anisotropic splitting constants a and b.

(d) Convert a and b to coupling constants $A_\parallel$ and $A_\perp$. Apply these to the relations

$$A_\parallel = A_0 + 2B \qquad A_0 = \frac{8\pi}{3}\,h^{-1}g\beta g_N\beta_N\,|\psi(0)|^2$$

and

$$A_\perp = A_0 - B \qquad B = h^{-1}g\beta g_N\beta_N\,\left\langle\frac{z^2 - \frac{1}{2}(x^2 + y^2)}{r^5}\right\rangle$$

to determine numerical values of A_0 and B.

What is the density of the unpaired electron at the nitrogen nucleus? Compare the values given here with the coupling constants for atomic nitrogen (Table C) to give an approximate description of the state of the unpaired electron.

What other experimental data would you ask for to give a more detailed answer to the previous question?

8-9. Figure 8-12 shows the successive shells of neighboring ions about an electron in a negative-ion vacancy in an alkali halide crystal having the rock salt structure.

(a) For such an F center in NaF, calculate the intensity distribution of hyperfine lines for **H** ∥ [111] if only the six first-shell neighbors interact significantly.

(b) What further splittings result from the 12 second-shell nuclei?

(c) Make an accurate line plot of the ESR spectrum of the F center in NaF, taking $a^{Na} = 105.6$ G for first-shell nuclei and $a^F = 61.6$ G for second-shell nuclei.

(d) Formulate conditions regarding the relative magnitudes of hyperfine splittings from three successive shells of neighbors such as to permit full resolution of hyperfine lines without overlapping of groups.

8-10. For the F center in MgO the relative abundance of ^{25}Mg with $I = \frac{5}{2}$ is 10.05 percent; the remaining magnesium nuclei are ^{24}Mg or ^{26}Mg with $I = 0$.

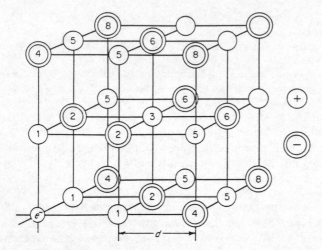

Fig. 8-12 Successive shells of neighbors of an F center in a crystal of the rock-salt structure. The shell number represents the sum of the squares of the Miller indices of a point relative to the origin at which an electron is trapped.

(a) Show that ~36 percent of the F centers will have one ^{25}Mg nucleus as a nearest neighbor.

(b) If the field **H** makes an angle θ with the $e^- - {}^{25}Mg$ axis, the splitting in gauss is given by

$$\Delta H = 3.94 + 0.48\,(3\cos^2\theta - 1)$$

Make a stick plot of the spectrum for $H \parallel [001]$, taking into account the several $e^- - {}^{25}Mg$ orientations. Show (with the correct intensity) the line arising from F centers with only ^{24}Mg or ^{26}Mg neighbors.

(c) The percentage of MgO F centers having two ^{25}Mg neighbors is ~10 percent. What is the maximum possible number of hyperfine components from such centers?

(d) Using an intensity scale 10 times as great as that for part b of this problem, make a stick plot of the hyperfine spectrum of F centers with two ^{25}Mg neighbors for $H \parallel [111]$.

(e) From the isotropic portion of the splitting, calculate the percentage s character of the trapped-electron wave function on each ^{25}Mg atom.

8-11. (a) Taking the relative abundances of the ^{35}Cl and ^{37}Cl nuclides as $3:1$, show that the relative abundances for the three Cl_2^- species are as follows:

$^{35}Cl—{}^{35}Cl^-$ 56.25 percent
$^{35}Cl—{}^{37}Cl^-$ 37.50 percent
$^{37}Cl—{}^{37}Cl^-$ 6.25 percent

(b) Enumerate the various spin states of a $^{35}Cl—{}^{37}Cl^-$ molecule. Are there any degeneracies? What should be the ESR spectrum of $^{35}Cl—{}^{37}Cl^-$?

(c) Using transparent graph paper overlaid on Fig. 8-7, indicate with lines the three separate Cl_2^- spectra.

8-12. Irradiation of LiF is said to give rise to the ion F_3^{--}; two hyperfine tensors are given, the first referring to the central fluorine atom[†]:

$$\begin{bmatrix} 715 & 0 & 0 \\ 0 & 3,146 & 0 \\ 0 & 0 & 504 \end{bmatrix} \quad \begin{bmatrix} 224 & 0 & 0 \\ 0 & 1,060 & 0 \\ 0 & 0 & 398 \end{bmatrix}$$

[†] M. H. Cohen, W. Känzig, and T. O. Woodruff, *J. Phys. Chem. Solids*, **11**:120 (1959).

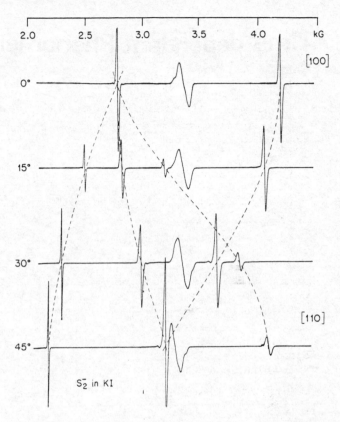

Fig. 8-13 ESR spectra of the S_2^- ion in KI at 4 K. with $\nu = 9.390$ GHz, for four orientations of the magnetic field in a (100) plane. Here 0° corresponds to $\mathbf{H} \parallel [100]$. [*Figure kindly supplied by Dr. J. R. Morton. See L. E. Vanotti and J. R. Morton, Phys. Rev.*, **161**:282 (1967).]

Sketch the spectrum for $\mathbf{H} \parallel X$, Y, and Z in succession. Draw conclusions about the structure of the ion radical. The Z axis will be taken as the axis of the linear F_3^{--} ion.

8-13. ESR spectra of the S_2^- ion in KI at 4 K are given at four orientations in a (100) plane in Fig. 8-13.† Here θ is the angle between the $\langle 100 \rangle$ axis and the magnetic field direction. The broad line near 3.4 kG arises from a different defect. The positions of the lines at $\theta = 0°$ with $\nu = 9.390$ GHz are approximately 2,732 and 4,128 G, whereas for $\theta = 45°$, they are at about 2,190, 3,221, and 4,099 G. The linewidths are highly anisotropic because of unresolved hyperfine interaction with host nuclei. At arbitrary orientations, six lines are observed.

   (*a*) From the number of lines and from their coincidences, determine the orientation of the S_2^- ion, i.e., specify the directions of the principal axes.

   (*b*) What are the principal g components?

(Note: The π-orbital direction is that corresponding to the minimum value of g. In KI, KBr, and NaI this orientation is identical. For KCl, RbBr, and RbI the orbital lies along a [110] direction as in the case of O_2^-.)

† L. E. Vannotti and J. R. Morton, *Phys. Rev.*, **161**:282 (1967).

9
Time-dependent Phenomena

9-1 INTRODUCTION

It is a common misconception that an ESR spectrum such as that of the naphthalene anion (Fig. 4-16) represents the spectrum that would be observed for a single radical ion. This is not true, since if the signal from a single naphthalene anion could be detected, it would consist of but *one* line! For example, the line occurring at the highest field in Fig. 4-16 corresponds to all eight protons having spin α (that is, $M_I = \frac{1}{2}$). Furthermore, if enough different molecules could be observed in succession, eventually all of the lines in the spectrum would be obtained. The observed ESR spectrum represents a statistical average over the ensemble of radicals. The fact that one line in the spectrum is four times as intense as another means that four times as many radicals are resonating at the field of the former line as at the field of the latter. It is tacitly assumed that all the radicals are completely independent and noninteracting. Such would be the case only at infinite dilution. This chapter is concerned with some of the effects of radicals interacting magnetically and chemically with each other and with their environment. The principal effect of these interactions is to give a finite width to the lines in the ESR spectrum. An analysis of linewidths in these spectra can give important information concerning time-dependent phenomena in solution. However, to understand the causes of line broadening one must first understand relaxation processes.

9-2 SPIN–LATTICE RELAXATION TIME

Although an isolated electronic magnetic dipole does not behave completely classically in a magnetic field (that is, μ can never be parallel to **H**), one may compute a resultant magnetic moment per unit volume $\mathscr{M}$ by taking the vector sum of all the individual magnetic moments. At equilibrium, there is a net magnetic moment only in the field direction (z direction) since the x and y components of μ average to zero over the spin ensemble.

It is of some interest to determine how the magnetization $\mathscr{M}$ approaches its equilibrium value when one makes a sudden change in the magnetic field. The situation is analogous to temperature-jump experiments in which a chemical system initially at equilibrium is subjected to a sudden increase in temperature. In this case one usually finds that the approach to the new equilibrium follows first-order kinetics. The approach of $\mathscr{M}$ to a new equilibrium value $\mathscr{M}^0$ behaves similarly. Thus

$$\frac{d\mathscr{M}_z}{dt} = -k(\mathscr{M}_z - \mathscr{M}_z{}^0) \tag{9-1}$$

Here $\mathscr{M}_z$ is used in place of $\mathscr{M}$ because $\mathscr{M}$ is parallel to **H**. The solution of this and of similar equations was first given by F. Bloch.[†] The solution to Eq. (9-1) is

$$\mathscr{M}_z = \mathscr{M}_z{}^0(1 - e^{-kt}) = \mathscr{M}_z{}^0(1 - e^{-t/T_1}) \tag{9-2}$$

Here $k = 1/T_1$, where T_1 is the time required for $\mathscr{M}_z$ to rise to within a fraction $1/e$ of the value $\mathscr{M}_z{}^0$.[‡] Equation (9-2) applies when $\mathscr{M}_z = 0$ at $t = 0$. Figure 9-1 depicts the behavior of $\mathscr{M}_z$ when the magnetic field is suddenly increased from zero to a value H'.

[†] F. Bloch, *Phys. Rev.*, **70**:460 (1946).
[‡] The reader may wince at using a capital T for time. This usage is universal and owes its origin to the great classic paper by F. Bloch, just cited.

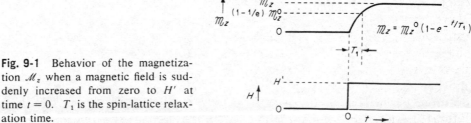

Fig. 9-1 Behavior of the magnetization $\mathscr{M}_z$ when a magnetic field is suddenly increased from zero to H' at time $t = 0$. T_1 is the spin-lattice relaxation time.

Equation (9-2) can be derived as follows: As seen earlier, the application of a magnetic field to a system of unpaired electrons results in a splitting of the energy levels into two components given by

$$W = \pm\tfrac{1}{2}g\beta H \tag{9-3}$$

as shown in Fig. 9-2. Here n_α and n_β are the populations of the upper and lower levels, respectively, and P. and $P_{\downarrow}$ correspond to the respective upward and downward lattice-induced transition probabilities. These are called lattice-induced transitions because it is assumed that the spin system is coupled in some way to the bulk sample or "lattice." For the spin system to be in thermodynamic equilibrium with the lattice, it must be at the same temperature, which for isolated spins† corresponds to that given by the Boltzmann distribution

$$\frac{n_\alpha}{n_\beta} = e^{-\Delta W/kT} = e^{-g\beta H/kT} \tag{9-4}$$

Let

$$N = (n_\beta + n_\alpha)$$

$$n = (n_\beta - n_\alpha)$$

$$n_0 = (n_\beta - n_\alpha) \text{ at equilibrium}$$

Also

$$n_\alpha = \tfrac{1}{2}(N - n)$$

and

$$n_\beta = \tfrac{1}{2}(N + n)$$

Since it has been assumed that the spins are isolated, it is expected that the rate of change of n will follow first-order kinetics; that is,

$$\frac{dn}{dt} = -2n_\beta P_{\downarrow} + 2n_\alpha P_{\downarrow} \tag{9-5}$$

† Strictly speaking, electrons obey Fermi-Dirac statistics, but when interactions between electron spins are weak, it can be shown that Boltzmann statistics apply.

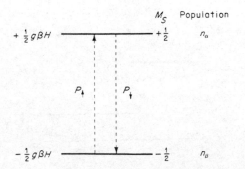

Fig. 9-2 Electron spin energy levels in a magnetic field H. $P_{\uparrow}$ and $P_{\downarrow}$ refer to the lattice-induced transition probabilities.

The factor 2 appears because an upward or downward transition changes n by 2. Equation (9-5) can be rewritten as

$$\frac{dn}{dt} = N(P_\downarrow - P_\uparrow) - n(P_\downarrow + P_\uparrow) = (P_\downarrow + P_\uparrow)\left(N\frac{P_\downarrow - P_\uparrow}{P_\downarrow + P_\uparrow} - n\right)$$

(9-6)

At the steady state, $dn/dt = 0$; hence, from Eq. (9-6)

$$n_0 = N\left(\frac{P_\downarrow - P_\uparrow}{P_\downarrow + P_\uparrow}\right)$$

(9-7)

Equation (9-6) then becomes

$$\frac{dn}{dt} = (n_0 - n)(P_\downarrow + P_\uparrow)$$

(9-8)

$1/(P_\downarrow + P_\uparrow)$ has the dimensions of time, and it is designated as T_1. The mean spin lifetime is then $2T_1$.

$$\frac{dn}{dt} = -\frac{n - n_0}{T_1}$$

(9-9)

Since $\mathcal{M}_z$ is proportional to n, Eq. (9-9) can be rewritten as

$$\frac{d\mathcal{M}_z}{dt} = -\frac{\mathcal{M}_z - \mathcal{M}_z^0}{T_1}$$

(9-10)

This is the same as Eq. (9-1).

At this point it is profitable to examine the significance of T_1. Since $\mathcal{M}_z$ is the sum of all μ_z components per unit volume, $\mathcal{M}_z$ can change only if some of the dipoles change their spin state, corresponding to a change in M_S. This can be accomplished (in the absence of an external radiation field) only by an exchange of energy with the environment (lattice). This exchange of energy is accomplished by lattice-induced transitions between spin levels. T_1 is the time which characterizes this "spin-lattice" interaction and hence is called the "spin-lattice relaxation time."

From the definition of T_1 as the inverse of the sum of the transition probabilities, it is clear that T_1 is characteristic of the mean lifetime of a given spin state. Consider the Heisenberg uncertainty principle written in terms of uncertainties in the energy and in the time,

$$\Delta W\, \Delta t \geqslant \frac{h}{2\pi}$$

(9-11)

If T_1 is used as an estimate of Δt, then a small T_1 value will lead to a large ΔW and hence to a smearing out of the energy levels. There will therefore be a broadening of the lines observed in the ESR spectrum. For example, if

$T_1 = 10^{-9}$ s, then $\Delta W \simeq 10^{-18}$ erg or $\Delta \nu \simeq 2 \times 10^8 \, s^{-1}$. This corresponds to a linewidth of about 60 G. For most transition-metal ions the linewidth is governed exclusively by their small values of T_1.

9-3 OTHER SOURCES OF LINE BROADENING

For stable free radicals the broadening due to a small value of T_1 is relatively unimportant, and hence other line-broadening mechanisms must be considered.† These may be divided into two major groups.

9-3a. Inhomogeneous broadening In this case the unpaired electron in various free radicals in the sample is subjected to slightly different effective magnetic fields; hence, at any time only a small fraction of the spins is in resonance as the external magnetic field is swept through the "line." The observed line is then a superposition of a large number of individual components (referred to as "spin packets"), each slightly shifted from the others. The resultant envelope has approximately a gaussian shape. (See Fig. 2-10.) The following are some causes of inhomogeneous broadening:

1. An inhomogeneous magnetic field.
2. Anisotropic interactions in randomly oriented systems in the solid state. Here the distribution of local magnetic fields resulting from the anisotropic g and hyperfine interactions gives rise to the inhomogeneity. In this case the line shape may be highly unsymmetrical. (See Sec. 7-9.)
3. Unresolved hyperfine structure, e.g., a trapped electron in KCl; the number of hyperfine components from nearby nuclei is so great that no structure is observed, i.e., one detects the envelope of a multitude of lines. These may be resolved by the technique of electron-nuclear double resonance (ENDOR, see Chap. 13).

9-3b. Homogeneous broadening In the inhomogeneous case, line broadening arises because the static or the average magnetic field at each dipole is *not* the same. For homogeneous broadening, the *static* plus the *time-average* magnetic fields can be considered to be the same at each dipole, but the instantaneous magnetic field is not. This means that the line shape (i.e., the transition probability as a function of magnetic field) will be the same for each dipole. The resulting line usually has a lorentzian shape. (See Fig. 2-9.) The width of this line is usually much greater than one expects from the value of T_1. However, it is useful to retain the concept of a relaxation time in discussing linewidth. One may define a new relaxation time T_2 based on the width of a normalized line, such as that depicted in Fig. 2-9a or Fig. 2-10a. T_2 may be defined as

† It is assumed that the microwave power level is adequately low, so that there is no broadening from saturation effects. (See Sec. D-2b.)

$$\frac{1}{T_2} = \kappa\gamma_e\Gamma \tag{9-12}$$

where Γ is half the linewidth at half-height, in the absence of microwave power saturation (see Sec. D-2c). κ is a constant which depends on the line shape, and γ_e is the electronic magnetogyric ratio. For lorentzian lines $\kappa = 1$, whereas for gaussian lines $\kappa = (\pi\ln2)^{\frac{1}{2}}.$†

Since $1/T_2$ is a linear function of the linewidth, it encompasses both lifetime broadening (characterized by the value of T_1) and other broadening processes which are usually homogeneous in nature. These other processes may be included by introducing a new relaxation time T_2' such that

$$\frac{1}{T_2} = \frac{1}{T_2'} + \frac{1}{2T_1} \tag{9-13}$$

T_2' is called the *spin-spin relaxation time*. For many systems, especially for stable free radicals, $T_1 >> T_2'$, so that for all practical purposes $T_2 \simeq T_2'$. The factor of 2 appears in Eq. (9-13), since the mean spin lifetime is $2T_1$. [See Eqs. (9-8) and (9-9).] The prime superscript will be deleted in the following discussion where various contributions to T_2' are considered. That is, the linewidth will be considered to be a measure of $1/T_2'.$‡

9-4 MECHANISMS CONTRIBUTING TO LINE BROADENING

9-4a. Electron spin–electron spin dipolar interactions For solids this is probably the most important single contribution to T_2; spin-spin interaction occurs in all but the most dilute samples. Associated with the electron magnetic moment there is a magnetic field. If another unpaired electron is at a distance r, the magnitude of this field at the electron can take any value between $\pm 2\mu_e/r^3$, depending on orientation. [See Eq. (3-2).] Consider a 0.1 mole percent solid solution of a paramagnetic species in a host with lattice spacing of 0.2 nm. At one electron the average dipolar field arising from another electron will be $\sim\pm 2$ G. Hence the deviations in the electron energy levels will result in a linewidth of ~ 4 G. In liquid solutions of low viscosity this effect is usually rather small, and it is masked by electron-exchange effects to be considered in Sec. 9-5b.

9-4b. Electron spin–nuclear spin interactions These interactions are analogous to the electron spin-spin interactions considered above except that the random local magnetic fields are produced by magnetic *nuclei* in the vicinity

---

† An alternative definition for T_2 in the case of gaussian lines sets $1/T_2$ equal to the standard deviation. Under this definition $\kappa = (2 \ln 2)^{\frac{1}{2}}$.

‡ It must be noted carefully that T_2 has been related to the inverse width of a *homogeneously* broadened line. The inverse width of an *in*homogeneously broadened line is not related to any relaxation time; however, the inverse width of each component spin packet should be a measure of T_2 for that packet.

of the radical. (Anisotropic interactions with magnetic nuclei *within* the radical are averaged to zero in a liquid of low viscosity.) These nuclei would typically be those of the host. Suppose that the host contains protons at an average distance of 0.3 nm. At such a distance the magnetic field at the unpaired electron would vary between $\sim\pm 1$ G. When the radical is tumbling slowly, as in a viscous liquid, the anisotropy of g and hyperfine interactions can contribute to T_2. This effect will be considered in more detail in Sec. 9-7.

9-5 CHEMICAL LINE–BROADENING MECHANISMS

9-5a. General model Chemical processes can give rise to ESR line broadening if they alter the magnetic environment of the unpaired electron. If the changes occur slowly, one observes lines assignable to distinct species. However, as the rate increases, the ESR lines broaden and finally coalesce to give single lines, the positions of which are weighted averages of the original line positions. This type of phenomenon was first observed for NMR spectra of interconverting species. The effect was explained by use of a modified form of the Bloch equations.[†],[‡]

Consider a free radical which can exist in two distinct forms (i.e., each has a distinctive ESR spectrum). For the sake of simplicity, consider equal concentrations of each form and assume that each gives rise to a single line in the ESR spectrum. If the rate of interconversion between form A and form B is slow, a simple two-line spectrum can be seen (Fig. 9-3a). As the rate of interconversion is increased, the first effect observed is a broadening of the two lines. The cause of the line broadening is the reduced lifetime of a given form, due to the increasing rate of the interconversion process. The minimum rate required to observe an effect is governed by the magnitude of the linewidth in the absence of interconversion between A and B. For free radicals in solution, the half half-width Γ is typically ~ 0.1 G. From the Heisenberg uncertainty principle, Eq. (9-11), the mean lifetime Δt would have to be approximately 5×10^{-7} s for line broadening to be significant. The relation between the linewidth Γ and the mean lifetime in the limit of slow interconversion is[‡]

$$\Gamma = \Gamma_0 + \frac{1}{2\tau\gamma_e} \tag{9-14}$$

Here Γ_0 is the width in gauss in the absence of interconversion, and 2τ is the mean lifetime of species A or B. It is standard practice to define τ in terms of the mean lifetimes τ_A and τ_B, viz.,

† H. S. Gutowsky, D. W. McCall, and C. P. Slichter, *J. Chem. Phys.*, **21**:279 (1953).

‡ J. A. Pople, W. G. Schneider, and H. J. Bernstein, "High Resolution Nuclear Magnetic Resonance," chap. 10, McGraw-Hill Book Company, New York, 1959.

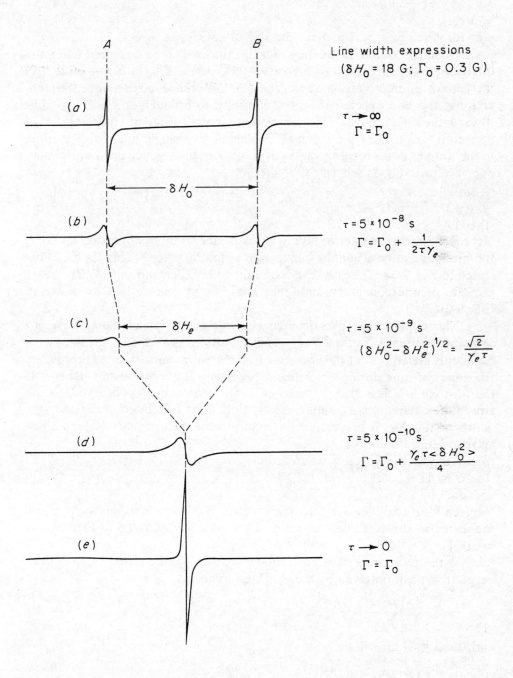

Fig. 9-3 Synthetic first-derivative spectra showing the effect of increasing rate of interconversion between form A and form B of a free radical.

$$\tau = \frac{\tau_A \tau_B}{\tau_A + \tau_B} \tag{9-15}$$

Hence, when $\tau_A = \tau_B$, the mean lifetime of species A or B is 2τ.

Figure 9-3b illustrates the spectrum that would be observed when the interconversion process has increased the linewidth by a factor of 2. As the rate of interconversion approaches the difference in resonance frequencies for the two species, the lines continue to broaden and begin to shift toward their midpoint. This can be anticipated, since at sufficiently high interconversion rates the spectrum coalesces to a single line corresponding to the time-average resonant-field position. For lines which do not originally overlap, this shift is related to the lifetime by[†]

$$(\delta H_0{}^2 - \delta H_e{}^2)^{\frac{1}{2}} = \frac{\sqrt{2}}{\gamma_e \tau} \tag{9-16}$$

Here δH_0 is the line separation in the absence of interconversion and δH_e the line separation when the conversion is taking place. This state is illustrated in Fig. 9-3c. The two lines continue to broaden and shift with increasing rate of interconversion until they coalesce to a single line, as shown in Fig. 9-3d.

The coalescence phenomenon is a manifestation of the uncertainty principle expressed in Eq. (9-11). If one writes this as $\Delta t \, \Delta \nu \approx 1/2\pi$, where $\Delta \nu$ is the separation of the two lines in frequency units, then Δt represents the smallest time during which the states A and B may be distinguished. If the lifetime τ is less than Δt, then only one central line is observed, as the two states cannot be distinguished. For very fast interconversion, the width of this line is proportional to τ and to the mean square of δH_0. The appropriate relation is[‡]

$$\Gamma = \Gamma_0 + \gamma_e \frac{\tau \langle (\delta H_0)^2 \rangle}{4} \tag{9-17}$$

Figures 9-3d and 9-3e illustrate the spectra. As the rate of interconversion increases to the fast interconversion limit, the line narrows to the limiting width Γ_0.

If the two forms have probabilities p_A and p_B, then the lines will converge to a position given by the weighted mean

$$\langle H \rangle = \frac{p_A H_A + p_B H_B}{p_A + p_B} \tag{9-18}$$

and the linewidth will be given by[‡]

$$\Gamma = \Gamma_0 + \gamma_e \tau p_A p_B \langle (\delta H_0)^2 \rangle \tag{9-19}$$

This synthetic example is characteristic of many of the phenomena

† H. S. Gutowsky and C. H. Holm, *J. Chem. Phys.*, **25**:1228 (1956).
‡ G. K. Fraenkel, *J. Phys. Chem.*, **71**:139 (1967).

observed when chemical processes occur at a rapid rate. Specific examples of such processes will now be considered.

9-5b. Electron spin exchange The term *electron spin exchange* is here reserved for a bimolecular reaction in which the unpaired electrons of two free radicals exchange their spin states.[†] If in dilute solution one observes a doublet spectrum (as for example in the case of hyperfine interaction with a single nucleus of spin $\frac{1}{2}$), then Eqs. (9-14), (9-16), and (9-17) are directly applicable. The only modification required is to set the actual electron spin exchange rate to be twice that indicated by the broadened spectra, since an exchange of electron spin states between two radicals with the same nuclear spin state will not change the resonant field.

Spin exchange was first studied for the $NO(SO_3)_2^{--}$ radical.[‡] Here we consider the electron spin exchange for the analogous case of the di-*t*-butyl nitroxide radical.[§] Figure 9-4a displays the spectrum observed at a very low concentration. This spectrum has already been encountered in Fig. 4-25. At a higher concentration in Fig. 9-4b, the lines are clearly broadened. From the additional linewidth one can calculate τ, using Eq. (9-14) multiplied by the statistical factor of $\frac{2}{3}$, since one-third of the encounters result in no field shift.

It is important to note that $1/2\tau$ is the electron spin exchange rate for a given *molecule*. One would expect that the exchange rate should be proportional to the concentration of radicals, and this is found to be the case. The second-order rate constant can be obtained from

$$k_2 = \frac{1}{2\tau[R]} \tag{9-20}$$

where [R] is the radical concentration. In dimethylformamide, $k_2 = 7.5 \times 10^9$ l mole^{-1} s^{-1}.[§] The high value of k_2 indicates that spin exchange must occur with a high probability since this rate constant approximates that of a diffusion-controlled reaction.

As the concentration continues to increase, the lines coalesce to a single line (Fig. 9-4c) which becomes narrower at even higher concentrations (Fig. 9-4d). The spectrum in Fig. 9-4d is often said to be *exchange-narrowed* since the electron spins are exchanging so fast that the time average of the hyperfine field is close to zero. A similar exchange-narrowed spectrum is observed for most pure solid free radicals. Here the strong exchange arises from the overlap of molecular wave functions.

Generally, electron spin exchange is to be avoided if narrow lines are desired. The electron spin exchange effect on linewidths is not the same as

---

[†] Electron exchange is detectable only if the colliding radicals have *different* electron spin states. There is no way of detecting exchanges if their initial spin states are identical.
[‡] J. P. Lloyd and G. E. Pake, *Phys. Rev.*, **94**:579 (1954).
[§] T. A. Miller and R. N. Adams, *J. Am. Chem. Soc.*, **88**:5713 (1966).

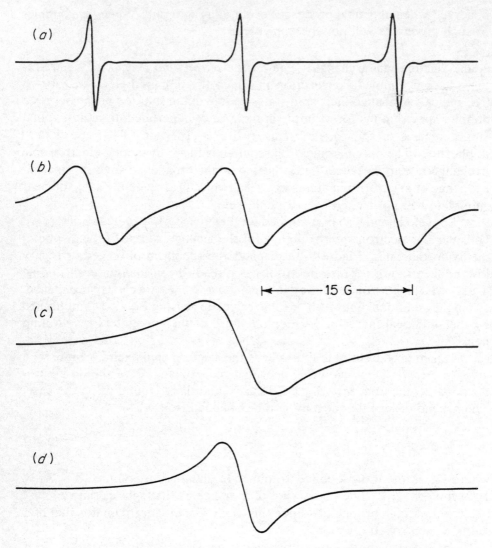

Fig. 9-4 First-derivative spectra of the di-*t*-butyl nitroxide radical in ethanol at room temperature at various radical concentrations. (*a*) $10^{-4}M$. (*b*) $10^{-2}M$. (*c*) $10^{-1}M$. (*d*) Pure liquid nitroxide.

the dipole-dipole effect considered in Sec. 9-4a. Electron spin exchange is a dynamic effect which in liquids produces a much greater broadening than the dipole-dipole effect. This can be shown by the following example. Suppose that the radical concentration is $10^{-3}M$ and the electron spin exchange rate constant is 10^{10} l mole^{-1} s^{-1}. From Eqs. (9-14) and (9-20) one calculates

$$\Gamma - \Gamma_0 = 0.57 \text{ G} \tag{9-21}$$

However, at the same concentration, the dipolar broadening would contribute only ~ 0.01 G to the linewidth.

9-5c. Electron transfer Electron transfer between a radical and a diamagnetic species is very similar to electron spin exchange in its effect on the spectrum. The first such electron transfer reaction studied was that between the naphthalene anion and neutral naphthalene molecules[†]

$$\text{naph}(1)^- + \text{naph}(2) \rightleftarrows \text{naph}(1) + \text{naph}(2)^- \tag{9-22}$$

At the beginning of this chapter the point was made that naphthalene molecules are distinguishable by virtue of the many different arrangements of the proton spins. In fact, the ESR spectrum shows 25 distinct resonant field positions, many of which are degenerate. Thus when an electron transfer reaction occurs, the electron resonance field is usually shifted. If the transfer rate is small compared to the separation between resonance lines (slow-transfer-rate region), the effect will be to cause a broadening of each resonance line in the spectrum. The broadening will, in general, not be the same for each hyperfine component. For example, there are 36 times as many molecules with a resonance field corresponding to the central line as for those molecules with a resonance field corresponding to one of the outermost lines. Since the probability of a jump between molecules with the same resonant field is much greater for molecules contributing to the central line, one might expect that this line would be narrower than lines toward the outsides of the spectrum. This phenomenon has been observed[‡] and indicates that the spin lifetime of an electron (characterized by T_1) is much longer than the average lifetime (2τ) of the electron on a given naphthalene molecule; hence, this broadening mechanism contributes mainly to T_2'. The broadening will be given by Eqs. (9-14) and (9-20). [Note that the concentration of neutral naphthalene must be used in Eq. (9-20).] Measurement of the linewidth as a function of the concentration of neutral naphthalene enables one to obtain the second-order electron transfer rate constant. For the naphthalene anion in tetrahydrofuran,[§] $k_2 = 5.7 \times 10^7$ l mole^{-1} s^{-1}. This value is almost 100 times smaller than the diffusion-controlled rate. Thus

[†] R. L. Ward and S. I. Weissman, *J. Am. Chem. Soc.*, **79**:2086 (1957).
[‡] P. J. Zandstra and S. I. Weissman, *J. Chem. Phys.*, **35**:757 (1961).
[§] T. A. Miller and R. N. Adams, *J. Am. Chem. Soc.*, **88**:5713 (1966).

one concludes that electron transfer occurs in only a small fraction of the collisions of a radical ion with neutral naphthalene molecules. This slow rate may be due to a transfer mechanism which involves the alkali-metal counterion. In a similar system, viz., the electron transfer between benzo-phenone and its anion, the spectrum coalesces to a quartet of equally intense lines at high concentrations of benzophenone.[†,‡] This observation indicates that in this case the transfer process involves an alkali-metal *atom*, instead of a single electron.

9-5d. Proton exchange The previous two examples have considered changes in the magnetic environment due to exchange of electron spin states or to transfer of an unpaired electron from one molecule to another. However, environmental changes can occur if a chemical reaction ex-changes one or more nuclei in the molecule with nuclei in the solvent. Usu-ally such reactions are too slow to have an effect on an ESR spectrum, al-though effects on NMR spectra can be very pronounced. In the case of proton exchange, reaction rates are sometimes large enough to exhibit detectable effects. A good example is that of the $\dot{C}H_2OH$ radical considered in Chap. 4. Figure 9-5 displays the spectrum of the $\dot{C}H_2OH$ radical at various pH values. At high pH, the OH doublet is clearly resolved, but as the pH is lowered, the doublet lines broaden, coalesce, and finally collapse into a single line. In this case the OH proton is rapidly exchanging with H^+ ions in solution. The proton-exchange rate may be computed from the field shift given by Eq. (9-16). The second-order rate constant is 1.76×10^8 l mole^{-1} s^{-1}.[§]

9-6 VARIATION OF LINEWIDTHS WITHIN AN ESR SPECTRUM

The ESR spectra which have been considered up to this point have had all lines of equal width. Hence the peak-to-peak amplitude of each derivative line is proportional to the relative intensity of the corresponding transition. Under certain conditions, linewidths can vary from one line to another in a given spectrum. The most obvious effect is a departure of the proportion-ality between the derivative amplitude and the line intensity, since the de-rivative amplitude is inversely proportional to the *square* of the linewidth. Thus small changes in linewidths can cause large changes in the relative amplitudes of various lines in the spectrum.

Variation in linewidths can arise from time-dependent hyperfine split-tings arising both from chemical processes and from internal rearrangements within a molecule. These variations can also result from the effects of a decrease in the tumbling rate of radicals in solution. This section qualita-tively surveys the types of effects which can be encountered and presents

† S. I. Weissman, *Z. Elektrochem.*, **64**:47 (1964).
‡ F. C. Adam and S. I. Weissman, *J. Am. Chem. Soc.*, **80**:1518 (1958).
§ H. Fischer, *Mol. Phys.*, **9**:149 (1965).

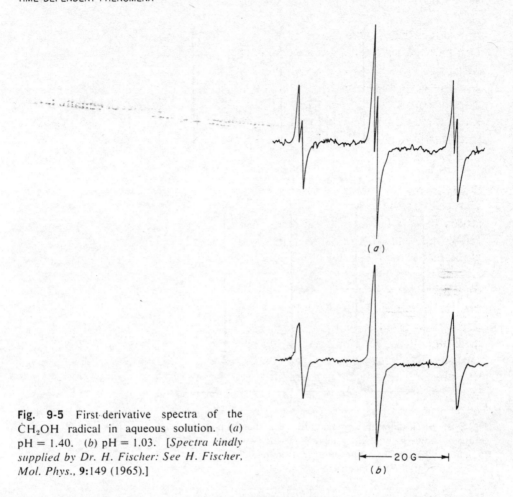

Fig. 9-5 First derivative spectra of the $\dot{C}H_2OH$ radical in aqueous solution. (a) pH = 1.40. (b) pH = 1.03. [*Spectra kindly supplied by Dr. H. Fischer: See H. Fischer, Mol. Phys.,* **9**:149 (1965).]

numerous examples. For a survey of the detailed theory, the reader is referred to a comprehensive review.†

9-6a. Time-dependent hyperfine splitting for a single nucleus Consider a radical which gives rise to hyperfine splitting from a single nucleus with spin I. Suppose that the radical can exist in two forms, A and B, which can interconvert. Let a_1 and a_2 be the hyperfine splittings for A and B, respectively. At slow interconversion rates, two superimposed spectra should be observed. Each spectrum would have $2I + 1$ lines corresponding to the possible values of M_I. In the limit of fast interconversion, a single spectrum would be observed; it would consist of $2I + 1$ lines, with a mean hyperfine splitting given by

$$\langle a \rangle = p_A a_1 + p_B a_2 \tag{9-23}$$

where p_A and p_B are the mole fractions of A and B, respectively.

† G. K. Fraenkel, *J. Phys. Chem.,* **71**:139 (1967).

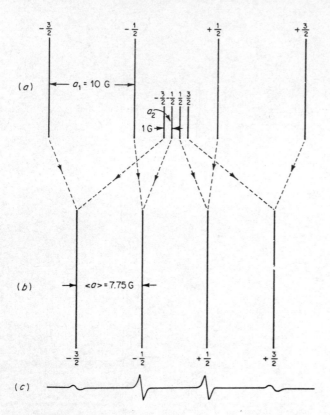

Fig. 9-6 Stick-diagram representation of a spectrum in the limits of slow (*a*) and fast (*b*) exchange for two forms of a radical exhibiting a four-line spectrum from a nucleus of spin $\frac{3}{2}$. (*c*) Simulated ESR spectrum.

Consider a specific example in which $I = \frac{3}{2}$, $a_1 = 10.0$ G, $a_2 = 1.0$ G, $p_A = 0.75$, and $p_B = 0.25$. Figure 9-6*a* displays a stick diagram of the spectrum which would be observed in the region of slow interconversion between A and B. Two distinct four-line spectra are apparent. Figure 9-6*b* displays the average spectrum in the fast region of interconversion. It is important to note that in going from form A to form B, the value of M_I *does not change*. Thus there is a one-to-one correspondence between lines in Fig. 9-6*a* and lines in Fig. 9-6*b*.

In Sec. 9-5*a* it was pointed out that in the fast region of interconversion, the linewidth is given by Eq. (9-19). Referring again to Fig. 9-6, it is clear that the $M_I = \pm\frac{1}{2}$ lines exhibit small shifts on conversion from form A to form B; however, the $M_I = \pm\frac{3}{2}$ lines exhibit much larger shifts. Thus the latter can be expected to be broader than the former. (See Fig. 9-6*c*.) In general, for a nuclear spin I†

$$\Gamma = \Gamma_0 + \gamma_e \tau p_A p_B (a_1 - a_2)^2 M_I^2 \tag{9-24}$$

† G. K. Fraenkel, *J. Phys. Chem.*, **71**:139 (1967).

A good example of this effect is shown by the sodium naphthalenide spectrum† (Fig. 9-7) in a mixed [tetrahydrofuran (25%) and diethyl ether (75%)] solvent. Ion pairing may be inferred, since a marked hyperfine splitting from ^{23}Na ($I = \frac{3}{2}$) is observed. At $-60°$ C, all four lines of a given ^{23}Na multiplet have roughly the same amplitude and hence nearly the same width (Fig. 9-7a). However, as the temperature is lowered, the $M_I = \pm\frac{3}{2}$ lines broaden relative to the $M_I = \pm\frac{1}{2}$ lines (Figs. 9-7b and c). These spectra have been interpreted in terms of a fast equilibrium between two ion pairs, one having a large and the other a small ^{23}Na hyperfine splitting. The differential broadening of the $M_I = \pm\frac{3}{2}$ and the $M_I = \pm\frac{1}{2}$ lines may be used to obtain a value of the rate constant for interconversion, using Eq. (9-24). As the temperature is lowered, the ^{23}Na hyperfine splitting decreases. The relative amounts of the two ion pairs and hence the equilibrium constant at each temperature can be obtained by using Eq. (9-23). Thus thermodynamic as well as kinetic information can be obtained from a study of these effects. (See Prob. 9-3 for a quantitative analysis of this system.)

9-6b. Time-dependent hyperfine splittings for systems with several nuclei
Consider a radical with two nuclei each having a nonzero nuclear spin. If these two identical nuclei have the same *time-average* hyperfine splitting, they are said to be *equivalent*. If they have the same *instantaneous* hyperfine splitting, they are said to be *completely equivalent*. As an example consider the hypothetical *cis*-1,2-dichloroethylene anion forming an ion

† N. Hirota, *J. Phys. Chem.*, **71**:127 (1967).

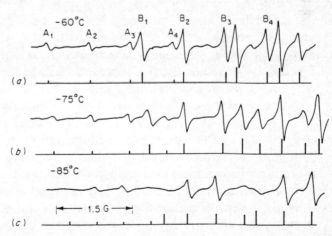

Fig. 9-7 First-derivative spectra of the low-field portion of the sodium naphthalenide spectrum at temperatures $-60°$ C. $-75°$ C, and $-85°$ C. The solvent is a mixture of tetrahydrofuran (25%) and diethyl ether (75%). Lines marked $A_1 \cdots A_4$ and $B_1 \cdots B_4$ are the ^{23}Na quartets for the two outermost proton line components. [*From N. Hirota, J. Phys. Chem.*, **71**:127 (1967).]

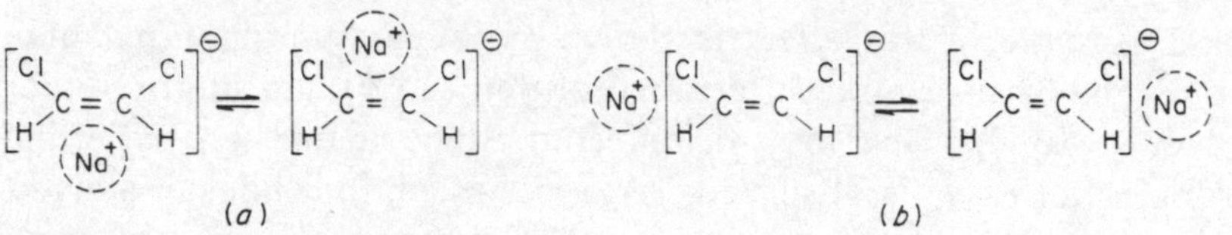

Fig. 9-8 (a) Ion-pair structures with the cation below and above the plane of the hypothetical cis-1,2-dichloroethylene anion. (b) Structures with the cation located at either end of the anion.

pair with Na^+ (Fig. 9-8). Two cases will be distinguished:

1. An in-phase modulation causes the hyperfine splittings to increase or decrease in unison. This is illustrated in Fig. 9-8a. In both structures the two protons are equivalent at any instant. Such nuclei are said to be *completely equivalent*.
2. The modulation of the two hyperfine splittings is exactly out-of-phase, i.e., when one increases the other decreases. This case is illustrated in Fig. 9-8b. The proton closer to the Na^+ will have a hyperfine splitting which is different from that of the other proton. When the Na^+ ion jumps to the other end of the molecule, the hyperfine splittings will be interchanged. On the *average* the hyperfine splittings for the two protons will have the same value. Such nuclei are then *equivalent* (as contrasted with *completely equivalent*).

First consider the case of in-phase modulation of hyperfine splittings (case 1 above). Since the nuclei are completely equivalent, the positions of the spectral lines at all times may be described by the total nuclear spin quantum number $M = \Sigma_i M_i$. This means that the widths of the lines can be treated as if there were one interacting nucleus with a total nuclear spin of $I = \Sigma_i I_i$. Consequently, this represents a case analogous to that given in Sec. 9-6a. That is, the linewidths will vary as M^2 [see Eq. (9-24)].

Next consider the case of out-of-phase modulation of hyperfine splittings (case 2 above). The fact that the two nuclei are not *instantaneously* equivalent leads to an interesting phenomenon in the ESR spectrum. This is commonly referred to as the *alternating-linewidth effect*, which was first observed in the spectrum of the dihydroxydurene cation[†] (Fig. 9-9) and the dinitrodurene anion[‡] (Fig. 9-10). The interpretation of this striking effect is aided by the consideration of the simpler example of two nuclei with $I = 1$. The model assumed is one in which alternately one hyperfine splitting is large, and the other is small. It is further assumed that the radical can exist in two thermodynamically equivalent states, A and B (Fig. 9-11), i.e., states of the same energy.

[†] J. R. Bolton and A. Carrington, *Mol. Phys.*, 5:161 (1962).
[‡] J. H. Freed and G. K. Fraenkel, *J. Chem. Phys.*, 37:1156 (1962).

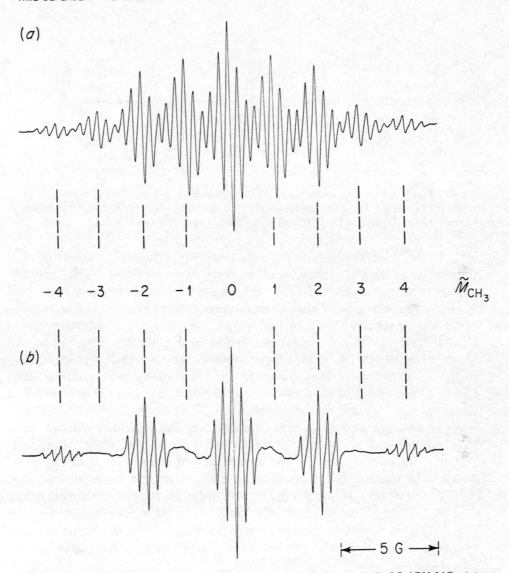

Fig. 9-9 ESR spectra of the positive ion of dideuteroxydurene in D_2SO_4/CH_3NO_2 (a) at $+60°$ C (b) at $-10°$ C. The hyperfine splittings are $a^H_{CH_3} = 2.05$ G and $a^D_{OD} = 0.42$ G. [*Taken from P. D. Sullivan and J. R. Bolton, Advan. Mag. Resonance,* **4**:39 (1970).]

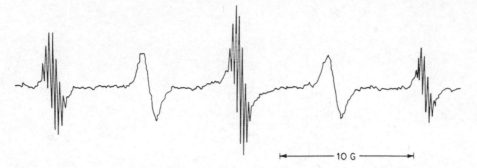

Fig. 9-10 First-derivative spectrum of the dinitrodurene anion in dimethyl formamide at room temperature. The major groups are due to hyperfine splitting from the two nitrogen nuclei. [*J. H. Freed and G. K. Fraenkel, J. Chem. Phys.*, **37**:1156 (1962).]

In this model the variation in the nitrogen hyperfine splitting is attributed to rotation of one nitro group out of the plane of the molecule. The hyperfine splittings are assumed values, but they are probably not far from the actual values. In the region of slow interconversion from A to B, one would observe the spectrum shown in Fig. 9-12a. As the rate of interconversion increases, the lines will coalesce, as shown in Fig. 9-12b. The widths will be given by Eq. (9-19). Three line components do not shift in going from A to B, and hence these lines remain sharp. These are the two outside lines and one component of the central line. The $M = \pm 1$ lines will appear broad because of the sizable field shifts involved. Two components of the $M = 0$ line will undergo a large magnetic field shift (see Fig. 9-12b), and hence these two components are usually not detected; instead, the single, sharp, unshifted component is seen. The appearance of the spectrum in Fig. 9-10 can now be understood. Knowledge of the two nitrogen hyperfine splittings (in the limit of slow exchange) and the use of Eq. (9-19) permits the exchange rate $(1/2\tau)$ to be obtained from the width of the $M = \pm 1$ components. In Fig. 9-12c the hyperfine splittings are completely averaged, and the five-line spectrum characteristic of two equivalent nuclei of spin 1 is obtained (see Fig. 4-26).

A second example of the alternating-linewidth effect is shown by the spectrum of the p-benzosemiquinone radical. The spectra are displayed as a

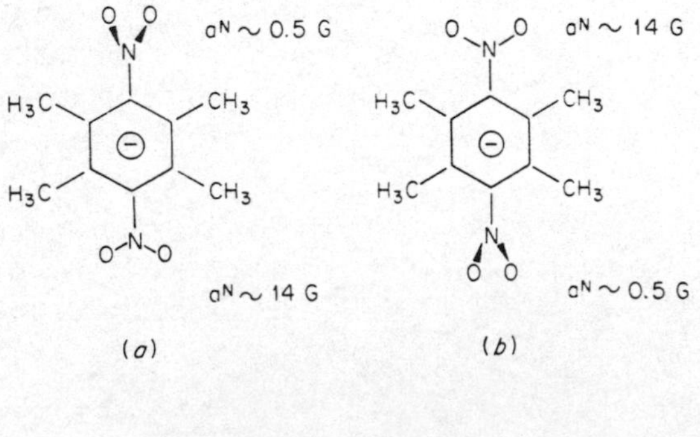

Fig. 9-11 Two structures for the dinitrodurene anion.

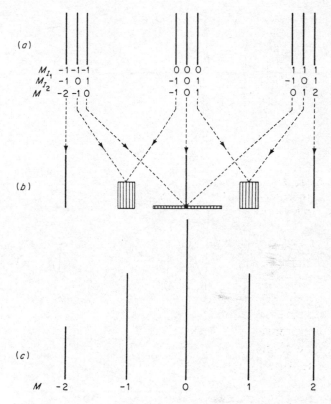

Fig. 9-12 Representation of the spectra of a radical containing two nitrogen nuclei subjected to an out-of-phase modulation. $M = M_{I_1} + M_{I_2}$. (a) Slow-interconversion limit. (b) Intermediate rate of interconversion; blocks represent linewidths and amplitudes. (c) Fast-interconversion limit.

function of pH in Fig. 9-13. At pH 8.3 (Fig. 9-13a) the radical exists predominantly as the anion (form A in Fig. 9-14).

Here the four protons are completely equivalent. With decreasing pH, the equilibrium shifts toward the neutral radical, forms B or C in Fig. 9-14. B and C are thermodynamically equivalent. In concentrated H_2SO_4, only the radical cation, shown as form D in Fig. 9-14, is observed. (See Fig. 4-34.) The spectrum of the neutral free radical has previously been given in Fig. 4-33. However, in the aqueous system, forms B and C are rapidly interconverting via forms A and D. The direct conversion of B to C can be ruled out as unimportant, since no OH proton hyperfine splitting is observed. This indicates that proton exchange with the solvent is the dominant mechanism. To interpret the spectra in Fig. 9-13 one first notes that in forms B and C, protons 2 and 6 are completely equivalent, as also are protons 3 and 5. The two *sets* together are not *completely* equivalent. The proposed mechanism results in an *out-of-phase* modulation of the hyperfine splittings. Hence this case is very similar to that considered in Fig. 9-12,

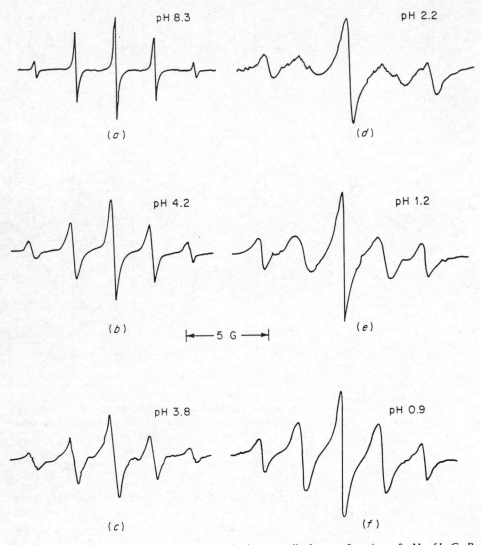

pH 8.3

(a)

pH 4.2

(b)

pH 3.8

(c)

pH 2.2

(d)

pH 1.2

(e)

pH 0.9

(f)

← 5 G →

Fig. 9-13 ESR spectra of the *p*-benzosemiquinone radicals as a function of pH. [*I. C. P. Smith and A. Carrington, Mol. Phys.*, **12**:439 (1967).]

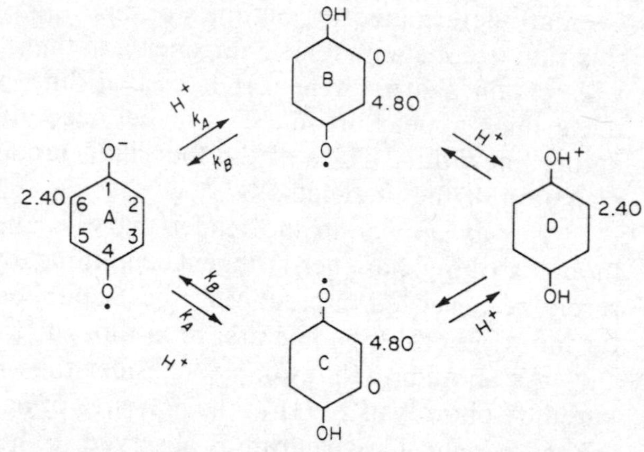

Fig. 9-14 Equilibria among the anionic (A), neutral (B and C), and cationic (D) forms of *p*-benzosemiquinone radicals. [*I. C. P. Smith and A. Carrington, Mol. Phys.*, **12**:439 (1967).]

since the two sets of completely equivalent protons can be treated as analogous to two nuclei of spin 1. The only difference is one of degeneracies of the individual lines; the expressions for the widths are identical.

It is now profitable to return to the spectrum of the *p*-dihydroxydurene cation displayed in Fig. 9-9. The spectrum was taken in CH_3NO_2 with added D_2SO_4. The small hyperfine splitting from the deuterons makes the effect more evident. The main structure in the spectrum is due to hyperfine interaction with the 12 methyl protons. A given methyl group can be considered to be rotating rapidly enough so that the three methyl protons are completely equivalent. A four-jump model depicted in Fig. 9-15 has been proposed[†] for semiquinone molecules. This mechanism assumes a *cis-trans* rotational isomerism involving the OH groups. Here no two methyl groups are completely equivalent. However, examination of spectra taken at low temperatures shows that the sums $a_2 + a_6$ and $a_3 + a_5$ are constant and equal for all forms. The positions of methyl proton hyperfine lines for which M is

[†] A. Carrington, *Mol. Phys.*, **5**:425 (1962).

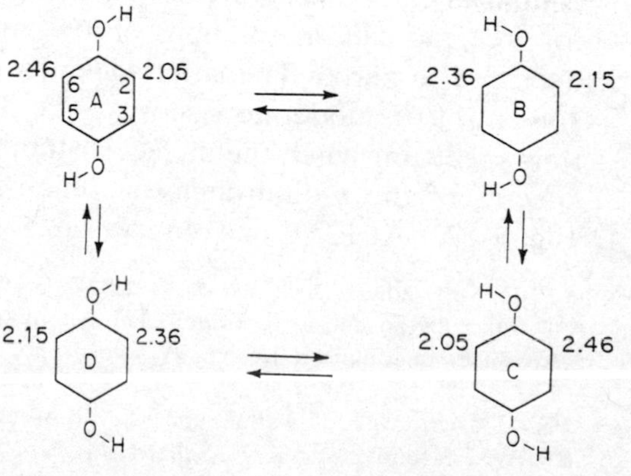

Fig. 9-15 Four-jump model for *cis-trans* rotational isomerism in the hydroquinone cation. The numbers indicate the proton hyperfine splittings in each form.

even are determined by multiples of this sum of hyperfine splittings. Since this sum is constant, some components of these lines will not shift and hence will remain sharp. When M is odd, all line components shift, and hence these lines appear broad. The model accounts qualitatively for the alternating linewidths. Use of the four-jump model leads to a quantitative explanation of the linewidths.[†]

It may appear to the reader that the alternating-linewidth effect, although striking, has such stringent conditions for its appearance that it would rarely be observed in practice. Recent publications show that the effect is far more general than had first been thought.[‡]

As an additional example, consider the vinyl radical. When it is prepared by photolysis of HI in the presence of acetylene in an argon matrix at 4 K,[§] an eight-line spectrum is observed. The spectrum is consistent with three nonequivalent hyperfine splittings of 68.5, 34.3 and 15.7 G. (The reason for these large proton hyperfine couplings is discussed in Sec. 8-3.) However, when the radical is produced by electron bombardment of a mixture of liquid ethane and ethylene,[¶] only a four-line spectrum is observed. It appears that only two hyperfine splittings of 102.8 and 15.7 G are present. This anomaly can be understood if it is assumed that in solution the radical is rapidly interconverting between two tautomers as shown in Fig. 9-16. This process results in an out-of-phase modulation of the hyperfine splittings of H_a and H_b. (See Fig. 9-16a.) If the nuclear-spin orientation for H_a and H_b is either $(\alpha_a \alpha_b)$ or $(\beta_a \beta_b)$, then no field shift of the lines corresponding to these spin states results from the interconversion. Thus, the outside lines should remain sharp, and they will be separated by $(68.5 + 34.3 \text{ G}) = 102.8$ G. The lines corresponding to the spin states $(\alpha_a \beta_b)$ or $(\beta_a \alpha_b)$ will shift, and at an intermediate interconversion rate they may appear so broad as not to be detected. This explanation accounts for the missing lines in the solution spectrum of the vinyl radical.[¶]

9-7 SPECTRAL EFFECTS OF SLOW MOLECULAR TUMBLING RATES

In Chaps. 7 and 8 it was shown that in solids the g factors and the hyperfine couplings can be very orientation-dependent. It has also been indicated that for free radicals in solutions of low viscosity, anisotropic interactions are averaged to zero. The latter statement is not always true, especially if the solvent has a moderate viscosity. To illustrate what may happen to a solution spectrum when the molecular tumbling rate is decreased, consider the spectra of the p-dinitrobenzene anion in dimethylformamide as shown in Fig. 9-17. At 12° C one observes a "normal" spectrum in which the relative

† P. D. Sullivan, *J. Am. Chem. Soc.*, **89**:4294 (1967).
‡ P. D. Sullivan and J. R. Bolton, The Alternating Line Width Effect, in J. S. Waugh (ed.), "Advances in Magnetic Resonance," vol. 4, pp. 39 to 85. Academic Press Inc., New York, 1970.
§ E. L. Cochran, F. J. Adrian, and V. A. Bowers, *J. Chem. Phys.*, **40**:213 (1963).
¶ R. W. Fessenden and R. H. Schuler, *J. Chem. Phys.*, **39**:2147 (1963).

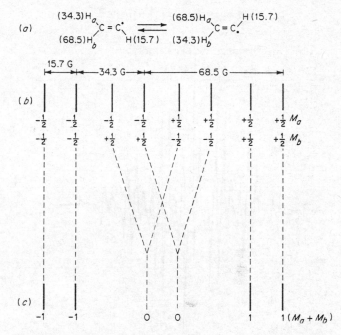

Fig. 9-16 Effect on the ESR spectrum of rapid interconversion between valence tautomers of the vinyl radical. (*a*) Hyperfine splittings (in gauss) in each tautomer. (*b*) Representation of the spectrum observed at low temperatures in a rigid matrix (slow interconversion rate limit). (*c*) Representation of the spectrum observed in solution. The center lines are broadened beyond observation by the interconversion process. The outer lines are not affected because they do not shift in the interconversion. (The figure is not drawn to scale.)

derivative amplitudes are proportional to the degeneracies of the corresponding energy levels. However, at $-55°$C the appearance of the spectrum has changed drastically, although the line positions are unaltered. The change in the spectrum results from variation in the widths of the several lines. Furthermore, this variation is not symmetric about the center line.

To understand the origin of these effects, first consider the di-t-butyl nitroxide radical which contains a single nucleus with spin $I = 1$. This is a simple case, since the g tensor and the hyperfine tensor have the same principal axes, and each tensor is approximately axially symmetric. Figure 9-18a illustrates the spectrum obtained for a randomly oriented solid. The parallel and perpendicular features of such spectra were considered in Chap. 7. Figure 9-18b illustrates the spectrum obtained from a solution of moderate viscosity. The tumbling rate is sufficiently rapid so that the line positions correspond to the completely averaged spectrum in Fig. 9-18c. However, the widths are different for each line. Recalling from Eq. (9-19) that the linewidth is proportional to the mean square of the shift in line positions, it is clear why the three lines have different widths and why the high-field line is broader than the low-field line. In interpreting Fig. 9-18, the following ex-

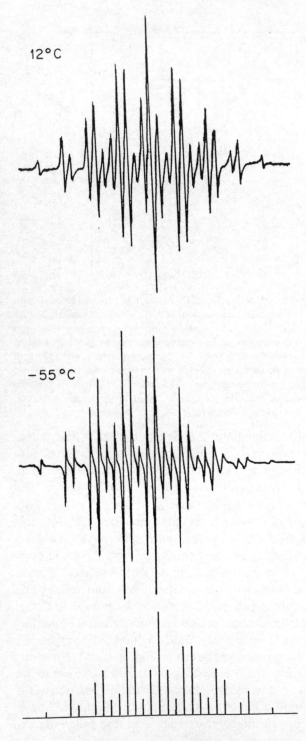

12°C

−55°C

Fig. 9-17 Electron spin resonance spectra of the negative ion of *p*-dinitrobenzene in dimethylformamide. The stick plot is based on the hyperfine splittings: $a^N = 1.51$ G and $a^H = 1.12$ G. [*Spectra taken from J. H. Freed and G. K. Fraenkel, J. Chem. Phys.*, **40**:1815 (1964).]

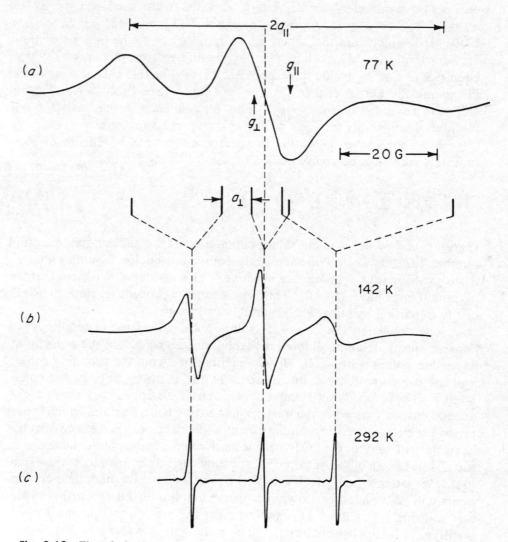

Fig. 9-18 First derivative spectra of the di-*t*-butyl nitroxide radical. (*a*) At 77 K (solid). (*b*) At 142 K (viscous ethanol solution). (*c*) At 292 K (low viscosity). The values of $a_\parallel$, $a_\perp$, $g_\parallel$, and $g_\perp$ are taken from data on single crystals. [*From O. H. Griffith, D. W. Cornell, and H. M. McConnell, J. Chem. Phys.*, **43**:2909 (1965).]

perimental parameters for di-t-butyl nitroxide were used[†]: $a_\parallel = +32$ G, $a_\perp = +6.3$ G, $a_{iso} = +15$ G, $h\nu/(g_\perp - g_\parallel)\beta = 10$ G, and $\frac{1}{3}(2g_\perp + g_\parallel) = g_{iso} = 2.00$. It is significant that when $a_{iso} > 0$ and $g_\perp > g_\parallel$, the high-field line is broader than the low-field line. Had a_{iso} been negative, then $a_\parallel$ would have been $+2$ G and $a_\perp$ would have been -23.7 G. In this case the high-field line would be narrower than the low-field line. (See Prob. 9-7.) The results are reversed when $g_\parallel > g_\perp$. This phenomenon is the basis for one method of measuring the signs of isotropic hyperfine splittings.[‡]

In general, the linewidths in spectra exhibiting the effects of slow tumbling rates can be approximated by the following relation:

$$\Gamma = A + \sum_i B_i \tilde{M}_i + \sum_i C_i \tilde{M}_i^2 + \sum_{i<j} E_{ij} \tilde{M}_i \tilde{M}_j \qquad (9\text{-}25)$$

Here $\tilde{M}_i$ and $\tilde{M}_j$ refer to the total z component of the nuclear spin quantum number for sets i and j of completely equivalent nuclei, assuming that all isotropic hyperfine splittings are negative. This assumption assures that the high-field lines have $\tilde{M}_i > 0$. This is an arbitrary procedure, since the signs of the hyperfine splittings are generally not known.

The coefficients in Eq. (9-25) will now be considered in detail. A is a constant term including all line-broadening effects which are the same for all hyperfine components. The B_i coefficients arise from the product of the g and the hyperfine tensors; in certain cases the B_i coefficients can be calculated.[§] These coefficients cause the spectrum to appear asymmetric. A special case occurs when the nucleus in question has a p orbital which is part of a π-electron system containing an unpaired electron. For example, it has been shown[‡] that for the ^{13}C hyperfine components in the naphthalene anion, the absolute sign of the isotropic ^{13}C splitting can be obtained from the asymmetric broadening (B_i coefficients). For such nuclei the high-field components will be broader (i.e., lower derivative amplitude) if the isotropic hyperfine splitting is positive. This assumes that the π-electron spin density ρ_i is positive. The opposite is true if ρ_i is negative. It also assumes that $g_\parallel < g_\perp$ which is true for most π-electron radicals.

As an example, consider the ESR spectrum of the 2,5-dihydroxy-p-benzosemiquinone anion shown in Fig. 9-19. The two ^{13}C splittings of $a_1^C = +2.63$ G and $a_3^C = -6.66$ G are indicated on the figure. Note that for each of the ^{13}C lines, the amplitudes are such that the high-field line is always broader than the corresponding low-field line. This implies that the product $a^C \rho^C > 0$. At position 1, it is reasonably certain that $\rho_1^C > 0$; hence $a_1^C > 0$. At position 3 the very small proton hyperfine splitting (0.79 G) implies that ρ_3^C is very small. The large magnitude of the ^{13}C splitting must then arise from spin density on neighboring carbon atoms. From Eq. (6-23)

[†] O. H. Griffith, D. W. Cornell, and H. M. McConnell, *J. Chem. Phys.*, **43**:2909 (1965).

[‡] E. de Boer and E. L. Mackor, *J. Chem. Phys.*, **38**:1450 (1963).

[§] G. K. Fraenkel, *J. Phys. Chem.*, **71**:139 (1967).

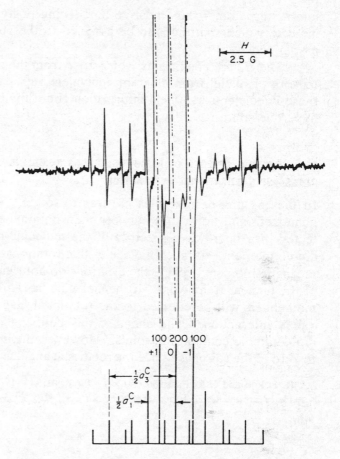

Fig. 9-19 First derivative of the ESR spectrum of 2,5-dioxy-1,4-benzosemiquinone in KOH solution. The lines arising from proton splittings are off scale in the spectrum. The quantum numbers $(+1, 0, -1)$ and the intensities $(100, 200, 100)$ of the proton lines are indicated to approximate scale relative to the smallest ^{13}C lines. [*Taken from M. R. Das and G. K. Fraenkel, J. Chem. Phys.*, **42**:1350 (1965).]

this contribution to $a_3{}^C$ is negative. Since this represents the largest contribution to $a_3{}^C$, $a_3{}^C < 0$. Hence $\rho_3{}^C < 0$. It should be emphasized that this type of argument does not apply to proton hyperfine splittings.

The C_i coefficients are a function only of the hyperfine anisotropy. Where they can be calculated, they provide information on the rotational correlation time for tumbling of the radical in the liquid. (This correlation time can be thought of as roughly the the time for rotation through ~ 1 radian about a principal axis. The assumption is made that the correlation time is isotropic.) In the special cases where the nucleus in question is part of the skeleton of a π-electron system, the coefficients can be used to make assignments of hyperfine splittings. In these cases it has been shown that the C_i coefficients are proportional to the square of the π-spin density on the interacting atom†; thus, relative broadening of hyperfine components can indicate which splitting is to be assigned to the position having the higher π-spin density.

The E_{ij} are coefficients which arise from the products of the hyperfine tensors of nuclei from different equivalent sets. It has been shown‡ that these coefficients can yield information about the relative signs of different hyperfine splittings.

9-8 SPECTRAL EFFECTS OF RAPID MOLECULAR TUMBLING RATES—SPIN–ROTATIONAL INTERACTION

In the gas phase, molecules are free to rotate. This rotational motion is quantized, and transitions between the rotational energy levels may be detected in a microwave spectrum if the molecule has a permanent electric dipole moment. In such molecules this rotational motion also generates a magnetic moment because the electrons do not rigidly follow the movement of the nuclear framework. If the molecule has a net electron *spin* magnetic moment, it will be coupled to the rotational magnetic moment by dipole-dipole interaction. The effect of this coupling is analogous to electron dipole-dipole couplings in solids. However, the interaction is *not* averaged to zero in the gas phase since the rotational angular momentum and magnetic

† J. R. Bolton and G.K. Fraenkel. *J. Chem. Phys.*, **41**:944 (1964).
‡ J. H. Freed and G. K. Fraenkel, *J. Chem. Phys.*, **41**:699 (1964).

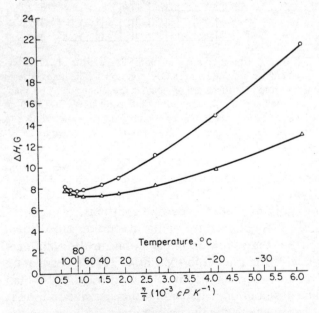

Fig. 9-20   Peak-to-peak widths of the $M_I = -\frac{7}{2}$ lines (○) and the $M_I = -\frac{1}{2}$ lines (△) of $5 \times 10^{-4}M$ vanadyl acetylacetonate in toluene. [*Taken from R. Wilson and D. Kivelson, J. Chem. Phys.*, **44**:154 (1966).]

moment vectors are colinear and fixed in space. Gas-phase ESR spectra are very complex as a result of this "spin-rotation" interaction. (See Sec. 12-6.)

For liquids of low viscosity (and usually at high temperatures), molecules in the liquid state may have an opportunity to undergo a few cycles of rotation before a collision occurs. Hence, the rotational magnetic moment generated can couple with the electron-spin magnetic moment.[†,‡] It has been shown that this effect broadens all lines equally.[‡] The linewidths generally vary as T/η, where η is the coefficient of viscosity and T is the absolute temperature. Broadening due to anisotropic hyperfine effects generally varies as η/T. Hence, one should generally expect an optimum temperature for best resolution of an ESR spectrum. Figure 9-20 shows the linewidth variation with temperature for the $M_I = -\frac{7}{2}$ and $-\frac{1}{2}$ lines in the ESR spectrum of vanadyl acetylacetonate in toluene. The former lines show a much greater linewidth variation with η/T than do the latter. The temperature corresponding to the minimum linewidth is also different for the two lines.

9-9 SUMMARY

It should now be apparent to the reader that a large number of different mechanisms can contribute to line broadening. In this chapter it has been possible only briefly to survey the origins of some of the important line-broadening effects. Though these effects contribute to the complexity of ESR spectra, analysis can yield valuable structural and kinetic information. In some cases, ignorance of these effects makes it virtually impossible to interpret an ESR spectrum (e.g., the vinyl radical in solution). For a quantitative interpretation of these line-broadening phenomena an understanding of the theory of relaxation processes is required. For further details and discussion the reader is urged to consult the literature.[§,¶¶]

PROBLEMS

9-1. Linewidths in ESR spectra may vary from 15 mG to 1,000 G or more.

(a) Compute the *minimum possible* value of T_1 for lines of width 15 mG, 1 G, and 1,000 G.

(b) Suggest possible methods which might be used to measure the spin-lattice relaxation time T_1.[¶,††]

(c) What are the distinguishing characteristics of the relaxation times T_1 and T_2?

† G. Nyberg, *Mol. Phys.*, **12**:69 (1967).

‡ R. Wilson and D. Kivelson, *J. Chem. Phys.*, **44**:154 (1966); P. W. Atkins and D. Kivelson, *J. Chem. Phys.*, **44**:169 (1966).

§J. H. Freed and G. K. Fraenkel, *J. Chem. Phys.*, **39**:326 (1963); **40**:1815 (1964); G. K. Fraenkel, *J. Phys. Chem.*, **71**:139 (1967); A. Carrington and H. C. Longuet-Higgins, *Mol. Phys.*, **5**:447 (1962). A. G. Redfield, The Theory of Relaxation Processes, p. 1, in vol. 1 of J. S. Waugh (ed.), "Advances in Magnetic Resonance," Academic Press Inc., New York, 1965.

¶ See K. J. Standley and R. A. Vaughan, "Electron Spin Relaxation Phenomena in Solids," chaps. 5 to 8, Adam Hilger, Ltd., London, 1969.

††J. Pescia, *J. Physique*, **27**:782 (1966).

¶¶A. Hudson and G.R. Luckhurst Chem. Rev. **69**:191 (1969)

9-2. From the linewidths in Fig. 9-4 at concentrations of $10^{-1}M$ and $10^{-2}M$ determine the second-order rate constant for electron spin exchange.

9-3. Consider the spectrum of sodium naphthalenide shown in Fig. 9-7. The sodium hyperfine splittings, in the limit of slow conversion between two distinct ion pairs, are $a_1 = 1.1$ G and $a_2 = 0$ G.

   (*a*) From the magnitude of the hyperfine splitting (computed using the scale in the figure) and using Eq. (9-23) compute the equilibrium constant for the reaction

$$A \underset{k_{-1}}{\overset{k_1}{\rightleftharpoons}} B$$

   (*b*) Plot ln K versus $1/T$ to obtain ΔH^0, the enthalpy change, for this reaction.

   (*c*) Using the scale in the figure and the relative amplitudes, compute approximate linewidths for lines B_1, B_2, B_3, and B_4 in Figs. 9-7*a*, *b*, and *c*.

   (*d*) Using Eq. (9-24) and noting that $\Gamma = (\sqrt{3}/2)\Delta H_{pp}$ for lorentzian lines where ΔH_{pp} is the peak-to-peak width shown in Fig. 2-9, compute the lifetime τ at the three temperatures.

   (*e*) From Eq. (9-15) and the equilibrium constant, compute τ_A and τ_B at the three temperatures.

   (*f*) Plot ln $1/\tau_A$ and ln $1/\tau_B$ versus $1/T$ to obtain the activation energies for the forward and reverse reactions.

9-4. The spectra of the *p*-benzosemiquinone radical are given as a function of pH in Fig. 9-13.

   (*a*) Compute the widths of the $M_I = \pm 1$ and $M_I = \pm 2$ lines from the scale on the spectra for *b*, *c*, and *d* of Fig. 9-13.

   (*b*) From the known hyperfine splittings of forms B or C of Fig. 9-14 and the linewidths of the $M = +1$ and -1 components, compute the approximate value of the lifetime of the neutral radical at pH values 4.2, 3.8, and 2.2.

   (*c*) From the data in *b*, compute the second-order rate constant for the conversion of form B to form A. (See Fig. 9-14.)

9-5. The spectrum of the VO_2^{--} ion is given in Fig. 9-21. The ^{51}V nucleus has $I = \frac{7}{2}$.

   (*a*) Determine relative linewidths for each component, using the fourth line as a reference.

   (*b*) Assuming that $g_{\parallel} < g_{\perp}$, determine the sign of the hyperfine splitting.

   (*c*) Determine the coefficients A, B, and C by application of Eq. (9-25) to this spectrum.

3329 G

$M = -\frac{7}{2}$ $M = -\frac{5}{2}$ $M = -\frac{3}{2}$ $M = -\frac{1}{2}$ $M = \frac{1}{2}$ $M = \frac{3}{2}$ $M = \frac{5}{2}$ $M = \frac{7}{2}$

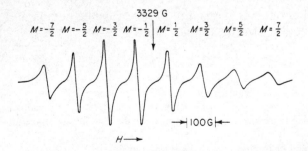

$\longrightarrow$ 100G $\longleftarrow$

$H \longrightarrow$

Fig. 9-21 First-derivative spectrum of the VO_2^{++} ion in toluene solution at 236 K. [*From R. Wilson and D. Kivelson, J. Chem. Phys.*, **44**:154 (1966).]

9-6. (*a*) From the spectrum of the *p*-dinitrobenzene anion at $-55°$C in Fig. 9-17, determine the relative linewidths of each component assuming the center line has a unit linewidth.

   (*b*) Assign a value of $\tilde{M}_H$ and $\tilde{M}_N$ to each line component.

   (*c*) Determine the sign of B_N and B_H (Eq. 9-25). (Hint: B_N should be determined from the line components for which $\tilde{M}_H$ is zero and vice versa.)

   (*d*) Determine the sign of E_{NH} (Eq. 9-25). (Hint: Compare the relative widths of the proton hyperfine components for which $\tilde{M}_N = -1$ and $\tilde{M}_N = +1$.)

9-7. Referring to Fig. 9-18, construct a stick diagram for the di-*t*-butyl nitroxide radical under the assumption that $a_{iso} = -15$ G, $a_{\parallel} = +2$ G, $a_{\perp} = -23.7$ G. From this diagram, predict the relative widths of the three hyperfine lines in a solution of moderate viscosity.

10

Energy-level Splitting
in Zero Magnetic Field.
The Triplet State

10-1 INTRODUCTION

For nearly all of the species considered in previous chapters the spin S is $\frac{1}{2}$; that is, we have considered the magnetic properties of only one electron in the system. In the absence of a magnetic field, the spin states for these systems are degenerate. In fact, for all systems with an *odd* number of electrons, a theorem due to Kramers† guarantees at least a twofold degeneracy which can only be removed by the application of a magnetic field. Thus, in principle, an ESR spectrum should be obtainable for any system with an odd number of electrons ($S = \frac{1}{2}, \frac{3}{2}, \frac{5}{2}, \ldots$).

For the case of two noninteracting electrons one may construct the four configurations

$$\alpha(1)\alpha(2) \qquad \alpha(1)\beta(2) \qquad \beta(1)\alpha(2) \qquad \beta(1)\beta(2)$$

In a molecule of finite size where interaction occurs, it is convenient to combine these configurations into states which are either *symmetric* or

---

† See M. Tinkham, "Group Theory and Quantum Mechanics," p. 143, McGraw-Hill Book Company, New York, 1964.

antisymmetric with respect to exchange of the electrons. These states are

$$\alpha(1)\alpha(2)$$

$$\frac{1}{\sqrt{2}}\left[\alpha(1)\beta(2) + \beta(1)\alpha(2)\right] \qquad \frac{1}{\sqrt{2}}\left[\alpha(1)\beta(2) - \beta(1)\alpha(2)\right]$$

$$\beta(1)\beta(2)$$

Symmetric   Antisymmetric

(Triplet state, $S = 1$)   (Singlet state, $S = 0$)

The multiplicity of the state with $S = 1$ is $(2S + 1) = 3$; hence it is called a triplet state; the state with $S = 0$ is analogously called a singlet state. If the two electrons occupy the same spatial orbital, then for this configuration only the antisymmetric or singlet state is possible because of the Pauli principle. However, if the two electrons occupy different orbitals, then both the singlet and triplet states exist.

For systems with two or more unpaired electrons, the degeneracy of these spin states may be removed even in the absence of a magnetic field. This phenomenon is referred to as *zero field splitting*. If the number of unpaired electrons is even ($S = 1, 2, \ldots$), the degeneracy may be *completely* lifted in zero magnetic field. (See Chap. 11.) If the ensuing separation of state energies is larger than the energy of the microwave quantum, it may be impossible to observe an ESR spectrum. In other cases, only some of the allowed transitions may be observed. If the zero field splitting is less than the energy of the microwave quantum, the resulting ESR spectra will show considerable anisotropy.

10-2 THE SPIN HAMILTONIAN FOR $S = 1$

In the absence of hyperfine splitting, the spin hamiltonian for $S = \frac{1}{2}$ is

$$\hat{\mathscr{H}} = \beta \mathbf{H} \cdot \boldsymbol{g} \cdot \hat{\mathbf{S}} \tag{10-1}$$

This cannot give rise to a splitting in zero field. Clearly, additional interactions must be considered.†

At small distances, two unpaired electrons will experience a strong dipole-dipole interaction.‡ This is analogous to the corresponding interaction between electron and nuclear magnetic dipoles which gives rise to the anisotropic hyperfine interaction. (See Fig. 3-3.) The electron spin–

---

† In the presence of hyperfine splitting, there will be a small zero field splitting (see Sec. C-8), but this is far too small to explain the magnitude of the zero field splitting found in most systems with $S \geq 1$.

‡ There is also a much stronger exchange interaction which separates the singlet and triplet states. (See Sec. 10-11.) Here we shall assume that the molecule is exclusively in the triplet state.

electron spin interaction is given by a spin-spin hamiltonian analogous to Eq. (7-23),

$$\hat{\mathcal{H}}_{ss} = \frac{\hat{\boldsymbol{\mu}}_{e_1} \cdot \hat{\boldsymbol{\mu}}_{e_2}}{r_{12}{}^3} - \frac{3(\hat{\boldsymbol{\mu}}_{e_1} \cdot \mathbf{r}_{12})(\hat{\boldsymbol{\mu}}_{e_2} \cdot \mathbf{r}_{12})}{r_{12}{}^5} \tag{10-2}$$

The magnetic moment operators may be replaced by the corresponding spin operators

$$\hat{\mathcal{H}}_{ss} = g^2\beta^2 \left[\frac{\hat{\mathbf{S}}_1 \cdot \hat{\mathbf{S}}_2}{r^3} - \frac{3(\hat{\mathbf{S}}_1 \cdot \mathbf{r})(\hat{\mathbf{S}}_2 \cdot \mathbf{r})}{r^5} \right] \tag{10-3}$$

Here $\mathbf{r}$ is defined as in Fig. 3-3 if $\boldsymbol{\mu}_N$ is replaced by $\boldsymbol{\mu}_e$. The scalar products in Eq. (10-3) are

$$\hat{\mathbf{S}}_1 \cdot \hat{\mathbf{S}}_2 = \hat{S}_{1x}\hat{S}_{2x} + \hat{S}_{1y}\hat{S}_{2y} + \hat{S}_{1z}\hat{S}_{2z} \tag{10-4a}$$
$$\hat{\mathbf{S}}_1 \cdot \mathbf{r} = \hat{S}_{1x}x + \hat{S}_{1y}y + \hat{S}_{1z}z \tag{10-4b}$$
$$\hat{\mathbf{S}}_2 \cdot \mathbf{r} = \hat{S}_{2x}x + \hat{S}_{2y}y + \hat{S}_{2z}z \tag{10-4c}$$

Substitution of Eqs. (10-4) into Eq. (10-3) yields

$$\hat{\mathcal{H}}_{ss} = \frac{g^2\beta^2}{r^5} \left[(r^2 - 3x^2)\hat{S}_{1x}\hat{S}_{2x} + (r^2 - 3y^2)\hat{S}_{1y}\hat{S}_{2y} + (r^2 - 3z^2)\hat{S}_{1z}\hat{S}_{2z} \right.$$
$$- 3xy(\hat{S}_{1x}\hat{S}_{2y} + \hat{S}_{1y}\hat{S}_{2x}) - 3xz(\hat{S}_{1x}\hat{S}_{2z} + \hat{S}_{1z}\hat{S}_{2x})$$
$$\left. - 3yz(\hat{S}_{1y}\hat{S}_{2z} + \hat{S}_{1z}\hat{S}_{2y}) \right] \tag{10-5}$$

It is more convenient to express $\hat{\mathcal{H}}_{ss}$ in terms of the total spin operator $\hat{\mathbf{S}}$, defined by

$$\hat{\mathbf{S}} = \hat{\mathbf{S}}_1 + \hat{\mathbf{S}}_2 \tag{10-6}$$

This is accomplished by expanding the appropriate operators

$$\hat{S}_x{}^2 = (\hat{S}_{1x} + \hat{S}_{2x})^2 = \hat{S}_{1x}{}^2 + \hat{S}_{2x}{}^2 + 2\hat{S}_{1x}\hat{S}_{2x} \tag{10-7}$$

Note that $\hat{\mathbf{S}}_1$ and $\hat{\mathbf{S}}_2$ commute. Hence

$$\hat{S}_{1x}\hat{S}_{2x} = \tfrac{1}{2}\hat{S}_x{}^2 - \tfrac{1}{4} \tag{10-8a}$$

since the eigenvalues of $\hat{S}_{1x}{}^2$ and $\hat{S}_{2x}{}^2$ are both $\tfrac{1}{4}$. (See Sec. B-6.) Similarly

$$\hat{S}_{1y}\hat{S}_{2y} = \tfrac{1}{2}\hat{S}_y{}^2 - \tfrac{1}{4} \tag{10-8b}$$
$$\hat{S}_{1z}\hat{S}_{2z} = \tfrac{1}{2}\hat{S}_z{}^2 - \tfrac{1}{4} \tag{10-8c}$$

Also

$$\hat{S}_x\hat{S}_y = (\hat{S}_{1x} + \hat{S}_{2x})(\hat{S}_{1y} + \hat{S}_{2y})$$
$$= \hat{S}_{1x}\hat{S}_{2y} + \hat{S}_{2x}\hat{S}_{1y} + \hat{S}_{1x}\hat{S}_{1y} + \hat{S}_{2x}\hat{S}_{2y} \tag{10-9a}$$

and

$$\hat{S}_{1x}\hat{S}_{1y} = \frac{i}{2}\hat{S}_{1z} \tag{10-9b}$$

Equation (10-9*b*) can be verified by taking the matrix product for the spin matrices with $S = \frac{1}{2}$. Then

$$\hat{S}_x\hat{S}_y = (\hat{S}_{1x}\hat{S}_{2y} + \hat{S}_{2x}\hat{S}_{1y}) + \frac{i}{2}(\hat{S}_{1z} + \hat{S}_{2z}) \tag{10-10a}$$

Similarly, since $\hat{S}_{1y}\hat{S}_{1x} = -i/2\hat{S}_{1z}$,

$$\hat{S}_y\hat{S}_x = (\hat{S}_{1x}\hat{S}_{2y} + \hat{S}_{2x}\hat{S}_{1y}) - \frac{i}{2}(\hat{S}_{1z} + \hat{S}_{2z}) \tag{10-10b}$$

Hence

$$(\hat{S}_{1x}\hat{S}_{2y} + \hat{S}_{2x}\hat{S}_{1y}) = \tfrac{1}{2}(\hat{S}_x\hat{S}_y + \hat{S}_y\hat{S}_x) \tag{10-11}$$

with similar expressions for the *xz* and *yz* components.

Substitution of these expressions in Eq. (10-5) yields

$$\begin{aligned}
\hat{\mathscr{H}}_{ss} = \frac{1}{2}\frac{g^2\beta^2}{r^5}\,[&(r^2 - 3x^2)\hat{S}_x{}^2 + (r^2 - 3y^2)\hat{S}_y{}^2 \\
&+ (r^2 - 3z^2)\hat{S}_z{}^2 - 3xy(\hat{S}_x\hat{S}_y + \hat{S}_y\hat{S}_x) \\
&- 3xz(\hat{S}_x\hat{S}_z + \hat{S}_z\hat{S}_x) - 3yz(\hat{S}_y\hat{S}_z + \hat{S}_z\hat{S}_y)]
\end{aligned} \tag{10-12}$$

since $r^2 = x^2 + y^2 + z^2$. Equation (10-12) can be written more conveniently in matrix form; that is

$$\hat{\mathscr{H}}_{ss} = \tfrac{1}{2}g^2\beta^2[\hat{S}_x,\hat{S}_y,\hat{S}_z]
\begin{bmatrix}
\dfrac{(r^2 - 3x^2)}{r^5} & \dfrac{-3xy}{r^5} & \dfrac{-3xz}{r^5} \\[2ex]
\dfrac{-3xy}{r^5} & \dfrac{(r^2 - 3y^2)}{r^5} & \dfrac{-3yz}{r^5} \\[2ex]
\dfrac{-3xz}{r^5} & \dfrac{-3yz}{r^5} & \dfrac{(r^2 - 3z^2)}{r^5}
\end{bmatrix}
\begin{bmatrix} \hat{S}_x \\[2ex] \hat{S}_y \\[2ex] \hat{S}_z \end{bmatrix} \tag{10-13a}$$

$$\hat{\mathscr{H}}_{ss} = \hat{\mathbf{S}} \cdot \mathbf{D} \cdot \hat{\mathbf{S}} \tag{10-13b}$$

The elements of **D** must be averaged over the electronic wave function, that is, $D_{xy} = (g^2\beta^2/2)\langle -3xy/r^5 \rangle$. etc. **D** is a second-rank tensor with a trace of zero. As with the tensors encountered in Chaps. 7 and 8, **D** can be diagonalized to $^d\mathbf{D}$. The diagonal elements of $^d\mathbf{D}$ are D_{XX}, D_{YY} and D_{ZZ}; in this axis system Eq. (10-12) becomes

$$\hat{\mathscr{H}}_{ss} = D_{XX}\hat{S}_x{}^2 + D_{YY}\hat{S}_y{}^2 + D_{ZZ}\hat{S}_z{}^2 \tag{10-14}$$

where $D_{XX} + D_{YY} + D_{ZZ} = 0.$† It is customary to replace Eq. (10-14) by

---

† The axis of quantization for the spin operators $\hat{S}_x$, $\hat{S}_y$, and $\hat{S}_z$ is always determined by the effective magnetic field. In many cases, this axis is the same as that of the external magnetic field. Hence, lowercase subscripts (referring to laboratory-fixed axes) are appropriate. However, in this case where the external field is zero, the axes of quantization are the principal axes. Thus, strictly speaking, one should use capitalized subscripts to refer to molecule-fixed axes. We have chosen to use the lowercase subscripts since later applications will involve an external magnetic field.

$$\hat{\mathscr{H}}_{ss} = -\mathscr{X}\hat{S}_x^{\,2} - \mathscr{Y}\hat{S}_y^{\,2} - \mathscr{Z}\hat{S}_z^{\,2} \tag{10-15}$$

where $\mathscr{X} = -D_{XX}$, etc. Since **D** is a traceless tensor, $\mathscr{Z} = -(\mathscr{X} + \mathscr{Y})$. $\mathscr{X}$, $\mathscr{Y}$, and $\mathscr{Z}$ are the respective energies of this system in zero magnetic field since they are the elements of the diagonal hamiltonian matrix. [See Eqs. (10-23).]

The dipole-dipole interaction between the two unpaired electrons is not the only one which can lead to a hamiltonian term of the form of Eq. (10-13b). Coupling between electron orbital and spin angular momenta gives rise to a term of the same form. (See Sec. 11-6.)

Irrespective of the origin of **D**, Eq. (10-13b) must be added to Eq. (10-1) to obtain the correct spin hamiltonian for an $S \geqslant 1$ system, that is,

$$\hat{\mathscr{H}} = \beta\mathbf{H} \cdot \mathbf{g} \cdot \hat{\mathbf{S}} + \hat{\mathbf{S}} \cdot \mathbf{D} \cdot \hat{\mathbf{S}} \tag{10-16a}$$

or

$$\hat{\mathscr{H}} = g\beta\mathbf{H} \cdot \hat{\mathbf{S}} - \mathscr{X}\hat{S}_x^{\,2} - \mathscr{Y}\hat{S}_y^{\,2} - \mathscr{Z}\hat{S}_z^{\,2} \tag{10-16b}$$

In Eq. (10-16b) an isotropic g factor has been assumed.

10-3   STATE ENERGIES FOR A SYSTEM WITH $S = 1$

It is convenient to use the functions $|1\rangle$, $|0\rangle$, and $|-1\rangle$ as a basis set; these are the eigenfunctions of $\hat{\mathscr{H}}$ (Eq. 10-16a) for $H \to \infty$. However, these are *not* eigenfunctions of $\hat{\mathscr{H}}_{ss}$; hence, it is necessary to compute the hamiltonian matrix. This is best accomplished by use of the spin matrices developed in Sec. B-6. Equation (10-16b) can be written in matrix form as follows:

$$\mathscr{H} = g\beta H_X \mathbf{S}_x + g\beta H_Y \mathbf{S}_y + g\beta H_Z \mathbf{S}_z - \mathscr{X}\mathbf{S}_x^{\,2} - \mathscr{Y}\mathbf{S}_y^{\,2} - \mathscr{Z}\mathbf{S}_z^{\,2} \tag{10-17}$$

The spin matrices are

$$\mathbf{S}_x = \begin{bmatrix} 0 & \dfrac{1}{\sqrt{2}} & 0 \\[2mm] \dfrac{1}{\sqrt{2}} & 0 & \dfrac{1}{\sqrt{2}} \\[2mm] 0 & \dfrac{1}{\sqrt{2}} & 0 \end{bmatrix} \tag{10-18a}$$

$$\mathbf{S}_y = \begin{bmatrix} 0 & \dfrac{-i}{\sqrt{2}} & 0 \\[2mm] \dfrac{i}{\sqrt{2}} & 0 & \dfrac{-i}{\sqrt{2}} \\[2mm] 0 & \dfrac{i}{\sqrt{2}} & 0 \end{bmatrix} \tag{10-18b}$$

$$\mathbf{S}_z = \begin{bmatrix} 1 & 0 & 0 \\ 0 & 0 & 0 \\ 0 & 0 & -1 \end{bmatrix} \tag{10-18c}$$

Substitution of the above spin matrices in Eq. (10-17) with subsequent matrix addition and multiplication yields

$$
\mathcal{H} = \begin{matrix} & |1\rangle & |0\rangle & |-1\rangle \\ \langle 1| & \left[g\beta H_Z - \tfrac{1}{2}(\mathcal{X} + \mathcal{Y}) - \mathcal{Z} & \frac{1}{\sqrt{2}}\,g\beta(H_X - iH_Y) & -\tfrac{1}{2}(\mathcal{X} - \mathcal{Y}) \right. \\ \langle 0| & \frac{1}{\sqrt{2}}\,g\beta(H_X + iH_Y) & -(\mathcal{X} + \mathcal{Y}) & \frac{1}{\sqrt{2}}\,g\beta(H_X - iH_Y) \\ \langle -1| & \left. -\tfrac{1}{2}(\mathcal{X} - \mathcal{Y}) & \frac{1}{\sqrt{2}}\,g\beta(H_X + iH_Y) & -g\beta H_Z - \tfrac{1}{2}(\mathcal{X} + \mathcal{Y}) - \mathcal{Z} \right] \end{matrix}
$$

$$(10\text{-}19)$$

The secular determinant is obtained from $\mathcal{H}$ by subtracting W from each diagonal element; this determinant is set equal to zero:

$$
\begin{vmatrix} g\beta H_Z - \tfrac{1}{2}\mathcal{Z} - W & \frac{1}{\sqrt{2}}\,g\beta(H_X - iH_Y) & -\tfrac{1}{2}(\mathcal{X} - \mathcal{Y}) \\ \frac{1}{\sqrt{2}}\,g\beta(H_X + iH_Y) & \mathcal{Z} - W & \frac{1}{\sqrt{2}}\,g\beta(H_X - iH_Y) \\ -\tfrac{1}{2}(\mathcal{X} - \mathcal{Y}) & \frac{1}{\sqrt{2}}\,g\beta(H_X + iH_Y) & -g\beta H_Z - \tfrac{1}{2}\mathcal{Z} - W \end{vmatrix} = 0
$$

$$(10\text{-}20)$$

Here the relation $\mathcal{Z} = -(\mathcal{X} + \mathcal{Y})$ has been used. The solutions to Eq. (10-20) are especially simple when $\mathbf{H} \parallel Z$. Then $H_X = H_Y = 0$, and Eq. (10-20) becomes

$$
\begin{vmatrix} g\beta H_Z - \tfrac{1}{2}\mathcal{Z} - W & 0 & -\tfrac{1}{2}(\mathcal{X} - \mathcal{Y}) \\ 0 & \mathcal{Z} - W & 0 \\ -\tfrac{1}{2}(\mathcal{X} - \mathcal{Y}) & 0 & -g\beta H_Z - \tfrac{1}{2}\mathcal{Z} - W \end{vmatrix} = 0 \qquad (10\text{-}21)
$$

The solution $W = \mathcal{Z}$ is obtained by inspection. Expansion of the remaining 2×2 determinant gives the other two energy values as

$$
W_{X,Y} = -\frac{\mathcal{Z}}{2} \pm \frac{1}{2}\,\sqrt{4g^2\beta^2 H_Z^2 + (\mathcal{X} - \mathcal{Y})^2} \qquad (10\text{-}22)
$$

In zero magnetic field, the degeneracy is completely removed if $\mathcal{X} \neq \mathcal{Y}$. Then

$$
W_X = -\frac{\mathcal{Z}}{2} + \frac{\mathcal{X} - \mathcal{Y}}{2} = \mathcal{X} \qquad (10\text{-}23a)
$$

$$
W_Y = -\frac{\mathcal{Z}}{2} - \frac{\mathcal{X} - \mathcal{Y}}{2} = \mathcal{Y} \qquad (10\text{-}23b)
$$

$$
W_Z = +\mathcal{Z} \qquad (10\text{-}23c)
$$

At this point it is apparent why $\hat{\mathcal{H}}_{ss}$ was written in the form of Eq. (10-15). Since the trace of $\mathbf{D}$ is zero, only two independent parameters are required. It is usual to designate these as D and E, defined as

$$D = \frac{-3\mathscr{Z}}{2} \tag{10-24a}$$

$$E = -\tfrac{1}{2}(\mathscr{X} - \mathscr{Y}) \tag{10-24b}$$

The hamiltonian operator, Eq. (10-16b), then becomes

$$\hat{\mathscr{H}} = g\beta \mathbf{H} \cdot \hat{\mathbf{S}} + D(\hat{S}_z{}^2 - \tfrac{1}{3}\hat{S}^2) + E(\hat{S}_x{}^2 - \hat{S}_y{}^2) \tag{10-25}$$

Hence Eqs. (10-23) become

$$W_X = \tfrac{1}{3}D - E \tag{10-26a}$$

$$W_Y = \tfrac{1}{3}D + E \tag{10-26b}$$

$$W_Z = -\tfrac{2}{3}D \tag{10-26c}$$

It is important to note that the values of D and E are not unique. They depend on which axis is chosen as Z. It is always possible to choose Z such that $|E| \leqslant |D/3|$. D and E may be positive or negative and may have opposite signs. The signs of D and E can be determined from relative intensity measurements of lines at low temperatures.[†] One usually is ignorant of the absolute signs of D and E since line positions depend only on their *relative* signs. The values quoted for D and E are usually given as absolute magnitudes. It is convenient to express D and E in units of magnetic field by defining the quantities $D' = D/g\beta$ and $E' = E/g\beta$.

The energies of the three states as a function of magnetic field are plotted in Fig. 10-1 for $\mathbf{H}$ parallel to Z. Here it is assumed that $D > 0$.

[†] A. W. Hornig and J. S. Hyde, *Mol. Phys.*, 6:33 (1963).

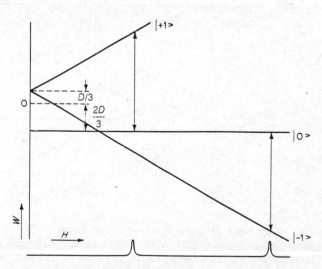

Fig. 10-1 State energies and transitions for a system of spin 1, with axial symmetry, and for $\mathbf{H} \parallel Z$.

For $D < 0$, the states at zero field are reversed. This figure is appropriate to systems with axial symmetry, since then $\mathscr{X} = \mathscr{Y}$ and hence $E = 0$. This means that two of the states will be degenerate at zero field.

When $E \neq 0$, as for systems with orthorhombic or lower symmetry, all three states are nondegenerate. The state energies and transitions are shown in Figs. 10-2a, b, and c for **H** parallel to X, Y, and Z, respectively.

Both D and E depend on the distance between two electrons with parallel spins in a simple model. In particular,

$$D = \tfrac{3}{4} g^2 \beta^2 \left\langle \frac{r^2 - 3z^2}{r^5} \right\rangle \tag{10-27a}$$

$$E = \tfrac{3}{4} g^2 \beta^2 \left\langle \frac{y^2 - x^2}{r^5} \right\rangle \tag{10-27b}$$

where the angular brackets imply an average over the electronic wave function. Experimental values of D and E can provide information on the mean separation between the two electrons if the averages in Eqs. (10-27) can be calculated. This procedure is valid only if the interaction is predominantly dipolar in nature, i.e., if there is not a significant contribution from spin-orbit coupling to the zero field splitting.

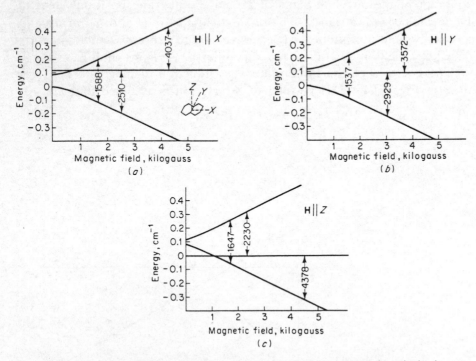

Fig. 10-2 State energies as a function of magnetic field for naphthalene in its lowest triplet state, with $\nu = 9.272$ GHz. The transition at lowest field in each case is allowed only for the microwave magnetic field parallel to the static field. (a) **H** $\|$ X. (b) **H** $\|$ Y. (c) **H** $\|$ Z.

10-4 THE SPIN EIGENFUNCTIONS FOR A SYSTEM WITH $S = 1$

The eigenfunctions of $\hat{\mathcal{H}}$ in Eq. (10-16) are linear combinations of the functions $|1\rangle$, $|0\rangle$, and $|-1\rangle$. The coefficients are obtained by substitution of the eigenvalues of Eq. (10-23) into the determinant (10-21). [See Eq. (5-24).] The coefficients will depend on the magnitude of $\mathbf{H}$. Two limiting cases can be delineated:

1. In the limit as $H \to 0$ with $\mathbf{H}$ parallel to the principal axis Z, the zero field eigenfunctions are[†],[‡]

$$|T_x\rangle = \frac{1}{\sqrt{2}} [|-1\rangle - |+1\rangle] \qquad (10\text{-}28a)$$

$$|T_y\rangle = \frac{i}{\sqrt{2}} [|-1\rangle + |+1\rangle] \qquad (10\text{-}28b)$$

$$|T_z\rangle = |0\rangle \qquad (10\text{-}28c)$$

Note that the functions $|T_x\rangle$, $|T_y\rangle$, and $|T_z\rangle$ are closely analogous to the orbital angular momentum wave functions $|p_x\rangle$, $|p_y\rangle$, and $|p_z\rangle$ for $l = 1$. (See Table 11-2.)

It is sometimes convenient to choose the functions in Eqs. (10-28) as a basis set since they are eigenfunctions of $\hat{\mathcal{H}}_{ss}$. In the presence of a magnetic field, the hamiltonian matrix then becomes (see Prob. 10-2)

$$
\mathcal{H} = \begin{array}{c} \\ \langle T_x| \\ \langle T_y| \\ \langle T_z| \end{array}
\overset{\displaystyle \begin{array}{ccc} |T_x\rangle & |T_y\rangle & |T_z\rangle \end{array}}{
\begin{bmatrix} \mathscr{X} & -ig\beta H_z & +ig\beta H_y \\ +ig\beta H_z & \mathscr{Y} & -ig\beta H_x \\ -ig\beta H_y & +ig\beta H_x & \mathscr{Z} \end{bmatrix}}
\qquad (10\text{-}29)
$$

In this form it is apparent that when $\mathbf{H}$ is parallel to X, Y, or Z, the corresponding state energy is independent of the magnitude of $\mathbf{H}$.

2. When $\mathbf{H}$ is very large and is parallel to Z, the off-diagonal elements in the hamiltonian matrix (10-19) can be neglected, and the eigenfunctions of $\hat{S}_z$ are then eigenfunctions of the hamiltonian of Eq. (10-16b). Hence at high fields, the state energies are usually labeled by the value of M_S.

For intermediate values of $\mathbf{H}$, if $\mathbf{H} \parallel Z$, the spin functions may be taken as

$$c_1|-1\rangle - c_2|1\rangle$$
$$c_2|-1\rangle + c_1|1\rangle$$

---

† H. F. Hameka and L. J. Oosterhoff, *Mol. Phys.*, 1:358 (1958).
‡ For arbitrary orientations of $\mathbf{H}$, these equations do not apply since the molecular axes do not correspond to the axes of quantization.

and

$|0\rangle$

These functions are of importance for the "$\Delta M = 2$" transitions considered in Sec. 10-7 and in Prob. 10-3.

10-5 ELECTRON SPIN RESONANCE OF TRIPLET–STATE MOLECULES

In large magnetic fields the substates of the triplet state are designated as $|+1\rangle$, $|0\rangle$, and $|-1\rangle$. It is important to note that there will always also be a state with $S = 0$ (singlet state). If the highest occupied level is nondegenerate and is doubly occupied by electrons, the ground state is a singlet (Fig. 10-3a). If one electron is excited from the highest occupied orbital to an unoccupied one by absorption of a quantum of the appropriate energy (Fig. 10-3b), the system is still in a singlet state, since allowed transitions occur without change of multiplicity. The molecule may then convert to a metastable triplet state with a change of spin (Fig. 10-3c). This occurs by a radiationless process. A triplet ground state (Fig. 10-3d) requires that this state have at least a twofold orbital degeneracy (or near degeneracy). Low-lying orbitals filled with electrons have been shown in Fig. 10-3 to emphasize that it is the highest *occupied* orbitals that are important in determining the multiplicity of the state.

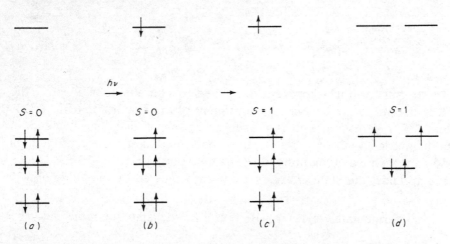

Fig. 10-3 (*a*) Configuration of a system in its singlet ground state. The horizontal lines represent orbital energies. (*b*) Configuration of the same system after excitation: the spins remain paired, and hence the state is a singlet. (*c*) Configuration of the same system after the molecule goes to a metastable triplet state via a radiationless process. The triplet state is usually lower in energy than the singlet state because of decreased interelectronic repulsion. (*d*) Configuration of a system with a triplet ground state. The lowest level is symbolic of filled orbitals below the degenerate pair; it is irrelevant whether the filled orbitals are degenerate or nondegenerate.

After irradiation with visible or ultraviolet light, many aromatic hydrocarbons in rigid solutions at low temperature exhibit excited states of unusually long lifetime—some of the order of minutes. This behavior is the result of the existence of a metastable state which is populated via other excited states. G. N. Lewis postulated that this long-lived state is a triplet state. Direct excitation or emission from this state is spin-forbidden, and hence the observed metastable behavior is to be expected. Following Lewis's prediction, magnetic susceptibility experiments on excited molecules in rigid media gave results in qualitative accord with the triplet nature of the state. That is, upon irradiation, there is an increase in paramagnetism; this decays upon cessation of irradiation with the same decay constant as the phosphorescence. One would expect these paramagnetic excited molecules to show an ESR absorption, but a number of early experiments failed to detect it. One reason for the early failures is the marked anisotropy of the ESR spectrum which results from the dipolar interaction of the two electrons coupling to give $S = 1$.[†] A second reason is the low sensitivity of the spectrometers at the time attempts were made.

Once the cause of the earlier failures was recognized, a successful observation of the first excited triplet state of naphthalene was achieved by irradiating single crystals of durene (1,2,4,5-tetramethyl benzene) containing a small fraction of naphthalene.[‡] Since the two molecules are similar in shape, the naphthalene directly replaces durene in the lattice. (Use of dilute single crystals instead of pure naphthalene greatly lengthens the lifetime of the excited triplet state in a particular molecule. In the pure crystal, rapid migration of the triplet excitation usually leads to effective quenching.)

The spectra observed for naphthalene are precisely in accord with expectation for a system with $S = 1$; $D/hc = 0.1003$ cm^{-1}, $E/hc = -0.0137$ cm^{-1}, and g (isotropic) $= 2.0030$. The axis system is shown in Fig. 10-2a. The positions of ESR lines for the several principal-axis orientations of the field are given in Figs. 10-4a to c. Figure 10-4d shows the expected line positions if it were possible to cause naphthalene molecules to rotate rapidly about their Z axes. The system would thus have Z as an axis of symmetry, and all directions in the XY plane would be equivalent. Thus the X and Y pairs of lines would merge to a single pair at the average positions. Upon setting $\mathscr{X} = \mathscr{Y}$, D remains unchanged, but the positions of the lines will be very slightly shifted. The shift will be greatest at low fields, because of the curvature of the state energies when $\mathscr{X} \neq \mathscr{Y}$. The lines in the vicinity of $h\nu/2g\beta$ will be considered in Sec. 10-7. The spectrum of triplet naphthalene in durene is more complicated than that shown in Fig. 10-4, since

[†] S. I. Weissman, cited in C. A. Hutchison Jr. and B. W. Mangum, *J. Chem. Phys.*, **34**:908 (1961).

[‡] This technique had been used in some fundamental studies of the excitation of oriented molecules by polarized light (D. S. McClure, *J. Chem. Phys.*, **22**:1668 (1954); **24**:1 (1956)).

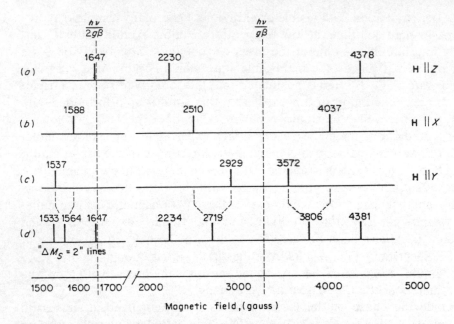

Fig. 10-4 ESR line positions for the triplet state of naphthalene for $\nu = 9.272$ GHz. (a) $H \parallel Z$. (b) $H \parallel X$. (c) $H \parallel Y$. (d) Hypothetical spectrum (total) for a naphthalene molecule in rapid rotation about the Z axis. The lines at 1,564, 2,719, and 3,806 G would be obtained for H in the plane of the rotating molecule; the lines at 2,234 and at 4,381 G would be obtained for H parallel to the axis of rotation. The line at 1,533 G arises when H is at a special orientation ("turning point," see Sec. 10-7). [*This figure was derived from M. S. de Groot and J. H. van der Waals, Mol. Phys.*, **6**:545 (1963).]

there are two types of sites per unit cell. Figure 10-5 shows the line positions for one type of site. The student is urged to interpret the angular-dependence curves and to extract the zero-field splitting parameters (Prob. 10-4).

With the field along the X or Y axes of the naphthalene molecule (in durene) it is possible to resolve a $1:4:6:4:1$ quintet at 77 K.[†] The analysis of hyperfine splittings of variously deuterated samples gives $a = 5.61$ G for the 1, 4, 5, 8 protons and 2.29 G for the 2, 3, 6, 7 protons. These values refer to $H \parallel Z$. It may be recalled that the splittings of the naphthalene anion in liquid solution are 4.90 and 1.83 G, respectively. The similarity of the ratio of the splittings may be explained as follows. The triplet molecule has one electron each in the highest bonding and the lowest antibonding orbitals. By the pairing theorem (Chap. 6), the electron distribution will be the same in each orbital; hence the relative unpaired electron density at the 1 and 2 positions should be the same in the triplet state as in the anion. Since the molecules are rigidly fixed in the crystal, there will be large anisotropic hyperfine coupling contributions. The interpretation is analogous to that for the C—H fragment. (See Sec. 8-2b.) When the magnetic

† N. Hirota, C. A. Hutchison Jr., and P. Palmer, *J. Chem. Phys.*, **40**:3717 (1964).

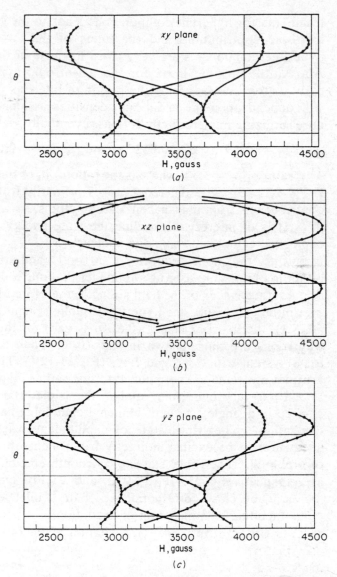

Fig. 10-5 Angular dependence of the resonant field for triplet naphthalene in durene, as a function of rotation with **H** in several planes of one type of durene molecule. (*a*) **H** in *xy* plane. (*b*) **H** in *xz* plane. (*c*) **H** in *yz* plane. [*Plots taken from C. A. Hutchison Jr. and B. W. Mangum, J. Chem. Phys.,* **34**:908 (1961).]

field is along the X axis of a naphthalene molecule, the field is perpendicular both to the C—H bond direction and to the $2p_z$ orbitals of the carbon atoms at positions 1, 4, 5, and 8. A treatment of the naphthalene molecule in the triplet state as a collection of fragments for hyperfine computations leads to the conclusion that both the isotropic contributions and the ratios of the three principal anisotropic hyperfine components are very close to those of the malonic acid radical. Allowance must be made for the anisotropic hyperfine contributions of carbon atoms not adjoining the proton of interest. The magnitude of these will depend upon the initial spin densities assumed. An iterative procedure, utilizing numerous corrections, eventually yields the spin densities

$$\rho_1 = 0.219 \qquad \rho_2 = 0.062 \qquad \text{and} \qquad \rho_9 = -0.063$$

The ratio $\rho_1/\rho_2 = 3.52$, whereas the ratio a_1/a_2 of the hyperfine splittings for $\mathbf{H} \parallel Z$ is 2.45. The difference arises primarily from the $2p_z$ dipole-dipole contributions from nonadjacent carbon atoms.

It is of interest to examine the state energy diagram for an excited triplet molecule containing one nucleus with $I = \frac{1}{2}$. Figure 10-6 shows the hyperfine transitions between an "$M_S = 1$" and an "$M_S = 0$" level. (The quotation marks are used as a reminder that one is dealing with mixed states at moderate values of the field strength; therefore M_S is not a "good" quantum number.) Since the hyperfine coupling hamiltonian involves the term $hA_0\hat{S}_z\hat{I}_z$, the matrix element $\langle 0|\hat{S}_z|0\rangle$ is zero for the "$M_S = 0$" level. The transitions are those shown in Fig. 10-6. Their difference is hA_0, just as for a system with $S = \frac{1}{2}$ and $I = \frac{1}{2}$ [Eq. (3-12)]. Thus the spin densities in a triplet molecule are normalized to one rather than two.

In a very interesting study,[†] the triplet state of phenanthrene was excited in a single crystal of biphenyl containing naphthalene at a low concentration. The triplet state of naphthalene was observed (by its ESR spectrum) to be excited indirectly from phenanthrene via the triplet state of biphenyl. The triplet states of phenanthrene, biphenyl, and naphthalene are, respectively, 21,410, 23,010, and 21,110 cm^{-1} above their ground states. The close correspondence of the triplet levels in the three molecules facilitates the migration of excitation.

In tribenzotriptycene (I), excitation of one of the naphthalene subunits

(I)

takes place in the primary process. Migration of triplet excitation among the three naphthalene-like subunits of the tribenzotriptycene is then found to be very rapid at 77 K.[‡] In this molecule one expects just the sort of

† N. Hirota and C. A. Hutchison Jr., *J. Chem. Phys.*, **42**:2869 (1965).

‡ M. S. de Groot and J. H. van der Waals, *Mol. Phys.*, **6**:545 (1963).

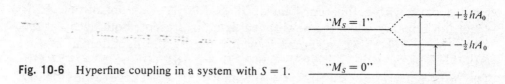

Fig. 10-6 Hyperfine coupling in a system with $S = 1$.

averaging over the X and Y directions that is illustrated in Fig. 10-4d. Some migration among the three units is still detectable at 20 K.

All of the excited triplet-state observations considered to this point have arisen from isolated molecules at low concentration in a rigid matrix. With increasing concentration of the guest molecules, one expects to have a significant number of *pairs* of guest molecules. If one prepares a single crystal of naphthalene-d_8 containing up to 10 percent naphthalene-h_8 (i.e., ordinary naphthalene), and excites with ultraviolet light, one observes a new pair of lines,[†] with $D/hc = -0.0059$ cm^{-1} and $E/hc = +0.0485$ cm^{-1}. The principal axes are *not* those of the molecules, but are closely related to the axes of the crystal. These lines are attributed to a triplet excitation which migrates rapidly between the two molecules of an isolated pair. Pure naphthalene represents the limiting case of a crystal made of noncrystallographically equivalent pairs of molecules. It had been predicted that for a rapidly migrating triplet excitation (referred to as a triplet exciton)[‡] among an array of paired molecules, one should observe in the ESR spectrum a pair of lines with a field separation of $2D/g\beta$ with the field parallel to a principal axis of the crystal.[§] Such a triplet-exciton ESR spectrum was observed for anthracene[¶] and for naphthalene.[††] Triplet-exciton spectra had been observed earlier in zero external field.[‡‡]

One of the simplest excited-triplet-state systems is the excited M center in the alkali halides.[§§] Each of two adjacent anion vacancies along a [110]-type direction contains one electron.

$$
\begin{array}{cccc}
K^+ & Cl^- & K^+ & Cl^- \\
Cl^- & K^+ & [e^-] & K^+ \\
K^+ & [e^-] & K^+ & Cl^- \\
Cl^- & K^+ & Cl^- & K^+
\end{array}
$$

In the ground state, the electrons pair to form a singlet state. Optical absorption bands have been assigned to this center. Irradiation in them is said to cause a "temporary bleaching." Simultaneously, a new band appears at a shorter wavelength. Upon cessation of the irradiation, this new

† M. Schwoerer and H. C. Wolf, *Molec. Crystals,* **3**:177 (1967).

‡ J. Franck and E. Teller, *J. Chem. Phys.,* **6**:861 (1938).

§ H. Sternlicht and H. M. McConnell, *J. Chem. Phys.,* **35**:1793 (1961).

¶ D. Haarer, D. Schmid, and H. C. Wolf, *Phys. Status Solidi,* **23**:633 (1967).

†† D. Haarer and H. C. Wolf, *Phys. Status Solidi,* **33**:K117 (1969).

‡‡ D. D. Thomas, A. W. Merkl, A. F. Hildebrandt, and H. M. McConnell, *J. Chem. Phys.,* **40**:2588 (1964).

§§ C. Z. van Doorn and Y. Haven, *Philips Res. Rept.,* **11**:479 (1956); **12**:309 (1957).

band decays with a time constant of 50 s at 90 K. An anisotropic ESR spectrum appears when the new optical band is present; both decay at the same rate.† The ESR spectrum is well fitted by the hamiltonian of Eq. (10-25), with $S = 1$, $g = 1.998 \pm 0.003$, $D' = \pm 161$ G, and $E' = \mp 54$ G.† The principal axes are $< 110 >$, $< \bar{1}10 >$ and $< 001 >$. These results clearly establish the nature of the excited triplet state of the M center.

10-6 LINE SHAPES FOR RANDOMLY ORIENTED SYSTEMS IN THE TRIPLET STATE

Few triplet systems have thus far been investigated in the oriented solid state. This is due largely to the difficulties of preparing single crystals of adequate size, with well-defined orientation of guest molecules at an appropriate concentration. The observation of a "$\Delta M_S = 2$" line in the region of $g \approx 4$ was the stimulus for the detection of triplet states in numerous nonoriented systems; some of these had not been investigated earlier by optical or other techniques. The relatively large amplitude of the "$\Delta M = 2$" lines is associated with their small anisotropy, i.e., with a small value of $dH/d\theta$ in Eq. (7-58). Subsequently, it was recognized that even for nonoriented systems there is a hope of seeing the ordinary $\Delta M_S = 1$ transitions at turning points. Lines at these positions were found by scanning the field under conditions of high sensitivity.‡

   For axially symmetric molecules, the calculated shapes of the $\Delta M_S = 1$ lines are given in Fig. 10-7. The separation of the outer vertical lines in Fig. 10-7a (which represents the theoretical line shape) is just $2|D'|$, while that of the two inner lines is $|D'|$. The high-field portion of the spectrum in Fig. 10-8 shows a satisfying correspondence with the derivative spectrum in Fig. 10-7b. The spectrum of Fig. 10-8 will be considered further in Sec. 10-8.

   The theoretical line shape for a randomly oriented triplet system with $E' \neq 0$ is given in Fig. 10-9a, and the derivative spectrum is given in Fig. 10-9b. The separation of outermost lines is again $2|D'|$, whereas that of the intermediate and inner pairs is $|D'| + 3|E'|$ and $|D'| - 3|E'|$, respectively. There is a close correspondence between Fig. 10-9b and Fig. 10-10, which gives the spectrum of the excited triplet state of naphthalene in a rigid, nonoriented ("glassy") matrix at 77 K. The compound used was actually $C_{10}D_8$ instead of $C_{10}H_8$, so as to minimize linewidth contributions from unresolved hyperfine splittings. The pairs of lines correspond with those given in Fig. 10-4 for a single crystal. In the $g = 2$ region an additional line is seen at high microwave power. This line has been identified as a double-quantum transition.§ For observations of the $\Delta M_S = 1$ lines in the nonoriented state, one requires a far greater ESR spectrometer sensi-

† H. Seidel, *Phys. Letters*, **7**:27 (1963).
‡ W. A. Yager, E. Wasserman, and R. M. R. Cramer, *J. Chem. Phys.*, **37**:1148 (1962).
§ M. S. de Groot and J. H. van der Waals, *Physica*, **29**:1128 (1963).

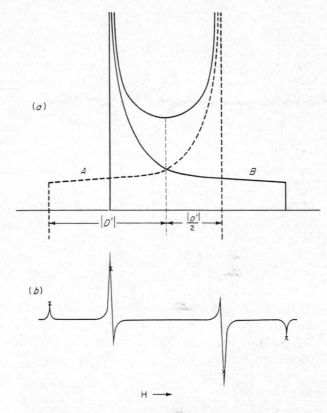

Fig. 10-7 (*a*) Theoretical ESR absorption spectrum for a randomly oriented triplet system for a given value of D' and $\nu(E' = 0)$. A zero linewidth is assumed. The solid curve B corresponds to the curve of Fig. 7-10a; the dotted curve A represents a reflection about the central-field value. (*b*) Derivative curve computed from (*a*) after assuming a finite linewidth. Only the field region corresponding to $\Delta M_S = \pm 1$ is shown. The points marked x correspond to the resonant field values when the magnetic field is oriented along the Z axis (bell-shaped lines) or perpendicular to Z. [*These figures are taken from E. Wasserman, L. C. Snyder, and W. A. Yager, J. Chem. Phys.*, **41**:1763 (1964).]

tivity than for an equivalent concentration in a single crystal. In the former case, only a small fraction of all molecules in the triplet state contribute to any of the observable derivative lines. The $\Delta M_S = 1$ lines are seen to be weak compared with the "$\Delta M = 2$" line.

For rigid media in which the geometries of host and guest molecules are markedly dissimilar, the linewidths in the triplet spectrum may be many times as broad as when host and guest are very similar. {Specifically, diphenylmethylene (C_6H_5—$\dot{C}$—C_6H_5) in diphenyldiazomethane shows a line-

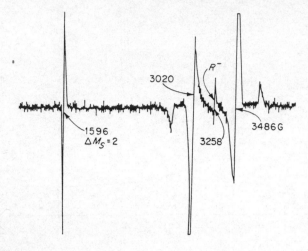

Fig. 10-8 ESR spectrum of a rigid solution of the dianion of triphenylbenzene. $\nu = 9.150$ GHz. The line R^- arises from the mononegative anion. [*Taken from R. E. Jesse, P. Biloen, R. Prins, J. D. W. van Voorst, and G. J. Hoijtink, Mol. Phys., 6:633 (1963).*]

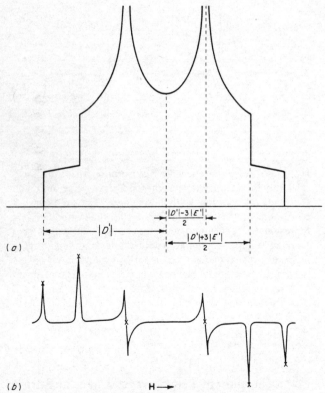

Fig. 10-9 (*a*) Theoretical ESR absorption spectrum for a randomly oriented triplet system for given values of D', E', and ν. A zero linewidth is assumed. (*b*) Derivative curve computed from (*a*) after assuming a finite linewidth. Only the field region corresponding to $\Delta M_S = \pm 1$ is shown. The points marked *x* correspond to resonant field values when the magnetic field is oriented along one of the principal axes of the system. [*These figures are taken from E. Wasserman, L. C. Snyder, and W. A. Yager, J. Chem. Phys., 41:1763 (1964).*]

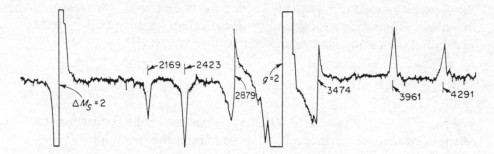

Fig. 10-10 ESR spectrum at 9.08 GHz of perdeuteronaphthalene ($C_{10}D_8$) in a glassy mixture of hydrocarbon solvents ("EPA") at 77 K. Lines in the region of $g \approx 2$ arise from free radicals ($S = \frac{1}{2}$) and from double quantum transitions. [*Taken from W. A. Yager, E. Wasserman, and R. M. R. Cramer, J. Chem. Phys.*, **37**:1148 (1962).]

width of 17 G; in *n*-pentane [$CH_3(CH_2)_3CH_3$] the linewidth is 94 G.†} In a dissimilar host-matrix system, it appears likely that a range of solute-solvent configurations is tolerated; the various configurations display a distribution of D and E values.

In nonrigid media, ESR absorption for triplet-state systems is not observed unless D and E are very small. If intramolecular spin-spin interactions are modulated at a rapid rate because of molecular reorientations, one expects a spread in the components of **D**. Since the trace of **D** is zero, the contribution of the term $\hat{S} \cdot \boldsymbol{D} \cdot \hat{S}$ may become negligible. Two limiting cases may be considered:

1. When D and E are large, the modulations of the spin-spin interaction will lead to so short a spin lifetime that the averaged spectrum would have undetectably broad lines.
2. When D and E are very small, the line-broadening effects in nonrigid media will also be small. In the absence of hyperfine splitting, one would see a single line, as if the spectrum were due to a system with $S = \frac{1}{2}$.

A system which may be an example of case 2 is the ion quartet shown below. Two ketyl anions (formed by reaction of carbonyl compounds and an alkali metal) may be bound by two alkali ions to form an ion quartet.‡

$$\begin{array}{ccc} R & Na^+ & R \\ \diagdown & & \diagup \\ C-O^- & {}^-O-C & \\ \diagup & Na^+ & \diagdown \\ R & & R \end{array}$$

These systems have very small D/hc values (0.007 to 0.015 cm^{-1}) at 77 K. At room temperature they show a seven-line spectrum from two equivalent

---

† A. M. Trozzolo, E. Wasserman, and W. A. Yager, *J. Chim. Phys.*, **1964**:1663.
‡ N. Hirota and S. I. Weissman, *J. Am. Chem. Soc.*, **86**:2538 (1964).

alkali-metal nuclei, just *as if* the second ketyl unit were not present. The two alkali atoms may be replaced by one divalent ion such as Be^{++} or Mg^{++}. The latter may also bind two α,α'-bipyridine anions (II)

(II)

The very low E/hc value (<0.0012 cm^{-1}) has led to the suggestion that the nitrogen atoms are tetrahedrally disposed about the metal ion.[†]

10-7 THE "$\Delta M_S = 2$" TRANSITIONS

At high fields, where the quantum numbers $M_S = +1, 0$, and -1 correspond to the eigenfunctions of the spin hamiltonian, a "$\Delta M_S = 2$" transition is not allowed if the microwave field is perpendicular to the static field. However, at low field, the wave functions become linear combinations of the high-field states (Sec. 10-4). The usual $\Delta M_S = \pm 1$ selection rule does not apply, since one cannot assign M_S values to the states. The "$\Delta M_S = 2$" transition is permitted for the component of microwave field *parallel* to the static field. This can be shown by taking the $\hat{S}_z$ matrix element for the states

$$c_1|-1\rangle - c_2|+1\rangle$$
$$c_2|-1\rangle + c_1|+1\rangle$$

between which the "$\Delta M_S = 2$" transition occurs. (See Prob. 10-3.) When **H** is at an arbitrary orientation to the principal axes, the $|1\rangle$, $|0\rangle$ and $|-1\rangle$ states are mixed by the spin-spin interaction. Hence "$\Delta M_S = 2$" transitions can be seen in a normal ESR cavity[‡] (i.e., with the microwave field perpendicular to the static field).

The position of the low-field side of the "$\Delta M_S = 2$" transition in randomly oriented solids does not correspond with that of the low-field X, Y, or Z components from Fig. 10-4 but occurs at a turning point $H_{\min}$.[§] It is possible to show that the angular dependence of the lines in the triplet spectrum for any frequency ν is given by[¶]

$$W(H,\nu) = \mathscr{X} \sin^2\theta \cos^2\phi + \mathscr{Y} \sin^2\theta \sin^2\phi + \mathscr{Z} \cos^2\theta$$
$$= \frac{\mathscr{X}\mathscr{Y}\mathscr{Z}}{(g\beta H)^2} \mp 3^{-\frac{3}{2}}\left[\frac{h^2\nu^2 + \mathscr{X}\mathscr{Y} + \mathscr{X}\mathscr{Z} + \mathscr{Y}\mathscr{Z}}{(g\beta H)^2} - 1\right]$$
$$\times \sqrt{4(g\beta H)^2 - h^2\nu^2 - 4(\mathscr{X}\mathscr{Y} + \mathscr{X}\mathscr{Z} + \mathscr{Y}\mathscr{Z})}. \quad (10\text{-}30)$$

Here θ and ϕ are the Euler angles of Fig. A-6 for **H** $\parallel Z$. The function $W(H,\nu)$ is plotted against **H** in Fig. 10-11a for the naphthalene triplet molecule at $\nu = 9.279$ GHz.

[†] I. M. Brown, S. I. Weissman and L. C. Snyder, *J. Chem. Phys.*, **42**:1105 (1965).
[‡] M. S. de Groot and J. H. van der Waals, *Mol. Phys.*, **3**:190 (1960).
[§] J. H. van der Waals and M. S. de Groot, *Mol. Phys.*, **2**:333 (1959); **3**:190 (1960).
[¶] P. Kottis and R. Lefebvre, *J. Chem. Phys.*, **39**:393 (1963).

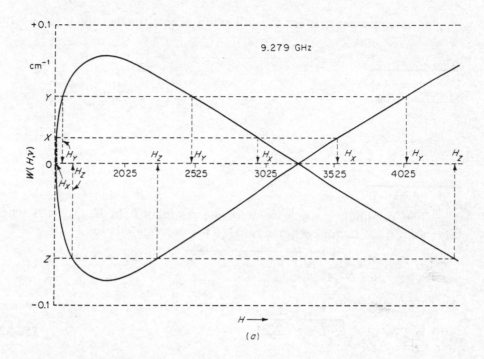

(a)

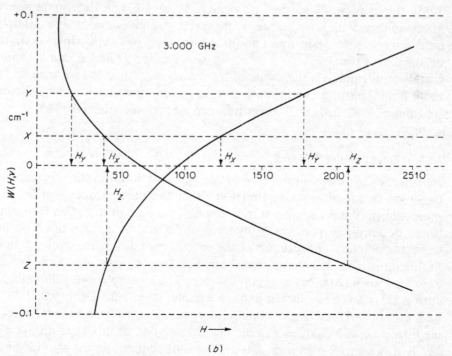

(b)

Fig. 10-11 The function $W(H, \nu)$ as a function of magnetic field for naphthalene in the triplet state. The dotted lines correspond to the resonant fields when the magnetic field is oriented along one of the principal axes of the system. (a) $\nu = 9.279$ GHz. (b) $\nu = 3.000$ GHz. [*Taken from P. Kottis and R. Lefebvre, J. Chem. Phys.*, **39**:393 (1963).]

It is apparent that for the following special angles, W takes on the values $\mathcal{X}$, $\mathcal{Y}$ and $\mathcal{Z}$.

W	θ	ϕ
$\mathcal{X}$	$\dfrac{\pi}{2}$	0
$\mathcal{Y}$	$\dfrac{\pi}{2}$	$\dfrac{\pi}{2}$
$\mathcal{Z}$	0	0

The minimum possible value of the resonant field $H_{\min}$ occurs when the square-root factor becomes zero. This corresponds to

$$H_{\min} = \frac{1}{g\beta}\sqrt{\frac{h^2\nu^2}{4} + \mathcal{XY} + \mathcal{XZ} + \mathcal{YZ}} \tag{10-31}$$

An equivalent form is

$$H_{\min} = \frac{1}{g\beta}\sqrt{\frac{h^2\nu^2}{4} - \frac{D^2 + 3E^2}{3}} \tag{10-32}$$

Thus the g factor (assumed isotropic), D, and E will determine the minimum field at which resonance is observed at a given value of ν. The low-field edge of the derivative line for a randomly oriented triplet system can be used to estimate $\sqrt{D^2 + 3E^2}$. In some cases D and E can be approximately determined if the shape of the "$\Delta M_S = 2$" line is analyzed.[†] If the zero field splitting is appropriately large compared with the microwave frequency, *no* "$\Delta M_S = 2$" transition can occur (see Fig. 10-11b).

10-8 TRIPLET GROUND STATES

In Fig. 10-3d, the lowest energy state (ground state) is that corresponding to single occupation with parallel spins of the highest occupied levels. A molecule may have a ground triplet state in the neutral, cationic, or anionic form, depending upon which involves one electron in each of a pair of degenerate orbitals. Inspection of the orbital energies of the cyclopentadiene molecule in Fig. 5-3 shows that the cation (having four electrons) would be expected to have a triplet ground state. The pentaphenylcyclopentadienyl cation (III) has been shown to have a triplet state with $D/hc = 0.1050$ cm^{-1} and $E = 0$.[‡] Degenerate orbital energy levels are found in molecules with an n-fold ($n \geq 3$) axis of symmetry. Molecules of this type do not *necessarily* have a triplet ground state. This is dependent on the sign of the elec-

[†] M. S. de Groot and J. H. van der Waals, *Mol. Phys.*, **3**:190 (1960).
[‡] R. Breslow, H. W. Chang and W. A. Yager, *J. Am. Chem. Soc.*, **85**:203 (1963).

$\phi = C_6H_5$

(III)

tron exchange integral. [See Eq. (10-34).] If this integral is positive the singlet state will lie lower. This is the case for the coronene dinegative ion.[†]

Occupancy by one electron in each of two degenerate orbitals of symmetrically substituted benzenes may be achieved if the dianion can be formed. Triplet ground states have been demonstrated for symmetrical molecules such as the triphenylbenzene (IV) or triphenylene dianions (V). These possess degenerate antibonding orbitals, analogous to those shown in Fig. 10-12 for the hypothetical benzene dianion.

(IV)   (V)

The D value for the triphenylbenzene dianion (ground triplet state, Fig. 10-8) is less ($D/hc = 0.042$ cm^{-1}) than that of the neutral excited triplet state molecule ($D/hc = 0.111$ cm^{-1}).[‡] The orbital occupation is very different for these two cases; calculations show that in the excited triplet molecule, there is a

[†] M. Glasbeek, J. D. W. van Voorst, and G. J. Hoijtink, *J. Chem. Phys.*, **45**:1852 (1966).
[‡] R. E. Jesse, P. Biloen, J. D. W. van Voorst, and G. J. Hoijtink, *Mol. Phys.*, **6**:633 (1963).

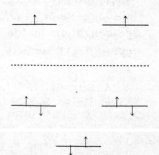

Fig. 10-12 Configuration of the hypothetical benzene dianion.

greater interaction (leading to a larger D value) between two electrons in the "paired" bonding and antibonding orbitals than between two electrons in the antibonding orbitals of the ground state dianion.

10-9 CARBENES AND NITRENES

Two of the simplest of ground-triplet-state molecules are O_2 and S_2; hence a molecule need be no larger than diatomic if it has an appropriate set of degenerate orbitals. The ultimate lower limit would be one atom (possibly with other nonparticipating attached atoms) with two degenerate orthogonal orbitals. The H—$\dot{\text{C}}$—H fragment (methylene or "carbene") is such a system, and its triplet ground state has been established spectroscopically. The ESR spectrum of methylene has been reported both for $\dot{\text{C}}\text{H}_2$ and $\dot{\text{C}}\text{D}_2$.[†] For the former, $D = 0.69$ cm^{-1} and $E = 0.003$ cm^{-1}; for the latter, $D = 0.75$ cm^{-1} and $E = 0.011$ cm^{-1}. Many substituted methylenes have also been studied. (See Table 10-1.) The nonzero value of E indicates that these molecules have no axis of symmetry of order 3 or greater; this indicates that the molecules are nonlinear. For such systems, the maximum number of lines (six $\Delta M_S = 1$ transition) is expected in the ESR spectrum, just as for naphthalene in the excited triplet state. When the system is nearly axial, the E parameter may be so small that one may only be able to set an upper limit. An increase in the extent of the conjugated system attached to the methylene carbon atom may lead to a decrease in the parameter D, as shown in Table 10-1. Conversely, it is of great interest to calculate the value of D when the two coupled π electrons are on the same carbon atom, viz., for $\dot{\text{C}}\text{H}_2$. Early calculations gave $D = 0.906$ cm^{-1}, $E = 0$[‡] or $D = 1.1$ cm^{-1}.[§] However, a later calculation gives $D = 0.71$ cm^{-1} and $E = 0.05$ cm^{-1}.[¶] Since the experimental values for $\dot{\text{C}}\text{D}_2$ are now at hand, the interpretation of the zero field parameters for the two methylenes is likely to be a field of continuing interest. In these small molecules, one must be concerned about a possible contribution to D from spin-orbit coupling. In O_2, this contribution is appreciable. However, calculations for $\dot{\text{C}}\text{H}_2$ suggest that the spin-orbit contribution to D and E is negligible. (In Chap. 11 we shall be dealing with the opposite extremes—the cases in which the zero field splittings may arise entirely from spin-orbit coupling.) The large value of D/hc in H—$\dot{\text{C}}$—C≡N, notwithstanding the possibility of delocalization in the C≡N group, is probably due to the existence of a negative spin density on the central atom of the C—C≡N group. This π

† R. A. Bernheim, H. W. Bernard, P. S. Wang, L. S. Wood, and P. S. Skell, *J. Chem. Phys.*, **53**:1280 (1970); **54**:3223 (1971); R. A. Bernheim, R. J. Kempf, and E. F. Reichenbecher, *J. Mag. Res.*, **3**:5 (1970). E. Wasserman, V. J. Kuck, R. S. Hutton, E. D. Anderson, and W. A. Yager, *J. Chem. Phys.*, **54**:4120 (1971).

‡ J. Higuchi, *J. Chem. Phys.*, **38**:1237 (1963).

§ R. D. Sharma, *J. Chem. Phys.*, **38**:2350 (1963); **41**:3259 (1964).

¶ J. F. Harrison (to be published).

Table 10-1   Zero field splitting parameters for ground-triplet-state molecules

Ground triplet molecule	$\dfrac{\|D\|}{hc}$ cm^{-1}	$\dfrac{\|E\|}{hc}$ cm^{-1}	Reference
H—Ċ—H	0.69	0.003	†
D—Ċ—D	0.75	0.011	†
H—Ċ—C≡N	0.8629	0	‡
H—Ċ—CF$_3$	0.72	0.021	§
H—Ċ—C$_6$H$_5$	0.5150	0.0251	¶
H—Ċ—C≡CH	0.6276	0	‡
H—Ċ—C≡C—CH$_3$	0.6263	0	‡
H—Ċ—C≡C—C$_6$H$_5$	0.5413	0.0035	‡
C$_6$H$_5$—Ċ—C$_6$H$_5$	0.4055	0.0194	¶
N≡C—Ċ—C≡N	1.002	<0.002	††
Ṅ—C≡N	1.52	<0.002	††

† R. A. Bernheim, H. W. Bernard, P. S. Wang, L. S. Wood and P. S. Skell, *J. Chem. Phys.*, **53**:1280 (1970); **54**:3223 (1971); R. A. Bernheim, R. J. Kempf and E. F. Reichenbecher, *J. Mag. Res.*, **3**:5 (1970); E. Wasserman, V. J. Kuck, R. S. Hutton, E. D. Anderson, and W. A. Yager, *J. Chem. Phys.*, **54**:4120 (1971).

‡ R. A. Bernheim, R. J. Kempf, J. V. Gramas, and P. S. Skell. *J. Chem. Phys.*, **43**:196 (1965).

§ E. Wasserman, cited in Ref.†

¶ E. Wasserman, A. M. Trozzolo, W. A. Yager, and R. W. Murray. *J. Chem. Phys.*, **40**:2408 (1964).

†† E. Wasserman, L. Barash, and W. A. Yager, *J. Am. Chem. Soc.*, **87**:2075 (1965).

system is akin to that of the allyl radical.   The expected negative spin density on the central atom would thus lead to an increased positive spin density on the left-hand carbon atom and hence to an increased value of D. Such an effect cannot occur with the H—Ċ—CF$_3$ molecule listed on the fourth line of Table 10-1.   The effect is probably operative in the molecule H—Ċ—C≡C—H also, and assuredly also in N≡C—Ċ—C≡N, where there are five π-electron centers.

Table 10-1 also gives the parameters for variously substituted nitrenes, R—Ṅ, which are isoelectronic with the carbenes.   The parent compound for the nitrenes is Ṅ—H.   It has been estimated that $D/hc = 1.86$ cm^{-1} for this fragment.†   For Ṅ—C≡N the reduction in the D value by delocalization is probably somewhat offset by the enhancement of the positive spin density on the nitrogen atoms due to a negative spin density on the carbon.

It is also possible to prepare dicarbenes such as

† J. A. R. Coope, J. B. Farmer, C. L. Gardner, and C. A. McDowell, *J. Chem. Phys.*, **42**:54 (1965).

For this molecule $D/hc = 0.0521$ and $E/hc < 0.002$ cm^{-1}. Here the ground state is presumably a triplet, with each electron located primarily on one of the two methylene carbon atoms. The low value of D/hc is consistent with a large interelectronic mean distance. Likewise, one may also observe triplet ESR spectra from dinitrenes:

$$\dot{N}=\!\!\!\!\bigcirc\!\!\!\!=\dot{N} \qquad \frac{D}{hc} = 0.0675, \; E \sim 0\dagger$$

Carbenes or nitrenes may be generated in a rigid matrix by photolysis. The starting material for the former is the appropriate diazo compound $R\!\!-\!\!C(N_2)\!\!-\!\!R'$, whereas for the latter it is the azide $R\!\!-\!\!CN_3$. (The evolved nitrogen does not interfere with the stability of the products.) Although irradiation is used to generate both excited- and ground-state triplets, the latter persist indefinitely after cessation of irradiation, whereas the former decay with a characteristic time. Of course the triplet stability is contingent upon the rigidity of the matrix which retards diffusion and subsequent reaction. It is especially important to retard the diffusion of oxygen to triplet-state molecules. A few ground-state triplet molecules have been studied at room temperature in a polymer. Here the diffusion of oxygen is no longer negligible.‡

If the zero field splitting D is large compared with $h\nu$ for a microwave quantum, only certain of the lines allowed by the selection rules will be observed. Figure 10-13 shows the energy-level diagrams for the fluorenylidene molecule (VI).§ The molecule is generated in its ground triplet state

$$\underset{(VI)}{\bigcirc\!\!\!\!\bigcirc}$$

by irradiation of diazofluorene. It is thus to be regarded as a derivative of methylene.

When **H** is parallel to the X or the Y axes, only one transition is observed for $\nu \approx 9.7$ GHz. Since the ordinate is expressed in gigahertz, the frequency required to cause a transition between adjacent levels is immediately apparent. For $\mathbf{H} \parallel Z$, three transitions are expected and are observed. Two of these are between the levels designated by $|0\rangle$ and $|-1\rangle$. Note that the "$\Delta M_S = 2$" transition occurs at an intermediate value of the magnetic field.

† A. M. Trozzolo, R. W. Murray, G. Smolinsky, W. A. Yager, and E. Wasserman, *J. Am. Chem. Soc.*, **85**:2526 (1963).

‡ C. Thomson, *J. Chem. Phys.*, **41**:1 (1964).

§ C. A. Hutchison Jr. and G. A. Pearson, *J. Chem. Phys.*, **47**:520 (1967).

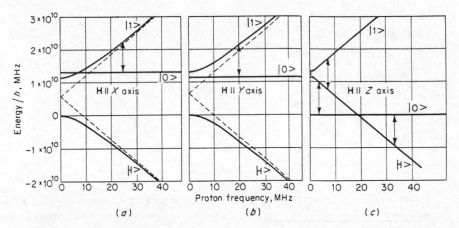

Fig. 10-13 Energy levels for fluorenylidene (VI) in the triplet ground state. (a) $H \parallel \langle X \rangle$. (b) $H \parallel \langle Y \rangle$. (c) $H \parallel \langle Z \rangle$. [Figures after C. A. Hutchison Jr. and G. A. Pearson, J. Chem. Phys., **47**:520 (1967).]

10-10 THERMALLY ACCESSIBLE TRIPLET STATES

Sections 10-5 and 10-8 of this chapter have dealt with systems in which the triplet state of interest is respectively an excited state or the ground state. In either case we have tacitly assumed that the separation of the triplet state from a nearby singlet state is large enough so that one need not consider mixing of the two states. An additional interesting case is that in which the singlet-triplet separation $W_T - W_S = \Delta W$ is small enough to make the triplet state thermally accessible but still without serious mixing of states. The single-triplet separation is just Jh, where the exchange interaction between two electrons is given by $Jh\hat{S}_1 \cdot \hat{S}_2$. (See Eqs. 10–33 and 10–34.) This is more commonly written as $J\hat{S}_1 \cdot \hat{S}_2$; however, in parallel with the notation for hyperfine coupling, we write Jh as an energy if J is in hertz. The population of the triplet state will be governed by the Boltzmann factor

$$3 \exp\left(-\frac{\Delta W}{kT}\right) = 3 \exp\left(\frac{-Jh}{kT}\right)$$

For a given population of a paramagnetic state, the intensity of ESR absorption will be given by a Curie law dependence, i.e., $\mathscr{I} \propto J/T$. Hence the intensity of ESR absorption arising from a thermally excited triplet state should be proportional to $(3/T) \exp(-Jh/kT)$. Thus a study of the temperature dependence of the ESR intensity permits a determination of the value of J. The energy levels for such a system are shown in Fig. 10-14. These energies will be derived in Sec. 10-11.

A clear-cut example of a thermally accessible triplet state is provided by copper acetate.† At 90 K, an ESR line is detected at zero magnetic field

† B. Bleaney and K. D. Bowers, Proc. Roy. Soc. (London), **A214**:451 (1952).

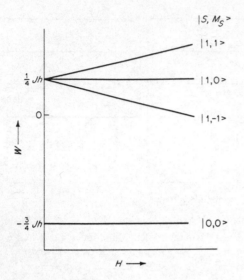

Fig. 10-14 The state energies of a system of two interacting electrons. When J is positive, the singlet state lies lower in energy. The states are labeled by their S, M_S values at high fields.

for a frequency of 3.3 GHz. A single Cu^{++} ion has a spin $S = \frac{1}{2}$, and it must hence retain a twofold Kramers degeneracy; therefore, no absorption from single Cu^{++} ions is possible at zero magnetic field. At room temperature the line at zero field is somewhat more intense than that at 90 K, but at 20 K no absorption is observed. However, at higher frequencies, the ESR spectrum is qualitatively similar to that of the Ni^{++} ion, for which the spin is known to be 1. These data are readily accounted for by assuming association of pairs of copper ions. (The x-ray determination[†] of the crystal structure has verified the occurrence of Cu^{++} pairs with a Cu—Cu separation of 0.264 nm.) From the variation of the line intensity with temperature, the value of $J/c = 260$ cm^{-1}. Alternatively, $Jh/k = 370$ K. The measured zero field splitting parameters D/hc and E/hc are, respectively, 0.34 and 0.01 cm^{-1}. The hyperfine coupling from the $1:2:3:4:3:2:1$ ^{63,65}Cu septet is given by $A/c \approx 0.008$ cm^{-1}. Thus

$$hJ \gg g\beta H_r \gg hA$$

10-11 BIRADICALS. EXCHANGE INTERACTION

A biradical is formed by the linkage of two molecular fragments, each containing one unpaired electron.[‡] The biradical feature of paramount interest is the electron exchange coupling between the two fragments. For fast exchange (i.e., strong coupling) the ground state of the biradical will be either a singlet or a triplet. However, when exchange is weak, the ESR spectrum of the biradical is analogous to that expected for a combination of two

† J. N. van Niekerk and F. K. L. Schoening, *Nature*, **171**:36 (1953); *Acta Cryst.*, **6**:501 (1953).
‡ S. H. Glarum and J. H. Marshall, *J. Chem. Phys.*, **47**:1374 (1967).

independent doublet radicals. In this case, the singlet and triplet states are almost degenerate.

Consider a biradical in which each molecular fragment contains one magnetic nucleus of spin I which gives rise to a hyperfine splitting. The spin hamiltonian appropriate to this system is[†]

$$\hat{\mathscr{H}} = g\beta H(\hat{S}_{1z} + \hat{S}_{2z}) + hA_0(\hat{S}_{1z}\hat{I}_{1z} + \hat{S}_{2z}\hat{I}_{2z}) + hJ\hat{\mathbf{S}}_1 \cdot \hat{\mathbf{S}}_2 \qquad (10\text{-}33)$$

Here $\hat{\mathbf{S}}_1$ and $\hat{\mathbf{S}}_2$ are the spin operators for electrons 1 and 2, and J is the exchange integral defined by

$$J = h^{-1}\langle\phi_A(1)\phi_B(2)|\, \frac{e^2}{r_{12}} \,|\phi_B(1)\phi_A(2)\rangle \qquad (10\text{-}34)$$

Either a positive or a negative sign has been used in the literature for the third term of Eq. (10-33). We choose a positive sign to be consistent with atomic spectroscopy. Then a negative value of J corresponds to the case of the triplet ground state. ϕ_A is a one-electron orbital on fragment A and ϕ_B is the orbital on fragment B. r_{12} is the separation of electrons 1 and 2. For a biradical in a solid, one must also consider an additional term, $\hat{\mathbf{S}}_1 \cdot \mathbf{D} \cdot \hat{\mathbf{S}}_2$ in Eq. (10-33); however, since the trace of $\mathbf{D}$ is zero, the effects of this term average out in a liquid, providing the elements of $\mathbf{D}$ are small.

First, consider the case where $|J| >> |A_0|$. The zero-order hamiltonian is then

$$\hat{\mathscr{H}}_0 = g\beta H(\hat{S}_{1z} + \hat{S}_{2z}) + hJ\hat{\mathbf{S}}_1 \cdot \hat{\mathbf{S}}_2 \qquad (10\text{-}35)$$

which can be expanded as (see Sec. C-1)

$$\hat{\mathscr{H}}_0 = g\beta H(\hat{S}_{1z} + \hat{S}_{2z}) + hJ[\hat{S}_{1z}\hat{S}_{2z} + \tfrac{1}{2}(\hat{S}_{1+}\hat{S}_{2-} + \hat{S}_{1-}\hat{S}_{2+})] \qquad (10\text{-}36)$$

The following set of functions are eigenfunctions of $\hat{\mathscr{H}}_0$

$$|T_1\rangle = |\tfrac{1}{2}, \tfrac{1}{2}\rangle \equiv |1, 1\rangle \qquad (10\text{-}37a)$$

$$|T_0\rangle = \frac{1}{\sqrt{2}}\,(|\tfrac{1}{2}, -\tfrac{1}{2}\rangle + |-\tfrac{1}{2}, \tfrac{1}{2}\rangle) \equiv |1, 0\rangle \qquad (10\text{-}37b)$$

$$|T_{-1}\rangle = |-\tfrac{1}{2}, -\tfrac{1}{2}\rangle \equiv |1, -1\rangle \qquad (10\text{-}37c)$$

$$|S_0\rangle = \frac{1}{\sqrt{2}}\,(|\tfrac{1}{2}, -\tfrac{1}{2}\rangle - |-\tfrac{1}{2}, \tfrac{1}{2}\rangle) \equiv |0, 0\rangle \qquad (10\text{-}37d)$$

The quantum numbers in the middle sets of functions refer to the eigenvalues of $\hat{S}_{1z}$ and $\hat{S}_{2z}$, whereas those on the right refer to the eigenvalues of $\hat{S}^2$ and $\hat{S}_z$, where $\hat{S}_z = \hat{S}_{1z} + \hat{S}_{2z}$.

† D. C. Reitz and S. I. Weissman, *J. Chem. Phys.*, **33**:700 (1960); G. R. Luckhurst, *Mol. Phys.*, **10**:543 (1966).

The eigenvalues of $\hat{\mathcal{H}}_0$ are

$$W^{(0)}_{1,1} = g\beta H + \frac{hJ}{4} \tag{10-38a}$$

$$W^{(0)}_{1,0} = +\frac{hJ}{4} \tag{10-38b}$$

$$W^{(0)}_{1,-1} = -g\beta H + \frac{hJ}{4} \tag{10-38c}$$

$$W^{(0)}_{0,0} = -\frac{3hJ}{4} \tag{10-38d}$$

These energies are plotted in Fig. 10-14 for $J > 0$. If one takes the remaining term of Eq. (10-33) as the perturbation hamiltonian

$$\hat{\mathcal{H}}' = hA_0(\hat{S}_{1z}\hat{I}_{1z} + \hat{S}_{2z}\hat{I}_{2z}) \tag{10-39}$$

the energies to first order are

$$W^{(1)}_{1,1} = g\beta H + \frac{hJ}{4} + \frac{hA_0}{2}M_I \tag{10-40a}$$

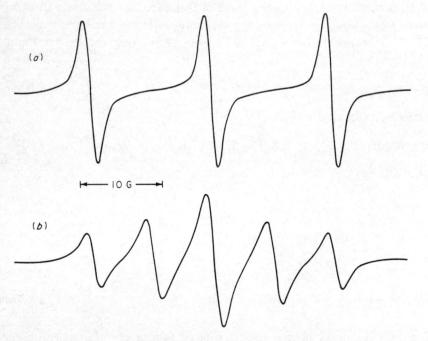

Fig. 10-15 (a) ESR spectrum of the radical (VII) in dimethylformamide; $a = 14.80$ G. (b) ESR spectrum of biradical (VIII) showing interaction with both nitroxide nitrogen atoms. The spacing between the lines is 7.40 G. This is an illustration of the case $|J| \gg |A_0|$; i.e., the line spacing is $\frac{1}{2}a$. [*Spectrum kindly supplied by A. Rassat. See R. M. Dupeyre, H. Lemaire, and A. Rassat, J. Am. Chem. Soc.*, **87**:3771 (1965).]

$$W_{1,0}^{(1)} = +\frac{hJ}{4} \tag{10-40b}$$

$$W_{1,-1}^{(1)} = -g\beta H + \frac{hJ}{4} - \frac{hA_0}{2} M_I \tag{10-40c}$$

$$W_{0,0}^{(0)} = -\frac{3hJ}{4} \tag{10-40d}$$

Here $M_I = M_I(1) + M_I(2)$. In the case of $I_1 = I_2 = 1$, the spectrum would consist of five lines separated by $\frac{1}{2}a$ ($a = hA_0/g\beta$) with intensity ratios $1:2:3:2:1$. This should be contrasted with the case of a radical ($S = \frac{1}{2}$) in which one nucleus has $I = 1$. In the latter case, pairs of lines would be separated by a. An example of the latter case is given in Fig. 10-15a, where the radical (VII) is

(VII)

The hyperfine splitting is 14.80 G.† However, in the case of the biradical (VIII)

(VIII)

the line separation is only 7.40 G (See Fig. 10-15b.)

$\hat{\mathscr{H}}'$ of Eq. (10-39) also causes a mixing of the triplet-state functions, Eqs. (10-37a) to (10-37c), and the singlet-state function, Eq. (10-37d). This can be seen by calculating the first-order corrections to the wave functions of Eqs. (10-37). (See Sec. A-7.) The only functions which are mixed are Eqs. (10-37b) and (10-37d). The corrected functions are then

$$|``T_0"\rangle = |1, 0, M_I(1), M_I(2)\rangle + \lambda|0, 0, M_I(1), M_I(2)\rangle \tag{10-41a}$$

$$|``S_0"\rangle = |0, 0, M_I(1), M_I(2)\rangle - \lambda|1, 0, M_I(1), M_I(2)\rangle \tag{10-41b}$$

† R. M. Dupeyre, H. Lemaire, and A. Rassat, *J. Am. Chem. Soc.*, **87:**3771 (1965).

where

$$\lambda = hA_0 \frac{\langle 1, 0, M_I(1), M_I(2)|(\hat{S}_{1z}\hat{I}_{1z} + \hat{S}_{2z}\hat{I}_{2z})|0, 0, M_I(1), M_I(2)\rangle}{W_{1,0}^{(0)} - W_{0,0}^{(0)}}$$

$$= +\frac{A_0}{2J} [\langle \tfrac{1}{2}, -\tfrac{1}{2}, M_I(1), M_I(2)|(\hat{S}_{1z}\hat{I}_{1z} + \hat{S}_{2z}\hat{I}_{2z})|\tfrac{1}{2}, -\tfrac{1}{2}, M_I(1), M_I(2)\rangle$$

$$- \langle -\tfrac{1}{2}, \tfrac{1}{2}, M_I(1), M_I(2)|(\hat{S}_{1z}\hat{I}_{1z} + \hat{S}_{2z}\hat{I}_{2z})|-\tfrac{1}{2}, \tfrac{1}{2}, M_I(1), M_I(2)\rangle]$$

$$= +\frac{A_0}{2J} [M_I(1) - M_I(2)] \qquad (10\text{-}42)$$

Thus the two states mix only when $M_I(1) \neq M_I(2)$. The physical reason is that the singlet and triplet states mix only when the two radical fragments can be identified. When $M_I(1) = M_I(2)$ the two fragments are indistinguishable.

Now consider the case of $|J| \ll |A_0|$. The zero-order spin hamiltonian is

$$\mathcal{H}_0 = g\beta H(\hat{S}_{1z} + \hat{S}_{2z}) + hA_0(\hat{S}_{1z}\hat{I}_{1z} + \hat{S}_{2z}\hat{I}_{2z}) \qquad (10\text{-}43)$$

$\mathcal{H}_0$ is separable into two parts,

$$\mathcal{H}_0(1) = g\beta H\hat{S}_{1z} + hA_0\hat{S}_{1z}\hat{I}_{1z} \qquad (10\text{-}44a)$$

$$\mathcal{H}_0(2) = g\beta H\hat{S}_{2z} + hA_0\hat{S}_{2z}\hat{I}_{2z} \qquad (10\text{-}44b)$$

The eigenfunctions of $\mathcal{H}_0(1)$ are

$$|\tfrac{1}{2}, M_I(1)\rangle \qquad |-\tfrac{1}{2}, M_I(1)\rangle$$

with eigenvalues

$$W_{\frac{1}{2}}^{(0)} = +\tfrac{1}{2}g\beta H + \tfrac{1}{2}hA_0M_I(1) \qquad (10\text{-}45a)$$

$$W_{-\frac{1}{2}}^{(0)} = -\tfrac{1}{2}g\beta H - \tfrac{1}{2}hA_0M_I(2) \qquad (10\text{-}45b)$$

One obtains identical results for $\mathcal{H}_0(2)$. Thus, this case may be considered as two noninteracting systems with $S = \tfrac{1}{2}$. When $I = 1$, the ESR spectrum consists of three lines separated by a gauss. An example of this case is shown in Fig. 10-16 where the biradical is (IX).

(IX)

The intermediate case of $|J| \sim |A_0|$ gives rise to a complex group of lines. The intensity and position of these lines are a strong function of $|J|/|A_0|$.[†,‡]

† S. H. Glarum and J. H. Marshall, *J. Chem. Phys.*, **47**:1374 (1967).
‡ R. Brière, R. M. Dupeyre, H. Lemaire, C. Morat, A. Rassat, and P. Rey, *Bull. Soc. Chim. France*, **11**:3290 (1965).

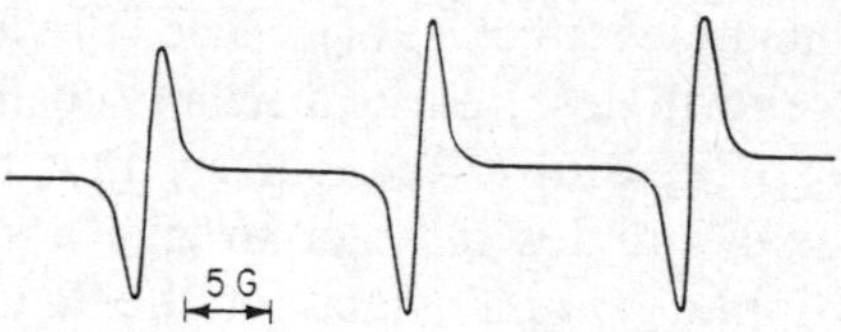

Fig. 10-16 ESR spectrum of the biradical (IX). The hyperfine splitting (15.6 G) is just the same as that of the corresponding mono-radical. This is an illustration of the case $|J| << |A_0|$. [*Spectrum taken from R. Brière, R. M. Dupeyre, H. Lemaire, C. Morat, A. Rassat, and P. Rey, Bull. Soc. Chim. France,* **11**:3290 (1965).]

One of the first biradicals to be studied was the compound (X) known as Chichibabin's hydrocarbon.† The solid-state ESR spectrum was origi-

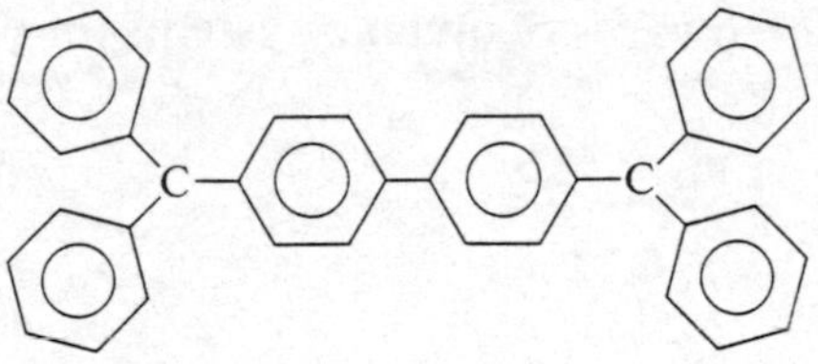

(X)

nally interpreted as an example of a thermally excited triplet state. This requires that $hJ \approx kT \approx 200$ cm^{-1}.

In solution, the ESR spectrum is very similar to that of triphenyl-methyl.‡ This suggests that $|J| << |A_0|$. Since $A_0 \sim 10^{-3}$ to 10^{-4} cm^{-1}, these two experiments appear to give inconsistent results. The great disparity between these J values has been termed the "biradical paradox."§ A clue to the resolution of this apparent paradox was given in 1964 when it was shown that the Chichibabin hydrocarbon polymerizes in solution.¶ This polymerization can be represented schematically in Fig. 10-17. Since the intermediate electrons are paired, it is the remote electrons which couple to give a very small value of J.

10-12 SYSTEMS WITH $S > 1$

Outside the realm of transition-metal or rare-earth ions, there are few documented examples of systems with $S > 1$. Systems with $S \geqslant \frac{3}{2}$ are also described by the zero field splitting hamiltonian, Eqs. (10-16). Perhaps

† C. A. Hutchison Jr., A. Kowalsky, R. C. Pastor, and G. W. Wheland, *J. Chem. Phys.,* **20**:1485 (1952).
‡ D. C. Reitz and S. I. Weissman, *J. Chem. Phys.,* **33**:700 (1960).
§ H. M. McConnell, *J. Chem. Phys.,* **33**:1868 (1960).
¶ R. K. Waring, Jr., and G. J. Sloan, *J. Chem. Phys.,* **40**:772 (1964).

Fig. 10-17 Coupling of biradicals to give a polymer biradical in which the two electrons are remote.

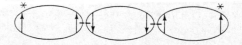

the most intriguing is the R center, briefly described as a point defect in Sec. 8-5c. The system of three interacting electrons, occupying adjacent vacancies in a (111) plane, obviously possesses a Kramers degeneracy. Nevertheless, early attempts to find an ESR signal from this center were unsuccessful. Recognizing that the two electrons in the M center in the alkali halides interact to give a singlet ground state, one would expect a doublet ground state for the R center. Irradiation in either of a pair of optical R bands gives an appreciable concentration of molecules in the quartet state ($S = \frac{3}{2}$). The assumed model (due to van Doorn and Haven†) predicts axial symmetry, and hence E should be zero. This prediction is verified, with $g_\parallel = g_\perp = 1.996$ and with $D' = 168.5$ G.‡ The "$\Delta M_S = 2$" transitions are also seen.

There have been a few documented cases of organic systems for which $S = 2$. An example is (XI).§

(XI)

The parameters $D/hc = 0.071$ cm^{-1} and $E/hc = 0.019$ cm^{-1} of this quintet-state molecule should be contrasted with the values 0.4055 and 0.0194 cm^{-1} for the triplet state of diphenyl carbene (see Table 10-1).

A general treatment of spin hamiltonians for systems with $S = 2$ has been given.¶ The observed transitions depend very strongly on the zero field splittings.

BIBLIOGRAPHY

1. Zahlen, A. B. (ed.). "The Triplet State—Beirut Symposium." Cambridge University Press, Cambridge, England, 1967.
2. McGlynn, S. P., T. Azumi, and M. Kinoshita, "Molecular Spectroscopy of the Triplet State," Prentice-Hall, Inc., Englewood Cliffs, N.J., 1969.
3. Wasserman, E., L. C. Snyder, and W. A. Yager, ESR of the Triplet States of Randomly Oriented Molecules, *J. Chem. Phys.*, **41**:1763 (1964).

PROBLEMS

10-1. By substitution of the appropriate spin matrices, derive the hamiltonian matrix of Eq. (10-19).

† C. Z. van Doorn and Y. Haven. *Philips Research Rept.*, **11**:479 (1956); **12**:309 (1957).
‡ H. Seidel, M. Schwoerer, and D. Schmid, *Z. Physik*, **182**:398 (1965).
§ E. Wasserman, R. W. Murray, W. A. Yager, A. M. Trozzolo, and G. Smolinsky, *J. Am. Chem. Soc.*, **89**:5076 (1967); K. Itoh, *Chem. Phys. Letters*, **1**:235 (1967).
¶ R. D. Dowsing, *J. Mag. Res.*, **2**:332 (1970).

10-2. (a) Obtain the spin matrices S_x, S_y, S_z, $S_x{}^2$, $S_y{}^2$, and $S_z{}^2$ using the eigenfunctions of Eqs. (10-28) as a basis set.

(b) Use spin matrices to obtain the hamiltonian matrix (10-29).

10-3. Show that the $\hat{S}_z$ operator will cause a transition between the low-field states $(c_1|-1\rangle - c_2|1\rangle)$ and $(c_2|-1\rangle + c_1|1\rangle)$. What is the relative intensity of this transition compared with a "$\Delta M_S = 1$" transition at high field? (Apply the $\hat{S}_x$ operator in the latter case.) Thus justify the statements made in Sec. 10-7 concerning the relative orientation of the static and microwave magnetic fields for observation of the "$\Delta M_S = 2$" transition.

10-4. Use the methods developed in Chap. 7 to extract the **D** tensor for the naphthalene triplet in a single crystal of durene from the angular-dependence curves given in Fig. 10-5. Diagonalize **D** and obtain the values of D and E.

10-5. The W center produced by x irradiation of MgO is an example of an $S = 1$ system in an electric field of tetragonal symmetry. This defect consists of two positive holes on opposite sides of a positive ion vacancy, i.e., the array $O^- \square\, O^-$ instead of $O^{--} \square O^{--}$. (The absence of an electron is referred to as a positive hole.) The spin hamiltonian for this system is Eq. (10-16a). For the case of $\mathbf{H} \parallel Z$, where Z is the tetragonal axis $\langle 001 \rangle$ of the defect, the energy-level scheme and the allowed transitions for a system of this type are given in Fig. 10-1.

The *proton* resonance frequencies at the magnetic fields corresponding to the two transitions shown occur at 13.3345 and 15.2680 MHz; $\nu = 9.4174$ GHz.

(a) Calculate the energies of the states in zero field and from these obtain the zero field splitting.

(b) Write expressions for the energies of the two transitions.

(c) From these, find the value of D/hc.

(d) Obtain the value of g (here $g_\parallel$).

(e) What feature of the spectrum could prove that in zero field the $M_S = \pm 1$ states lie below the $M_S = 0$ state, rather than vice versa?

(f) The separation of the pair of lines is given by $(3\mu/2)\langle r^{-3}\rangle(3\cos^2\theta - 1)$, as one expects for interacting dipoles. From this calculate $\langle r^{-3}\rangle$ and hence the separation of the two dipoles of spin $\frac{1}{2}$. (The magnetic moment of a positive hole has the same absolute magnitude as that for the electron.)

10-6. Show that the eigenvalues in Eq. (10-38) follow from the application of $\hat{\mathscr{H}}_0$ of Eq. (10-36) to the eigenfunctions in Eqs. (10-37).

10-7. Verify the final form of Eq. (10-42).

10-8. The zero-field splitting parameters for the triplet exciton in anthracene are given as $D/hc = -0.0058$ cm^{-1} and $E/hc = 0.0327$ cm^{-1}. The low value of D is deceptive, since (for this crystal axis system) $E > D$. After ascertaining the direction cosines of the axes of the anthracene molecules relative to the crystal axes [see V. C. Sinclair, J. M. Robertson and A. McL. Mathieson, *Acta Cryst.*, 3:251 (1950)], show that the parameters ascribable to the individual molecules are $D/hc = 0.0688$ cm^{-1} and $E/hc = -0.0081$ cm^{-1}. [See D. Haarer, D. Schmid and H. C. Wolf, *Phys. Stat. Solidi*, 23:633 (1967).]

11
Transition-metal Ions. I

The transition-metal, rare-earth, and actinide ions, i.e., the members of the $3d$, $4d$, $5d$, $4f$, and $5f$ groups, have been the subject of a host of ESR investigations. Observation of the ESR spectra of these ions in low concentrations is generally no more difficult than for free radicals, though linewidths may be much greater, and low temperatures may be required to see a spectrum at all. However, *even a qualitative understanding of the spectrum of an ion requires a detailed consideration both of the individual ion and of its environment.*[†] The results of such a detailed analysis may provide:

1. Identification of the element, its specific valence state and configuration
2. The symmetry of the crystalline electric field to which an ion is subjected
3. Numerical values for the parameters in the spin hamiltonian

[†] A very clear exposition of the behavior of transition-metal ions is given in "Electron Paramagnetic Resonance," by J. W. Orton, Iliffe Books, Ltd., London, 1968. A more advanced exposition is the chapter Electron Spin Resonance by B. R. McGarvey, in R. L. Carlin (ed.), "Transition Metal Chemistry," Marcel Dekker, Inc., New York, 1966, pp. 89 to 201. The most detailed presentation—yet very readable—is the treatise "Electron Paramagnetic Resonance of Transition Ions" by A. Abragam and B. Bleaney, Oxford University Press, London, 1970.

The behavior of free gaseous ions in a magnetic field may be deduced from a study of the Zeeman effect. However, when these ions are placed in condensed media, their behavior in a magnetic field is profoundly altered. The alteration arises because the symmetry of the electric field is lowered from spherical to cubic, axial, rhombic, or even lower symmetry. Hence orbitals which were degenerate in the free ion become widely separated in energy. The qualitative array of the energy levels resulting from the inter-action of an ion with the electric field of its neighbors is predictable from the symmetry of the field. Therefore, our primary concern is not with the an-gular momentum nor the magnetic moment of the free ion, but with those properties *of the low-lying states which are thermally populated.*

11-1 STATES OF GASEOUS TRANSITION-METAL IONS

The "spectroscopic alphabet" s, p, d, f, etc., is used to represent the orbital angular momentum (in units of $\hbar$) of a *single* electron in an atom, whereas the capital letters S, P, D, F, . . . represent corresponding integral values for the whole atom. Lamentably, s is also used to represent the spin quantum number of a single electron, while S is used for the total spin quantum number of an atom or a paramagnetic sys-tem. The student must live with this state of affairs, which does not introduce excessive confusion. The simplest ions of the $3d$ series are Sc^{++} and Ti^{3+}; these have one $3d$ electron, and hence

$$s = \tfrac{1}{2} \qquad l = 2 \qquad M_l(\text{max}) = 2$$
$$S = \tfrac{1}{2} \qquad L = 2 \qquad M_L(\text{max}) = 2$$

The symbol for the ground state represents the value of L; a left super-script indicates the multiplicity ($2S + 1$). Thus, for a $3d^1$ ion, the ground state is indicated as ^2D.

The next trivalent ion is V^{3+}, which has two $3d$ electrons; as-signment of the L value is simplified by tabulating the quantum num-bers as follows:

Electron	s	l	$M_l(\text{max})$	S	L
1	$\tfrac{1}{2}$	2	2		
2	$\tfrac{1}{2}$	2	1		
Total quantum numbers:				1	3

Here the Pauli principle has been invoked to avoid identical quantum numbers for the two electrons. The ground state has the maximum possible multiplicity, i.e., $2S + 1 = 3$. The L value is taken as the sum of the maximum allowed M_l values, i.e., 3. Hence the ground state of V^{3+} is designated as ^3F.

In elements of low-to-moderate atomic weight, the total spin angular momentum vector $\mathbf{S}\hbar$ [of magnitude $\sqrt{S(S+1)}\hbar$] and the total orbital angular momentum vector $\mathbf{L}\hbar$ [of magnitude $\sqrt{L(L+1)}\hbar$] are coupled vectorially to form the resultant vector $\mathbf{J}\hbar$ [of magnitude $\sqrt{J(J+1)}\hbar$]. The possible values of J range from $L+S$ to $|L-S|$ in integral steps. For example, the Ti^{3+} ion, which has a ground-state configuration of ^2D, has two possible values of J; these are $2+\frac{1}{2}=\frac{5}{2}$ and $2-\frac{1}{2}=\frac{3}{2}$. The J value is indicated on the term symbol by a right subscript. In the above example, the two states are designated by $^2D_{\frac{3}{2}}$ and $^2D_{\frac{5}{2}}$. They are separated in energy because of the interaction between spin- and orbital angular momenta. This interaction is called "spin-orbit coupling." In the Ti^{3+} example the $^2D_{\frac{3}{2}}$ state lies lower in energy. In general, if the d shell is less than half full, the state with the minimum value of J will lie lowest. The reverse is true if the d shell is more than half full.

The ground states of $3d^n$ ("iron group") ions have only three distinct values of the total orbital-angular-momentum quantum number L, namely, 0, 2, and 3, as shown in Table 11-1. They are thus labeled, respectively, as S-, D-, and F-state ions.

The energy of a gaseous ion changes from the value W_0 in the absence of a magnetic field to W in the presence of the field. Here

$$W = W_0 + g'\beta M_J H \tag{11-1}$$

where β and M_J are, respectively, the Bohr magneton and the component of $\mathbf{J}$ in the direction of the field $\mathbf{H}$. The factor

$$g' = 1 + \frac{J(J+1) + S(S+1) - L(L+1)}{2J(J+1)} \tag{11-2}$$

Table 11-1   Ground-state properties of $3d^n$ ions

No. of d electrons	S	L	J	Orbital degeneracy	Spectroscopic symbol	Examples
	(of the ground state)					
1	$\frac{1}{2}$	2	$\frac{3}{2}$	5	$^2D_{\frac{3}{2}}$	Sc^{++}, Ti^{3+}, VO^{++}, Cr^{5+}
2	1	3	2	7	$^3F_2$	Ti^{++}, V^{3+}, Cr^{4+}
3	$\frac{3}{2}$	3	$\frac{3}{2}$	7	$^4F_{\frac{3}{2}}$	Ti^+, V^{++}, Cr^{3+}, Mn^{4+}
4	2	2	0	5	$^5D_0$	Cr^{++}, Mn^{3+}
5	$\frac{5}{2}$	0	$\frac{5}{2}$	1	$^6S_{\frac{5}{2}}$	Cr^+, Mn^{++}, Fe^{3+}
6	2	2	4	5	$^5D_4$	Fe^{++}
7	$\frac{3}{2}$	3	$\frac{9}{2}$	7	$^4F_{\frac{9}{2}}$	Fe^+, Co^{++}, Ni^{3+}
8	1	3	4	7	$^3F_4$	Co^+, Ni^{++}, Cu^{3+}
9	$\frac{1}{2}$	2	$\frac{5}{2}$	5	$^2D_{\frac{5}{2}}$	Ni^+, Cu^{++}

is the Landé splitting factor.† Note that this factor will be different for each value of J. The allowed free-ion transitions are those for which $\Delta M_J = \pm 1$. *One of the primary tasks in Chaps. 11 and 12 is to show why one must use expressions other than the Landé formula (11-2) for the ground state to get g factors for ions in crystalline electric fields.*

11-2 REMOVAL OF ORBITAL DEGENERACY IN CRYSTALLINE ELECTRIC FIELDS

If a number of different kinds of interactions occur in a molecule, the task of finding the energy levels is far too complex to consider all effects simultaneously. Instead, one tries to establish a hierarchy of interactions, ranging from the strongest to the weakest. First the energy-level array is determined from the strongest interactions (i.e., those which account for perhaps 90 percent of the total energy); then the displacement of energy levels is computed by introducing—one at a time if possible—the much weaker interactions. The student by now should be convinced that the Zeeman term (i.e., $g\beta\mathbf{H} \cdot \hat{\mathbf{S}}$) is well down the list. For an ion in a crystalline electric field, the magnitude of the crystal field energy relative to the spin-orbit coupling energy is crucial. (Note the wording "ion in a crystalline electric field," rather than "ion in a crystalline solid." Attention is hereby directed to the fact that transition-metal ions in solution are able to organize ligands—solvent molecules or negative ions—into an ordered array not unlike that of the nearest neighbors in a crystalline solid.) The three cases to be considered are: (1) weak, (2) medium, and (3) strong crystal fields.‡

The case 1 of crystal fields weak compared with the spin-orbit interaction is typified by rare-earth or actinide ions in most solids. Both $4f$ and $5f$ electrons are well shielded from the crystal field by other electrons. In most $4f$ cases this field will have trigonal symmetry; this is in contrast with the usual octahedral or tetrahedral symmetry (sometimes distorted) found for most ions of the $3d$ or $4d$ groups. Because of the strong coupling of $\mathbf{L}$ and $\mathbf{S}$ to give the resultant total angular-momentum vector $\mathbf{J}$, one is concerned with the ordering of the $(2J + 1)$ M_J states. For these ions, M_L and M_S are *not* good quantum numbers. The splittings arising from the spin-orbit interaction may be of the order of 5,000 cm^{-1}, whereas the separation of the M_J states may be about 100 cm^{-1} in the crystal field. The crystal field splits the M_J states into doublets $\pm M_J$ [and a singlet $(M_J = 0)$ if J is integral]. Because of this small splitting, many rare-earth ions in solids or in solutions give magnetic susceptibility values not much different from those of the free ion.

† G. Herzberg, "Atomic Spectra and Atomic Structure," pp. 106ff., Dover Publications, Inc., New York, 1944.

‡ This designation is not universally accepted. Some workers prefer the designations rare-earth scheme, weak-field scheme, and strong-field scheme for the respective terms weak, medium, and strong field used above.

The case 2 of medium crystal fields implies that the crystal field energy exceeds the spin-orbit coupling energy. This is typically the case for ions of the first transition series. The d electrons range over the exterior regions of the atom and interact strongly with the crystalline electric field. Some d electrons will interact more strongly than others with this field, and as a result, there will be a splitting of the energies of the d orbitals. The splitting apart of orbital states is usually so great that ordinarily only the lowest-lying states will be thermally populated. While the spin-orbit interaction is typically ~ 50 to 850 cm^{-1} for $3d^n$ ions, the crystalline electric field commonly causes splittings of the order of 10^4 cm^{-1}. Hence the magnetic susceptibility of a $3d$ ion in a crystalline electric field will be *very* different from that of the free ion because only the lowest-lying states will be populated. In fact, if the lowest state has no orbital degeneracy, the susceptibility will be close to the contribution of the spin angular momentum (spin-only susceptibility).

The case 3 of strong crystal fields involves covalent bonding between the $3d$ ion and its diamagnetic ligands. The electrostatic interaction between $3d$ electrons is weak compared with the interaction of each electron with the crystal field. It is now no longer valid to consider only the $3d$ electrons of the central ion, even as a first approximation. Rather, the ligand electrons must be considered explicitly.[†],[‡]

In this chapter we shall be concerned primarily with $3d$ ions in crystal fields corresponding to case 2. The method of dealing with the other two cases will be considered briefly in Secs. 12-3, 12-4, and 12-5; however, detailed calculations are beyond the scope of this book.

The orbital-angular-momentum functions of an ion in the gaseous state are usually taken as the eigenfunctions of $\hat{L}^2$ and of $\hat{L}_z$ (i.e., the complex functions listed in the second column of Table 11-2). However, in a crystalline electric field some or all of the orbital angular momentum may be "quenched," i.e., the orbital degeneracy is removed. If some of the orbital degeneracy remains, the wave functions of the degenerate states may be chosen to be complex, i.e., of the form $\psi = Ae^{iM\phi}$. However, if all orbital degeneracy is removed, it is necessary to use the real form of the orbital angular-momentum wave functions given in the third column of Table 11-2.

That an orbitally nondegenerate level should have zero associated angular momentum (and hence zero orbital magnetic moment) is shown as follows: Suppose that an eigenfunction (corresponding to some

---

† For a review of this subject see J. Owen and J. H. M. Thornley, *Rept. Prog. Phys.*, **29**:675 (1966).

‡ It is to be noted that for ions in medium or strong crystalline electric fields, one does not append a subscript (corresponding to J) in the spectroscopic designation. When the interaction with the crystalline electric field exceeds the spin-orbit coupling, the various functions which would correspond to the possible values of J for given values of L and S are mixed by the crystal field. In other words, J is not a good quantum number for these two cases.

eigenstate of the system) is complex. In this case there always exists at least one other independent eigenfunction having the same energy, namely the complex conjugate. Consequently any state which can be represented by a complex eigenfunction must be at least *doubly* degenerate. Conversely, if the state is nondegenerate, the eigenfunction must be real (at least in the case where the potential energy is purely electrostatic).

The operator $\hat{L}_z$ for the z component of the orbital angular momentum is the purely imaginary operator

$$\hat{L}_z = i\hbar \left(x \frac{\partial}{\partial y} - y \frac{\partial}{\partial x} \right) \tag{11-3}$$

$\hat{L}_z$ is a hermitian operator; this means that its eigenvalues must be real. If the wave function of an orbitally nondegenerate ground state of a system is designated by $|n\rangle$, operation by $\hat{L}_z$ will give

$$\hat{L}_z|n\rangle = M_l|n\rangle \tag{11-4}$$

where M_l must be real. However, if $|n\rangle$ corresponds to an orbitally nondegenerate state, then $|n\rangle$ must also be real. $\hat{L}_z$ is purely imaginary, and hence the result of operating with an imaginary operator on a real eigenfunction must be imaginary or zero. Since M_l must be real, the only possibility is to have $M_l = 0$. Q.E.D.

11-3 THE CRYSTAL FIELD POTENTIAL

Calculation of the energy levels of an ion in a crystalline electric field requires explicit consideration of the form of the crystal-field potential. In "crystal-field theory," the ligands are not considered explicitly, except as the source of negative point charges. The crystalline electric fields of interest are produced by a regular array of point negative charges about the central magnetic ion. Some arrays that are frequently encountered are the following:

Number of charges	Location at corners of	Symmetry designation
4	Tetrahedron	Tetrahedral
6	Octahedron	Octahedral
8	Cube	Eightfold, cubal

Since all of these arrays are based on a cube (the tetrahedron and the octahedron may be inscribed into a cube) one must use terms more explicit than "cubic." Another, less common, eightfold symmetry is based on the dodecahedron, for example, $Mo(CN)_8^{-3}$. (See Fig. 12-14.)

Table 11-2 Real and complex forms of the orbital angular momentum wave functions

l	Complex forms†	Real forms‡	Octahedral group representation§
0 (s orbitals or S states)	—	$\dfrac{1}{\sqrt{4\pi}}$	A_1
1 (p orbitals or P states)	$-\sqrt{\dfrac{3}{8\pi}}\,\sin\theta\,e^{i\phi} = \lvert 1, 1\rangle$ $\sqrt{\dfrac{3}{4\pi}}\,\cos\theta = \lvert 1, 0\rangle$ $\sqrt{\dfrac{3}{8\pi}}\,\sin\theta\,e^{-i\phi} = \lvert 1, -1\rangle$	$p_x = \sqrt{\dfrac{3}{4\pi}}\,\sin\theta\,\cos\phi = \dfrac{1}{\sqrt{2}}(\lvert 1, -1\rangle - \lvert 1, +1\rangle)$ $p_y = \sqrt{\dfrac{3}{4\pi}}\,\sin\theta\,\sin\phi = \dfrac{i}{\sqrt{2}}(\lvert 1, -1\rangle + \lvert 1, +1\rangle)$ $p_z = \sqrt{\dfrac{3}{4\pi}}\,\cos\theta = \lvert 1, 0\rangle$	T_1
2 (d orbitals or D states)	$-\sqrt{\dfrac{15}{32\pi}}\,\sin^2\theta\,e^{2i\phi} = \lvert 2, 2\rangle$ $\sqrt{\dfrac{15}{8\pi}}\,\sin\theta\,\cos\theta\,e^{i\phi} = \lvert 2, 1\rangle$ $\sqrt{\dfrac{5}{16\pi}}\,(3\cos^2\theta - 1) = \lvert 2, 0\rangle$ $\sqrt{\dfrac{15}{8\pi}}\,\sin\theta\,\cos\theta\,e^{-i\phi} = \lvert 2, -1\rangle$ $\sqrt{\dfrac{15}{32\pi}}\,\sin^2\theta\,e^{-2i\phi} = \lvert 2, -2\rangle$	$d_{xy} = \sqrt{\dfrac{15}{16\pi}}\,\sin^2\theta\,\sin 2\phi = -\dfrac{i}{\sqrt{2}}(\lvert 2, 2\rangle - \lvert 2, -2\rangle)$ $d_{xz} = \sqrt{\dfrac{15}{16\pi}}\,\sin 2\theta\,\cos\phi = -\dfrac{1}{\sqrt{2}}(\lvert 2, -1\rangle - \lvert 2, +1\rangle)$ $d_{yz} = \sqrt{\dfrac{15}{16\pi}}\,\sin 2\theta\,\sin\phi = +\dfrac{i}{\sqrt{2}}(\lvert 2, -1\rangle + \lvert 2, +1\rangle)$ $d_{x^2-y^2} = \sqrt{\dfrac{15}{16\pi}}\,\sin^2\theta\,\cos 2\phi = \dfrac{1}{\sqrt{2}}(\lvert 2, 2\rangle + \lvert 2, -2\rangle)$ $d_{z^2} = \sqrt{\dfrac{5}{16\pi}}\,(3\cos^2\theta - 1) = \lvert 2, 0\rangle$	T_2 E

$$-\sqrt{\frac{35}{64\pi}}\sin^3\theta\, e^{3i\phi} \quad = |3, 3\rangle$$

$$\sqrt{\frac{105}{32\pi}}\sin^2\theta\cos\theta\, e^{2i\phi} \quad = |3, 2\rangle$$

$$-\sqrt{\frac{21}{64\pi}}\sin\theta\,(5\cos^2\theta - 1)e^{i\phi} \quad = |3, 1\rangle$$

$$\sqrt{\frac{7}{16\pi}}\,(5\cos^3\theta - 3\cos\theta) \quad = |3, 0\rangle$$

$$\sqrt{\frac{21}{64\pi}}\sin\theta\,(5\cos^2\theta - 1)e^{-i\phi} \quad = |3, -1\rangle$$

$$\sqrt{\frac{105}{32\pi}}\sin^2\theta\cos\theta\, e^{-2i\phi} \quad = |3, -2\rangle$$

$$\sqrt{\frac{35}{64\pi}}\sin^3\theta\, e^{-3i\phi} \quad = |3, -3\rangle$$

3 (f orbitals or F states)

$$f_{xyz} = -\frac{i}{\sqrt{2}}\,(|3, 2\rangle - |3, -2\rangle) \quad\Big\} \ A_2$$

$$f_{x(y^2-z^2)} = \frac{1}{4}\left[\sqrt{5}(|3, -1\rangle - |3, 1\rangle) - \sqrt{3}(|3, 3\rangle - |3, -3\rangle)\right]$$

$$f_{y(z^2-x^2)} = -\frac{i}{4}\left[\sqrt{5}(|3, -1\rangle + |3, 1\rangle) - \sqrt{3}(|3, 3\rangle + |3, -3\rangle)\right]$$

$$f_{z(x^2-y^2)} = \frac{1}{\sqrt{2}}\,(|3, 2\rangle + |3, -2\rangle) \quad\Big\} \ T_2$$

$$f_{x(2x^2-3y^2-3z^2)} = \frac{1}{4}\left[\sqrt{3}(|3, 1\rangle - |3, -1\rangle) + \sqrt{5}(|3, -3\rangle - |3, 3\rangle)\right]$$

$$f_{y(2y^2-3x^2-3z^2)} = -\frac{i}{4}\left[\sqrt{3}(|3, 1\rangle + |3, -1\rangle) + \sqrt{5}(|3, -3\rangle + |3, 3\rangle)\right]$$

$$f_{z(2z^2-3x^2-3y^2)} = |3, 0\rangle \quad\Big\} \ T_1$$

† These functions are eigenfunctions of $\hat{L}_z$.

‡ These functions have the same symmetry characteristics as the subscripts on each symbol.

§ The symbols A, E, and T designate nondegenerate, doubly degenerate, and triply degenerate states or their wave functions. Subscripts are used to indicate symmetry properties. (See F. A. Cotton, "Chemical Applications of Group Theory," Interscience Publishers, a division of John Wiley & Sons, Inc., New York, 1963.)

At an arbitrary point near the origin with radius vector $\mathbf{r}$, the potential V_j due to a negative point charge q_j at $\mathbf{R}_j$ will be given by

$$V_j = \frac{q_j}{|\mathbf{R}_j - \mathbf{r}|} \tag{11-5}$$

Hence the total potential at $\mathbf{r}$ due to all of the point charges is

$$V = \sum_j \frac{q_j}{|\mathbf{R}_j - \mathbf{r}|} \tag{11-6}$$

The interaction of unpaired electrons in particular orbitals of the magnetic ion with this potential is of principal interest. If the charge due to these electrons at the point defined by the vector $\mathbf{r}_i$ is q_i and the potential at this point is V_i, the potential energy W_{cryst} is given by

$$W_{cryst} = \sum_i q_i V_i = \sum_i \sum_j \frac{q_i q_j}{|\mathbf{R}_j - \mathbf{r}_i|} \tag{11-7}$$

The form of the potential due to the point charges is determined by their configuration and by their distance d from the central magnetic ion. From Fig. 11-1 it may be seen that the contribution V_y to the potential at $\mathbf{r}_i$ from the two point charges at $y = \pm d$ is given by

$$V_y = q[(r^2 + d^2 - 2dy)^{-\frac{1}{2}} + (r^2 + d^2 + 2dy)^{-\frac{1}{2}}] \tag{11-8}$$

Here y is one of the three components of $\mathbf{r}$. Expressions for the potential-energy contributions V_x and V_z from the charges along the x and the z axes are precisely analogous to Eq. (11-8). Extensive algebraic manipulation and series expansions of the sum $V_x + V_y + V_z = V(x,y,z)$ leads eventually to the result[†],[‡]

$$V_{oct}(x,y,z) = \frac{6q}{d} + \frac{35q}{4d^5}\left[(x^4 + y^4 + z^4) - \frac{3}{5}r^4\right] - \frac{21q}{2d^7}\left[(x^6 + y^6 + z^6)\right.$$
$$\left. + \frac{15}{4}(x^2y^4 + x^2z^4 + y^2x^4 + y^2z^4 + z^2x^4 + z^2y^4) - \frac{15}{14}r^6\right] \tag{11-9}$$

For ions with $3d$ electrons, only the fourth-power terms are required; the full expression is needed for ions with $4f$ electrons. Since the constant term shifts all levels equally, it need not be considered further.

For an ion in tetragonal symmetry (see the distorted configuration in Fig. 11-1) the crystalline potential is

$$V_{ttgl} = A_t\left[(3z^2 - r^2) + \frac{5}{4d^2}\left(\frac{35}{3}z^4 - 10r^2z^2 + r^4\right)\right]$$
$$+ B_c\left(x^4 + y^4 + z^4 - \frac{3}{5}r^4\right) \tag{11-10}$$

† M. T. Hutchings, in F. Seitz and D. Turnbull (eds.), "Solid State Physics," vol. 16, p. 227, Academic Press Inc., New York, 1965; see also T. M. Dunn, D. S. McClure, and R. G. Pearson, p. 10, "Some Aspects of Crystal Field Theory," Harper & Row, Publishers, Incorporated, 1965.
‡ J. W. Orton, "Electron Paramagnetic Resonance," pp. 92ff., Iliffe Books, Ltd., London, 1968.

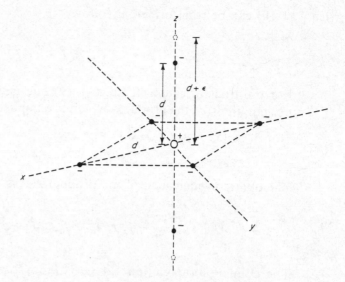

Fig. 11-1 Octahedral arrangement of six negative ions at a distance d from a central positive ion (solid circles). A tetragonal distortion resulting from an increased separation ϵ along the z axis is shown by open circles.

Here $A_t = -3q\epsilon/d^4$, where $d + \epsilon$ is the distance from the origin to the negative charges on the z axis and $B_c = 35q/4d^5$. It is assumed that $\epsilon << d$. Note that this tetragonal potential can be regarded as an octahedral potential [last term of Eq. (11-10)] plus a tetragonal distortion term (first two terms).

11-4 THE CRYSTAL FIELD OPERATORS

The effect of the crystal field is to modify the potential energy of the d electrons. One requires calculation of matrix elements of the type

$$\langle J, M_J | \hat{\mathcal{H}}_{cr} | J, M_J \rangle$$

Here J and M_J refer to the *total* angular momentum of the ion, and $\hat{\mathcal{H}}_{cr}$ is the crystal-field hamiltonian. Since $\hat{\mathcal{H}}_{cr}$ involves only the potential-energy operator, one requires evaluation of matrix elements involving V_{oct} or V_{ttgl} in Eq. (11-9) or (11-10).

The operator-equivalent approach uses the fact that in computing the crystal field state energies, the functions of x, y, and z in Eqs. (11-9) and (11-10) may be replaced by the general angular-momentum operators $\hat{J}_x$, $\hat{J}_y$, and $\hat{J}_z$. The justification for this replacement is as follows: For a set of eigenfunctions of $\hat{J}^2$ and $\hat{J}_z$, all having the same eigenvalue J, the matrix elements of $\hat{x}$ and $\hat{J}_x$, $\hat{y}$ and $\hat{J}_y$, and $\hat{z}$ and $\hat{J}_z$ are respectively proportional to one another. In this transformation, one must take due account of the non-commuting properties of these operators. The factor in the second term of

Eq. (11-10) can be transformed as follows:

$$B_c(x^4 + y^4 + z^4 - \tfrac{3}{5}r^4) \longrightarrow$$
$$\beta_c\{\hat{J}_x^4 + \hat{J}_y^4 + \hat{J}_z^4 - \tfrac{1}{5}J(J+1)[3J(J+1)-1]\} \quad (11\text{-}11)$$

For the transformation in Eq. (11-11) one must consider all possible permutations of the products which are implied in r^4. That is,

$$r^4 = (x^2 + y^2 + z^2)(x^2 + y^2 + z^2)$$
$$= x^4 + y^4 + z^4 + 2(x^2y^2 + x^2z^2 + y^2z^2) \quad (11\text{-}12)$$

The operator equivalent of the product x^2z^2 is

$$x^2z^2 \longrightarrow$$
$$\tfrac{1}{6}(\hat{J}_x^2\hat{J}_z^2 + \hat{J}_z^2\hat{J}_x^2 + \hat{J}_x\hat{J}_z^2\hat{J}_x + \hat{J}_z\hat{J}_x^2\hat{J}_z + \hat{J}_x\hat{J}_z\hat{J}_x\hat{J}_z + \hat{J}_z\hat{J}_x\hat{J}_z\hat{J}_x)$$
$$(11\text{-}13)$$

The commutation relation $\hat{J}_z\hat{J}_x - \hat{J}_x\hat{J}_z = i\hat{J}_y$ may be used to evaluate $\hat{J}_x\hat{J}_z\hat{J}_x\hat{J}_z$ as

$$\hat{J}_x\hat{J}_z\hat{J}_x\hat{J}_z = \hat{J}_x(i\hat{J}_y + \hat{J}_x\hat{J}_z)\hat{J}_z$$
$$= i\hat{J}_x\hat{J}_y\hat{J}_z + \hat{J}_x^2\hat{J}_z^2 \quad (11\text{-}14)$$

The products $\hat{J}_x\hat{J}_z^2\hat{J}_x$, $\hat{J}_z\hat{J}_x^2\hat{J}_z$, and $\hat{J}_z\hat{J}_x\hat{J}_z\hat{J}_x$ are transformed with the use of the commutation relations into forms such that the only imaginary term is $i\hat{J}_x\hat{J}_y\hat{J}_z$. The final result is†

$$x^2z^2 \longrightarrow \tfrac{1}{6}[3\hat{J}_x^2\hat{J}_z^2 + 3\hat{J}_z^2\hat{J}_x^2 - 2\hat{J}_x^2 + 3\hat{J}_y^2 - 2\hat{J}_z^2]. \quad (11\text{-}15)$$

This transformation is further considered in Prob. 11-2. The operator equivalents of the terms y^2z^2 and x^2y^2 are given in Table 11-3. Substitution of these operator equivalents into Eq. (11-12) gives

$$r^4 = J^2(J+1)^2 - \tfrac{1}{3}J(J+1) \quad (11\text{-}16)$$

The relation (11-11) is thus justified.

It is desirable to replace $\hat{J}_x^4 + \hat{J}_y^4$ of Eq. (11-11) by its equivalent in terms of the raising and lowering operators $\hat{J}_+$ and $\hat{J}_-$. This can be done by noting that one may write

$$x^4 - 6x^2y^2 + y^4 = \tfrac{1}{2}[(x+iy)^4 + (x-iy)^4] \quad (11\text{-}17)$$

The operator equivalent of the right-hand side is $\tfrac{1}{2}(\hat{J}_+^4 + \hat{J}_-^4)$. On addition of the operator equivalent of $6x^2y^2$. the right side of Eq. (11-11) becomes

$$\frac{\beta_c}{20}[35\hat{J}_z^4 - 30J(J+1)\hat{J}_z^2 + 25\hat{J}_z^2$$

$$- 6J(J+1) + 3J^2(J+1)^2] + \frac{\beta_c}{8}(\hat{J}_+^4 + \hat{J}_-^4) \quad (11\text{-}18)$$

---

† M. T. Hutchings. *Solid State Phys.*. **16**:227 (1964).

Table 11-3 Operator equivalents

Function	Operator equivalents
x^n, y^n, z^n	$(\hat{J}_x)^n$, $(\hat{J}_y)^n$. $(\hat{J}_z)^n$
r^2	$J(J+1)$
$x^2 z^2$	$\frac{1}{6}[3\hat{J}_x^2\hat{J}_z^2 + 3\hat{J}_z^2\hat{J}_x^2 - 2\hat{J}_x^2 + 3\hat{J}_y^2 - 2\hat{J}_z^2]$
$y^2 z^2$	$\frac{1}{6}[3\hat{J}_y^2\hat{J}_z^2 + 3\hat{J}_z^2\hat{J}_y^2 + 3\hat{J}_x^2 - 2\hat{J}_y^2 - 2\hat{J}_z^2]$
$x^2 y^2$	$\frac{1}{6}[3\hat{J}_x^2\hat{J}_y^2 + 3\hat{J}_y^2\hat{J}_x^2 - 2J(J+1) + 5\hat{J}_z^2]$
r^4	$[J^2(J+1)^2 - \frac{1}{3}J(J+1)]$
$35z^2 - 30r^2z^2 + 3r^4$	$\{35\hat{J}_z^4 - [30J(J+1) - 25]\hat{J}_z^2 - 6J(J+1) + 3J^2(J+1)^2\}$

The term for the tetragonal distortion in Eq. (11-10) becomes

$$A_t(3z^2 - r^2) \rightsquigarrow \alpha_t[3\hat{J}_z^2 - J(J+1)] \tag{11-19}$$

The second term of Eq. (11-10) transforms to terms analogous to those in Eq. (11-18). These terms will be incorporated into the octahedral constant β_c.

Equations (11-18) and (11-19) are appropriate for case 1 of weak crystal field. For ions in crystal fields of intermediate strength (case 2), Eqs. (11-18) and (11-19) are replaced by

$$\hat{\mathscr{H}}_{oct} = \frac{\beta_c}{20} [35\hat{L}_z^4 - 30L(L+1)\,\hat{L}_z^2$$

$$+ 25\hat{L}_z^2 - 6L(L+1) + 3L^2(L+1)^2] + \frac{\beta_c}{8} [\hat{L}_+^4 + \hat{L}_-^4] \tag{11-20}$$

$$\hat{\mathscr{H}}_{ttgl} = \hat{\mathscr{H}}_{oct} + \alpha_t[3\hat{L}_z^2 - L(L+1)] \tag{11-21}$$

In a tetrahedral field the crystal field hamiltonian is the same as Eq. (11-20) except for the magnitude and sign of β_c. There β_c has a sign opposite to that of an octahedral field.

11-5 CRYSTAL FIELD SPLITTINGS OF STATES FOR P-, D-, AND F-STATE IONS

As a first example, consider an ion in a P state ($L = 1$) placed in an octahedral crystal field. Computation of the energy levels is most expeditious in terms of the angular-momentum matrices given in Sec. B-8, using the complex functions of Table 11-2. The terms in $\hat{L}_\pm^4$ of Eq. (11-20) do not contribute to $\mathscr{H}_{oct}$, since for $L = 1$ there are no nonzero matrix elements. (The reader can check this by matrix multiplication of $\mathbf{L}_+$.) Also for $L = 1$, $\mathbf{L}_z^4 = \mathbf{L}_z^2$, which is given by

$$\mathbf{L}_z^2 = \begin{bmatrix} 1 & 0 & 0 \\ 0 & 0 & 0 \\ 0 & 0 & 1 \end{bmatrix} \tag{11-22}$$

Substitution of $\mathbf{L}_z^2$ and $\mathbf{L}_z^4$ into Eq. (11-20) shows that

$$
\mathscr{H}_{\text{oct}} = \begin{array}{c} \\ \langle 1| \\ \langle 0| \\ \langle -1| \end{array}
\begin{array}{ccc} |1\rangle & |0\rangle & |-1\rangle \\ \left[\begin{array}{ccc} 0 & 0 & 0 \\ 0 & 0 & 0 \\ 0 & 0 & 0 \end{array}\right] \end{array}
\tag{11-23}
$$

Consequently, an octahedral field does not remove any orbital degeneracy for an ion in a P state.†

The effect of a tetragonal distortion can be computed by adding the term $\alpha_t[3\mathbf{L}_z^2 - L(L+1)\mathbf{1}]$ to $\mathscr{H}_{\text{oct}}$. In this case the hamiltonian matrix becomes

$$
\mathscr{H}_{\text{ttgl}} = \begin{array}{c} \\ \langle 1| \\ \langle 0| \\ \langle -1| \end{array}
\begin{array}{ccc} |1\rangle & |0\rangle & |-1\rangle \\ \left[\begin{array}{ccc} \alpha_t & 0 & 0 \\ 0 & -2\alpha_t & 0 \\ 0 & 0 & \alpha_t \end{array}\right] \end{array}
\tag{11-24}
$$

Thus if $\alpha_t > 0$, the $|\pm 1\rangle$ states are raised in energy by α_t and the $|0\rangle$ state is lowered in energy by $2\alpha_t$. (See Fig. 11-2.) This situation corresponds to a distortion which elongates the octahedron along the z axis. (See Fig. 11-1.)

In the case of a D-state ion, the fivefold orbital degeneracy of the ion in the gaseous state is partially removed in an octahedral or a tetrahedral

† The reader who is familiar with the properties of spherical harmonics will regard this result as obvious, since matrix elements such as $\langle M_L|\hat{L}^4|M_L\rangle$ will vanish unless $L \geqslant 2$. However, assuming that this property is not obvious to all, we have given this "proof," since we wish to make the extension in Eq. (11-24) to the case of octahedral symmetry plus a tetragonal distortion.

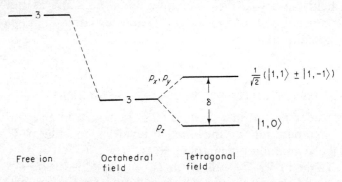

Fig. 11-2 Displacement of the energy levels of a P-state ion in an octahedral crystal field with a subsequent splitting in a tetragonal field. The real wave functions p_x and p_y corresponding to these states are indicated. For purposes of computation, it is more expedient to make use of the complex forms $|+1\rangle$ and $|-1\rangle$ of Table 11-2.

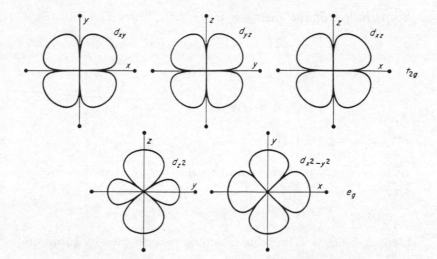

Fig. 11-3 Representation of $3d$ orbitals, showing the relation of the orbital lobes to the x, y, and z axes. Each of the subscripts should be multiplied by r^{-2}; the orbital usually referred to as d_{z^2} is given more fully as $(3z^2 - r^2)r^{-2}$. The orbitals indicated in the figure are representations of real wave functions. These are obtained by taking linear combinations of the imaginary wave functions which are eigenfunctions of $\hat{L}_z$. (See Table 11-2.) The symbols t_{2g} and e_g are defined in the text.

crystal field. Consider a D-state positive ion, e.g., Ti^{3+} in a field of octahedral symmetry resulting from six equidistant point negative charges located along the x, y, and z axes. (See Fig. 11-1.) Figure 11-3 shows the orientation of the lobes of the various d orbitals relative to the negative charges. An electron in an orbital in the upper row of this figure (d_{xy}, d_{yz}, and d_{xz}) avoids to a maximum extent a proximity to the negative charges. However, the lobes of the $d_{x^2-y^2}$ and d_{z^2} orbitals point directly at the negative charges. The latter two orbitals can be shown to be degenerate, though this is not obvious. Thus one should expect the latter two orbitals to have a higher energy than the former set.

The hamiltonian matrix for a D-state ion in an octahedral field can be computed in a fashion analogous to that for the P-state ion. One requires the angular-momentum matrices for $L = 2$. These may be computed using the relations (B-51) to (B-56), followed by matrix multiplication. For example,

$$\mathbf{L}_+ = \begin{array}{c} \\ \langle 2| \\ \langle 1| \\ \langle 0| \\ \langle -1| \\ \langle -2| \end{array} \begin{array}{ccccc} |2\rangle & |1\rangle & |0\rangle & |-1\rangle & |-2\rangle \\ \left[\begin{array}{ccccc} 0 & 2 & 0 & 0 & 0 \\ 0 & 0 & \sqrt{6} & 0 & 0 \\ 0 & 0 & 0 & \sqrt{6} & 0 \\ 0 & 0 & 0 & 0 & 2 \\ 0 & 0 & 0 & 0 & 0 \end{array}\right] \end{array} \qquad (11\text{-}25)$$

Substitution of the matrices L_z^4, L_z^2, L_+^4, and L_-^4 into Eq. (11-20) yields

$$\mathscr{H}_{\text{oct}} = \begin{array}{c} \\ \langle 2| \\ \langle 1| \\ \langle 0| \\ \langle -1| \\ \langle -2| \end{array} \begin{array}{ccccc} |2\rangle & |1\rangle & |0\rangle & |-1\rangle & |-2\rangle \\ \left[\begin{array}{ccccc} \frac{1}{10}\Delta & 0 & 0 & 0 & \frac{1}{2}\Delta \\ 0 & -\frac{2}{5}\Delta & 0 & 0 & 0 \\ 0 & 0 & \frac{3}{5}\Delta & 0 & 0 \\ 0 & 0 & 0 & -\frac{2}{5}\Delta & 0 \\ \frac{1}{2}\Delta & 0 & 0 & 0 & \frac{1}{10}\Delta \end{array} \right] \end{array} \qquad (11\text{-}26)$$

Here $\Delta = 6\beta_c$.[†] The secular determinant formed by subtraction of W from

[†] The quantity Δ for D-state ions is sometimes indicated as $10Dq$. (See J. W. Orton, "Electron Paramagnetic Resonance," p. 98, Iliffe Books, Ltd., London, 1968.)

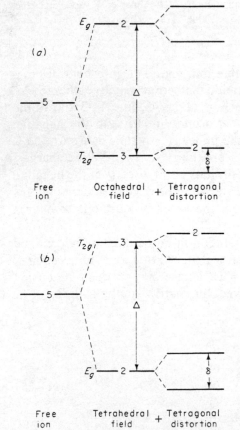

Fig. 11-4 Splitting of the D-state energies in a crystal field for $3d^1$ or $3d^6$ ions. (a) Octahedral crystal field. (b) Tetrahedral crystal field. For $3d^4$ and $3d^9$ ions, (a) applies to the tetrahedral case and (b) to the octahedral case.

the diagonal elements yields the eigenvalues†

$$W(T_{2g}) = -\frac{2}{5}\Delta \qquad \text{triply degenerate} \qquad\qquad (11\text{-}27a)$$

$$W(E_g) = +\frac{3}{5}\Delta \qquad \text{doubly degenerate} \qquad\qquad (11\text{-}27b)$$

The splitting of the D-state levels in an octahedral crystal field is shown in Fig. 11-4a for $3d^1$ and $3d^6$ ions. Since $3d^4$ and $3d^9$ ions have one electron missing from a complete half- or full shell, one may treat them as having one positive hole. This is equivalent to reversing the sign of the crystal field potential; hence, the levels in Fig. 11-4a will be inverted for the $3d^4$ and $3d^9$ ions in an octahedral field.

The application of the tetragonal distortion term $\alpha_t[3L_z^2 - L(L+1)\mathbf{1}]$ causes a further removal of orbital degeneracy. The hamiltonian matrix then becomes

$$\mathcal{H}_{ttgl} = \begin{array}{c} \\ \langle+2| \\ \langle+1| \\ \langle 0| \\ \langle-1| \\ \langle-2| \end{array} \begin{array}{ccccc} |+2\rangle & |+1\rangle & |0\rangle & |-1\rangle & |-2\rangle \\ \left[\begin{array}{ccccc} \frac{1}{10}\Delta + 6\alpha_t & 0 & 0 & 0 & \frac{1}{2}\Delta \\ 0 & -\frac{2}{5}\Delta - 3\alpha_t & 0 & 0 & 0 \\ 0 & 0 & \frac{3}{5}\Delta - 6\alpha_t & 0 & 0 \\ 0 & 0 & 0 & -\frac{2}{5}\Delta - 3\alpha_t & 0 \\ \frac{1}{2}\Delta & 0 & 0 & 0 & \frac{1}{10}\Delta + 6\alpha_t \end{array}\right] \end{array}$$

$$(11\text{-}28)$$

Expansion of the corresponding secular determinant yields the energies

$$W_1 = +\frac{3}{5}\Delta + \frac{2}{3}\delta \qquad\qquad (11\text{-}29a)$$

$$W_2 = +\frac{3}{5}\Delta - \frac{2}{3}\delta \qquad\qquad (11\text{-}29b)$$

$$W_3 = -\frac{2}{5}\Delta + \frac{2}{3}\delta \qquad\qquad (11\text{-}29c)$$

$$W_{4,5} = -\frac{2}{5}\Delta - \frac{\delta}{3} \qquad \text{doubly degenerate} \qquad (11\text{-}29d)$$

where $\delta = 9\alpha_t$ is the tetragonal splitting in Fig. 11-4a.

† The symbols in parentheses of Eq. (11-27) are standard group-theoretical designations. a, e, and t refer to nondegenerate, doubly degenerate, and triply degenerate orbitals. The corresponding *states* will be indicated by A, E, and T. (See Sec. 5-2 for a discussion of the distinction between *orbitals* and *states*.) The subscripts g and u are used to respectively denote *even* and *odd* symmetry with respect to inversion through a center of symmetry. Subscripts 1 and 2 respectively denote even and odd symmetry with respect to a 90° rotation about x, y, or z.

If α_t is negative, corresponding to a compression along the z axis, the state with energy W_3 will lie lowest, and hence in this case all orbital degeneracy of the lowest state will be removed. (See Fig. 11-4a.)

At this point it is appropriate to comment on the wave functions used to describe the states of an ion in a crystalline electric field. The eigenfunctions of $\hat{L}_z$, namely $|L,M_L\rangle$, have been used as a basis set. These are the complex functions (except for $M_L = 0$) tabulated in Table 11-2. However, because of the off-diagonal elements in Eqs. (11-26) and (11-28), the states $|L,M_L\rangle$ are not eigenfunctions of $\hat{\mathscr{H}}_{oct}$ or $\hat{\mathscr{H}}_{ttgl}$. On the other hand, the real forms of the wave functions given in column 3 of Table 11-2 are eigenfunctions of $\hat{\mathscr{H}}_{oct}$ and $\hat{\mathscr{H}}_{ttgl}$.

Consider next the effect of a tetrahedral rather than an octahedral field on these same ions. One may inscribe a tetrahedron in a cube by drawing a set of face diagonals through nonadjacent vertices of the cube (Fig. 11-5). Negative charges are now placed at each of the vertices of the tetrahedron, and the five d orbitals are inscribed into the original cube. It is the $d_{x^2-y^2}$ and d_{z^2} orbitals which have the lower energy for $3d^1$ and $3d^6$, whereas the d_{xy}, d_{xz}, and d_{yz} orbitals have the lower energies for $3d^4$ and $3d^9$ ions. Thus, there is an inversion in the sequence of levels in tetrahedral, as compared with octahedral crystal field symmetry for a given $3d^n$ ion.

In a tetrahedral field the sign of β_c, and hence of Δ, will be reversed. Thus the energy levels in a tetrahedral field will be those in Fig. 11-4b for $3d^1$ and $3d^6$ ions. For $3d^4$ and $3d^9$ ions, the levels will be inverted.

We come now to a consideration of F-state ions in an octahedral field. The hamiltonian matrix for such ions is computed in a fashion analogous to that used to obtain Eq. (11-28).

$$
\mathscr{H}_{oct} = \begin{array}{c} \langle 3| \\ \langle 2| \\ \langle 1| \\ \langle 0| \\ \langle -1| \\ \langle -2| \\ \langle -3| \end{array}
\begin{array}{ccccccc}
|3\rangle & |2\rangle & |1\rangle & |0\rangle & |-1\rangle & |-2\rangle & |-3\rangle \\
\frac{3}{10}\Delta & 0 & 0 & 0 & \frac{\sqrt{15}}{10}\Delta & 0 & 0 \\
0 & -\frac{7}{10}\Delta & 0 & 0 & 0 & \frac{\Delta}{2} & 0 \\
0 & 0 & \frac{1}{10}\Delta & 0 & 0 & 0 & \frac{\sqrt{15}}{10}\Delta \\
0 & 0 & 0 & \frac{3}{5}\Delta & 0 & 0 & 0 \\
\frac{\sqrt{15}}{10}\Delta & 0 & 0 & 0 & \frac{1}{10}\Delta & 0 & 0 \\
0 & \frac{\Delta}{2} & 0 & 0 & 0 & -\frac{7}{10}\Delta & 0 \\
0 & 0 & \frac{\sqrt{15}}{10}\Delta & 0 & 0 & 0 & \frac{3}{10}\Delta
\end{array}
$$

$$(11\text{-}30a)$$

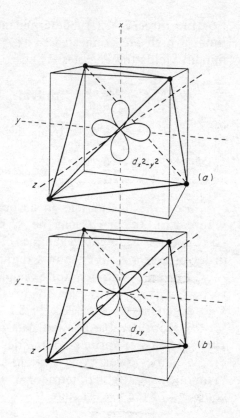

Fig. 11-5 (a) The $d_{x^2-y^2}$ orbital inscribed in a tetrahedron within a cube. For both the $d_{x^2-y^2}$ and the d_{z^2} orbitals the lobes point to the centers of edges of the tetrahedron, i.e., midway between the four ions at the corners. These same points of intersection would be four corners of an inscribed octahedron. (b) The d_{xy} orbital is similarly inscribed. The lobes of this orbital are directed away from the axes of an octahedron, but they are in closer proximity to tetrahedron vertices than the d orbital in (a). (*After D. S. McClure, "Electronic Spectra of Molecules and Ions in Crystals,"* Academic Press Inc., New York, 1959.)

Here $\Delta = 30\beta_c$. The secular determinant corresponding to this matrix may be rearranged into diagonal blocks as follows:

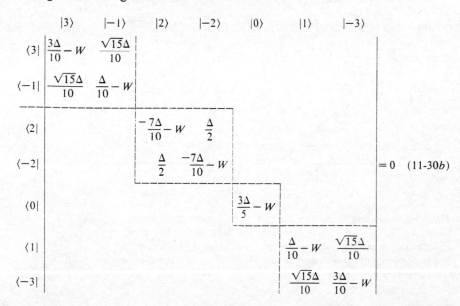

For the three 2×2 subdeterminants, it is apparent that the elements are those which are connected by $(\hat{L}_+{}^4 + \hat{L}_-{}^4)$. Solution of the secular determinant yields the energies

$$W(T_{1g}) = \frac{3}{5}\Delta \qquad \text{triply degenerate} \qquad (11\text{-}31a)$$

$$W(T_{2g}) = -\frac{1}{5}\Delta \qquad \text{triply degenerate} \qquad (11\text{-}31b)$$

$$W(A_{2g}) = -\frac{6}{5}\Delta \qquad \text{nondegenerate} \qquad (11\text{-}31c)$$

For $3d^2$ or $3d^7$ ions in an octahedral field, the T_{1g} states lie lowest, whereas for $3d^3$ or $3d^8$ ions the A_{2g} state lies lowest. (See Fig. 11-6a and b.) Cr^{3+} and Ni^{++} are good examples of the latter case. As in a D-state ion, the order of levels is inverted in a tetrahedral field.

For an S-state ion (for example, Mn^{++}) there is no orbital degeneracy to be removed. However, the crystal field can cause some removal of the spin degeneracy. (See Sec. 11-8.)

A considerable body of data relating orbital and spin degeneracies in electric fields of various symmetries is collected in Table 11-4. The degeneracy of the lowest-lying state in an octahedral field is indicated with the symbol ¶. A trigonal, tetragonal, or rhombic field reduces the orbital degeneracy of the lowest state.

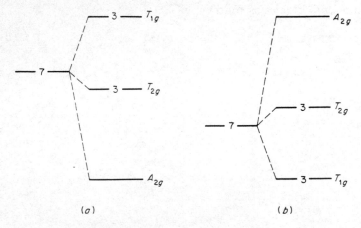

(a) (b)

Fig. 11-6 Splitting of the state energies of an F-state ion in a crystal field. (a) $3d^3$ or $3d^8$ ions in an octahedral field. (b) $3d^2$ or $3d^7$ ions in an octahedral field. In a tetrahedral field, splittings for $3d^3$ or $3d^8$ ions are obtained by inversion of (a). For $3d^2$ or $3d^7$ ions they are obtained by inversion of (b).

Table 11-4 Ground states, quantum numbers, and degeneracies† in various fields of ions of the iron group

Configuration	d^1	d^2	d^3	d^4	d^5	d^6	d^7	d^8	d^9
	^2D	^3F	^4F	^5D	^6S	^5D	^4F	^3F	^2D
Examples	Sc^{++} Ti^{3+} VO^{++} Cr^{5+}	Ti^{++} V^{3+} Cr^{4+}	Ti$^+$ V^{++} Cr^{3+} Mn^{4+}	Cr^{++} Mn^{3+}	Cr$^+$ Mn^{++} Fe^{3+}	Fe^{++}	Fe$^+$ Co^{++} Ni^{3+}	Co$^+$ Ni^{++}	Ni$^+$ Cu^{++}
S	$\frac{1}{2}$	1	$\frac{3}{2}$	2	$\frac{5}{2}$	2	$\frac{3}{2}$	1	$\frac{1}{2}$
L	2	3	3	2	0	2	3	3	2
J (free ion)	$\frac{3}{2}$	2	$\frac{3}{2}$	0	$\frac{5}{2}$	4	$\frac{9}{2}$	4	$\frac{5}{2}$
λ (cm^{-1})	154(Ti^{3+})	104(V^{3+})	56(V^{++})	58(Cr^{++})		-103	-178	-325	-829
(Free ion)‡	248(V^{4+})		91(Cr^{3+})	88(Mn^{3+})		(Fe^{++})	(Co^{++})	(Ni^{++})	(Cu^{++})

Orbital degeneracy in fields of various symmetries†

	d^1	d^2	d^3	d^4	d^5	d^6	d^7	d^8	d^9
Free ion	5	7	7	5	1	5	7	7	5
Octahedral§	2, 3¶	1, 2 · 3	1¶, 2 · 3	2¶, 3	1	2, 3¶	1, 2 · 3	1¶, 2 · 3	2¶, 3
Trigonal	1, 2 · 2	3 · 1, 2 · 2	3 · 1, 2 · 2	1, 2 · 2	1	1, 2 · 2	3 · 1, 2 · 2	3 · 1, 2 · 2	1, 2 · 2
Tetragonal	3 · 1, 2	3 · 1, 2 · 2	3 · 1, 2 · 2	3 · 1, 2	1	3 · 1, 2	3 · 1, 2 · 2	3 · 1, 2 · 2	3 · 1, 2
Rhombic	5 · 1	7 · 1	7 · 1	5 · 1	1	5 · 1	7 · 1	7 · 1	5 · 1

Spin degeneracy in fields of various symmetries for a single orbital level

	d^1	d^2	d^3	d^4	d^5	d^6	d^7	d^8	d^9
Free ion	2	3	4	5	6	5	4	3	2
Octahedral	2	3	4	2, 3	2, 4	2, 3	4	3	2
Trigonal	2	1, 2	2 · 2	1, 2 · 2	3 · 2	1, 2 · 2	2 · 2	1, 2	2
Tetragonal	2	1, 2	2 · 2	3 · 1, 2	3 · 2	3 · 1, 2	2 · 2	1, 2	2
Rhombic	2	3 · 1	2 · 2	5 · 1	3 · 2	5 · 1	2 · 2	3 · 1, 3 · 1	2

(After W. Gordy, W. V. Smith and R. F. Trambarulo, "Microwave Spectroscopy," p. 225, John Wiley & Sons, Inc., New York, 1963.)

† $a \cdot b$ means that there are a sets of states of b-fold degeneracy.

‡ Some authors have used Griffith's spin-orbit coupling parameter ζ in place of λ. For one d electron, $\lambda = \zeta$. However, if more than one d electron is present, $\lambda = \pm \zeta/2S$. The positive sign applies to ions with less than a half-filled shell, and the negative sign to ions with more than a half-filled shell. (See J. S. Griffith, "The Theory of Transition Metal Ions," p. 111, Cambridge University Press, 1961.)

§ Fields of tetrahedral symmetry will invert the order of these states.

¶ Lower or lowest state.

11-6 SPIN–ORBIT COUPLING AND THE SPIN HAMILTONIAN

The intrinsic spin angular momentum of a free electron is associated with a g factor of 2.00232. An electron in an atom or molecule may also possess orbital angular momentum. The corresponding orbital magnetic moment adds vectorially to the spin magnetic moment. For free atoms, one employs the Landé formula (Eq. 11-2) to obtain the g factor for the Zeeman splitting appropriate to each of the atomic energy levels in a magnetic field. Since the ground state of most molecules (including radicals) has zero orbital angular momentum, one might hope that in these cases the g factor would have precisely the free-electron value. However, the interaction of a presumably "pure spin" ground state with certain excited states admixes a small amount of orbital angular momentum into the ground state. This interaction will be inversely proportional to the energy separation of the states. One of the results is a change in the *effective* g factor.

In any atom, the orbital and spin angular momenta will be coupled through the spin-orbit interaction term, which for the present purposes may be given as

$$\hat{\mathscr{H}}_{so} = \lambda \hat{\mathbf{L}} \cdot \hat{\mathbf{S}} = \lambda [\hat{L}_x \hat{S}_x + \hat{L}_y \hat{S}_y + \hat{L}_z \hat{S}_z] \tag{11-32}$$

This term must be added to the Zeeman terms in the hamiltonian, that is,

$$\hat{\mathscr{H}} = \hat{\mathscr{H}}_{mag} + \hat{\mathscr{H}}_{so} = \beta \mathbf{H} \cdot (\hat{\mathbf{L}} + g_e \hat{\mathbf{S}}) + \lambda \hat{\mathbf{L}} \cdot \hat{\mathbf{S}} \dagger \tag{11-33}$$

It is assumed that the ground state which will be represented by $|G, M_S\rangle$ is orbitally nondegenerate. The energy (to first order) is given by the diagonal matrix element

$$W_G^{(1)} = \langle G, M_S | g_e \beta H_z \hat{S}_z | G, M_S \rangle + \langle G, M_S | (\beta H_z + \lambda \hat{S}_z) \hat{L}_z | G, M_S \rangle \tag{11-34}$$

The first term gives the "spin-only" electron Zeeman energy. The second term may be expanded as

$$\langle M_S | \beta H_z + \lambda \hat{S}_z | M_S \rangle \langle G | \hat{L}_z | G \rangle$$

We have shown in Sec. 11-2 that an orbitally nondegenerate state has zero orbital angular momentum. Hence for this case $\langle G | \hat{L}_z | G \rangle = 0$. The second-order correction to each *element* in the hamiltonian matrix is given by (see Sec. A-7)

$$(\mathscr{H})_{M_S, M'_S} = \frac{-\sum_n' |\langle G, M_S | (\beta \mathbf{H} + \lambda \hat{\mathbf{S}}) \cdot \hat{\mathbf{L}} + g_e \beta \mathbf{H} \cdot \hat{\mathbf{S}} | n, M'_S \rangle|^2}{W_n^{(0)} - W_G^{(0)}} \tag{11-35}$$

---

† The use of the terms $\beta \mathbf{H} \cdot (\hat{\mathbf{L}} + g_e \hat{\mathbf{S}})$ and $\lambda \hat{\mathbf{L}} \cdot \hat{\mathbf{S}}$ to proceed from the isotropic factor g_e to the symmetrical tensor **g** implicitly requires the unpaired electrons to be in a field of central symmetry. When this condition is not fulfilled, the tensor **g** may be asymmetric. [See F. K. Kneubühl, *Phys. kondens Materie*, 1:410 (1963).]

The prime designates summation over all states except the ground state. The matrix elements of $g_e \beta \mathbf{H} \cdot \hat{\mathbf{S}}$ will vanish since $\langle G|n \rangle$ is zero. $(\mathscr{H})_{M_S M'_S}$ can then be expanded as

$$(\mathscr{H})_{M_S, M'_S} = - \sum_n' \frac{[\langle M_S|(\beta \mathbf{H} + \lambda \hat{\mathbf{S}})|M'_S \rangle \cdot \langle G|\hat{\mathbf{L}}|n \rangle]}{W_n^{(0)} - W_G^{(0)}} \times [\langle n|\hat{\mathbf{L}}|G \rangle \cdot \langle M'_S|(\beta \mathbf{H} + \lambda \hat{\mathbf{S}})|M_S \rangle]}{W_n^{(0)} - W_G^{(0)}} \quad (11\text{-}36)$$

It is convenient to factor out of Eq. (11-36) the quantity

$$- \sum_n' \frac{\langle G|\hat{\mathbf{L}}|n \rangle \langle n|\hat{\mathbf{L}}|G \rangle}{W_n^{(0)} - W_G^{(0)}} = \begin{bmatrix} \Lambda_{xx} & \Lambda_{xy} & \Lambda_{xz} \\ \Lambda_{xy} & \Lambda_{yy} & \Lambda_{yz} \\ \Lambda_{xz} & \Lambda_{yz} & \Lambda_{zz} \end{bmatrix} = \Lambda \quad (11\text{-}37a)$$

The product of the two vector matrix elements is called an "outer product" and yields the second-rank tensor Λ. (See Sec. A-4.) The ijth element of this tensor is given by

$$\Lambda_{ij} = - \sum_n' \frac{\langle G|\hat{L}_i|n \rangle \langle n|\hat{L}_j|G \rangle}{W_n^{(0)} - W_G^{(0)}} \quad (11\text{-}37b)$$

$\hat{L}_i$ and $\hat{L}_j$ are orbital angular momentum operators appropriate to the x, y, or z directions. Substitution of Eq. (11-37a) into Eq. (11-36) yields

$$(\mathscr{H})_{M_S, M'_S} = \langle M_S| \beta^2 \mathbf{H} \cdot \Lambda \cdot \mathbf{H} + 2\lambda \beta \mathbf{H} \cdot \Lambda \cdot \hat{\mathbf{S}} + \lambda^2 \hat{\mathbf{S}} \cdot \Lambda \cdot \hat{\mathbf{S}} |M'_S \rangle \quad (11\text{-}38)$$

The first operator in Eq. (11-38) represents a constant contribution to the temperature-independent paramagnetism; it need not be considered further. The second and third terms in the matrix element of Eq. (11-38) constitute a hamiltonian which operates only on spin variables. When combined with the operator $g_e \beta \mathbf{H} \cdot \hat{\mathbf{S}}$ from Eq. (11-33), it is called the "spin hamiltonian" $\hat{\mathscr{H}}_S$. This may be written as

$$\hat{\mathscr{H}}_S = \beta \mathbf{H} \cdot (g_e \mathbf{1} + 2\lambda \Lambda) \cdot \hat{\mathbf{S}} + \lambda^2 \hat{\mathbf{S}} \cdot \Lambda \cdot \hat{\mathbf{S}} \quad (11\text{-}39a)$$

$$= \beta \mathbf{H} \cdot \boldsymbol{g} \cdot \hat{\mathbf{S}} + \hat{\mathbf{S}} \cdot \boldsymbol{D} \cdot \hat{\mathbf{S}} \quad (11\text{-}39b)$$

where

$$\boldsymbol{g} = g_e \mathbf{1} + 2\lambda \Lambda \quad (11\text{-}39c)$$

and

$$\boldsymbol{D} = \lambda^2 \Lambda \quad (11\text{-}39d)$$

and $\mathbf{1}$ is the unit tensor.

$\hat{\mathbf{S}}$ in Eq. (11-39a) is an operator which corresponds to the apparent or *effective* spin of the ground state. This need not be the actual spin, as will be illustrated in Sec. 12-1.

If the angular momentum of a system is *solely* due to spin angular momentum, the tensor $\boldsymbol{g}$ should be isotropic, with the value 2.00232. Any anisotropy or deviation from this value results from the tensor Λ which involves only contributions of the orbital angular momentum from excited states. [See Eq. (11-37).]

Equation (11-39c) indicates that one may immediately obtain the tensor $\boldsymbol{g}$ when the tensor Λ is known. As a simple example, consider a P-state ion in a tetragonal crystal field such that the state $|1, 0\rangle$ lies lowest. (See Fig. 11-2.) For the degenerate upper states one may exercise the prerogative of using the complex forms $|L, M_L\rangle$ of the functions (see Table 11-2), viz., $|1, 1\rangle$ and $|1, -1\rangle$. Since in each case $L = 1$, it is sufficient to represent the three levels as $|0\rangle$, $|1\rangle$, and $|-1\rangle$.

As the symmetry is tetragonal, one principal axis is the tetragonal axis Z. The other two axes (X and Y) are equivalent and are perpendicular to Z. In this principal-axis system, the only nonzero elements of the tensor Λ will be the diagonal elements. From Eq. (11-37b), the matrix elements $\langle 0|\hat{L}_z|M_L\rangle$ and $\langle M_L|\hat{L}_z|0\rangle$ will vanish since $\hat{L}_z$ couples only states of the same M_L value. Hence

$$\Lambda_{ZZ} = 0 \quad \text{and} \quad g_{ZZ} = g_{\parallel} = g_e \tag{11-40a}$$

The value of $g_\perp$ is obtained from either Λ_{XX} or Λ_{YY} as follows:

$$\Lambda_{XX} = -\frac{[\langle 0|\hat{L}_x|1\rangle\langle 1|\hat{L}_x|0\rangle + \langle 0|\hat{L}_x|-1\rangle\langle -1|\hat{L}_x|0\rangle]}{\delta}$$

$$= -\frac{\langle 0|\tfrac{1}{2}\hat{L}_-|1\rangle\langle 1|\tfrac{1}{2}\hat{L}_+|0\rangle + \langle 0|\tfrac{1}{2}\hat{L}_+|-1\rangle\langle -1|\tfrac{1}{2}\hat{L}_-|0\rangle}{\delta} = \frac{-1}{\delta} \tag{11-40b}$$

Here Eqs. (B-51) and (B-52) have been used. Hence from Eq. (11-39c)

$$g_{XX} = g_e - \frac{2\lambda}{\delta} = g_\perp \tag{11-41}$$

The V_1 point-defect center (O^- ion next to a cation vacancy) in MgO serves as an excellent example of a P-state ion in a tetragonal crystal field. This center was considered in Sec. 7-1. We are now in a position to interpret the g factors. According to Eq. (11-40a), $g_{\parallel}$ should be very close to the free-electron value. In fact, $g_{\parallel}$(observed) = 2.00327. Since λ for a positive hole on oxygen is negative, Eq. (11-41) predicts that $g_\perp > g_e$. This is again in agreement with experiment† since $g_\perp$ (observed) = 2.03859.

This procedure for calculating $\boldsymbol{g}$ for ions with orbitally nondegenerate ground states is extremely simple, and it demonstrates in a very clear way the source of the deviations of g from the value of g_e.

An alternative, much more tedious procedure, is as follows:

1. Determine the wave function of both of the ground spin states after admixture of orbital angular momentum from excited states. If the

† J. E. Wertz, P. Auzins, J. H. E. Griffiths, and J. W. Orton, *Discussions Faraday Soc.*, **28**:136 (1959); W. C. O'Mara, J. J. Davies, and J. E. Wertz, *Phys. Rev.*, **179**:816 (1969).

perturbed states are called $|\alpha'\rangle$ and $|\beta'\rangle$, the former may be written as

$$|\alpha'\rangle = |G, \tfrac{1}{2}\rangle + \sum_{M_L}' \sum_{M_S}' |M_L, M_S\rangle$$

$$\times \frac{\langle M_L, M_S|\lambda[\tfrac{1}{2}(\hat{L}_+\hat{S}_- + \hat{L}_-\hat{S}_+) + \hat{L}_z\hat{S}_z]|G, \tfrac{1}{2}\rangle}{W(G,\tfrac{1}{2}) - W(M_L,M_S)} \quad (11\text{-}42)$$

Evaluation of the matrix elements gives the admixture coefficients c_1 in the expressions

$$|\alpha'\rangle = |G, \tfrac{1}{2}\rangle + c_1|M_L, M_S\rangle + c_2|M_L', M_S'\rangle + \cdots +$$

$$(11\text{-}43a)$$

and

$$|\beta'\rangle = |G, -\tfrac{1}{2}\rangle + c_1|M_L, M_S\rangle + c_2|M_L', M_S'\rangle + \cdots +$$

$$(11\text{-}43b)$$

2. Use the functions $|\alpha'\rangle$ and $|\beta'\rangle$ with the hamiltonian $\hat{\mathcal{H}}_{\text{mag}}$ of Eq. (11-33), which may be rewritten as

$$\hat{\mathcal{H}}_{\text{mag}} = \beta\hat{\mathbf{L}} \cdot \mathbf{H} + g_e\beta\hat{\mathbf{S}} \cdot \mathbf{H} = \beta H_z(\hat{L}_z + g_e\hat{S}_z)$$
$$+ \frac{1}{2}\beta H_x[(\hat{L}_+ + \hat{L}_-) + g_e(\hat{S}_+ + \hat{S}_-)]$$
$$+ \frac{1}{2i}\beta H_y[(\hat{L}_+ - \hat{L}_-) + g_e(\hat{S}_+ - \hat{S}_-)] \quad (11\text{-}44)$$

3. The 2×2 hamiltonian matrix involving the functions $|\alpha'\rangle$ and $|\beta'\rangle$ is then constructed for each principal-axis direction. Solution of the secular determinant gives the energies.
4. These energies may be expressed in the form $W = g_i\beta H_i M_S$, where g_i is the effective g factor in the direction of the field component H_i. This procedure is applied in Prob. 11-4.

The second term of Eq. (11-39b) is effective only in systems with $S \geq 1$. One notes that this term in the spin hamiltonian is analogous to that derived for the spin-spin hamiltonian of Eq. (10-13b). Experimentally, it is not possible to separate the spin-spin contribution to $\mathbf{D}$ from the anisotropic part of the spin-orbit coupling contribution.

The $\hat{\mathbf{S}} \cdot \mathbf{D} \cdot \hat{\mathbf{S}}$ term can be written in the alternative form,

$$\hat{\mathbf{S}} \cdot \mathbf{D} \cdot \hat{\mathbf{S}} = D_{XX}\hat{S}_x{}^2 + D_{YY}\hat{S}_y{}^2 + D_{ZZ}\hat{S}_z{}^2 \quad (11\text{-}45)$$

$$= D[\hat{S}_z{}^2 - \tfrac{1}{3}S(S + 1)] + E(\hat{S}_x{}^2 - \hat{S}_y{}^2)$$
$$+ \tfrac{1}{3}(D_{XX} + D_{YY} + D_{ZZ})S(S + 1) \quad (11\text{-}46a)$$

where

$$D = D_{ZZ} - \frac{D_{XX} + D_{YY}}{2} \quad (11\text{-}46b)$$

and

$$E = \frac{D_{XX} - D_{YY}}{2} \qquad (11\text{-}46c)$$

The last term in (11-46a) is a constant which shifts all components of the ground state equally. The coefficient of this term is proportional to the trace of D. This term is not usually included in the spin hamiltonian; it would be zero for pure spin-spin coupling.

The spin hamiltonian of Eqs. (11-39) is incomplete for ions with nuclei of nonzero nuclear spin. The hyperfine and nuclear Zeeman interactions can be treated by the addition of the extra terms $h\hat{S} \cdot A \cdot \hat{I} - g_N \beta_N H \cdot \hat{I}$. The hyperfine interaction was treated in Sec. 7-6. The spin hamiltonian then becomes[†],[‡]

$$\hat{\mathscr{H}}_S = \beta H \cdot g \cdot \hat{S} + \hat{S} \cdot D \cdot \hat{S} + h\hat{S} \cdot A \cdot \hat{I} - g_N \beta_N H \cdot \hat{I} \qquad (11\text{-}47)$$

The hamiltonian matrix will now contain off-diagonal terms, since the functions $|M_S, M_I\rangle$ are not eigenfunctions of this hamiltonian. In such cases the hamiltonian matrix cannot be factored.

The complication of changes in the axes of quantization has been considered briefly in Sec. 7-8. For a system with axial symmetry and for a small hyperfine interaction, one may diagonalize the hamiltonian matrix by rotation of the coordinate system. (See Sec. A-5e.) For the electrons, the axis of quantization is the direction of μ_e, and for the nuclei, it is that of μ_N. Calculation of the energy levels to the approximation required for consideration of transition-metal ion spectra requires second-order perturbation theory. The hyperfine transition energy in the absence of zero-field splitting is given by [§]

$$\Delta W = g_{\text{eff}} \beta H + K M_I + \frac{h^2 A_\perp^2}{4 g_{\text{eff}} \beta H} \left[\frac{h^2 A_\parallel^2 + K^2}{K^2} \right] [I(I+1) - M_I^2]$$

$$+ \frac{h^2 A_\perp^2}{2 g_{\text{eff}} \beta H} \left(\frac{h A_\parallel}{K} \right) (2 M_S - 1) M_I$$

$$+ \frac{h^4}{2 g_{\text{eff}} \beta H} \left[\frac{A_\parallel^2 - A_\perp^2}{K} \right]^2 \left(\frac{g_\parallel g_\perp}{g_{\text{eff}}^2} \right)^2 \sin^2 \theta \cos^2 \theta \, M_I^2 \qquad (11\text{-}48)$$

Here

$$K^2 = \frac{h^2 A_\parallel^2 g_\parallel^2}{g_{\text{eff}}^2} \cos^2 \theta + \frac{h^2 A_\perp^2 g_\perp^2}{g_{\text{eff}}^2} \sin^2 \theta \qquad (11\text{-}49)$$

Here θ is the angle between the symmetry axis and the direction of H; g_{eff} is

---

† Equation (11-47) is valid only for ions with orbitally nondegenerate ground states.‡

‡ B. Bleaney, *Phil. Mag.*, **42**:441 (1951); an explanation of the details of calculation of the hyperfine energy is given by J. W. Orton, in "Electron Paramagnetic Resonance," pp. 64, 124, Iliffe Books, Ltd., London, 1968.

§ B. Bleaney, *Phil. Mag.*, **42**:441 (1951).

given by Eq. (7-3). A in Eqs. (11-48) and (11-49) is in units of hertz; however, it is customary to express hyperfine couplings in reciprocal centimeters, that is, A/c for transition-metal ions.

The spin hamiltonian of Eq. (11-39b) (plus hyperfine terms, if necessary) is adequate for systems with $S = 1$. If $S = \frac{3}{2}$, as for $3d^7$ ions, then a term involving $\hat{S}^3 \cdot \mathbf{H}$ is allowed.[†] If there is a nucleus contributing hyperfine splitting, an additional term of the form $\hat{S}^3 \cdot \hat{\mathbf{I}}$ may be required. For octahedral or tetrahedral symmetries, the additional terms have the form

$$u\beta[H_x\hat{S}_x{}^3 + H_y\hat{S}_y{}^3 + H_z\hat{S}_z{}^3 - \tfrac{1}{5}\mathbf{H} \cdot \hat{\mathbf{S}}(3S^2 - 1)]$$
$$+ U[\hat{I}_x\hat{S}_x{}^3 + \hat{I}_y\hat{S}_y{}^3 + \hat{I}_z\hat{S}_z{}^3 - \tfrac{1}{5}\hat{\mathbf{I}} \cdot \hat{\mathbf{S}}(3S^2 - 1)] \quad (11\text{-}50)$$

Conversely, if description of an experimental spectrum requires such a term, this confirms the identification of an $S = \frac{3}{2}$ state. For $S \geqslant 2$, a term in $\hat{S}^4$ is allowed. This will be illustrated in Sec. 11-8 for $3d^5$ ions.

11-7 D- AND F-STATE IONS WITH ORBITALLY NONDEGENERATE GROUND STATES

The rest of this chapter will be devoted to a consideration of specific ions in a variety of crystal fields such that the ground state is orbitally nondegenerate. Table 11-5 summarizes the D- and F-state ions considered in this section.

[†] B. Bleaney, *Proc. Phys. Soc.*, **A73**:939 (1959); G. F. Koster and H. Statz, *Phys. Rev.*, **113**:445 (1959).

Table 11-5 Ions with orbitally nondegenerate ground states

Subsection	Ion	Crystal field symmetry	Symbol
11-7a	D-state ions		
	$3d^1$	Tetrahedral plus tetragonal distortion	$3d^1$ (ttdl + ttgl)
	$3d^7$ (low spin)	Octahedral plus tetragonal distortion	$3d^7$ (ls)(oct + ttgl)
	$3d^9$	Octahedral plus tetragonal distortion	$3d^9$ (oct + ttgl)
11-7b	F-state ions		
	$3d^8$	Octahedral	$3d^8$ (oct)
	$3d^2$	Tetrahedral	$3d^2$ (ttdl)
	$3d^8$	Octahedral plus tetragonal distortion	$3d^8$ (oct + ttgl)
	$3d^2$	Tetrahedral plus tetragonal distortion	$3d^2$ (ttdl + ttgl)
	$3d^3$	Octahedral	$3d^3$ (oct)
	$3d^7$ (high spin)	Tetrahedral	$3d^7$ (hs)(ttdl)
	$3d^3$	Octahedral plus tetragonal distortion	$3d^3$ (oct + ttgl)
11-8	S-state ions		
	$3d^5$	Octahedral	$3d^5$ (oct)

The designation "low spin" or "high spin" refers to electron distributions which lead to minimum or maximum total spin respectively. In an octahedral crystal field the $3d$ levels split so as to leave the triply degenerate t_{2g} orbitals lower in energy than the e_g orbitals. For example, in $3d^5$ ions two situations arise, depending on the magnitude of the crystal-field splitting. In the high-spin case, Hund's rule applies, i.e., the state with maximum spin multiplicity has the lowest energy. In the low-spin case the crystal-field splitting is so large that electrons occupy only the lower group of levels. These cases are illustrated in Fig. 11-7. Low- and high-spin cases are also found for $3d^4$, $3d^5$, $3d^6$, and $3d^7$ ions.

For a number of reasons it is important to know precisely how the energy of each atomic state of the free ion varies as a function of the magnitude of the crystalline electric field. In general, the change of energy with crystal field strength is different for each state, and the dependence is usually nonlinear. Thus far we have completely ignored the excited states of the free ion. In the crystal field, some components of these excited states may often come close enough to the ground state so that there may be significant contributions both to g factors and to the zero field splitting parameters D and E. Further, for some d^n cases (d^4, d^5, d^6, d^7) the decrease in energy with increasing crystal field of at least one of the excited states is much greater than that of the initial ground state. (See Fig. 11-8a.) Hence, for magnitudes of the crystal field beyond some critical value, there will be a new ground state. This is represented symbolically in Fig. 11-8a for some of the states of a d^4 ion.

Since it is the separation of energies of excited states from that of the ground state which is usually of greatest interest, it is convenient to redraw the diagram, keeping the energy of the ground state horizontal. This causes the energy of the ^5E state to rise relative to that of the $^3T_1$ ground state beyond the crossover point. (See Fig. 11-8b.) The areas at the left- and right-hand sides of this diagram are thus the high-spin and low-spin regions, respectively.

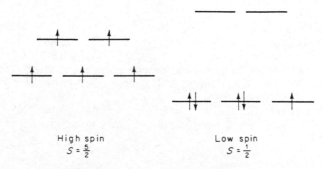

High spin
$S = \frac{5}{2}$

Low spin
$S = \frac{1}{2}$

Fig. 11-7 Configurations of a $3d^5$ ion in the high-spin and the low-spin states.

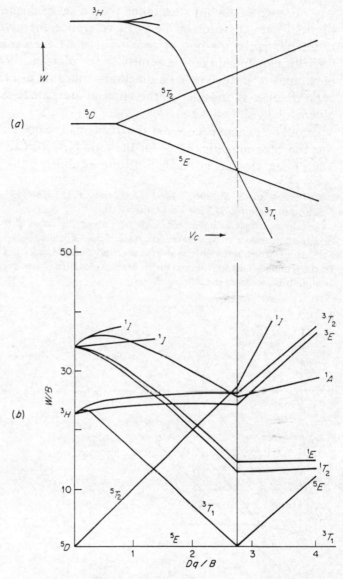

Fig. 11-8 (a) Splitting of the states of a $3d^4$ ion as a function of octahedral crystal field strength. The dashed line separates the medium and strong field regions. (b) The $3d^4$ splitting diagram of Fig. 11-8a redrawn for an assumed constant ground state energy. A number of levels which increase monotonically with octahedral crystal field have not been depicted. This representation is due to Tanabe and Sugano; it is applicable to all $3d^4$ ions, since both W and Dq are scaled by an energy characteristic of the particular ion. Note that $Dq = \Delta/10$. [After Y. Tanabe and S. Sugano, J. Phys. Soc. Japan, **9**:753, 766 (1954).]

Figure 11-8*b* has been taken from a set of diagrams of energy vs. crystal field strength, each divided by a reference energy B which is appropriate to the particular ion-host system.   These are the Tanabe-Sugano diagrams,[†] well known to inorganic chemists.   By plotting W/B versus Dq/B[‡] for a particular d^n case, one has a diagram applicable to all ions of that type.[§] As each d^n case is considered, the appropriate Tanabe-Sugano diagram will be given.

As an example, consider the d^8 case; here one may use a single Tanabe-Sugano diagram (Fig. 11-9) for the ions Ni^{++} or Cu^{3+}.   (Both of these ions have been studied in Al_2O_3.)   If one establishes the location of the optical

[†] Y. Tanabe and S. Sugano, *J. Phys. Soc. Japan*, **9**:753, 766 (1954).

[‡] See footnote on p. 272 for a definition of Dq.

[§] Certain assumptions have been made here about the ratios of free-ion parameters being applicable to the ions in an octahedral crystal field.   For details see D. S. McClure, "Electronic Spectra of Molecules and Ions in Crystals," pp. 58ff. Academic Press Inc., New York, 1959. In the Tanabe-Sugano diagrams given here, the excited states corresponding to different electron distributions are usually not shown.

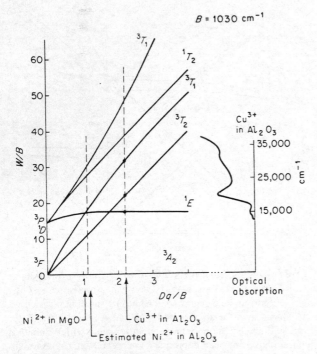

Fig. 11-9 Tanabe-Sugano diagram for a $3d^8$ ion in an octahedral field. The absorption spectrum of Cu^{3+} in Al_2O_3 is plotted vertically on the right. The location of the first three absorption bands serves to establish the value of the crystal field for this system. [*Taken from S. Geschwind and J. P. Remeika, J. Appl. Phys.*, **33 Suppl.**:370 (1962).]

absorption band of lowest energy for a specific ion, for example, Cu^{3+}, this establishes the value of Dq/B for that ion in that host. One can then predict (or verify) the absorption bands next higher in energy, i.e., the $^3T_2 \leftarrow {}^3A_2$ and the $^3T_1 \leftarrow {}^3A_2$ transitions for Cu^{3+}. The observed absorption bands for Cu^{3+} are shown at the right of Fig. 11-9.

11-7a D-state ions

$3d^1$ *(ttdl + ttgl) or* $3d^1$ *(cubal + ttgl)*   In a tetrahedral crystal field a $3d^1$ ion will have the E_g state lowest. A further tetragonal distortion will completely remove the orbital degeneracy of the ground state, as shown in Fig. 11-4b. If α_t of Eq. (11-21) is positive, the $|0\rangle$ or d_{z^2} state will lie lower; conversely if α_t is negative, the $(1/\sqrt{2})(|2\rangle + |-2\rangle)$ or $d_{x^2-y^2}$ state will lie lower. The former case will be considered first, since it is simple to treat.

The distortion axis will be taken as the Z direction. The values of the g components are obtained by using Eqs. (11-37a) and (11-39c) as follows:

$$g_{zz} = g_\parallel = g_e + 2\lambda\Lambda_{zz} = g_e \tag{11-51}$$

$\Lambda_{zz} = 0$ since $\hat{L}_z$ couples only states with the same M_L components.

For $\mathbf{H} \parallel X$

$$g_{xx} = g_\perp = g_e + 2\lambda\Lambda_{xx} = g_e - \frac{2\lambda}{\Delta}\sum_{M_l}{}' \langle 0|\hat{L}_x|M_L\rangle\langle M_L|\hat{L}_x|0\rangle \tag{11-52a}$$

The only states coupled to $|0\rangle$ by $\hat{L}_x$ are $|+1\rangle$ and $|-1\rangle$, and hence

$$g_\perp = g_e - \frac{2\lambda}{\Delta}[\langle 0|\tfrac{1}{2}\hat{L}_-|1\rangle\langle 1|\tfrac{1}{2}\hat{L}_+|0\rangle + \langle 0|\tfrac{1}{2}\hat{L}_+|-1\rangle\langle -1|\tfrac{1}{2}\hat{L}_-|0\rangle] \tag{11-52b}$$

The matrix elements in Eq. (11-52b) are the same as those in Eq. (11-40b), but now $L = 2$. Application of Eqs. (B-51) and (B-52) to these elements gives

$$g_\perp = g_e - \frac{6\lambda}{\Delta} \tag{11-53}$$

The energies in a magnetic field are then obtained by inserting the appropriate g factor into Eqs. (1-16).

A good example of a $3d^1$ (ttdl + ttgl) ion is Cr^{5+} in CrO_4^{3-} doped in Ca_2PO_4Cl single crystals.[†] Here $g_\parallel = 1.9936$ and $g_\perp = 1.9498$. Thus $\alpha_t > 0$ and the unpaired electron resides primarily in the d_{z^2} orbital. This corresponds to a compressed tetrahedron.

† E. Banks, M. Greenblatt and B. R. McGarvey. *J. Chem. Phys.*, 47:3772 (1967); B. R. McGarvey, in T. F. Yen (ed.), "Electron Spin Resonance of Metal Complexes," p. 1, Plenum Press, Plenum Publishing Corporation, New York, 1969.

In the case that α_t is negative, the g factors are (see Prob. 11-6)

$$g_{\|} = g_e - \frac{8\lambda}{\Delta} \tag{11-54a}$$

$$g_{\perp} = g_e - \frac{2\lambda}{\Delta} \tag{11-54b}$$

One might have expected that substitution of V^{4+} for W in the WO_4^{--} ion of $CaWO_4$ would provide a good example of a $3d^1$ (ttdl + ttgl) ion. The WO_4^{--} ions are indeed compressed tetrahedra with their unique axes aligned with the c axis of the tetragonal unit cell which contains four molecules. The hyperfine coupling tensor reflects this symmetry, with $A_{\|}/c = 0.00179$ cm^{-1} and $A_{\perp}/c = 0.00190$ cm^{-1} at 77 K. However the g factor is isotropic and has the value 2.0245.[†] The positive deviation of g from the free-electron value is opposite to the prediction of the point-charge model. The W—O bond is known to be strongly covalent, and the V—O bond will presumably be also; it will be noted in Sec. 12-5 that whereas small amounts of covalent bonding may diminish the negative contributions of excited states to **g**, large amounts can actually give positive g shifts, as observed here. A diminished magnitude of central-atom hyperfine coupling is a strong indication of the importance of covalent bonding. More typically, the hyperfine coupling for V^{4+} lies in the range 0.0070 to 0.0100 cm^{-1}. Thus the much smaller values in this case also attest to a strong covalent bonding.

In $CaWO_4$, V^{4+} may also substitute for Ca^{++} instead of W. There are then eight surrounding oxygen atoms, forming one elongated and one compressed tetrahedron of oxygen atoms. This is not a simple $3d^1$ (ttdl + ttgl) case; however, in these more ionic surroundings, the g shifts are *negative*, as expected, and $A_0/c = 0.0087$ cm^{-1}.

$3d^7$ *(ls)(oct + ttgl); $3d^9$ (oct + ttgl)*   Equations (11-51) and (11-53) also apply to $3d^7$ (low spin) and $3d^9$ ions in an octahedral field with tetragonal distortion. For these ions $\lambda < 0$ (see Table 11-4), and $g_{\perp} > g_{\|}$ for $\alpha_t > 0$. Two examples of the $3d^7$ (ls)(oct + ttgl) case will be considered. The first is cobalt phthalocyanine, in which the unpaired electron is essentially in the d_{z^2} orbital and hence in an orbital perpendicular to the plane of this "square-planar" complex. In solvents such as pyridine, the two axial coordination sites are occupied by solvent molecules. (This is evidenced by hyperfine splitting from two equivalent nitrogen atoms.) For this system, $g_{\|} = 2.016$ and $g_{\perp} = 2.268$.[‡] Thus $g_{\|} \approx g_e$ as predicted from Eq. (11-51).

A second example is $Fe(CN)_5NOH^{--}$ in the diamagnetic host $Na_2Fe(CN)_5NO \cdot 2H_2O$.[§],[¶] For this system, $g_{\|} = 2.0069$ and $g_{\perp} = 2.0374$.

† N. Mahootian, C. Kikuchi and W. Viehmann, *J. Chem. Phys.*, **48**:1097 (1968).
‡ J. M. Assour, *J. Am. Chem. Soc.*, **87**:4701 (1965).
§ J. Danon, R. P. A. Muniz, and H. Panepucci, *J. Chem. Phys.*, **41**:3651 (1964).
¶ J. D. W. van Voorst and P. Hemmerich, *J. Chem. Phys.*, **45**:3914 (1966).

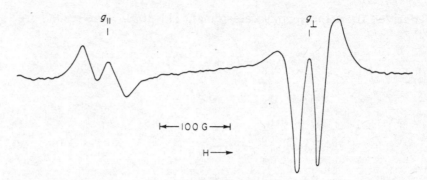

Fig. 11-10 Second-derivative ESR spectrum of Ag^{++} in (frozen) HNO$_3$ solution at 77 K, with $\nu = 9.35$ GHz. [*Taken from J. A. McMillan and B. Smaller, J. Chem. Phys.*, **35**:1698 (1961).]

Thus the g components indicate that the unpaired electron in this compound is in the d_{z^2} orbital as assumed for Eqs. (11-51) and (11-53).

The ESR spectrum of the $4d^9$ ion Ag^{++} in solutions of HNO$_3$ or H$_2$SO$_4$ shows a single line with $g = 2.133$ at room temperature.† The second-derivative spectrum in HNO$_3$ at 77 K is shown in Fig. 11-10. This spectrum is typical of a system with axial symmetry and with $(g_\| - g_\perp) \beta H >> hA$. Computer analysis of the spectrum yields $g_\| = 2.265$ and $g_\perp = 2.065$. The hyperfine splitting arises from ^{107}Ag and ^{109}Ag, each with about the same abundance, $I = \frac{1}{2}$, and very similar magnetic moments. Since $g_\| > g_\perp$, it appears that in this case $\alpha_t < 0$, and hence the unpaired electron resides in the $d_{x^2-y^2}$ orbital. Thus Eqs. (11-54) are the appropriate ones to use.

It is interesting that substitution of the value $\lambda = -1{,}840$ cm^{-1} (from the free ion) into Eqs. (11-54) leads to the values $\Delta = 53{,}500$ cm^{-1} for $g_\|$ and $\Delta = 48{,}750$ cm^{-1} for $g_\perp$. An absorption band is observed at 25,560 cm^{-1}. It is evident that either λ is much smaller in the complex or the assumption of a positive hole localized exclusively in the $d_{x^2-y^2}$ orbital of the silver ion is a poor approximation. Probably both statements are true.

A $3d^4$ (oct + ttgl) ion will exhibit the same splitting of orbital levels as a $3d^1$ (ttdl + ttgl) ion.

11-7b *F*-state ions

$3d^8$ *(oct)* The sevenfold orbital degeneracy of the ^3F ground state of gaseous $3d^8$ ions splits into two triply degenerate states and one nondegenerate state in an octahedral field. This was demonstrated in Sec. 11-5. In this case the nondegenerate $^3A_{2g}$ state lies lowest. (See Figs. 11-6a and 11-9.) The eigenfunctions corresponding to the states in an octahedral field can be found by substituting the energies of Eqs. (11-31) into the equations

† J. A. McMillan and B. Smaller, *J. Chem. Phys.*, **35**:1698 (1961).

derived from the secular determinant (11-30b). The results are

$$
T_{1g}: \quad
\begin{cases}
|t'_{-1}\rangle \equiv \sqrt{\tfrac{3}{8}}\,|-1\rangle + \sqrt{\tfrac{5}{8}}\,|3\rangle \\[2mm]
|t'_{+1}\rangle \equiv \sqrt{\tfrac{3}{8}}\,|1\rangle + \sqrt{\tfrac{5}{8}}\,|-3\rangle \\[2mm]
|t'_0\rangle \equiv |0\rangle
\end{cases}
$$

$$
T_{2g}: \quad
\begin{cases}
|t''_{+1}\rangle \equiv \sqrt{\tfrac{5}{8}}\,|-1\rangle - \sqrt{\tfrac{3}{8}}\,|3\rangle \\[2mm]
|t''_{-1}\rangle \equiv \sqrt{\tfrac{5}{8}}\,|1\rangle - \sqrt{\tfrac{3}{8}}\,|-3\rangle \\[2mm]
|t''_0\rangle \equiv \tfrac{1}{\sqrt{2}}\,(|2\rangle + |-2\rangle)
\end{cases}
$$

$$
A_{2g}: \quad |a\rangle \equiv \tfrac{1}{\sqrt{2}}\,(|2\rangle - |-2\rangle) \tag{11-55}
$$

The subscript on the t designates the expectation value of the fictitious operator $\hat{L}'_z$. (See Sec. 12-1.)

To zero order, one may ignore the effects of spin-orbit coupling on the ground-state wave function. The energies in a magnetic field are calculated using $\hat{\mathscr{H}}_{\text{mag}}$ of Eq. (11-33). For $\mathbf{H} \parallel \mathbf{Z}$

$$
\hat{\mathscr{H}}_{\text{mag}} = \beta H_z(\hat{L}_z + g_e \hat{S}_z) \tag{11-56}
$$

Since the ground state is orbitally nondegenerate, the contribution of $\hat{L}_z$ is zero. (See Prob. 11-5.) The ground-state wave function including spin may be written

$$
\frac{1}{\sqrt{2}}\,(|2, M_S\rangle - |-2, M_S\rangle)
$$

where $M_S = 1, 0, -1$. Thus the energies in a magnetic field are (see Fig. 11-11)

$$
W_{\pm 1} = \pm 2\beta H_z \tag{11-57a}
$$

$$
W_0 = 0 \tag{11-57b}
$$

Hence in this approximation $g_{ZZ} = 2$. Also, $g_{XX} = g_{YY} = 2$ as is required by the octahedral symmetry.

The spin-orbit coupling operator $\hat{\mathscr{H}}_{\text{so}}$ will cause an admixture of excited states into the ground state. Since $S = 1$, there may also be a splitting of the spin degeneracy in fields of lower than octahedral symmetry. The calculation of these effects is most conveniently approached through the tensor Λ of Eqs. (11-39). Since the field is octahedral, only one component of Λ need be calculated, e.g., Λ_{ZZ}. The only excited state which contributes to Λ_{ZZ} is the $|t''_0\rangle$ function of the T_{2g} state

$$
\frac{1}{\sqrt{2}}\,(|2\rangle + |-2\rangle)
$$

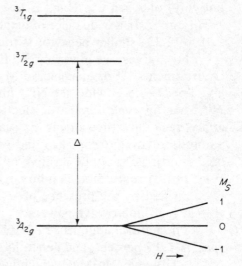

Fig. 11-11 Splitting of the states of a $3d^8$ ion in an octahedral field and in an added magnetic field. The same diagram applies to a d^2 ion in a tetrahedral field.

Thus

$$\Lambda_{zz} = -\frac{\left[\dfrac{1}{\sqrt{2}}\left(\langle 2| - \langle -2|\right)\hat{L}_z\dfrac{1}{\sqrt{2}}\left(|2\rangle + |-2\rangle\right)\right]^2}{\Delta}$$

$$= -\frac{4}{\Delta} \tag{11-58}$$

Recall that Δ is the energy separation between the T_{2g} states and the A_{2g} state. (See Fig. 11-11.)

Substitution of Eq. (11-58) into Eq. (11-39c) gives

$$g_{zz} = g_e - \frac{8\lambda}{\Delta} = g_{\text{isotropic}} \tag{11-59}$$

Since there are no excited states close to the ground state, ESR spectra of $3d^8$ ions can be seen at room temperature or at 77 K. The appropriate spin hamiltonian which allows for distortions from octahedral symmetry and for hyperfine coupling is

$$\hat{\mathscr{H}} = g_{\parallel}\beta H_z\hat{S}_z + g_{\perp}\beta(H_x\hat{S}_x + H_y\hat{S}_y) + D(\hat{S}_z^2 - \tfrac{2}{3})$$
$$+ E(\hat{S}_x^2 - \hat{S}_y^2) + hA_{\parallel}\hat{S}_z\hat{I}_z + hA_{\perp}(\hat{S}_x\hat{I}_x + \hat{S}_y\hat{I}_y) \tag{11-60}$$

The Ni^{++} ion is the most important example of a $3d^8$ (oct) ion. The g factor may be estimated from Eq. (11-59) if optical absorption data are available. For example in the $Ni(NH_3)_6^{++}$ ion an optical band (assigned to the $^3T_2 \leftarrow {}^3A_2$ transition) is observed at 10,700 cm^{-1}. If one takes the free-ion value of λ (-325 cm^{-1}), g is calculated to be 2.245. The experimental value is 2.162[†]; the discrepancy is probably due to some covalent bonding. Alter-

[†] T. Garofano, M. B. Palma-Vittorelli, M. U. Palma and F. Persico, *Paramagnetic Resonance, Proc. Intern. Cong.*, 1st, Jerusalem, **2:**582 (1962).

natively, one can use g and Δ to compute an effective value of λ, that is, -211 cm^{-1}. In a wide range of octahedral environments, g varies from about 2.10 to 2.33; similar behavior is found for the Cu^{3+} ion.

Ni^{++} gives rise to very broad ESR lines, even in presumably octahedral environments. For example, in MgO, where other substitutional ions may give lines ~ 0.5 G wide, the Ni^{++} linewidth may be as much as 40 G. Since Ni^{++} has an *even* number of electrons and the Kramers theorem does not apply, residual lattice strains may cause the $|0\rangle$ state to be shifted by varying amounts relative to the $|+1\rangle$ and $|-1\rangle$ states. (See Fig. 11-11.) The zero field splitting D may be positive or negative. Thus the observed spectrum (Fig. 11-12) represents an inhomogeneously broadened line. This analysis is confirmed by the presence of a sharp double-quantum transition observable at high microwave powers.†

The natural abundance of ^{61}Ni, the only nuclide with nonzero nuclear spin, is 1.25 percent, and hence hyperfine splitting has rarely been seen for Ni^{++}. However, in MgO which had been enriched with ^{61}Ni, the hyperfine quartet is readily seen; the coupling constant is 0.00083 cm^{-1}.‡

$3d^2$ *(ttdl)* The sevenfold orbital degeneracy of the ^3F states of a gaseous $3d^2$ ion is split by a tetrahedral crystal field so as to leave an orbital singlet lowest, as in Fig. 11-6a. The system is thus analogous to a $3d^8$ ion in octahedral symmetry considered above. In a *purely* tetrahedral field **g** will

† J. W. Orton, P. Auzins, J. H. E. Griffiths, and J. E. Wertz, *Proc. Phys. Soc.*, **78**:554 (1961).
‡ J. W. Orton, J. E. Wertz, and P. Auzins, *Phys. Rev. Letters*, **6**:339 (1963).

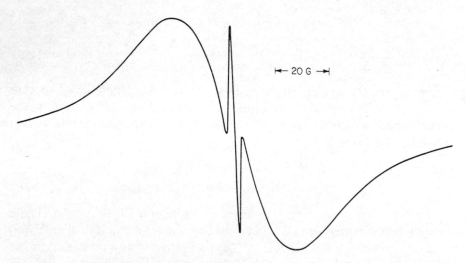

← 20 G →

Fig. 11-12 ESR spectrum of Ni^{++} in MgO at 115 K, with $\nu = 9.155$ GHz. The broad line is the superposition of the transitions $|-1\rangle \leftrightarrow |0\rangle$ and $|0\rangle \leftrightarrow |+1\rangle$. The narrow central line is the transition $|-1\rangle \leftrightarrow |+1\rangle$ effected by the absorption of two quanta at moderate microwave power.

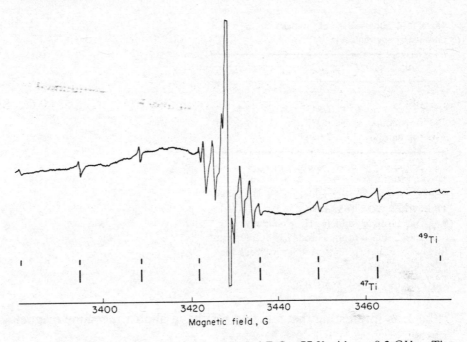

Fig. 11-13 ESR spectrum of Ti^{++} in tetrahedral ZnS at 77 K with $\nu = 9.2$ GHz. The broad line is analogous to that in Fig. 11-12: the narrow lines are double-quantum hyperfine lines. The central line corresponds to $^{46,48,50}Ti^{++}$. The widely spaced set of eight lines is a superposition of hyperfine lines from ^{47}Ti ($I = \frac{5}{2}$) and ^{49}Ti ($I = \frac{7}{2}$). The central sextet arises from one ^{67}Zn nucleus (4.12% natural abundance, $I = \frac{5}{2}$) as a nearest neighbor. [*J. Schneider and A. Räuber, Phys. Letters*, **21**:380 (1966).]

be isotropic and will be given by Eq. (11-59), since the Λ tensor will be identical to that of $3d^8$ ions in octahedral symmetry.

Since the ground state of the $3d^2$ (ttdl) ions is an orbital singlet, and there are no low-lying excited states, ESR spectra are observed at 77 K or even at room temperature. Figure 11-13 gives the spectrum of substitutional Ti^{++} ions in the "cubic" (i.e., tetrahedral) form of ZnS.† The appropriate spin hamiltonian ($S = 1$) is

$$\hat{\mathcal{H}} = g\beta\mathbf{H} \cdot \hat{\mathbf{S}} + hA_0\hat{\mathbf{S}} \cdot \hat{\mathbf{I}} \tag{11-61}$$

The hyperfine term is necessary, since ^{47}Ti and ^{49}Ti have nuclear spins of $\frac{5}{2}$ and $\frac{7}{2}$, respectively. For $^{46,48,50}Ti$, $I = 0$ and the first term suffices. The energy levels (omitting hyperfine splitting) are analogous to those in Fig. 11-11 with the g factor given by Eq. (11-59). It is not surprising that this system should behave in a very similar fashion to Ni^{++} in an octahedral environment. For Ti^{++} in ZnS, analogous broad lines are observed as also are sharp double-quantum transitions. The g factor (1.9280) is less than g_e since $\lambda > 0$.

† J. Schneider and A. Räuber, *Phys. Letters*, **21**:380 (1966).

Table 11-6 Comparison of g factors for $3d^2$ ions in tetrahedral environments

Ion	g factor	Reference
Ti^{++} in ZnS	1.9280	†
V^{3+} in ZnS	1.9433	‡
Cr^{4+} in silicon	1.9962	§
Mn^{5+} in silicon	2.0259	§

† J. Schneider and A. Räuber, *Phys. Letters,* **21**:380 (1966).
‡ W. C. Holton, J. Schneider, and T. L. Estle, *Phys. Rev.,* **133A**:1638 (1964).
§ G. W. Ludwig and H. H. Woodbury, in F. Seitz and D. Turnbull (eds.), "Solid State Physics," vol. 1, p. 265, Academic Press Inc., New York, 1962.

It is interesting that ^{47}Ti and ^{49}Ti have almost the same magnetogyric ratios:

$$\gamma(^{47}\text{Ti}) = 1.5102 \times 10^3 \text{ rad s}^{-1}\text{ G}^{-1}$$

$$\gamma(^{49}\text{Ti}) = 1.5106 \times 10^3 \text{ rad s}^{-1}\text{ G}^{-1}$$

Thus the hyperfine splittings for each isotope will be almost precisely the same, as is seen in Fig. 11-13. The outside pair of lines is weaker since only ^{49}Ti contributes to these lines. The central sextet arises from one ^{67}Zn nucleus (4.1% natural abundance) in the 12-atom second shell.

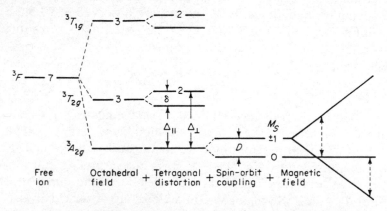

Fig. 11-14 Splitting of the states of a free d^8 ion in an octahedral field with an added tetragonal distortion. The combined effects of the tetragonal field and spin-orbit coupling partially remove the threefold spin degeneracy which exists in an octahedral field.

The isoelectronic V^{3+}, Cr^{4+}, and Mn^{5+} ions have also been observed in tetrahedral environments. The g factors show a progressive increase, as is shown in Table 11-6. This increase has been ascribed to an increasing contribution of covalent bonding along the series.

$3d^8$ *(oct + ttgl)* Suppose that a tetragonal distortion is applied along the Z axis for a $3d^8$ ion in an octahedral field. There will be some splitting of the T_{1g} and T_{2g} states by an amount δ which will be assumed to be small compared to $\Delta_\parallel$ or $\Delta_\perp$. (See Fig. 11-14.)

g_{ZZ} of Eq. (11-59) now becomes

$$g_{ZZ} = g_\parallel = g_e - \frac{8\lambda}{\Delta_\parallel} \tag{11-62}$$

Since this is a case of axial symmetry, only Λ_{XX} need be calculated. $\hat{L}_x$ couples the T_{2g} functions $|t_1''\rangle$ and $|t_{-1}''\rangle$ of Eq. (11-55). Thus

$$\Lambda_{XX} = - \frac{\left[\frac{1}{\sqrt{2}}\left(\langle 2| - \langle -2|\right)\hat{L}_x\left(\sqrt{\frac{5}{8}}|-1\rangle - \sqrt{\frac{3}{8}}|3\rangle\right)\right]^2}{\Delta_\perp}$$

$$- \frac{\left[\frac{1}{\sqrt{2}}\left(\langle 2| - \langle -2|\right)\hat{L}_x\left(\sqrt{\frac{5}{8}}|1\rangle - \sqrt{\frac{3}{8}}|-3\rangle\right)\right]^2}{\Delta_\perp}$$

$$= -\frac{4}{\Delta_\perp} \tag{11-63}$$

Thus

$$g_{XX} = g_\perp = g_e - \frac{8\lambda}{\Delta_\perp} \tag{11-64}$$

If $\delta << \Delta_\parallel$ or $\Delta_\perp$ (i.e., if $\Delta_\parallel \approx \Delta_\perp$), **g** will still be nearly isotropic. However, there will be a splitting of the threefold spin degeneracy in zero magnetic field.

From Eq. (11-46b)

$$D = D_{ZZ} - D_{XX}$$
$$= \lambda^2(\Lambda_{ZZ} - \Lambda_{XX}) \tag{11-65a}$$

for axial symmetry. Thus

$$D = -4\lambda^2\left(\frac{1}{\Delta_\parallel} - \frac{1}{\Delta_\perp}\right) = -4\lambda^2\frac{\Delta_\perp - \Delta_\parallel}{\Delta_\perp\Delta_\parallel}$$

$$\approx \frac{-4\lambda^2\delta}{\Delta^2} \tag{11-65b}$$

$E = 0$ since $D_{XX} = D_{YY}$. It is now clear why there will be no zero field splitting for a $3d^8$ ion in a purely octahedral field, since in that case $\delta = 0$.

In fields of orthorhombic or lower symmetry, the spin hamiltonian of Eq. (11-46a) will be required. Experimentally one observes that for most $3d^8$ ions, g is almost isotropic. Hence to simplify consideration of the second and third terms of Eq. (11-46a) we shall assume that g is isotropic, that is,

$$\hat{\mathscr{H}}_z = g\beta \mathbf{H} \cdot \hat{\mathbf{S}} + D(\hat{S}_z^2 - \tfrac{1}{3}\hat{S}^2) + E(\hat{S}_x^2 - \hat{S}_y^2) \qquad (11\text{-}66)$$

This hamiltonian is the same as Eq. (10-25). Hence the energy expressions of Eqs. (10-26) will apply here.

Ni^{++} has been observed in sites which give a strong zero field splitting (i.e., symmetry lower than octahedral), although the lines are very broad due to local variations in the magnitude of D. Measured values of D/hc range from 0.043 cm^{-1} in $Zn_3La_2(NO_3)_{12}\cdot24H_2O$ (with $E = 0$)† to -8.3 cm^{-1} in TiO_2 ($E/hc = 0.137$ cm^{-1}).‡ Provided that there is no mixing of the ground state with states of *different* L and S, the following expressions may be used to estimate D and E from the g components:

$$D = \tfrac{1}{2}\lambda[g_{ZZ} - \tfrac{1}{2}(g_{XX} + g_{YY})] \qquad (11\text{-}67a)$$

$$E = \tfrac{1}{4}\lambda(g_{XX} - g_{YY}) \qquad (11\text{-}67b)$$

These expressions are obtained from Eqs. (11-39) and (11-46). The student

† R. H. Hoskins, R. C. Pastor, and K. R. Trigger, *J. Chem. Phys.*, **30**:1630 (1959); J. W. Culvahouse, *J. Chem. Phys.*, **36**:2720 (1962).

‡ H. J. Gerritsen and E. S. Sabisky, *Phys. Rev.*, **125**:1853 (1962).

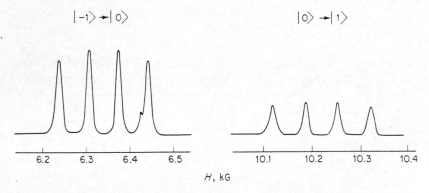

Fig. 11-15 The ESR spectrum of Cu^{3+} in Al_2O_3 at 1.8 K, with $\nu \approx 24$ GHz and H parallel to the trigonal axis. This is an example of a $3d^8$ ion which shows a zero-field splitting due to a trigonal field plus spin-orbit coupling. Hyperfine quartets from ^{63}Cu and ^{65}Cu are superimposed, due to similarity of their nuclear moments. There is no central line since there is no stable copper nuclide of zero nuclear spin. The greater intensity of the $|-1\rangle \leftrightarrow |0\rangle$ transitions shows that the parameter D has a negative value, i.e., for this ion in zero magnetic field the $|0\rangle$ level lies above the $|\pm1\rangle$ levels. (See Fig. 11-14.) The extraneous line at 6.43 kG is due to an iron impurity. [*Figure kindly supplied by Dr. S. Geschwind. See W. E. Blumberg, J. Eisinger, and S. Geschwind, Phys. Rev.*, **130**:900 (1963).]

is given an opportunity in Prob. 11-9 to examine the validity of these expressions.

Another d^8 ion is Cu^{3+}, which has been observed in Al_2O_3.[†] The ESR spectrum at 1.8 K of this system is given in Fig. 11-15 for **H** parallel to the trigonal axis. The hyperfine lines represent superimposed contributions of the ^{63}Cu and ^{65}Cu nuclides, both of which have $I = \frac{3}{2}$, and which have nuclear moments of 2.226 and 2.386 nuclear magnetons, respectively. At 1.8 K the upper levels are sufficiently depopulated such that the several transitions will have different intensities. Thus the greater intensity of the lines of the low-field group (corresponding to the $|-1\rangle \rightarrow |0\rangle$ transitions) compared with the lines of the high-field group indicates that the sign of D is negative. That is, at zero magnetic field the $|0\rangle$ level lies above the $|+1\rangle$ or the $|-1\rangle$ levels. The parameters which apply to Cu^{3+} in Al_2O_3 are

$$g_{\parallel} = 2.0788$$

$$g_{\perp} = 2.0772$$

$$\frac{D}{hc} = -0.1884 \text{ cm}^{-1}$$

$$\Delta(^3T_2) = 21,000 \text{ cm}^{-1}$$

$$^{63}A_{\parallel} = -0.00644 \text{ cm}^{-1} \qquad ^{63}A_{\perp} = -0.00601 \text{ cm}^{-1}$$

$$^{65}A_{\parallel} = -0.00689 \text{ cm}^{-1} \qquad ^{65}A_{\perp} = -0.00644 \text{ cm}^{-1}$$

The negative sign of the hyperfine coupling constants is considered in Prob. 11-10. ENDOR measurements (Sec. 13-4) allow observation of separate lines from the ^{63}Cu and ^{65}Cu nuclides.

$3d^2$ (ttdl + ttgl) For a d^2 ion in tetrahedral symmetry with an added tetragonal distortion the g components and zero field splitting will be given by Eqs. (11-62), (11-64), and (11-65b). An apparent example is the V^{3+} ion in CdS,[‡] for which $g_{\parallel} = 1.934$, $g_{\perp} = 1.932$, and $D/hc = 0.1130 \text{ cm}^{-1}$. From the g components one estimates the value of $\Delta_{\parallel}/hc$ as approximately 8,600 cm^{-1} if λ is taken as about 70 percent of the free-ion value. The corresponding value of $\Delta_{\parallel}/hc$ for V^{3+} in Al_2O_3 is 18,000 cm^{-1}. The ratio of these two splittings is very roughly 4:9, the expected ratio of tetrahedral to octahedral crystal field splittings. Thus the departure of the g components from g_e will be roughly twice as great as for octahedral ions.

For "Fe^{6+}" in K_2CrO_4, $g = 2.000$, $D/hc = 0.103 \text{ cm}^{-1}$, and $E/hc = 0.016 \text{ cm}^{-1}$.[§] The nonzero value of E reflects a significant distortion from axial symmetry.

† W. E. Blumberg, J. Eisinger, and S. Geschwind, *Phys. Rev.*, **130**:900 (1963).
‡ F. S. Ham and G. W. Ludwig, in W. Low, ed., "Paramagnetic Resonance," vol. 1, p. 130, Academic Press Inc., New York, 1963.
§ A. Carrington, D. J. E. Ingram, K. A. K. Lott, D. S. Schonland, and M. C. R. Symons, *Proc. Roy. Soc. (London)*, **A254**:101 (1960).

$3d^3$ *(oct)*   The ^4F ground state of gaseous $3d^3$ ions will split in an octahedral field in the same manner as for the $3d^8$ ions; i.e., the nondegenerate
A_{2g} state will again lie lowest. (See Fig. 11-6a or 11-16.) The principal
differences between $3d^3$ and $3d^8$ ions arise from the spin multiplicity. For
$3d^3$ ions, $S = \frac{3}{2}$, and the zero-order energies in a magnetic field will be given
as

$$W_{\pm\frac{3}{2}} = \pm\tfrac{3}{2} g_e \beta H_z \tag{11-68a}$$

$$W_{\pm\frac{1}{2}} = \pm\tfrac{1}{2} g_e \beta H_z \tag{11-68b}$$

Since the orbital wave functions are exactly the same as in the $3d^8$
case, the analytical forms of the tensor Λ will be identical. The expression
for g in an octahedral field [Eq. (11-59)] will also apply to the $3d^3$ (oct) case.

Numerous ESR studies of $3d^3$ ions or complexes in octahedral symmetry have been made; the lines are generally narrow, and hence spectra are
easily obtained at room temperature. Of the $3d^3$ ions, Cr^{3+} and its complexes have been studied more than any other ion. Figure 11-17 shows a
typical ESR spectrum of Cr^{3+} in octahedral symmetry in MgO. The spec-

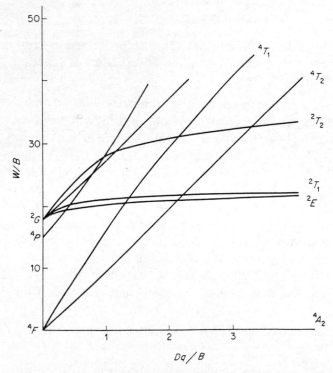

Fig. 11-16 The Tanabe-Sugano diagram for a d^3 ion in an
octahedral field.

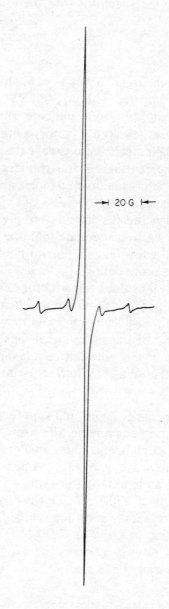

→ 20 G ←

Fig. 11-17 ESR spectrum of the $3d^3$ ion Cr^{3+} in octahedral symmetry in MgO at room temperature. The hyperfine quartet arises from $^{53}Cr^{3+}$ ($I = \frac{3}{2}$; natural abundance 9.54%).

trum is isotropic with $g = 1.9796$ at 290 K.[†] The central line arises from the nuclides of atomic mass 50, 52, and 54, all with $I = 0$; the hyperfine quartet arises from ^{53}Cr ($I = \frac{3}{2}$, natural abundance 9.54%) with $A_0/c = 0.00163$ cm^{-1}. One can describe the spectrum with the following spin hamiltonian ($S = \frac{3}{2}$, $I = \frac{3}{2}$):

$$\hat{\mathscr{H}} = g\beta\mathbf{H} \cdot \hat{\mathbf{S}} + hA_0\hat{\mathbf{S}} \cdot \hat{\mathbf{I}} \qquad (11\text{-}69)$$

For Cr^{3+}, $\lambda = 91$ cm^{-1}; in MgO, $\Delta(^4T_{2g}) = 16,900$ cm^{-1}; hence from Eq. (11-59), $g_{calc} = 1.96$ as compared with $g_{exp} = 1.98$. Similar calculations for the isoelectronic V^{++} and Mn^{4+} ions indicate an increasing discrepancy between the calculated and observed g factors as the nominal charge on the ion increases. More detailed calculations reveal a considerable transfer of charge from ligands to the central ion; this transfer increases as the charge on the ion increases.[‡],[§] Thus the discrepancies appear to arise from covalent bonding with the ligands. These effects are cited to give warning that crystal field theory (which regards ligands merely as point charges) *may*, under certain conditions, be so inadequate that one must explicitly consider the ligand electrons and energy levels (ligand field theory). Section 12-5 considers this problem in more detail.

For none of the $3d^3$ ions does the g factor fall below ~ 1.95. This indicates that excited states are far removed and hence there is a small spin-orbit interaction. It is a well-documented fact that these are the conditions under which ESR spectra of transition-metal ions are observable at room temperature. The reason is that spin-lattice relaxation in these ions usually proceeds via spin-orbit coupling. Where this interaction is small, T_1 is large, and hence the lines are narrow.

3d^7 (hs)(ttdl) The energy levels and spin for a $3d^7$ (high-spin) ion in a tetrahedral field are the same as for the $3d^3$ case considered above. Thus the expressions for the g components and zero field splitting parameters will be the same. Since λ is negative, $g > g_e$. The behavior of a $3d^7$ ion in cubal coordination is the same as in tetrahedral symmetry.

The Co^{++} ion in "cubic" ZnS or ZnTe shows the behavior expected of a high-spin d^7 ion in tetrahedral coordination.[¶] This is likewise true for Co^{++} in cubal coordination in CaF_2 or CdF_2.[††] Though the orbital singlet ground state lies lowest, the separation of the T_{1g} state from the A_{2g} ground state (see Fig. 11-6a) is small (4,200 cm^{-1} for Co^{++} in CdF_2); hence observa-

---

[†] ESR parameters for a number of ions in various simple and complex oxides are tabulated by W. Low and E. L. Offenbacher, in F. Seitz and D. Turnbull (eds.), "Solid State Physics," vol. 17, p. 135, Academic Press Inc., New York, 1965.

[‡] B. R. McGarvey, *J. Chem. Phys.*, **41**:3743 (1964).

[§] R. Lacroix and G. Emch, *Helv. Phys. Acta*, **35**:592 (1962).

[¶] F. S. Ham, G. W. Ludwig, G. D. Watkins, and H. H. Woodbury, *Phys. Rev. Letters*, **5**:468 (1960).

[††] T. P. P. Hall and W. Hayes, *J. Chem. Phys.*, **32**:1871 (1960).

tions are limited to temperatures below 20 K. For Co^{++} in CdF_2, $g = 2.278$. Setting $\lambda = -180$ cm^{-1} and $\Delta \approx 4300$ cm^{-1} in Eq. (11-59) gives $g_{calc} = 2.34$. As usual, the calculated g factor overestimates the departure from g_e; however, the result is sufficiently close to aid in the identification of the ground state of the ion. Two other facts confirm that $S = \frac{3}{2}$ for this system:

1. The terms $\hat{S}^3 \cdot \mathbf{H}$ and $\hat{S}^3 \cdot \hat{\mathbf{I}}$ in Eq. (11-50) are required to explain fine details of the spectrum.
2. The main transitions correspond to the transition $|-\frac{1}{2}\rangle \rightarrow |\frac{1}{2}\rangle$; however, at one-third the normal field value a weak line corresponding to the transition $|-\frac{3}{2}\rangle \rightarrow |\frac{3}{2}\rangle$ is observed.

$3d^3$ *(oct + ttgl)*   For a $3d^3$ ion in orthorhombic or lower symmetry, the energy levels in a magnetic field may be obtained by rewriting Eq. (11-66) in matrix form, noting that $S = \frac{3}{2}$. The necessary angular-momentum matrices correspond to those considered in Prob. B-4. The states to be used with the spin hamiltonian are $|\frac{3}{2}\rangle$, $|\frac{1}{2}\rangle$, $|-\frac{1}{2}\rangle$, and $|-\frac{3}{2}\rangle$, corresponding to the effective spin $S = \frac{3}{2}$. For $\mathbf{H} \parallel Z$, the hamiltonian matrix is

$$
\mathcal{H} = \begin{array}{c} \langle\frac{3}{2}| \\ \langle\frac{1}{2}| \\ \langle-\frac{1}{2}| \\ \langle-\frac{3}{2}| \end{array}
\begin{bmatrix}
\frac{3}{2}g\beta H_z + D & 0 & \sqrt{3}E & 0 \\
0 & \frac{1}{2}g\beta H_z - D & 0 & \sqrt{3}E \\
\sqrt{3}E & 0 & -\frac{1}{2}g\beta H_z - D & 0 \\
0 & \sqrt{3}E & 0 & -\frac{3}{2}g\beta H_z + D
\end{bmatrix}
$$

$$\begin{array}{cccc} |\frac{3}{2}\rangle & |\frac{1}{2}\rangle & |-\frac{1}{2}\rangle & |-\frac{3}{2}\rangle \end{array}$$

$$\tag{11-70a}$$

This matrix may be rearranged to give diagonal blocks; its conversion to the secular determinant leads to

$$
\begin{array}{c} \langle\frac{3}{2}| \\ \langle-\frac{1}{2}| \\ \langle\frac{1}{2}| \\ \langle-\frac{3}{2}| \end{array}
\begin{vmatrix}
\frac{3}{2}g\beta H_z + D - W & \sqrt{3}E & 0 & 0 \\
\sqrt{3}E & -\frac{1}{2}g\beta H_z - D - W & 0 & 0 \\
0 & 0 & \frac{1}{2}g\beta H_z - D - W & \sqrt{3}E \\
0 & 0 & \sqrt{3}E & -\frac{3}{2}g\beta H_z + D - W
\end{vmatrix} = 0
$$

$$\begin{array}{cccc} |\frac{3}{2}\rangle & |-\frac{1}{2}\rangle & |\frac{1}{2}\rangle & |-\frac{3}{2}\rangle \end{array}$$

$$\tag{11-70b}$$

Solution of the two 2×2 determinants gives

$$W_{\frac{3}{2}} = \frac{1}{2}g\beta H_z + [(g\beta H_z + D)^2 + 3E^2]^{\frac{1}{2}}$$
$$W_{-\frac{1}{2}} = \frac{1}{2}g\beta H_z - [(g\beta H_z + D)^2 + 3E^2]^{\frac{1}{2}}$$
$$W_{\frac{1}{2}} = -\frac{1}{2}g\beta H_z + [(g\beta H_z - D)^2 + 3E^2]^{\frac{1}{2}}$$
$$W_{-\frac{3}{2}} = -\frac{1}{2}g\beta H_z - [(g\beta H_z - D)^2 + 3E^2]^{\frac{1}{2}} \tag{11-71}$$

The subscripts on W refer to the eigenvalues of $\hat{S}_z$ as $H \rightarrow \infty$. In zero magnetic field there are pairs of degenerate energy levels at $\pm(D^2 + 3E^2)^{\frac{1}{2}}$. These pairs, which reflect the Kramers degeneracy of an ion with an odd number of electrons, are called Kramers doublets.

The simplest case is that of tetragonal symmetry. Since $E = 0$, the energy levels will be given by

$$W_{\pm\frac{1}{2}} = \pm\tfrac{1}{2}g\beta H_z - D$$
$$W_{\pm\frac{3}{2}} = \pm\tfrac{3}{2}g\beta H_z + D \qquad (11\text{-}72)$$

These levels are shown in Fig. 11-18. The transitions occur at fields

$$\frac{h\nu}{g\beta} - \frac{2D}{g\beta}, \quad \frac{h\nu}{g\beta}, \quad \text{and} \quad \frac{h\nu}{g\beta} + \frac{2D}{g\beta}$$

The determination of the energy levels for $\mathbf{H} \parallel X$ or $\mathbf{H} \parallel Y$ is simplified by examining the corresponding coefficients of the squares of the spin operators in the identity

$$D[\hat{S}_z^2 - \tfrac{1}{3}(\hat{S}_x^2 + \hat{S}_y^2 + \hat{S}_z^2)] + E(\hat{S}_x^2 - \hat{S}_y^2)$$
$$\equiv D_{XX}\hat{S}_x^2 + D_{YY}\hat{S}_y^2 + D_{ZZ}\hat{S}_z^2 \qquad (11\text{-}73)$$

The relations are

$$D_{ZZ} = \frac{2D}{3} \qquad D_{XX} = \frac{3E - D}{3} \qquad D_{YY} = -\frac{(D + 3E)}{3} \qquad (11\text{-}74)$$

Thus D in Eq. (11-70) should be replaced by $\tfrac{3}{2}D_{XX} = \tfrac{1}{2}(3E - D)$ for $\mathbf{H} \parallel X$, and by $-\tfrac{1}{2}(D + 3E)$ for $\mathbf{H} \parallel Y$. Further, E should be replaced by $-\tfrac{1}{2}(D + E)$ for $\mathbf{H} \parallel X$ and by $\tfrac{1}{2}(D - E)$ for $\mathbf{H} \parallel Y$. For $\mathbf{H} \parallel X$ and $E = 0$ one

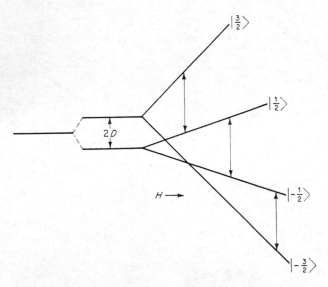

Fig. 11-18 Energy levels of a d^3 ion in octahedral-plus-tetragonal symmetry: the zero-field splitting gives two Kramers doublets. Application of a magnetic field makes possible three transitions if $2D$ is not large compared with $h\nu$.

may write

$$\hat{\mathscr{H}} = g\beta H_x \hat{S}_x - \frac{D}{2} \left[\hat{S}_z^2 - \frac{5}{4} \right] - \frac{D}{4} (\hat{S}_+^2 + \hat{S}_-^2) \tag{11-75}$$

For the tetragonal case, the energies are

$$W_{\frac{3}{2}} = \tfrac{1}{2} g\beta H_x + [(g\beta H_x)^2 + D^2 - g\beta H_x D]^{\frac{1}{2}}$$
$$W_{\frac{1}{2}} = -\tfrac{1}{2} g\beta H_x + [(g\beta H_x)^2 + D^2 + g\beta H_x D]^{\frac{1}{2}}$$
$$W_{-\frac{1}{2}} = \tfrac{1}{2} g\beta H_x - [(g\beta H_x)^2 + D^2 - g\beta H_x D]^{\frac{1}{2}}$$
$$W_{-\frac{3}{2}} = -\tfrac{1}{2} g\beta H_x - [(g\beta H_x)^2 + D^2 + g\beta H_x D]^{\frac{1}{2}} \tag{11-76}$$

The subscripts on W refer to the eigenvalues of $\hat{S}_z$ as $H \to \infty$.

These more complicated expressions for the energies result from the effect of the last term in Eq. (11-75). $\hat{S}_+^2$ mixes the $|+\tfrac{3}{2}\rangle$ with the $|-\tfrac{1}{2}\rangle$ functions, whereas $\hat{S}_-^2$ mixes the $|+\tfrac{1}{2}\rangle$ and $|-\tfrac{3}{2}\rangle$ functions.

It is of interest to note that the intensities of the three transitions shown in Fig. 11-18 are not $1:1:1$ but $3:4:3$. This may be seen by taking the square of the matrix elements of $\hat{S}_x = \tfrac{1}{2}(\hat{S}_+ + \hat{S}_-)$ between adjacent levels. (See Sec. C-4.) Thus

$$|\langle -\tfrac{1}{2} | \hat{S}_+ | -\tfrac{3}{2} \rangle|^2 = S(S+1) - M_S(M_S+1) = 3 \tag{11-77a}$$
$$|\langle \tfrac{1}{2} | \hat{S}_+ | -\tfrac{1}{2} \rangle|^2 \qquad\qquad\qquad = 4 \tag{11-77b}$$
$$|\langle \tfrac{3}{2} | \hat{S}_+ | \tfrac{1}{2} \rangle|^2 \qquad\qquad\qquad = 3 \tag{11-77c}$$

An example of a $3d^3$ (oct + ttgl) ion is considered in Prob. 11-11.

11-8 S–STATE IONS

$3d^5$ (hs)(oct)   The half-filled shell of the high-spin $3d^5$ ions implies that the ground state will be an orbital singlet, that is, $^6S_{\frac{5}{2}}$. (See Fig. 11-19.) For moderate crystal fields, the ground state is still essentially an orbital singlet, and hence the g factors for these ions are very close to the free-spin value. However, the presence of five unpaired electrons requires an additional term in the spin hamiltonian, since now the octahedral crystal-field operator can couple states with M_S values differing by ± 4. Reference to Eq. (11-18) shows that the new term must be of the form

$$\hat{\mathscr{H}}_c = \frac{a'}{120} [35\hat{S}_z^4 - 30S(S+1)\hat{S}_z^2 + 25\hat{S}_z^2 - 6S(S+1) + 3S^2(S+1)^2]$$

$$+ \frac{a'}{48} (\hat{S}_+^4 + \hat{S}_-^4) \tag{11-78}$$

Here $a' = \beta_c/6$.

For $S = \tfrac{5}{2}$, Eq. (11-78) reduces to

$$\hat{\mathscr{H}}_c = \frac{a'}{384} (112\hat{S}_z^4 - 760\hat{S}_z^2 + 567) + \frac{a'}{48} (\hat{S}_+^4 + \hat{S}_-^4) \tag{11-79}$$

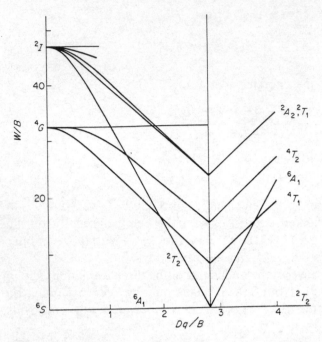

Fig. 11-19 Tanabe-Sugano diagram for a d^5 ion in octahedral symmetry. The transition between the low-field and high-field cases occurs for $Dq/B \approx 2.8$. In stronger crystal fields, the ground state is $^2T_2$ instead of $^6A_1$.

In zero magnetic field, the hamiltonian matrix will be

$$
\mathscr{H} =
\begin{array}{c}
 \\
\langle \tfrac{5}{2} | \\
\langle \tfrac{3}{2} | \\
\langle \tfrac{1}{2} | \\
\langle -\tfrac{1}{2} | \\
\langle -\tfrac{3}{2} | \\
\langle -\tfrac{5}{2} |
\end{array}
\begin{array}{c}
| \tfrac{5}{2} \rangle \quad | \tfrac{3}{2} \rangle \quad | \tfrac{1}{2} \rangle \quad | -\tfrac{1}{2} \rangle \quad | -\tfrac{3}{2} \rangle \quad | -\tfrac{5}{2} \rangle \\
\left[
\begin{array}{cccccc}
\dfrac{a'}{2} & 0 & 0 & 0 & \dfrac{\sqrt{5}}{2}\,a' & 0 \\
0 & -\dfrac{3}{2}\,a' & 0 & 0 & 0 & \dfrac{\sqrt{5}}{2}\,a' \\
0 & 0 & a' & 0 & 0 & 0 \\
0 & 0 & 0 & a' & 0 & 0 \\
\dfrac{\sqrt{5}}{2}\,a' & 0 & 0 & 0 & -\dfrac{3}{2}\,a' & 0 \\
0 & \dfrac{\sqrt{5}}{2}\,a' & 0 & 0 & 0 & \dfrac{a'}{2}
\end{array}
\right]
\end{array}
\qquad (11\text{-}80)
$$

The angular-momentum matrices for $S = \tfrac{5}{2}$ have been used to obtain (11-80).

The $| \pm\tfrac{1}{2} \rangle$ functions factor out immediately to give an energy of a'. The remaining matrix can then be reduced to

$$
\mathcal{H} =
\begin{array}{c}
\\
\langle\pm\tfrac{5}{2}| \\
\\
\langle\mp\tfrac{3}{2}|
\end{array}
\begin{array}{cc}
|\pm\tfrac{5}{2}\rangle & |\mp\tfrac{3}{2}\rangle \\
\left[\begin{array}{cc}
\dfrac{a'}{2} & \dfrac{\sqrt{5}}{2}\,a' \\[2mm]
\dfrac{\sqrt{5}}{2}\,a' & -\dfrac{3}{2}\,a'
\end{array}\right]
\end{array}
\tag{11-81}
$$

The corresponding secular determinant has the roots a' and $-2a'$. Thus at zero magnetic field the $|\pm\tfrac{5}{2}\rangle$ and $|\mp\tfrac{3}{2}\rangle$ states are strongly mixed. However, in a strong magnetic field such that $g\beta H \geqslant a'$, the states can be characterized by the kets $|M_S\rangle$. This is illustrated in Fig. 11-20.

In a magnetic field the electron Zeeman term must be added to the hamiltonian of Eq. (11-79). Operation on the six M_S states and solution of the resulting secular determinant leads to the energy levels

$$
W_{\pm\frac{1}{2}} = a' \pm \tfrac{1}{2}\,g\beta H \tag{11-82a}
$$

$$
W_{\frac{3}{2},-\frac{3}{2}} = \tfrac{1}{2}(g\beta H - a') \pm \tfrac{1}{2}(16g^2\beta^2 H^2 + 16a'g\beta H + 9a'^2)^{\frac{1}{2}} \tag{11-82b}
$$

$$
W_{\frac{3}{2},-\frac{5}{2}} = -\tfrac{1}{2}(g\beta H + a') \pm \tfrac{1}{2}(16g^2\beta^2 H^2 - 16a'g\beta H + 9a'^2)^{\frac{1}{2}} \tag{11-82c}
$$

when $\mathbf{H} \parallel Z$, i.e., one of the principal axes of the octahedron. If H is very large, the subscripts on W refer to the eigenvalues of $\hat{S}_z$. These levels are plotted as a function of magnetic field in Fig. 11-20.

The ESR spectra of high-spin $3d^5$ ions have been studied over a wide temperature range in a variety of hosts. For example, Mn^{++} has been studied in NaCl at ~ 870 K! This unusually favorable case is the result of the very small spin-orbit coupling with excited states, all of which are far removed from the $^6A_1$ ground state. Further, the Kramers degeneracy makes it possible for at least some lines to be observed at X band even when zero field splittings are large.

Fe^{3+} has been observed in octahedral symmetry in $SrTiO_3$, which has the perovskite structure. In this, the Fe^{3+} ion substitutes for a Ti^{4+} ion, which is surrounded by an undistorted octahedron of oxygen ions. Figure 11-21a illustrates the transitions shown in Fig. 11-20 for a $3d^5$ ion with $\mathbf{H} \parallel [001]$.[†] The outer lines correspond to the transitions $|\pm\tfrac{3}{2}\rangle \leftrightarrow |\pm\tfrac{1}{2}\rangle$; the intermediate lines to $|\pm\tfrac{5}{2}\rangle \leftrightarrow |\pm\tfrac{3}{2}\rangle$; and the central line to $|+\tfrac{1}{2}\rangle \leftrightarrow |-\tfrac{1}{2}\rangle$. In Prob. 11-12, one finds that the expected intensity ratios for these three groups of lines are $8:5:9$, respectively. The derivative amplitudes are not in this exact ratio due to linewidth differences; however, the integrated intensities are correct. For hosts in which the Fe^{3+} lines are sufficiently narrow, one may observe a doublet about the central line due to ^{57}Fe ($I = \tfrac{1}{2}$); $A/c = 0.00105$ cm^{-1}.

† E. S. Kirkpatrick, K. A. Müller, and R. S. Rubins, *Phys. Rev.*, **135**:A86 (1964).

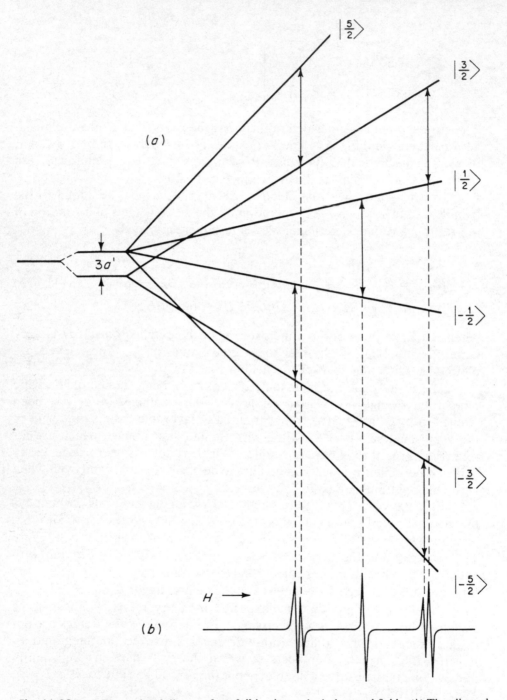

Fig. 11-20   (*a*) Energy level diagram for a $3d^5$ ion in octahedral crystal field.   (*b*) The allowed ESR spectrum is shown for $h\nu \gg 3a'$.   The diagram applies only for **H** parallel to a principal axis of the octahedron.

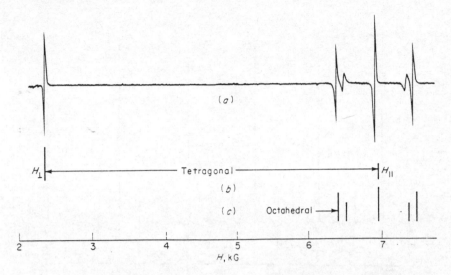

Fig. 11-21 (*a*) ESR spectrum of Fe^{3+} in sites of tetragonal and of octahedral symmetry in $SrTiO_3$ for **H** ∥ [001]. The spectrometer gain was greatly reduced in the region of the octahedral spectrum. $T = 300$ K and $\nu = 19.445$ GHz. (*b*) Stick plot of the tetragonal spectrum for the magnetic field along one of the octahedral axes. (*c*) Stick plot of the spectrum of Fe^{3+} ions in sites of octahedral symmetry. [*Spectrum kindly supplied by Dr. K. A. Müller. See E. S. Kirkpatrick, K. A. Müller, and R. S. Rubins, Phy. Rev., 135:A86 (1964).*]

For comparison, the spectrum of Mn^{++}, also in MgO, is given in Fig. 11-22. The spectrum is seen to be a sixfold replication of the Fe^{3+} spectrum, with some line coincidences occurring in the fifth group. This is due to the ^{55}Mn ($I = \frac{5}{2}$) hyperfine coupling. The parameter a' ($a'/hc = 0.001901$ cm^{-1}) for Mn^{++} is much smaller than A ($A/c = 0.008111$ cm^{-1}) and is also much smaller than a' for Fe^{3+} ($a'/hc = 0.02038$ cm^{-1}). As a consequence, the sextet of lines corresponding to the $|-\frac{1}{2}\rangle \leftrightarrow |\frac{1}{2}\rangle$ transition has a very small angular dependence. This particular sextet can thus readily be seen in a powder or in liquid solution. The other sets of lines are usually not observed under these conditions. For Fe^{3+}, one may observe the $|-\frac{1}{2}\rangle \leftrightarrow |\frac{1}{2}\rangle$ transition in powders of some octahedral hosts. The field variation between extrema is so great that only a single broad ESR line is seen for Fe^{3+} in liquid solution where the ion is presumably in tetrahedral coordination, i.e., $[FeCl_4]^-$.[†]

Due to the remoteness of the excited states, the g factors for $3d^5$ (hs) ions lie very close to the free-electron value. For example, $g = 2.0009$ for Mn^{++} in CaO,[‡] $g = 2.0052$ for Fe^{3+} in CaO,[‡] and $g = 1.9995$ for Cr^+ in ZnS.[§]

[†] G. R. Hertel and H. M. Clark, *J. Phys. Chem.*, **65**:1930 (1961).
[‡] A. J. Shuskus, *Phys. Rev.*, **127**:1529 (1962).
[§] R. S. Title, *Phys. Rev.*, **131**:623 (1963).

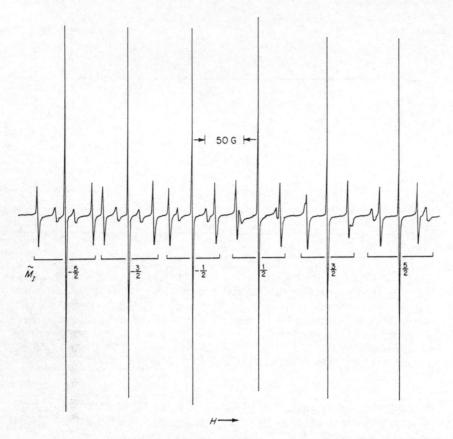

Fig. 11-22 ESR spectrum of Mn^{++} in MgO at room temperature, with $\nu \approx 9.3$ GHz. Since the hyperfine splitting is much greater than the zero-field splitting (Fig. 11-20), the spectrum consists of six sets of quintets such as those shown in Figs. 11-20 and 11-21. The numbers under each bracket represent the $\tilde{M}_I$ values for the given quintet.

3d^5 (hs)(oct + ttgl)   In fields of lower symmetry the terms

$$D(\hat{S}_z^2 - \tfrac{1}{3}\hat{S}^2) + E(\hat{S}_x^2 - \hat{S}_y^2)$$

must be added to Eq. (11-79). Usually D and E are much larger than a'; hence, the spin hamiltonian of Eq. (11-46a) may be used. g is usually isotropic and very close to 2. In a tetragonal crystal field ($E = 0$) the energy levels would appear as in Fig. 11-23 for $D/hc \approx 0.1$ cm^{-1} with **H** along the tetragonal axis. The energy-level expressions become complicated when **H** is not directed along a principal axis.

If $D >> h\nu$, only transitions between the $|\pm\tfrac{1}{2}\rangle$ states will be observed. For **H** $\parallel Z$, this case can be treated in terms of a spin hamiltonian with a fictitious spin $S' = \tfrac{1}{2}$. The energies of the $|\pm\tfrac{1}{2}\rangle$ states will then be $\pm\tfrac{1}{2} g_e \beta H_z$, since the $|\pm\tfrac{1}{2}\rangle$ functions are eigenfunctions of $\hat{S}_z$. Thus $g_\parallel = g_e$. For **H** $\parallel X$, one must recognize that the true spin is $\tfrac{5}{2}$ in computing the hamiltonian matrix in terms of the kets $|S, M_S\rangle$

$$\mathcal{H}_{\text{mag}} = \begin{array}{c} \langle \frac{5}{2}, \frac{1}{2}| \\ \langle \frac{5}{2}, -\frac{1}{2}| \end{array} \begin{bmatrix} 0 & \frac{3}{2}g_e\beta H_X \\ \frac{3}{2}g_e\beta H_X & 0 \end{bmatrix} \qquad (11\text{-}83)$$

The corresponding secular determinant has the roots $\pm\frac{3}{2}g_e\beta H_X$. Thus $g_\perp \approx 6$. The energy levels are plotted in Fig. 11-24.

By appropriate treatment, the octahedral Fe^{3+} ion in $SrTiO_3$ can be converted to a tetragonal system in which $g_\parallel = 2.0054$ and $g_\perp = 5.993$ (Fig. 11-21b).[†] This is thus an example for which $D >> h\nu$; in fact $|D/hc| = 1.42$ cm^{-1}. (Here D is obtained indirectly from the value of $g_\perp$.) It is likely that an adjacent oxygen vacancy is the origin of the tetragonal field. A rather important biological example of Fe^{3+} in a tetragonal field will be considered in Sec. 14-5.

[†] E. S. Kirkpatrick, K. A. Müller, and R. S. Rubins, *Phys. Rev.*, **135**:A86 (1964).

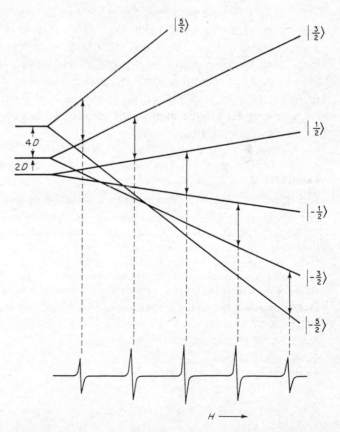

Fig. 11-23 Energy levels and allowed transitions for a d^5 ion in a weak tetragonal field, with **H** parallel to the tetragonal axis.

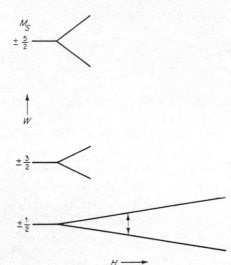

Fig. 11-24 Energy levels of a d^5 ion in a strong tetragonal field, with H parallel to the tetragonal axis. The zero-field splitting is so large compared with $h\nu$ that only the $|+\frac{1}{2}\rangle \leftrightarrow |-\frac{1}{2}\rangle$ transitions are observed.

If one is able to detect the ESR line of Fe^{3+} in solution, the linewidth is usually of the order of 1,000 G. This extreme width is presumably due to a nonoctahedral instantaneous distribution of ligands, with a consequently large and varying zero field splitting. However, upon addition of F^- ions (at pH > 5) to form the stable octahedral ion FeF_6^{3-}, one observes a septet of lines (of width 11 G) with a binomial distribution of intensities.[†] This complexing technique might well be applied to other ions which are difficult to observe in solution.

PROBLEMS

11-1. Compute the Landé g factors for each of the $3d^n$ gaseous ions in Table 11-1.

11-2. Make use of the appropriate commutation relations [see Eqs. (B-19)] to transform the products $\hat{J}_x\hat{J}_z^2\hat{J}_x$, $\hat{J}_z\hat{J}_x^2\hat{J}_z$, and $\hat{J}_z\hat{J}_x\hat{J}_z\hat{J}_x$ into forms in which the only imaginary term is $i\hat{J}_x\hat{J}_y\hat{J}_z$ (more than one commutator is required). Thus verify Eq. (11-15).

11-3. (a) Obtain the matrices L_z^4, L_z^2, L_+^4 and L_-^4 for $L = 2$ and thus verify $\mathscr{H}_{oct}$ of Eq. (11-26).

(b) Solve the secular determinant derived from Eq. (11-26) to obtain Eqs. (11-27). (Note that three of the energies factor out immediately, leaving only a 2 × 2 determinant.)

11-4. (a) For a P-state ion in a tetragonal crystal field, use Eq. (11-42) to obtain

$$|\alpha'\rangle = \left|0, \frac{1}{2}\right\rangle - \frac{1}{\sqrt{2}}\frac{\lambda}{\delta}\left|1, -\frac{1}{2}\right\rangle$$

$$|\beta'\rangle = \left|0, -\frac{1}{2}\right\rangle - \frac{1}{\sqrt{2}}\frac{\lambda}{\delta}\left|-1, \frac{1}{2}\right\rangle$$

(b) Then use $\mathscr{H}_{mag}$ of Eq. (11-44) to obtain $g_{\parallel} = g_e$ and $g_{\perp} = g_e - 2\lambda/\delta$.

[†] H. Levanon, G. Stein, and Z. Luz, *J. Am. Chem. Soc.*, **90**:5292 (1968).

11-5. Show that the eigenvalue of $\hat{L}_z$ for the $(1/\sqrt{2})(|2\rangle - |-2\rangle)$ function is zero.

11-6. For the $3d^1$ ion in a tetrahedral field with a tetragonal distortion such that $\alpha_t < 0$, the $(1/\sqrt{2})(|2\rangle + |-2\rangle)$ state is the lowest. Use the Λ-tensor method of Sec. 11-6 to show that in this case

$$g_{\|} = 2 - \frac{8\lambda}{\Delta}$$

$$g_{\perp} = 2 - \frac{2\lambda}{\Delta}$$

11-7. The wave functions and energies for a $3d^1$ ion in an octahedral field with trigonal distortion are given in Fig. 11-25. Derive the expressions for $g_{\|}$ and $g_{\perp}$ in this case and thus show that these expressions are the same as for the tetragonal distortion case, that is,

$$g_{\|} \cong g_e$$

$$g_{\perp} \cong g_e - \frac{2\lambda}{\delta}$$

where terms involving Δ^{-1} have been neglected.

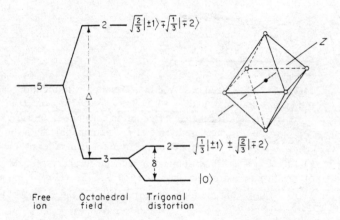

Fig. 11-25 Splitting of state energies of a $3d^1$ (oct + trgl) ion. The trigonal axis is a body diagonal of the circumscribing cube.

11-8. Derive Eqs. (11-67a) and (11-67b) from the dependence of **g** and **D** on Λ.

11-9. For Ni^{++} in Al_2O_3, $g_{\|} = 2.1957$ and $g_{\perp} = 2.1859$. Compute D/hc using Eq. (11-67a) and compare with the experimental value of -1.375 cm^{-1}. Alternatively, for Cr^{3+} in $MgWO_4$, $g_{\|} = 1.966$, $g_{\perp} = 1.960$, and $D/hc = +0.794$ cm^{-1}. Again calculate D/hc. Comment on any differences between calculated and experimental D/hc values.

11-10. The spin hamiltonian for a $3d^8$ ion in an axial field may be written in the following form:

$$\hat{\mathscr{H}} = D[\hat{S}_z^2 - \tfrac{1}{3}S(S+1)]$$
$$+ g_{\|}\beta H \hat{S}_z \cos\theta + \tfrac{1}{2}g_{\perp}\beta H(\hat{S}_+ + \hat{S}_-)\sin\theta$$
$$+ hA_{\|}\hat{S}_z\hat{I}_z + \tfrac{1}{2}hA_{\perp}(\hat{S}_+\hat{I}_- + \hat{S}_-\hat{I}_+)$$

where θ is the angle between **H** and the symmetry axis.

(a) For $\theta = 0$, use second-order perturbation theory (see Secs. A-7 and C-7) to show that transitions occur at fields given by

$$H_{-1 \to 0} = H' + \frac{D}{g_\parallel \beta} - \frac{hA_\parallel}{g_\parallel \beta} M_I - \frac{h^2 A_\perp^2}{2g_\parallel g_\perp \beta^2 H'} \left(\frac{15}{4} - M_I^2 - M_I \right)$$

$$H_{0 \to 1} = H' - \frac{D}{g_\parallel \beta} - \frac{hA_\parallel}{g_\parallel \beta} M_I - \frac{h^2 A_\perp^2}{2g_\parallel g_\perp \beta^2 H'} \left(\frac{15}{4} - M_I^2 + M_I \right)$$

(b) Given that $D < 0$, $S = 1$, $I = \frac{3}{2}$, use the field positions given below to confirm the sign and magnitude of the parameters (p. 297) for the $^{63}Cu^{3+}$ ion in Al_2O_3.

Frequency $\nu = 9.042$ GHz	
Transition	Field, G
$\lvert 0 \rangle \to \lvert 1 \rangle$	4947.7
	5013.4
	5080.3
	5148.5
$\lvert -1 \rangle \to \lvert 0 \rangle$	1067.0
	1131.4
	1197.1
	1264.1

11-11. Cr^{3+} has been observed in association with a cation vacancy in CaO. The local symmetry is tetragonal, with $g_\parallel = 1.9697$, $g_\perp = 1.9751$, and $D/hc = 0.13606$ cm^{-1}.

(a) Use λ for the Cr^{3+} free ion to calculate $\Delta_\parallel$ and $\Delta_\perp$.

(b) Is the measured value of D/hc compatible with the observed g factors?

(c) Sketch the expected spectrum for **H** parallel to one of the axes of the cubic crystal.

11-12. For a $3d^5$ (high spin) ion in an octahedral crystal field, show that the relative intensities of the five fine-structure lines are $5:8:9:8:5$. (Hint: Compute the matrix elements of $\hat{S}_+$.)

11-13. Show that the following expression for the energy of a state $\lvert M_S, M_I \rangle$,

$$\left(A_\parallel \frac{g_\parallel}{g} \cos^2 \theta + A_\perp \frac{g_\perp}{g} \sin^2 \theta \right) M_S M_I$$

which may be used for a system with axial symmetry if $A_\parallel \gg A_\perp$, may be transformed into the more usual expression for angular variation of hyperfine splitting

$$a + b(3 \cos^2 \theta - 1)$$

if the anisotropy of the g factor may be neglected.

12

Transition-metal Ions. II.
Electron Resonance in
the Gas Phase

12-1 IONS IN ORBITALLY DEGENERATE GROUND STATES

All of the ions considered in Chap. 11 were those for which the ground
state is orbitally nondegenerate. The deviations of the g factors from g_e
are adequately described by the use of second-order perturbation theory.
In Chap. 12 we encounter examples in which the crystal field does not
completely remove the orbital degeneracy of the free ion. Hence, the
ground state will have a net *orbital* angular momentum. It is not possible
to treat the spin and orbital angular momenta independently; hence, the
approach used in Chap. 11 is *no longer applicable.* Instead one must consider the effect of the spin-orbit coupling operator in terms of the *total* angular momentum of the ground state. Since the resulting splitting of states is
usually much larger than kT, one can deal primarily with the lowest-lying
state. It is essential to determine the degeneracy and eigenfunctions for
this state as these determine the characteristics of the ESR spectrum. The
ions in the various crystal fields to be considered in this chapter are listed
in Table 12-1.

Table 12-1   Ions in orbitally degenerate ground states

Section	Ion	Crystal field symmetry	Symbol
12-1a	D-state ions		
	$3d^1$	Octahedral	$3d^1$ (oct)
	$3d^1$	Octahedral + tetragonal distortion	$3d^1$ $(oct + ttgl)$
	$3d^1$	Octahedral + trigonal distortion	$3d^1$ $(oct + trgl)$
	$3d^5$ (low spin)	Octahedral + tetragonal distortion	$3d^5$ $(ls)(oct + ttgl)$
	$3d^9$	Tetrahedral + tetragonal distortion	$3d^9$ $(ttdl + ttgl)$
	$3d^6$ (high spin)	Octahedral	$3d^6$ $(hs)(oct)$
12-1b	F-state ions		
	$3d^2$	Octahedral	$3d^2$ (oct)
	$3d^2$	Octahedral + trigonal	$3d^2$ $(oct + trgl)$
	$3d^7$ (high spin)	Octahedral	$3d^7$ $(hs)(oct)$
12-1c	Jahn-Teller splitting		
	$3d^9$	Octahedral	$3d^9$ (oct)
	$3d^7$ (low spin)	Octahedral	$3d^7$ $(ls)(oct)$

12-1a   D-state ions

$3d^1$ (oct) The ^2D state of gaseous $3d^1$ ions splits in an octahedral field so as to leave the T_{2g} state lowest. (See Fig. 11-4a.) Thus the ground state will retain some orbital angular momentum even in zero order. It is convenient to consider the three degenerate T_{2g} states as the three components of a state with an orbital angular momentum quantum number $L' = 1$. (A prime has been used since L applies strictly only to systems with spherical symmetry.) Consideration of the ground state as a system with $L' = 1$ requires use of the following modified hamiltonian[†] analogous to Eq. (11-33)

$$\hat{\mathcal{H}}' = \beta(-\alpha\hat{\mathbf{L}}' + g_e\hat{\mathbf{S}}) \cdot \mathbf{H} - \alpha\lambda\hat{\mathbf{L}}' \cdot \hat{\mathbf{S}}$$

$$= \hat{\mathcal{H}}'_{\text{mag}} + \hat{\mathcal{H}}'_{\text{so}} \tag{12-1}$$

For D-state ions $\alpha \cong 1$, and for F-state ions $\alpha \cong \frac{3}{2}$ in fields of octahedral symmetry. The latter result can be obtained by applying $\hat{L}_z$ to the T_{1g} functions in Eq. (11-55).

Since $S = \frac{1}{2}$ for $3d^1$ ions, there will always be at least a twofold Kramers degeneracy. In order to evaluate the effect of $\hat{\mathcal{H}}'_{\text{so}}$ on the ground state, it is convenient to use the functions $|M_{L'}, M_S\rangle$ as a basis set even though most of these functions are not eigenfunctions of $\hat{\mathcal{H}}'_{\text{so}}$. Of the six possible $|M_{L'}, M_S\rangle$ functions, only $|\pm 1, \pm\frac{1}{2}\rangle$ are eigenfunctions of $\hat{\mathcal{H}}'_{\text{so}}$.

[†] A. Abragam and M. H. L. Pryce, *Proc. Roy. Soc.* (*London*), **A205**:135 (1951).

At this point it is convenient to introduce the quantum number J' for the *total* angular momentum of the system; J' is analogous to the J quantum number for atoms. The mixed states can be characterized by the eigenvalues of $\hat{J}_z'$, namely, $M_{J'} = M_{L'} + M_S$. Thus the functions $|\pm 1, \pm\frac{1}{2}\rangle$ will be characterized as eigenfunctions with $M_{J'} = \pm\frac{3}{2}$. Applications of $\hat{\mathscr{H}}_{so}'$ of Eq. (12-1) to the $|\pm 1, \pm\frac{1}{2}\rangle$ functions yields the energy

$$\langle \pm 1, \pm\tfrac{1}{2} | \hat{\mathscr{H}}_{so}' | \pm 1, \pm\tfrac{1}{2} \rangle = -\alpha\lambda \langle \pm 1, \pm\tfrac{1}{2} | [\hat{L}_z'\hat{S}_z + \tfrac{1}{2}(\hat{L}_+'\hat{S}_- + \hat{L}_-'\hat{S}_+)] | \pm 1, \pm\tfrac{1}{2} \rangle$$

$$= -\frac{\alpha\lambda}{2} \quad (12\text{-}2)$$

The functions with $M_{J'} = \pm\frac{1}{2}$ are $|0, \pm\frac{1}{2}\rangle$ and $|\pm 1, \mp\frac{1}{2}\rangle$. These are not eigenfunctions of $\hat{\mathscr{H}}_{so}'$, and hence one must solve a secular determinant. The hamiltonian matrix is

$$\mathscr{H}_{so}' = \begin{array}{cc} & \begin{array}{cc} |\pm 1, \mp\tfrac{1}{2}\rangle & |0, \pm\tfrac{1}{2}\rangle \end{array} \\ \begin{array}{c} \langle \pm 1, \mp\tfrac{1}{2}| \\ \langle 0, \pm\tfrac{1}{2}| \end{array} & \left[\begin{array}{cc} \tfrac{1}{2}\alpha\lambda & -\dfrac{1}{\sqrt{2}}\alpha\lambda \\ -\dfrac{1}{\sqrt{2}}\alpha\lambda & 0 \end{array} \right] \end{array} \quad (12\text{-}3)$$

Solution of the corresponding secular determinant yields the two energies $\alpha\lambda$ and $-\frac{1}{2}\alpha\lambda$. Thus only two distinct energies are found. These may be characterized by the two possible values of J', namely, $\frac{1}{2}$ and $\frac{3}{2}$. That is,

$$W_{\frac{3}{2}} = -\tfrac{1}{2}\alpha\lambda \quad (12\text{-}4a)$$

and

$$W_{\frac{1}{2}} = \alpha\lambda \quad (12\text{-}4b)$$

For $3d^1$ ions, $\lambda > 0$; hence the level $W_{\frac{3}{2}}$ will lie lower. If $\alpha\lambda \gg kT$, then this state will be the only one which is appreciably populated. Thus the ground state will be fourfold degenerate.

In order to compute the effect of $\hat{\mathscr{H}}_{mag}'$ of Eq. (12-1), one requires the eigenfunctions of the ground state. These are obtained by insertion of the energy of the $J' = \frac{3}{2}$ state into the secular equations derived from (12-3). The ground-state wave functions are

$$\left| \frac{3}{2}, \pm\frac{3}{2} \right\rangle = \left| \pm 1, \pm\frac{1}{2} \right\rangle \quad (12\text{-}5a)$$

$$\left| \frac{3}{2}, \pm\frac{1}{2} \right\rangle = \frac{1}{\sqrt{3}} \left| \pm 1, \mp\frac{1}{2} \right\rangle \pm \sqrt{\frac{2}{3}} \left| 0, \pm\frac{1}{2} \right\rangle \quad (12\text{-}5b)$$

where the kets on the left refer to $|J', M_{J'}\rangle$.

The energies and eigenfunctions in this problem can be obtained more directly by recognizing that

$$\hat{\mathbf{L}}' \cdot \hat{\mathbf{S}} = \tfrac{1}{2}[(\hat{\mathbf{L}}' + \hat{\mathbf{S}})^2 - \hat{\mathbf{L}}'^2 - \hat{\mathbf{S}}^2]$$
$$= \tfrac{1}{2}(\mathbf{J}'^2 - \hat{\mathbf{L}}'^2 - \hat{\mathbf{S}}^2) \tag{12-6}$$

Thus $\hat{\mathcal{H}}'_{so}$ becomes

$$\hat{\mathcal{H}}'_{so} = -\frac{\alpha\lambda}{2}(\mathbf{J}'^2 - \hat{\mathbf{L}}'^2 - \hat{\mathbf{S}}^2)$$

$$= -\frac{\alpha\lambda}{2}[J'(J'+1) - L'(L'+1) - S(S+1)] \tag{12-7}$$

The functions $|J', M_{J'}\rangle$ are eigenfunctions of $\hat{\mathcal{H}}'_{so}$, and hence the energies in Eqs. (12-4) are obtained directly by applying Eq. (12-7) to $|J', M_{J'}\rangle$.

The eigenfunctions in terms of $|M_{L'}, M_S\rangle$ may be obtained by successive application of

$$\hat{J}'_\pm = \hat{L}'_\pm + \hat{S}_\pm \tag{12-8}$$

using the methods outlined in Sec. B-7. For example, application of $\hat{J}'_-$ to Eq. (12-5a) gives

$$(\hat{L}'_- + \hat{S}_-)|1, \tfrac{1}{2}\rangle = \sqrt{2}|0, \tfrac{1}{2}\rangle + |1, -\tfrac{1}{2}\rangle \tag{12-9}$$

Renormalization gives the eigenfunction corresponding to $J' = \tfrac{3}{2}$, $M_{J'} = \tfrac{1}{2}$, i.e., Eq. (12-5b).

For $\mathbf{H} \parallel Z$, $\hat{\mathcal{H}}'_{mag}$ becomes

$$\hat{\mathcal{H}}'_{mag} = \beta(-\alpha\hat{L}'_z + g_e\hat{S}_z)H_Z \tag{12-10}$$

and hence the energies in a magnetic field are

$$W_{\frac{3}{2},\pm\frac{3}{2}} = \pm(-\alpha + 1)\beta H_Z = \tfrac{2}{3}(-\alpha + 1)\beta H_Z(\pm\tfrac{3}{2}) \tag{12-11a}$$
$$W_{\frac{3}{2},\pm\frac{1}{2}} = 2[\tfrac{1}{3}(-\alpha - 1) + \tfrac{2}{3}(+1)]\beta H_Z(\pm\tfrac{1}{2})$$
$$= \tfrac{2}{3}(-\alpha + 1)\beta H_Z(\pm\tfrac{1}{2}) \tag{12-11b}$$

Since $\alpha \approx 1$ for D-state ions, $g \approx 0$. Thus it is not surprising that ESR absorption is not detected for $3d^1$ ions in purely octahedral symmetry.

$3d^1$ (oct + ttgl), $\Delta \gg \delta \gg \lambda$ If spin-orbit coupling were absent, a tetragonal field would split the levels of a $3d^1$ ion as shown by the dotted lines in Fig. 12-1. Depending on the sign of the tetragonal splitting, either an orbital doublet or a singlet lies lower. In the latter case, the ground state is a spin doublet (Kramers doublet), with $g_\parallel = g_\perp = g_e$, since the spin-orbit coupling has been assumed to be zero.

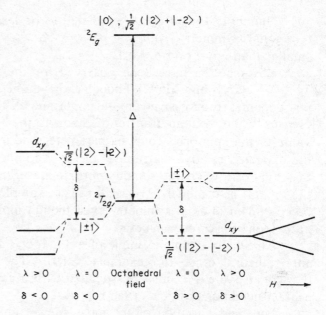

Fig. 12-1 Splitting of the states of a $3d^1$ ion in an octahedral field with an added tetragonal distortion large compared with the spin-orbit coupling. The right-hand and left-hand sides respectively refer to positive and to negative values of δ.

However, even a small spin-orbit coupling completely lifts the orbital degeneracy, as shown in Fig. 12-1. For the case of $0 < \delta >> \lambda$, one may calculate approximate g components from the corresponding Λ-tensor components. The distortion axis will be taken as the Z axis. Here

$$\Lambda_{ZZ} = -\frac{[(1/\sqrt{2})(\langle 2| - \langle -2|)\hat{L}_z(1/\sqrt{2})(|2\rangle + |-2\rangle)]^2}{\Delta + 2\delta/3} \approx \frac{-4}{\Delta} \quad (12\text{-}12)$$

Hence

$$g_{\parallel} = g_{ZZ} = g_e - \frac{8\lambda}{\Delta} \quad (12\text{-}13)$$

The $|\pm 1\rangle$ states, which are degenerate in an octahedral field, are coupled by $\hat{L}_-$ and $\hat{L}_+$, respectively, to the ground state. The approximate contributions to Λ_{XX} are as follows:

$$\Lambda_{XX} \approx -\frac{[(1/\sqrt{2})(\langle 2| - \langle -2|)\frac{1}{2}\hat{L}_+|1\rangle]^2 + [(1/\sqrt{2})(\langle 2| - \langle -2|)\frac{1}{2}\hat{L}_-|-1\rangle]^2}{\delta}$$

$$= -\frac{1}{\delta} \quad (12\text{-}14)$$

Hence

$$g_{\perp} = g_{XX} = g_{YY} \approx g_e - \frac{2\lambda}{\delta} \quad (12\text{-}15)$$

Thus both $g_\parallel$ and $g_\perp$ are predicted to be less than g_e for a $3d^1$ ion in tetragonal symmetry with $0 < \delta >> \lambda$, $\Delta >> \delta$. However, $g_\perp$ will be smaller than $g_\parallel$. For $\lambda/\delta = 0.1$, $g_\perp \approx 1.8$.

ESR spectra have been observed for the following $3d^1$ ions: Sc^{++}, Ti^{3+}, V^{4+}, Cr^{5+}, and Mn^{6+}. Though all have one d electron, covalent bonding becomes progressively more important as the charge on the ion increases. The resultant transfer of charge to the ligands lowers the effective value of the spin-orbit coupling parameter and in some cases may reverse the expected relative magnitudes of the g components. One thus expects that ions of low charge should approximate best the predictions based on the crystal field model. We shall select examples of $3d^1$ ions which have been studied in axial symmetry even though most of the reported spectra correspond to lower symmetry.

An ideally simple example of a $3d^1$ (oct + ttgl) ion is the Ti^{3+} ion in an octahedral oxide (for example, CaO). The symmetry is lowered to tetragonal if the Ti^{3+} ion is associated with a cation vacancy in the next-nearest-neighbor position. (See Fig. 12-2a.)

The field variation of the most intense lines for rotation in a (001) plane is given in Fig. 12-2c. The invariant line arises from those Ti^{3+} ions with a <001> tetragonal axis perpendicular to the field direction. The position of this line gives $g_\perp$, whereas that of the lines for the [100] or [010] direction yields $g_\parallel$. From the positions of the hyperfine lines, one may obtain $a_\parallel$ or a_{45} directly. The $A_\perp$ value (which is difficult to measure because of broadened lines) may then be calculated from Eqs. (11-48) and (11-49), ignoring second-order terms. The results are as follows. (See Prob. 12-1.)

$$g_\parallel = 1.9427 \qquad a_\parallel = 30.2 \text{ G} \qquad \frac{A_\parallel}{c} = 0.00274 \text{ cm}^{-1}$$

$$g_{45} = 1.9403 \qquad a_{45} = 22.7 \text{ G} \qquad \frac{A_\perp}{c} \text{ (calc)} = 0.00099 \text{ cm}^{-1}$$

$$g_\perp = 1.9380$$

The complexes of VO^{++} are the most widely studied examples of $3d^1$ ions; these ions have a very large tetragonal component superimposed upon octahedral symmetry. When the unpaired electron resides in the d_{xy} orbital (see Fig. 12-1) in such a strong tetragonal field, δ is so large that T_1 is long enough to permit observation of ESR spectra at room temperature. (A similar tetragonal component is present in the complexes of CrO^{3+} and MoO^{3+}.)

The $VO(CN)_5^{3-}$ ion in KBr exhibits a spectrum (see Fig. 12-3) with two groups of lines (intensity ratio 1:2) for **H** respectively parallel and perpendicular to the V—O axis.[†] Here $g_\parallel = 1.9711$ and $g_\perp = 1.9844$. Although as expected both g factors are less than g_e, the crystal-field theory

† H. A. Kuska and M. T. Rogers, in E. T. Kaiser and L. Kevan (eds.), "Radical Ions," p. 601, Interscience Publishers, a division of John Wiley & Sons, Inc., New York, 1968.

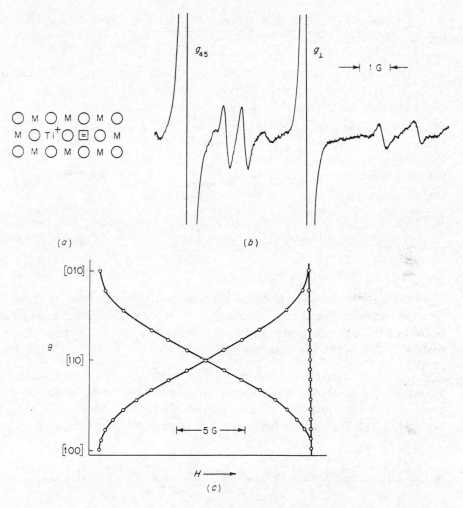

Fig. 12-2 (*a*) Model of a Ti^{3+} ion trapped at a next-nearest-neighbor cation vacancy in CaO. In this representation, only the *deviations* from normal site charge are shown. (*b*) ESR spectrum of Ti^{3+} at sites of tetragonal symmetry in CaO. $T = 77$ K and $\nu = 9.5$ GHz. The $\langle 110 \rangle$ crystal axis is oriented along the magnetic field. The two pairs of weaker lines arise from Ti^{3+} ions in two other sites of tetragonal symmetry for which the defects inducing the tetragonal distortion are unknown. [*Taken from J. J. Davies and J. E. Wertz, J. Magn. Resonance,* **1**:500 (1969).] (*c*) Variation in field position of the ESR lines of Ti^{3+} in CaO for rotation of the crystal in an (001) plane. The stationary line arises from those Ti^{3+} ions for which the axis of symmetry is always perpendicular to the magnetic field.

predicts that $g_\parallel > g_\perp$. [See Eqs. (12-13) and (12-15).] The correct ordering of the g factors is obtained when the ligands are considered explicitly in a molecular orbital calculation.† This calculation indicates that $g_\parallel$ is

† B. R. McGarvey, in R. L. Carlin (ed.), "Transition Metal Chemistry," vol. 3, p. 150, Marcel Dekker, New York, 1966.

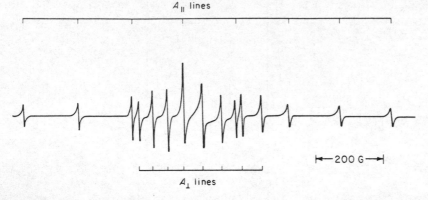

Fig. 12-3 ESR spectrum of the $VO(CN)_5^{3-}$ ion in a single crystal of KBr; the magnetic field is parallel to the [001] direction. (*After a figure kindly supplied by H. A. Kuska.*)

affected only by bonding in the XY (equatorial) plane, while bonding to the oxygen atom on the Z axis affects $g_\perp$. Further evidence for direct participation of the ligands is provided by the observation of ^{13}C hyperfine splitting for the $VO(CN)_5^{3-}$ ion in solution.[†]

$3d^1$ (*oct* + *ttgl*), $\Delta \gg \lambda \approx \delta$ The more complicated case in which λ and δ have a similar order of magnitude will now be considered. In this case, it is not permissible to separate the effects of λ and δ. To obtain the energies in zero magnetic field, one must apply the hamiltonian

$$\hat{\mathscr{H}} = \hat{\mathscr{H}}_{ttgl} + \hat{\mathscr{H}}_{so} \tag{12-16}$$

to the following six unperturbed functions which characterize the T_{2g} states:

$$\frac{1}{\sqrt{2}} \left(|2, \pm\tfrac{1}{2}\rangle - |-2, \pm\tfrac{1}{2}\rangle \right) \equiv |G, \pm\tfrac{1}{2}\rangle$$

$$|+1, \pm\tfrac{1}{2}\rangle$$

and

$$|-1, \pm\tfrac{1}{2}\rangle$$

$\hat{\mathscr{H}}_{ttgl}$ is given by Eq. (11-21); however, since $\hat{\mathscr{H}}_{oct}$ does not cause any splitting of the T_{2g} states, it need not be included. Thus Eq. (12-16) may be written as

$$\hat{\mathscr{H}} = \frac{-\delta}{9} (3\hat{L}_z{}^2 - 6) + \lambda[\hat{L}_z\hat{S}_z + \tfrac{1}{2}(\hat{L}_+\hat{S}_- + \hat{L}_-\hat{S}_+)] \tag{12-17}$$

Here $\delta = -9\alpha_t$, and $\hat{\tilde{\mathscr{H}}}_{so}$ is obtained from Eq. (11-32).

[†] H. A. Kuska and M. T. Rogers. *Inorg. Chem.*, **5**:313 (1966).

Since there are six basis functions, a 6×6 hamiltonian matrix may be constructed. Two of the functions, $|1, -\tfrac{1}{2}\rangle$ and $|-1, +\tfrac{1}{2}\rangle$, are already eigenfunctions of $\mathscr{H}$. The remaining portion of the matrix factors into the following 2×2 matrices

$$\mathscr{H}_1 = \begin{array}{c} \\ \langle +1, +\tfrac{1}{2}| \\ \\ \langle G, -\tfrac{1}{2}| \end{array} \overset{\displaystyle |+1,+\tfrac{1}{2}\rangle \qquad |G,-\tfrac{1}{2}\rangle}{\left[\begin{array}{cc} +\dfrac{\delta}{3}+\dfrac{\lambda}{2} & \dfrac{\lambda}{\sqrt{2}} \\[2ex] \dfrac{\lambda}{\sqrt{2}} & -\dfrac{2\delta}{3} \end{array}\right]} \tag{12-18a}$$

$$\mathscr{H}_2 = \begin{array}{c} \\ \langle -1, -\tfrac{1}{2}| \\ \\ \langle G, +\tfrac{1}{2}| \end{array} \overset{\displaystyle |-1,-\tfrac{1}{2}\rangle \qquad |G,+\tfrac{1}{2}\rangle}{\left[\begin{array}{cc} \dfrac{\delta}{3}+\dfrac{\lambda}{2} & -\dfrac{\lambda}{\sqrt{2}} \\[2ex] -\dfrac{\lambda}{\sqrt{2}} & -\dfrac{2\delta}{3} \end{array}\right]} \tag{12-18b}$$

It is evident that when $\lambda = 0$, the functions do not mix; the $|\pm 1, \pm\tfrac{1}{2}\rangle$ and $|\pm 1, \mp\tfrac{1}{2}\rangle$ functions are degenerate and are separated in energy by δ from the $|G, \pm\tfrac{1}{2}\rangle$ functions.

The energy eigenvalue of the $|\pm 1, \mp\tfrac{1}{2}\rangle$ functions is

$$W_0 = \delta(\tfrac{1}{3} - \tfrac{1}{2}\eta) \tag{12-19a}$$

where $\eta = \lambda/\delta$.

The above two matrices may be diagonalized using the coordinate rotation matrix method outlined in Sec. A-5e. The use of Eqs. (A-72e) and (A-72f) yields

$$W_\pm = \frac{\delta}{2}\left[\left(\frac{1}{2}\eta - \frac{1}{3}\right) \pm \mathscr{S}\right] \tag{12-19b}$$

where

$$\mathscr{S} = [(1 + \tfrac{1}{2}\eta)^2 + 2\eta^2]^{\frac{1}{2}} \tag{12-19c}$$

The three energies are plotted as W/δ versus η in Fig. 12-4. In the limit as $\delta \to 0$, one of these states goes to $W = \lambda$, while the other two go to $W = -\lambda/2$. This is the result that was obtained in Eqs. (12-4).

Eigenfunctions $|\psi_+\rangle$, $|\psi'_+\rangle$, $|\psi_-\rangle$, and $|\psi'_-\rangle$ corresponding to W_+ and W_- are also obtained by the methods of Sec. A-5. $|\psi_+\rangle$ and $|\psi'_+\rangle$ form a Kramers doublet which in zero magnetic field has energy W_+; similarly, the second Kramers doublet $|\psi_-\rangle$ and $|\psi'_-\rangle$ has energy W_- in zero field.

$$|\psi_+\rangle = \cos\omega|G, \tfrac{1}{2}\rangle - \sin\omega|-1, -\tfrac{1}{2}\rangle \tag{12-20a}$$

$$|\psi'_+\rangle = \cos\omega|G, -\tfrac{1}{2}\rangle + \sin\omega|1, \tfrac{1}{2}\rangle \tag{12-20b}$$

$$|\psi_-\rangle = \cos\omega|G, \tfrac{1}{2}\rangle + \sin\omega|-1, -\tfrac{1}{2}\rangle \tag{12-20c}$$

$$|\psi'_-\rangle = \cos\omega|G, -\tfrac{1}{2}\rangle - \sin\omega|1, \tfrac{1}{2}\rangle \tag{12-20d}$$

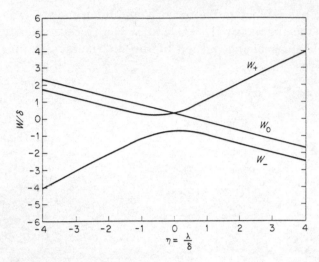

Fig. 12-4 Splitting of the T_{2g} state energies of a $3d^1$ (*oct* + *ttgl*) ion as a function of the ratio λ/δ.

From Eqs. (A-72)

$$\tan 2\omega = \frac{\sqrt{2}\eta}{1 + \frac{1}{2}\eta} \tag{12-21a}$$

$$\sin^2 \omega = \frac{1}{2}\left(1 - \frac{1 + \frac{1}{2}\eta}{\mathscr{S}}\right) \tag{12-21b}$$

$$\cos^2 \omega = \frac{1}{2}\left(1 + \frac{1 + \frac{1}{2}\eta}{\mathscr{S}}\right) \tag{12-21c}$$

$$\sin \omega \cos \omega = \frac{\eta}{\sqrt{2}\mathscr{S}} \tag{12-21d}$$

To obtain $g_\parallel$ and $g_\perp$ for the ground state (energy W_-), one must apply $\hat{\mathscr{H}}_{\text{mag}}$ of Eq. (11-44) to $|\psi_-\rangle$ and $|\psi'_-\rangle$. One associates a fictitious spin $S' = \frac{1}{2}$ with the ground state. Thus $M_{S'} = \frac{1}{2}$ is associated with $|\psi_-\rangle$ and $M_{S'} = -\frac{1}{2}$ with $|\psi'_-\rangle$. Hence

$$\langle\tfrac{1}{2}|g_\parallel\beta H_z\hat{S}'_z|\tfrac{1}{2}\rangle = \langle\psi_-|\beta H_z(\hat{L}_z + g_e\hat{S}_z)|\psi_-\rangle \tag{12-22a}$$

$$\langle\tfrac{1}{2}|g_\perp\beta H_x\frac{\hat{S}'_+}{2}|-\tfrac{1}{2}\rangle = \langle\psi_-|\beta H_x\frac{(\hat{L}_+ + g_e\hat{S}_+)}{2}|\psi'_-\rangle \tag{12-22b}$$

Substitution of Eqs. (12-20) and (12-21) into (12-22) gives

$$g_\parallel = g_e(\cos^2 \omega - \sin^2 \omega) - 2\sin^2 \omega$$

$$\approx \frac{1}{\mathscr{S}}\left[3(1 + \tfrac{1}{2}\eta) - \mathscr{S}\right] \tag{12-23a}$$

and

$$g_\perp = g_e \cos^2 \omega - 2\sqrt{2}\, \sin \omega \cos \omega$$

$$\approx \frac{1}{\mathscr{S}}\left(\mathscr{S} + 1 - \frac{3\eta}{2}\right) \qquad (12\text{-}23b)$$

Note that as $\delta \to 0$, $\mathscr{S} \to \frac{3}{2}\eta$ [see Eq. (12-19c)], and both $g_\parallel$ and $g_\perp$ approach 0 if $\lambda > 0$. This agrees with the conclusion derived in Sec. 12-1 for pure octahedral symmetry. However for $\lambda < 0$ and $\delta = 0$, the ground state would correspond to $J' = \frac{1}{2}$. In this case $g_\parallel = -2$ and $g_\perp = 2$, which correspond to the limiting values as $\eta \to -\infty$ in Fig. 12-5. This example illustrates a case for which g may be negative. As $\eta \to 0$, $\mathscr{S} \to 1$ and both $g_\parallel$ and $g_\perp$ approach 2. The behavior of these g components as a function of η is illustrated in Fig. 12-5. (Problem 12-4 considers the case $\delta < 0$.)

In the above discussion the E_g states have been neglected. Since λ/Δ is small, their effect may be accounted for by use of perturbation theory. Then (see Prob. 12-3) the following terms must be added to Eqs. (12-23):

$$\Delta g_\parallel = -\frac{8\lambda}{\Delta}\left(\cos^2 \omega + \frac{1}{\sqrt{2}}\sin \omega \cos \omega\right) \qquad (12\text{-}24a)$$

$$\Delta g_\perp = \frac{2\sqrt{2}\lambda}{\Delta}\left(\sin \omega \cos \omega + \frac{1}{\sqrt{2}}\sin^2 \omega\right) \qquad (12\text{-}24b)$$

Ti^{3+} in $CsTi(SO_4)_2\cdot 12H_2O$ exhibits an ESR spectrum with $g_\parallel = 1.25$ and $g_\perp = 1.14$. Although the distortion is trigonal, Eqs. (12-23) are ap-

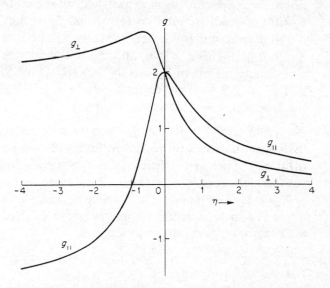

Fig. 12-5 Values of the g components of a $3d^1$ ($oct + trgl$) ion as a function of $\eta = \lambda/\delta$.

plicable to first order.[†] From Fig. 12-5 one estimates that $\eta \approx 0.6$. $\lambda = 154$ cm^{-1} for the free ion; hence $\delta \approx 250$ cm^{-1}.

$3d^1$ $(oct + trgl)$ There are a very large number of $3d$ ions which have primarily octahedral crystal fields, with a small trigonal distortion. A body diagonal of the cube enclosing an octahedron is a trigonal axis; compression or extension of the octahedron along this direction adds a trigonal component to the crystal field. The $3d^1$ case represents the simplest example. The choice of Z as the trigonal axis gives the simplest form to the crystal field operators and to the wave functions.[‡] The energy levels and wave functions for a $3d^1$ (oct + trgl) ion are given in Fig. 11-25. The splittings of energy levels and the g components (see Prob. 11-7) are the same as those for the case of octahedral-plus-tetragonal symmetry.

Titanium acetylacetonate in aluminum acetylacetonate is a good example of a $3d^1$ (oct + trgl) ion. Here $g_{\parallel} = 2.000_2$, $g_{\perp} = 1.921_1$, $A_{\parallel}/c = 0.00063$ cm^{-1}, and $A_{\perp}/c = 0.00175$ cm^{-1}.[§] The uncertainty in $g_{\parallel}$ would allow Δ values in the range 2,000 to 30,000 cm^{-1}, but no optical absorption is detected in the range 5,000 to 14,000 cm^{-1}. The temperature dependence of T_1 indicates that Δ lies in the range 2,000 to 5,000 cm^{-1}.

$3d^5$ $(ls)(oct + ttgl)$ A low-spin $3d^5$ ion in octahedral symmetry has the same ordering of energy levels and sets of orbitals as a $3d^1$ (oct) ion. Since there are five d electrons in the t_{2g} levels, the system can be regarded as having one positive hole in the t_{2g} set. Thus $\lambda < 0$.

A number of examples in this case are ions of the form $MX_6{}^{n-}$, which might be expected to have octahedral symmetry. However, they usually exhibit a small rhombic or tetragonal distortion. Since the magnitudes of δ and λ are comparable, the g factors will behave according to Eqs. (12-23). Since δ is positive, $\eta < 0$, and the left-hand side of Fig. 12-5 is applicable. For example $\overset{-}{Fe}{}^{3+}$ in $K_3Co(CN)_6$ exhibits an ESR spectrum with $g_{\parallel} = 0.915$ and $g_{\perp} \approx 2.2$.[¶] This corresponds to $\eta \approx -0.5$; since $\lambda = -103$ cm^{-1} for the free ion, $\delta \approx 200$ cm^{-1}.

$3d^9$ $(ttdl + ttgl)$ A $3d^9$ ion in a tetrahedral field with a tetragonal distortion can be considered in terms of one positive hole in the t_{2g} set of orbitals. The energy levels of this system will split in a fashion analogous to the $3d^1$ (oct + ttgl) case. The same equations will apply; however $\lambda < 0$. Unfortunately, there are few examples of d^9 ions which have been observed in tetrahedral symmetry; none of these approximates the case of a simple tetragonal distortion.

---

† B. R. McGarvey, in R. L. Carlin (ed.), "Transition Metal Chemistry," vol. 3, p. 111, Marcel Dekker, New York, 1966.
‡ See J. W. Orton, "Electron Paramagnetic Resonance," p. 104, Iliffe Books, Ltd., London, 1968.
§ B. R. McGarvey, *J. Chem. Phys.*, 38:388 (1963).
¶ B. Bleaney and M. C. M. O'Brien, *Proc. Phys. Soc.*, B69:1216 (1956).

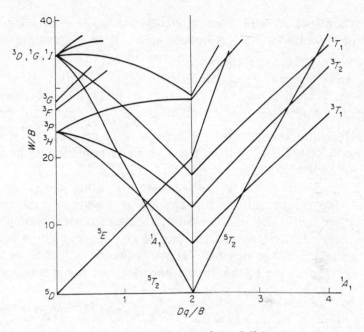

Fig. 12-6   Tanabe-Sugano diagram for a $3d^6$ ion in octahedral symmetry.

$3d^6$ $(hs)(oct)$ In an octahedral crystal field of intermediate strength, the ^5D state of a gaseous $3d^6$ ion will split so as to leave a $^5T_{2g}$ state lowest. (See Figs. 12-6 and 12-7.)   The strong-field case is not of interest to us since it corresponds to a diamagnetic ground state.   For the weak-field case, spin-orbit coupling in the ground state causes a splitting according to the

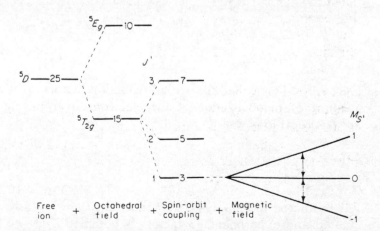

Fig. 12-7   Successive splitting of state energies of a $3d^6$ $(hs)(oct)$ ion. In an exactly octahedral field, the two transitions shown at the right would occur at the same field.

values of J', which may have the values 3, 2, and 1. Application of $\hat{\mathcal{H}}'_{so}$ of Eq. (12-7) to the eigenfunctions $|J', M_{J'}\rangle$ gives the energies

$$W_3 = -2\alpha\lambda \approx -2\lambda \qquad\qquad (12\text{-}25a)$$

$$W_2 = \alpha\lambda \approx \lambda \qquad\qquad (12\text{-}25b)$$

$$W_1 = 3\alpha\lambda \approx 3\lambda \qquad\qquad (12\text{-}25c)$$

Here the subscripts on W refer to the value of J'. If spin-orbit coupling with states arising from the E_g state is considered, the term $-\frac{18}{5}\lambda^2/\Delta$ must be added to each energy in Eqs. (12-25).

Since $\lambda < 0$ for $3d^6$ ions, the W_1 state lies lowest. (See Fig. 12-7.) If the temperature is low enough, this state will be the only one with an appreciable population. Since the state corresponding to W_1 is triply degenerate, the ground state may be characterized by a fictitious spin $S' = 1$. In terms of the original $|M_{L'}, M_S\rangle$ functions, the wave functions $|J', M_{J'}\rangle$ corresponding to the three components of the ground state are

$$|1, -1\rangle = \frac{1}{\sqrt{10}}|-1, 0\rangle - \sqrt{\frac{3}{10}}|0, -1\rangle + \sqrt{\frac{3}{5}}|1, -2\rangle \qquad (12\text{-}26a)$$

$$|1, 0\rangle = \sqrt{\frac{2}{5}}|0, 0\rangle - \sqrt{\frac{3}{10}}[|1, -1\rangle + |-1, 1\rangle] \qquad (12\text{-}26b)$$

$$|1, 1\rangle = \frac{1}{\sqrt{10}}|1, 0\rangle - \sqrt{\frac{3}{10}}|0, 1\rangle + \sqrt{\frac{3}{5}}|-1, 2\rangle \qquad (12\text{-}26c)$$

These functions are obtained by the methods outlined in Sec. B-7 (Prob. 12-5). The energy levels in a magnetic field are, respectively,†

$$W_{-1} = \beta H\left(-\frac{3}{2}g_e - \frac{k'}{2} + \frac{18\lambda}{5\Delta}\right) \qquad (12\text{-}27a)$$

$$W_0 = 0 \qquad\qquad (12\text{-}27b)$$

$$W_{+1} = \beta H\left(\frac{3}{2}g_e + \frac{k'}{2} - \frac{18\lambda}{5\Delta}\right) \qquad (12\text{-}27c)$$

Here k' (< 1) is called the orbital reduction factor. It is introduced to take account of delocalization of the electrons over the ligands. Hence for $3d^6$ (hs)(oct) ions with $k' \approx 1$

$$g = \frac{7}{2} - \frac{18\lambda}{5\Delta} \qquad\qquad (12\text{-}28)$$

For the free Fe^{++} ion, $\lambda = -103$ cm^{-1}. In MgO, $\Delta \cong 10,000$ cm^{-1}. Hence g is predicted to be 3.494. Experimentally $g = 3.428$.† Agreement of calculated and experimental g factors is obtained if $k' \approx 0.8$ in Eqs. (12-27).

† W. Low and M. Weger, *Phys. Rev.*, **118**:1119, 1130 (1960).

Spin-orbit coupling within the $^5T_{2g}$ state leads to so short a spin-lattice relaxation time that observations of the ESR spectrum can only be made at 20 K or lower. The Fe^{++} ion is coupled to its environment perhaps more strongly than any other $3d$ ion. Small departures from exact octahedral symmetry will lead to large zero field splittings. In MgO, in which other $3d$ ions (except $3d^8$ ions) may have ESR lines ~ 0.5 G wide, the Fe^{++} line may display a width of 500 G. A distribution in residual strains causes the $|0\rangle$ state to be shifted by varying amounts relative to the $|+1\rangle$ and the $|-1\rangle$ states. The sign of the zero field splitting D may be either positive or negative, i.e., the $|0\rangle$ state may lie either above or below the $|\pm 1\rangle$ states at zero magnetic field. This case is analogous to that of Ni^{++} considered in Sec. 11-7b.

Figure 12-8 shows the broad line of Fe^{++} in MgO observed at 4.2 K with a microwave frequency of 25 GHz. The great width of the line is attributed to random distortions of the crystal field from exact octahedral symmetry. The sharp line at the center is a transition involving two quanta (double-quantum transition) which was discussed for the case of Ni^{++} in Sec. 11-7b. Assuming that the g factor of the broad line may be determined from the center of the double-quantum transition, $g = 3.4277$. There is also a distorted line at approximately half the field value of the broad line; this is a $\Delta M = 2$ transition, which would be forbidden in exact octahedral symmetry. Its occurrence is further evidence for the existence of distortions from octahedral symmetry.

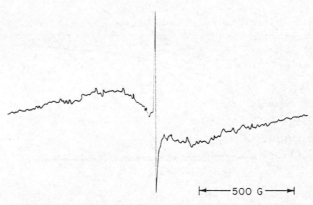

Fig. 12-8 ESR spectrum of Fe^{++} in octahedral symmetry in MgO at 4.2 K. $\nu = 25$ GHz. The central narrow line is a double-quantum transition which is observed at high microwave power levels. (*After W. Low, in F. Seitz and D. Turnbull, eds., "Solid State Physics", suppl. 2, p. 87, Academic Press Inc., New York, 1960.*)

12-1b   F-state ions

$3d^2$ (oct) For $3d^2$ ions in an octahedral field, the T_{1g} state lies lowest (see Figs. 12-9 and 12-10); in this case $S = 1$.   The spin-orbit coupling again results in a splitting of levels according to the value of the quantum number J', which here has the values 2, 1, and 0.   The energies of these three states are computed by the methods outlined in Sec. 12-1a as

$$W_2 = -\alpha\lambda \tag{12-29a}$$

$$W_1 = \alpha\lambda \tag{12-29b}$$

and

$$W_0 = 2\alpha\lambda \tag{12-29c}$$

Since $\lambda > 0$ for $3d^2$ ions, the W_2 state will lie lowest.   The eigenfunctions for $J' = 2$ are (see Sec. B-7)

$$|2, \pm 2\rangle = |\pm 1, \pm 1\rangle \tag{12-30a}$$

$$|2, \pm 1\rangle = \frac{1}{\sqrt{2}} \left(|\pm 1, 0\rangle + |0, \pm 1\rangle \right) \tag{12-30b}$$

$$|2, 0\rangle = \frac{1}{2} |1, -1\rangle + \frac{1}{\sqrt{2}} |0, 0\rangle + \frac{1}{2} |-1, 1\rangle \tag{12-30c}$$

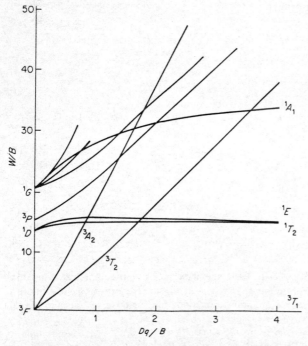

Fig. 12-9   Tanabe-Sugano diagram for d^2 ions in an octahedral field.

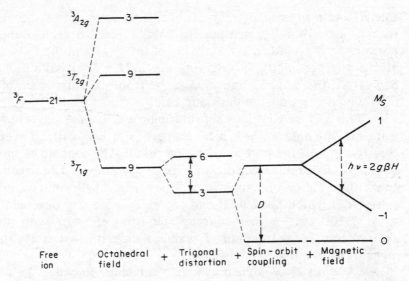

Fig. 12-10 Splitting of the states of a ^3F ion (d^2) in an octahedral field; additional splittings arise from trigonal distortion and from spin-orbit coupling. In a magnetic field, one may observe transitions between the $|\pm 1\rangle$ states at room temperatures, but the absorption intensity falls at temperatures such that $kT \approx h\nu$.

The kets on the left are labeled as $|J', M_{J'}\rangle$ and those on the right as $|M_{L'}, M_S\rangle$. The energies in a magnetic field are computed using $\hat{\mathcal{H}}'_{\text{mag}}$ of Eq. (12-1) as

$$W_{2,\pm 2} = \pm\beta(-\alpha + 2)H_z \approx (\tfrac{1}{4})\beta H_z(\pm 2) \qquad (12\text{-}31a)$$

$$W_{2,\pm 1} = \pm\beta\left(-\frac{\alpha}{2} + 1\right)H_z \approx (\tfrac{1}{4})\beta H_z(\pm 1) \qquad (12\text{-}31b)$$

$$W_{2,0} = 0 \qquad (12\text{-}31c)$$

Thus for this case $g \approx \tfrac{1}{4}$. The energies of Eqs. (12-31) are not quite correct, since there are additional small terms due to spin-orbit coupling with the T_{2g} states. No examples of this case appear to have been reported.

$3d^2$ *(oct + trgl)* The only known examples of a $3d^2$ ion in an octahedral field exhibit a strong trigonal distortion. This causes the T_{1g} states to split so as to leave an orbital singlet lowest. There is a large zero field splitting due to spin-orbit coupling with the upper components of the T_{1g} states. This leaves a nondegenerate state lowest, as shown in Fig. 12-10. Magnetic resonance has been detected between the $M_S = \pm 1$ states.

Since the two electrons couple to give $S = 1$, there is a likelihood of zero field splitting. From magnetic susceptibility experiments, the value of D/hc is found to be about 5 cm^{-1}. This means that the states, which at high field are represented by $|+1\rangle$ and $|-1\rangle$, will be too far removed from the $|0\rangle$ state in ordinary magnetic fields to permit observation of $\Delta M_S = 1$ transitions. In a distorted trigonal field, a nonzero value of the asymmetry parameter E would reflect a mixing of the $|+1\rangle$ and the $|-1\rangle$ states. This allows the possibility of observing the "$\Delta M = 2$" transitions. As noted in Sec. 10-7, the $\Delta M = 2$ transitions arise from the component of microwave field which is parallel to the static field.

The V^{3+} or Cr^{4+} ions substituting for Al^{3+} in Al_2O_3 (corundum) illustrate nicely some of the typical properties of ions with an even number of electrons.[†] In the corundum structure, all Al^{3+} ions lie along the trigonal axis of a distorted octahedron of six oxygen ions. The trigonal distortion splits the $^3T_{1g}$ ground state of V^{3+} (see Fig. 11-6b) into an orbital singlet state (A_{2g}) which lies lowest and an orbitally doubly degenerate state (E_g) some 1,200 cm^{-1} higher. Strong spin-orbit coupling with this low-lying excited state leads to a short T_1 value; hence, it is necessary to make ESR measurements at 4 K or lower.

^{51}V with $I = \frac{7}{2}$ is the only stable vanadium nuclide. Its magnetogyric ratio is large, and hence hyperfine splitting is usually obvious. The spin hamiltonian for $S = 1$ may be written

$$\hat{\mathscr{H}} = \beta[g_{\|}H_Z\hat{S}_z + g_{\perp}(H_X\hat{S}_x + H_Y\hat{S}_y)] + D(\hat{S}_z^2 - \tfrac{2}{3}) + hA_{\|}\hat{S}_z\hat{I}_z$$
$$+ \tfrac{1}{2}hA_{\perp}(\hat{S}_+\hat{I}_- + \hat{S}_-\hat{I}_+) \quad (12\text{-}32)$$

If $\mathbf{H} \parallel Z$, where Z is the trigonal axis, one may approximate Eq. (12-32) by an effective hamiltonian which neglects the x and y components. The $D\hat{S}_z^2$ term, which affects the $|\pm 1\rangle$ states equally, is also neglected. Then

$$\hat{\mathscr{H}}_{\text{eff}} = g_{\|}\beta H_Z\hat{S}_z + hA_{\|}\hat{S}_z\hat{I}_z \quad (12\text{-}33)$$

Application of $\hat{\mathscr{H}}_{\text{eff}}$ to the $|\pm 1\rangle$ wave functions leads to the expression

$$\Delta W = h\nu = 2g_{\|}\beta H_Z + 2hA_{\|}M_I \quad (12\text{-}34)$$

If $\mathbf{H}$ makes an angle θ with the Z axis, $g_{\|}$ in Eq. (12-34) must be replaced by $g_{\|} \cos \theta$.

The resonant field $H_r(M_I)$ for the hyperfine component corresponding to M_I is

$$H_r(M_I) = \frac{\tfrac{1}{2}h\nu_0 - hA_{\|}M_I}{g_{\|}\beta} \quad (12\text{-}35)$$

The spectrum in Fig. 12-11 shows a remarkably uniform spacing of ~ 110 G. Ordinarily, for such a large hyperfine splitting one would observe nonuniform spacings due to second-order shifts. Sections C-7 and C-9 indicate that it is the $(\hat{S}_+\hat{I}_- + \hat{S}_-\hat{I}_+)$ terms which cause these shifts. However, the above operators do not couple the $|\pm 1\rangle$ functions. They do couple the $|\pm 1\rangle$ functions with the $|0\rangle$ function, but the latter is too far removed for

[†] J. Lambe and C. Kikuchi, *Phys. Rev.*, **118**:71 (1960); G. M. Zhverev and A. M. Prokhorov, *Soviet Phys.—JETP*, **7**:707 (1958).

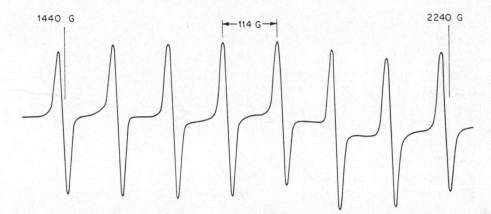

Fig. 12-11 ESR spectrum of the V^{3+} ion in Al_2O_3 at 4.2 K. The magnetic field H is along the trigonal axis. [*Figure kindly supplied by C. Kikuchi. See J. Lambe and C. Kikuchi, Phys. Rev.,* **118**:71 (1960).]

transitions to occur under ordinary conditions. The use of pulsed magnetic fields up to 100 kG has permitted observation of transitions between the $|-1\rangle$ and $|0\rangle$ states.† From these observations D/hc is computed to be 7.85 ± 0.4 cm^{-1}.

In Al_2O_3 the properties of the Cr^{4+} ion are very similar to those of the isoelectronic ion V^{3+}. For Cr^{4+}, $g_\| = 1.90$, $D/hc \approx 7$ cm^{-1}, and $E/hc < 0.05$ cm^{-1}.‡

$3d^7$ $(hs)(oct)$ The ^4F state of gaseous $3d^7$ ions splits in an octahedral field so as to leave the $^4T_{1g}$ state lowest. (See Figs. 11-6b and 12-12.) This situation is analogous to the $3d^1$ (oct) case; thus the ground state may be regarded as having $L' = 1$. The energy levels are obtained by the same procedure as that of Sec. 12-1a, except that now $S = \frac{3}{2}$. Thus J' takes on the values $\frac{5}{2}, \frac{3}{2}, \frac{1}{2}$, with corresponding energies

$$W_{\frac{5}{2}} = -\tfrac{3}{2}\alpha\lambda \approx -\tfrac{9}{4}\lambda \tag{12-36a}$$

$$W_{\frac{3}{2}} = \alpha\lambda \approx \tfrac{3}{2}\lambda \tag{12-36b}$$

$$W_{\frac{1}{2}} = \tfrac{5}{2}\alpha\lambda \approx \tfrac{15}{4}\lambda \tag{12-36c}$$

These state energies are illustrated in Fig. 12-13.

For $3d^7$ ions $\lambda < 0$; hence, $W_{\frac{1}{2}}$ will correspond to the state of lowest energy. If $\alpha\lambda \gg kT$, then the state with $J' = \frac{1}{2}$ will be the only one which will be appreciably populated. Thus the ground state will be doubly degenerate in the absence of a magnetic field.

The eigenfunctions for $J' = \frac{1}{2}$ are obtained by the methods outlined

† S. Foner and W. Low, *Phys. Rev.,* **120**:1585 (1960).
‡ R. H. Hoskins and B. H. Soffer. *Phys. Rev.,* **133**:A490 (1964).

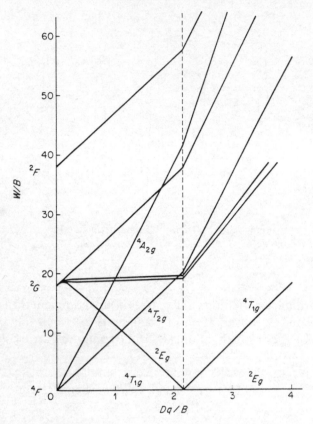

Fig. 12-12 Tanabe-Sugano diagram for a $3d^7$ ion in an octahedral field.

in Sec. B-7. The results are

$$|\tfrac{1}{2}, \pm\tfrac{1}{2}\rangle = \frac{1}{\sqrt{2}}\, |\mp 1, \pm\tfrac{3}{2}\rangle - \frac{1}{\sqrt{3}}\, |0, \pm\tfrac{1}{2}\rangle + \frac{1}{\sqrt{6}}\, |\pm 1, \mp\tfrac{1}{2}\rangle \qquad (12\text{-}37)$$

where the ket on the left refers to $|J', M_{J'}\rangle$.

Application of $\hat{\mathscr{H}}'_{\text{mag}}$ of Eq. (12-1) for $\mathbf{H} \parallel Z$ gives

$$W_{\tfrac{1}{2}, \pm\tfrac{1}{2}} = \pm\beta H_Z[\alpha(\tfrac{1}{2} - \tfrac{1}{6}) + (\tfrac{3}{2} + \tfrac{1}{3} - \tfrac{1}{6})] = (\tfrac{2}{3}\alpha + \tfrac{10}{3})\beta H_Z(\pm\tfrac{1}{2}) \quad (12\text{-}38)$$

Since $\alpha \approx \tfrac{3}{2}$ for F-state ions, $g \approx 4.33$. Second-order spin-orbit coupling gives a contribution to g of $-15\lambda/2\Delta$. Since the ground-state $W_{\tfrac{1}{2}}$ is doubly degenerate, $3d^7$ ions in an octahedral field are often described in terms of a spin hamiltonian in which the effective spin $S' = \tfrac{1}{2}$ even though the true spin of the system is $\tfrac{3}{2}$.

The best known example of a $3d^7$ (hs)(oct) ion is Co^{++}, which has been observed in a variety of environments.† Since the only naturally occurring nuclide is ^{59}Co ($I = \tfrac{7}{2}$, with a large magnetogyric ratio), a hyperfine octet is usually resolved. Since a quartet and a sextet state are only a few hundred

† J. H. M. Thornley, C. G. Windsor, and J. Owen, *Proc. Roy. Soc. (London)*, **A284**:252 (1965).

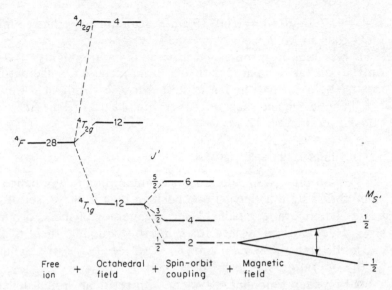

Fig. 12-13 Splitting of the state energies of a $3d^7$ (oct) ion due to spin-orbit interaction in the ground state.

reciprocal centimeters above the ground state (see Fig. 12-13), T_1 is very short at normal temperatures; hence observations must be made at 20 K or lower. Even in a distorted octahedral field, the following spin hamiltonian ($S' = \frac{1}{2}$) is appropriate:

$$\hat{\mathcal{H}} = \beta \mathbf{H} \cdot \boldsymbol{g} \cdot \hat{\mathbf{S}} + h\hat{\mathbf{S}} \cdot \boldsymbol{A} \cdot \hat{\mathbf{I}} \qquad (12\text{-}39)$$

In a purely octahedral field, the spectrum is isotropic, but marked second-order shifts are apparent. (See Fig. 1-9.) Equation (4-2) may be employed to account for these shifts.

The g factor to first order is predicted to be 4.33. [See Eq. (12-38).] This is reasonably close to the observed values of 4.2785 in MgO[†] and 4.372 in CaO.[‡] For these two hosts A/c is, respectively, 0.009779 and 0.01322 cm^{-1}. The lattice constants are, respectively, 0.420 and 0.482 nm; hence, the Co^{++} ion experiences a weaker crystalline electric field in CaO as compared to MgO. This explains why the hyperfine coupling is larger in CaO; there will likely be less covalent bonding in this host.

In distorted octahedral fields both spin-orbit coupling and the non-octahedral component of the crystal field will lead to markedly different g components. (See Sec. 12-1a.) An example is Co^{++} in TiO_2 for which $g_{xx} = 2.090$, $g_{YY} = 3.725$, and $g_{zz} = 5.860$.

The isoelectronic ion Fe^+ has been mentioned. (See Sec. 1-6.) In MgO,[§] $g = 4.1307$ and A/c (^{57}Fe) = 0.00101 cm^{-1}, whereas in CaO,[¶]

† W. Low, *Phys. Rev.*, **109**:256 (1958).
‡ W. Low and R. S. Rubins, *Phys. Letters (Netherlands)*, **1**:316 (1962).
§ J. W. Orton, P. Auzins, J. H. E. Griffiths, and J. E. Wertz, *Proc. Phys. Soc.*, **78**:554 (1961).
¶ W. Low and J. T. Suss, *Bull. Am. Phys. Soc.*, **9**:36 (1964).

$g = 4.1579$ and $A/c = 0.00339$ cm^{-1}; again the hyperfine coupling is larger in CaO than in MgO.

Ni^{3+} has been observed in various environments. The extra charge on the ion (as compared with Co^{++}) may cause a sufficient increase in the crystal field so that the ion is shifted to the strong-field (low spin) side of the Tanabe-Sugano diagram. (See Fig. 12-12.) This $3d^7$ $(ls)(oct)$ case is considered in Sec. 12-1c.

12-1c Jahn-Teller splittings

$3d^9$ (oct) A $3d^9$ ion has the configuration $t_{2g}^6 e_g^3$; hence one may consider the system as having one positive hole in the e_g set of orbitals. In a *purely* octahedral crystal field, the degeneracy of the e_g orbitals would persist. Even with allowance for spin-orbit coupling, there is no splitting because the states $|0\rangle$ ($\equiv d_{z^2}$) and $(1/\sqrt{2})(|2\rangle + |-2\rangle)(\equiv d_{x^2-y^2})$ are not coupled by the operators of $\mathcal{H}_{so}$ in Eq. (11-32).

For such a nonlinear system with residual orbital degeneracy, a theorem due to Jahn and Teller states that the system will be strongly coupled to those lattice vibrations which remove the degeneracy and lower the energy of the ground state.[†] Either a tetragonal or a rhombic distortion will serve to remove the degeneracy; a trigonal distortion will not. Thus even in crystals (such as Al$_2$O$_3$) which have a strong trigonal component, the crystal field does not remove the degeneracy. Depending on the sign of the distortion, either of the two resulting Kramers doublets may lie lower. If the state corresponding to $|0\rangle$ lies lower, one must consider the system as analogous to that treated in Sec. 11-7a, for which it was shown that

$$g_{\parallel} = g_e$$

$$g_{\perp} = g_e - \frac{6\lambda}{\Delta} \tag{12-40}$$

However, if the state corresponding to $(1/\sqrt{2})(|2\rangle + |-2\rangle)$ lies lower, then from Prob. (11-6)

$$g_{\parallel} = g_e - \frac{8\lambda}{\Delta_{\parallel}} \tag{12-41a}$$

and

$$g_{\perp} = g_e - \frac{2\lambda}{\Delta_{\perp}} \tag{12-41b}$$

If the Jahn-Teller distortion is large, then one expects to observe an ESR spectrum with tetragonal symmetry and with g components given by Eqs. (12-40) or (12-41). If the distortion is small, the distortion axis may

[†] For a review see F. S. Ham, in S. Geschwind (ed.), "Electron Paramagnetic Resonance." Plenum Press, Plenum Publishing Corporation, New York, 1969.

shift among the three equivalent octahedral axes. At temperatures such that the interconversion is sufficiently rapid, the system will exhibit an isotropic g factor given by

$$g = g_e - \frac{4\lambda}{\Delta} \tag{12-42}$$

One of the clearest examples of a Jahn-Teller splitting is Cu^{++} in MgO. At 77 K an isotropic g factor of 2.192[†] and hyperfine coupling $A/c = 0.0019$ cm^{-1} are observed. However at 1.2 K, very anisotropic behavior is observed.[‡] This has been explained[‡],[§] in terms of a tunneling among equivalent distorted Jahn-Teller configurations.

$3d^7$ $(ls)(oct)$ The same considerations, including the Jahn-Teller splitting, apply to a $3d^7$ $(ls)(oct)$ ion since the configuration is $t_{2g}{}^6 e_g{}^1$. Ni^{3+} in Al_2O_3[¶] is an example of this case. The orbital degeneracy is not removed either by spin-orbit coupling or by the trigonal field. However, a Jahn-Teller distortion will remove the degeneracy. Above 50 K, an isotropic g factor of 2.146 is found. On cooling to 4.2 K, the spectrum becomes strongly anisotropic, presumably because each of the statically distorted configurations contributes individually.

12-2 ELEMENTS OF THE 4d AND 5d GROUPS (PALLADIUM AND PLATINUM GROUPS)

The elements of the $4d$ and $5d$ groups may be treated in a fashion analogous to that used for the $3d$ ions, with some added complications. For $4d^3$ and $5d^3$ ions, both weak-field (high-spin) and strong-field (low-spin) examples are observed; for $4d^4$ to $4d^9$ and $5d^4$ to $5d^9$ ions only the strong-field examples are found. For some ions, for example, $5d^3$ (oct), the *additional* terms of Eq. (11-50) are required in the spin hamiltonian.

Because of the fact that spin-orbit coupling constants for the $4d$ and $5d$ groups are much larger than for the $3d$ group, spin-lattice relaxation times are usually very short. Thus it is often difficult to observe a spectrum at all except at very low temperatures. Only in cases where an orbital singlet lies much lower than any other state is one likely to be able to observe an ESR spectrum at temperatures above 20 K. Such a case is $4d^3$ (oct); for example, Tc^{4+} in K_2PtCl_6 was detected at 77 K with $g = 2.050$.[††]

Another unusual case is the d^1 ion in symmetry corresponding to an Archimedean antiprism.[‡‡] An orbital singlet lies lowest, and ESR spectra are observed for the $Mo(CN)_8{}^{3-}$ and $W(CN)_8{}^{3-}$ ions in aqueous solution

† J. W. Orton, P. Auzins, J. H. E. Griffiths, and J. E. Wertz, *Proc. Phys. Soc.*, **78**:554 (1961).
‡ R. E. Coffman, *J. Chem. Phys.*, **48**:609 (1968).
§ I. B. Bersuker, *Soviet Phys.—JETP*, **16**:933 (1963); **17**:836 (1964).
¶ S. Geschwind and J. P. Remeika, *J. Appl. Phys.*, **33**:370 (1962).
†† W. Low and P. M. Llewellyn, *Phys. Rev.*, **110**:842 (1958).
# B. R. McGarvey, *Inorg. Chem.*, **5**:476 (1966); R. G. Hayes, *J. Chem. Phys.*, **44**:2210 (1966).

at room temperature! The spectrum of $Mo(CN)_8^{3-}$ is displayed in Fig. 12-14. Here 75 percent of the Mo isotopes have $I = 0$; $^{95,97}Mo$ comprise the remaining 25 percent. Each has $I = \frac{5}{2}$, and their nuclear moments are almost identical. Thus a sextet of lines[†] with $A/c = 0.00298$ cm^{-1} is observed. The small doublet on the wings of the central line arises from ^{13}C in the CN$^-$ ligands.

12-3 THE RARE–EARTH IONS

The rare-earth ions are distinguished from the 3, 4, $5d^n$ ions by their very weak interaction with a crystal field. The $4f$ electrons are very well shielded from their environments by filled $5s$ and $5p$ electron shells. For some rare-earth ions in solution or in crystals one observes very narrow optical absorption or emission lines. These are only slightly displaced from the corresponding atomic transitions. The spin-orbit coupling is strong; λ ranges from 640 to 2,940 cm^{-1}. The angular momenta **L** and **S** combine to give a total angular momentum **J** of magnitude $\sqrt{J(J + 1)}$. The ground

[†] S. I. Weissman and M. Cohn. *J. Chem. Phys.*, **27**:1440 (1957).

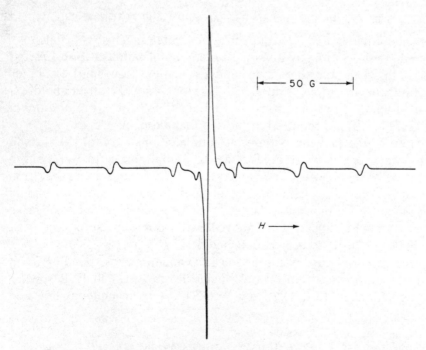

50 G

$H \longrightarrow$

Fig. 12-14 ESR spectrum of $Mo(CN)_8^{3-}$ in aqueous solution at room temperature. The sextets arise from $^{95,97}Mo$ ($I = \frac{5}{2}$). The doublet flanking the central line is due to ^{13}C. (*Taken from B. R. McGarvey, in R. L. Carlin, ed., "Transition Metal Chemistry," p. 92, Marcel Dekker, New York, 1966.*)

Table 12-2 Properties of the ground states of free rare-earth ions

Number of 4f electrons	1	2	3	4	5	6	7	8	9	10	11	12	13
Representative ions	Ce^{3+}	Pr^{3+}	Nd^{3+}	Pm^{3+}	Sm^{3+}	Eu^{2-}	Gd^{3-}	Tb^{3+}	Dy^{3+}	Ho^{3+}	Er^{3+}	$.Tm^{3+}$	Yb^{3+}
Ground state	$^2F_{5/2}$	$^3H_4$	$^4I_{9/2}$	$^5I_4$	$^6H_{5/2}$	$^7F_0$	$^8S_{7/2}$	$^7F_6$	$^6H_{15/2}$	$^5I_8$	$^4I_{15/2}$	$^3H_6$	$^2F_{7/2}$
Landé g factor	$\frac{6}{7}$	$\frac{4}{5}$	$\frac{8}{11}$	$\frac{3}{5}$	$\frac{2}{7}$	—	2	$\frac{3}{2}$	$\frac{4}{3}$	$\frac{5}{4}$	$\frac{6}{5}$	$\frac{7}{6}$	$\frac{8}{7}$
λ, cm⁻¹	640	800	900	1,070	1,200	1,410	1,540	−1,770	−1,860	−2,000	−2,350	−2,660	−2,940

state corresponds to $J = |L - S|$ for ions with less than seven electrons and $J = L + S$ for those with more than seven. The calculation of the spin-orbit coupling $\lambda \hat{\mathbf{L}} \cdot \hat{\mathbf{S}}$ within the ground state becomes very simple with a hamiltonian analogous to Eq. (12-7); that is,

$$\hat{\mathcal{H}}_{so} = \lambda \hat{\mathbf{L}} \cdot \hat{\mathbf{S}} = \frac{\lambda}{2} \left[J(J+1) - L(L+1) - S(S+1) \right] \qquad (12\text{-}43)$$

The values of L, S, and J for the $4f$ ions in their ground states can be obtained from the ground state given in Table 12-2.

The splittings arising from spin-orbit coupling in a $4f^1$ ion are shown in Fig. 12-15. The calculation of g components will be illustrated for this case. Here $J = \frac{7}{2}$ or $\frac{5}{2}$; thus from Eq. (12-43) the energies W_J are

$$W_{\frac{7}{2}} = \frac{3\lambda}{2} \qquad (12\text{-}44a)$$

$$W_{\frac{5}{2}} = -2\lambda \qquad (12\text{-}44b)$$

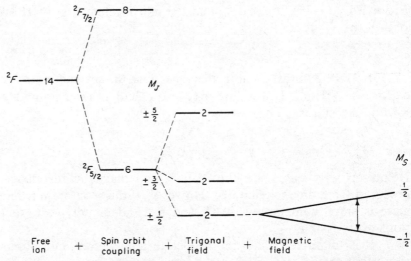

Fig. 12-15 Splitting of the states of a $4f^1$ ion by the combined action of spin-orbit coupling, a trigonal crystal field and an external magnetic field. The degeneracy of each state is indicated.

Since $\lambda > 0$, the $J = \frac{5}{2}$ state is the ground state.

The eigenfunctions in terms of the $|M_L, M_S\rangle$ functions for $J = \frac{5}{2}$ are obtained by application of the methods outlined in Sec. B-7. The results are

$$|\tfrac{5}{2}, \pm\tfrac{5}{2}\rangle = \sqrt{\tfrac{6}{7}}|\pm 3, \mp\tfrac{1}{2}\rangle - \sqrt{\tfrac{1}{7}}|\pm 2, \pm\tfrac{1}{2}\rangle \qquad (12\text{-}45a)$$

$$|\tfrac{5}{2}, \pm\tfrac{3}{2}\rangle = \sqrt{\tfrac{5}{7}}|\pm 2, \mp\tfrac{1}{2}\rangle - \sqrt{\tfrac{2}{7}}|\pm 1, \pm\tfrac{1}{2}\rangle \qquad (12\text{-}45b)$$

$$|\tfrac{5}{2}, \pm\tfrac{1}{2}\rangle = \sqrt{\tfrac{4}{7}}|\pm 1, \mp\tfrac{1}{2}\rangle - \sqrt{\tfrac{3}{7}}|0, \pm\tfrac{1}{2}\rangle \qquad (12\text{-}45c)$$

These functions may now be used as operands for a crystal-field operator appropriate to trigonal symmetry. This results in the splitting of the $M_J = \pm\frac{5}{2}, \pm\frac{3}{2}$, and $\pm\frac{1}{2}$ states into a set of three Kramers doublets.† Of these, the $M_J = \pm\frac{1}{2}$ states lie lowest.

The g components are obtained by the application of

$$\mathcal{H}_{\text{mag}} = \beta\mathbf{H} \cdot (\hat{\mathbf{L}} + g_e\hat{\mathbf{S}})$$

of (Eq. 11-33) on the $|\frac{5}{2}, \pm\frac{1}{2}\rangle$ functions. For the field parallel to the trigonal axis and for $M_J = +\frac{1}{2}$

$$W_{\frac{1}{2}} = (\sqrt{\tfrac{4}{7}}\langle 1, -\tfrac{1}{2}| - \sqrt{\tfrac{3}{7}}\langle 0, \tfrac{1}{2}|)\beta H(\hat{L}_z + g_e\hat{S}_z)(\sqrt{\tfrac{4}{7}}|1, -\tfrac{1}{2}\rangle - \sqrt{\tfrac{3}{7}}|0, +\tfrac{1}{2}\rangle)$$
$$\cong 0 + \tfrac{3}{7}\beta H \quad (12\text{-}46a)$$

Similarly

$$W_{-\frac{1}{2}} \cong -\tfrac{3}{7}\beta H \qquad (12\text{-}46b)$$

Hence

$$g_{\parallel} \cong \tfrac{6}{7} \qquad (12\text{-}47a)$$

From Eq. (11-44) one may obtain the matrix elements for $\mathbf{H} \parallel X$ and thus obtain the value

$$g_{\perp} \cong \tfrac{18}{7} \qquad (12\text{-}47b)$$

It is of interest to note that $g_{\parallel}$ is just the ground-state Landé factor g' for $J = \frac{5}{2}$ [Eq. (11-2)]. One may also calculate the g components for the $\pm\frac{3}{2}$ and $\pm\frac{5}{2}$ states. These are

$$M_J = \pm\tfrac{3}{2}: \qquad g_{\parallel} = \tfrac{18}{7} \qquad g_{\perp} = 0 \qquad (12\text{-}48a)$$

$$M_J = \pm\tfrac{5}{2}: \qquad g_{\parallel} = \tfrac{30}{7} \qquad g_{\perp} = 0 \qquad (12\text{-}48b)$$

Relatively few cases are known in which magnetic resonance has been detected from an excited state. However, if the separation from the lowest state is small compared with kT, and T_1 is adequately long, such observations are possible.

It is characteristic of the rare-earth ions with an odd number of electrons that the effects of spin-orbit coupling and an axial crystal field leave

† The effect of the crystal field in this case is analogous to the Stark splitting of the states of a gaseous atom.

a Kramers doublet lowest in energy. However, the large spin-orbit coupling to excited states implies a strong coupling with the environment. Hence even in this case. ESR observations usually require temperatures of 20 K or lower.

For rare-earth ions with an even number of electrons, a crystal field of C_{3v} or C_{3h} symmetry will split each J state into singlets and doublets. One might expect a Jahn-Teller splitting to remove the residual degeneracies. However such splittings are usually small, and the C_{3h} or C_{3v} symmetry is still appropriate. The observed effect of a magnetic field on the (non-Kramers) doublets is a first-order splitting if the field is directed along the symmetry axis. No splitting is obtained if the field is in a plane perpendicular to this axis. One may regard this system as having $S' = 1$ and a large negative zero field splitting parameter D such that the states corresponding to $|\pm 1\rangle$ are the only ones populated. There are additionally small splitting contributions attributed to crystalline imperfections; their effects are accounted for by the terms $\Delta_x \hat{S}_x + \Delta_y \hat{S}_y$ in the spin hamiltonian, which is usually written

$$\hat{\mathscr{H}} = g_\| \beta \hat{S}_z H_z + \Delta_x \hat{S}_x + \Delta_y \hat{S}_y + hA_\| \hat{S}_z \hat{I}_z \tag{12-49}$$

The transitions in a magnetic field are given by

$$(h\nu)^2 = (g_\| \beta H_z + hA_\| M_I)^2 + \Delta^2 \tag{12-50}$$

where $\Delta^2 = \Delta_x^2 + \Delta_y^2$.

12-4 THE ACTINIDE IONS

The ions of the actinide ($5f^n$) group exhibit many of the properties characteristic of the lanthanide ($4f^n$) group. The most common valency is $+3$; however, unlike the $4f$ electrons, the $5f$ electrons do participate in bonding with the ligands. Table 12-3 lists the common ions, along with their ground-state configurations.

The trivalent ions have magnetic properties very similar to the corresponding ions in the $4f$ group except that the spin-orbit coupling constants

Table 12-3 Magnetic properties of the ground states of actinide ions

Number of magnetic electrons	0	1	2	3	4	5	6	7
Representative ions	UO_2^{++}	NpO_2^{++}	PuO_2^{++}	AmO_2^{++}				
			Pa^{3+}	U^{3+}				
					Np^{3+}	Pu^{3+}	Am^{3+}	Cm^{3+}
	Th^{4+}	Pa^{4+}	U^{4+}	Np^{4+}	Pu^{4+}	Am^{4+}		
		U^{5+}	Np^{5-}	Pu^{5+}				
		Np^{6+}	Pu^{6+}	Am^{6+}				
Ground state		$^2F_{5/2}$	$^3H_4$	$^4I_{9/2}$	$^5I_4$	$^6H_{5/2}$	$^7F_0$	$^8S_{7/2}$

are somewhat larger. The ions of the type $(O\!-\!M\!-\!O)^{n+}$ are peculiar to the actinide group. These are linear complexes in which the axial interaction with oxygen is dominant. The ions are anomalous in that the "crystal field" is so strong that it is much larger than the spin-orbit coupling. The U^{++} ion has the configuration $5f6d7s^2$. These four electrons are used to form covalent bonds with the oxygen atoms, and thus the UO_2^{++} ion has no magnetic electrons. The electrons of NpO_2^{++}, PuO_2^{++}, and AmO_2^{++} are similarly involved; thus these three ions have, respectively, 1, 2, and 3 magnetic electrons.

The strong axial field in the NpO_2^{++} ion splits the sevenfold orbital degeneracy of the ^2F ground state into states with $M_L = \pm3, \pm2, \pm1, 0$. The state of lowest energy has $M_L = \pm3$. Spin-orbit coupling splits this state into two Kramers doublets with $M_J = \pm\frac{7}{2}$ and $\pm\frac{5}{2}$; the latter lies lower. The ground state can thus be characterized by $S' = \frac{1}{2}$. The g factors are obtained from

$$g_{\parallel} = 2|\langle 3, -\tfrac{1}{2}|\hat{L}_z + g_e\hat{S}_z|3, -\tfrac{1}{2}\rangle| \cong 4 \qquad (12\text{-}51a)$$

$$g_{\perp} = 2|\langle 3, -\tfrac{1}{2}|\hat{L}_x + g_e\hat{S}_x|-3, \tfrac{1}{2}\rangle| \cong 0 \qquad (12\text{-}51b)$$

The experimental values are $g_{\parallel} = 3.405$ and $g_{\perp} = 0.205$.[†] The deviations from the expected values can be ascribed to interactions with excited states and delocalization of electrons onto the ligands. Additional examples of actinide ions are covered in a review by Low.[‡]

12-5 DEFICIENCIES OF THE POINT–CHARGE CRYSTAL FIELD MODEL; LIGAND FIELD THEORY

In Chaps. 11 and 12 we have used the point-charge crystal field model in order to obtain a simple picture of the state energies of transition-metal ions in fields of various symmetries. Some of the deficiencies of this model are outlined below.

1. *Neglect of excited atomic states.* All of the calculations in these chapters have considered only the splitting of the ground atomic state in the crystal field (for example, ^3F for $3d^2$ ions). In some cases the crystal field splitting of excited atomic states may leave some states close enough to the ground state so that mixing via the spin-orbit coupling operator may occur. Although this type of interaction is often not important in the calculation of g factors and zero field splitting parameters, it is a very important consideration when treating optical spectra of transition-metal ions. (See, e.g., Fig. 11-9.)

† B. Bleaney, P. M. Llewellyn, M. H. L. Pryce, and G. R. Hall, *Phil. Mag.*, **45**:992 (1954).
‡ W. Low, "Paramagnetic Resonance in Solids," pp. 137ff., Academic Press Inc., New York, 1960.

2. *Neglect of ligand electrons.* We have considered the ligands only as a source of the crystal-field potential. The effects of ligand electrons have not been included explicitly. We have already indicated several observations which show that these effects must be included if quantitative (and in some cases qualitative) agreement with experiment is desired. Probably the most striking observation is the presence of ligand hyperfine splittings for some transition-metal ion spectra. Ligand hyperfine interaction was first seen from the chlorine nuclei of $IrCl_6^{--}$.[†]

The most satisfactory method for inclusion of ligand effects is to construct molecular orbitals made up from central-ion atomic orbitals and appropriate combinations of ligand atomic orbitals. As an example, consider an octahedral complex in which the ligands are F^- ions. The $2p$ orbitals of the F^- ions will combine with the $3d$, $4s$, and $4p$ orbitals of the central ion to form a set of delocalized molecular orbitals. In order to obtain the correct linear combinations of atomic orbitals, one must apply techniques of group theory.[‡] However, some of the qualitative features may be illustrated pictorially. Two types of molecular orbitals will be considered: (1) σ molecular orbitals, for which the charge distribution along the axis between a ligand atom and the central ion has cylindrical symmetry, and (2) π molecular orbitals, for which this axis contains one nodal plane. Figure 12-16 illustrates some of the combinations of $3d$ and $2p$ orbitals which are used to construct molecular orbitals. Only the *bonding* orbitals (i.e., those in which there is no node between a ligand atom and the central ion) are shown. There will be a corresponding set of antibonding molecular orbitals.

Consider the $3d_{x^2-y^2}$ orbital, which combines with the $2p$ orbitals of F^- ions in the xy plane. In an octahedral complex, all F^- ions are equivalent; hence, the $2p$ orbitals of F^- must each contribute equally. If the signs are chosen so as to generate a bonding orbital between the central ion and a ligand, the corresponding molecular orbital is written[‡]

$$\psi_{x^2-y^2} = N'_\sigma[d_{x^2-y^2} + \tfrac{1}{2}\Lambda'_\sigma(\phi_1 - \phi_2 + \phi_3 - \phi_4)] \qquad (12\text{-}52a)$$

When the signs are taken so as to generate an antibonding orbital between the central ion and a ligand, the following molecular orbital is obtained:

$$\psi^*_{x^2-y^2} = N_\sigma[d_{x^2-y^2} - \tfrac{1}{2}\Lambda_\sigma(\phi_1 - \phi_2 + \phi_3 - \phi_4)] \qquad (12\text{-}52b)$$

N_σ is a normalization constant and Λ_σ is a measure of the extent of ligand-orbital participation.

In octahedral symmetry, d_{z^2} and $d_{x^2-y^2}$ are degenerate. The molecular

[†] J. H. E. Griffiths and J. Owen, *Proc. Roy. Soc. (London)*, **A213**:459 (1952); **A226**:96 (1954); *Proc. Phys. Soc.*, **A65**:951 (1952).

[‡] For the details of these calculations, see C. J. Ballhausen and H. B. Gray, "Molecular Orbital Theory," chap. 8, W. A. Benjamin, Inc., New York, 1964.

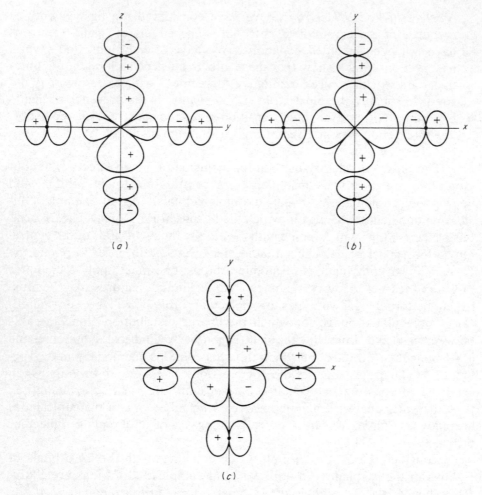

Fig. 12-16 Representation of the combination of central-ion d orbitals with ligand p orbitals. (a) d_{z^2}. (b) $d_{x^2-y^2}$. (c) d_{xy}.

orbitals containing d_{z^2} are

$$\psi_{z^2} = N'_\sigma \left[d_{z^2} - \frac{1}{\sqrt{12}} \Lambda'_\sigma (\phi_1 + \phi_2 + \phi_3 + \phi_4 - 2\phi_5 - 2\phi_6) \right] \quad (12\text{-}53a)$$

$$\psi_{z^2}^* = N_\sigma \left[d_{z^2} + \frac{1}{\sqrt{12}} \Lambda_\sigma (\phi_1 + \phi_2 + \phi_3 + \phi_4 - 2\phi_5 - 2\phi_6) \right] \quad (12\text{-}53b)$$

If overlap is neglected, the probability that an electron in $\psi_{z^2}^*$ will be found in ϕ_1 is $\frac{1}{12} N_\sigma^2 \Lambda_\sigma^2$; for an electron in $\psi_{x^2-y^2}^*$, the corresponding probability is $\frac{1}{4} N_\sigma^2 \Lambda_\sigma^2$. The average probability is $\frac{1}{6} N_\sigma^2 \Lambda_\sigma^2$. Thus $N_\sigma^2 \Lambda_\sigma^2$ measures the probability that an electron in the degenerate $\psi_{z^2}^*$, $\psi_{x^2-y^2}^*$ orbitals will be found on one of the ligand atoms.

It is apparent from Fig. 12-16c that the d_{xy} orbitals can only form π-type molecular orbitals with the ligands. The same applies to the d_{xz} and d_{yz} orbitals. The resulting antibonding molecular orbitals are

$$\psi_{xy}^* = N_\pi[d_{xy} - \tfrac{1}{2}\Lambda_\pi(\phi_1 + \phi_2 + \phi_3 + \phi_4)] \qquad (12\text{-}54a)$$

$$\psi_{xz}^* = N_\pi[d_{xz} - \tfrac{1}{2}\Lambda_\pi(\phi_2 + \phi_4 + \phi_5 + \phi_6)] \qquad (12\text{-}54b)$$

$$\psi_{yz}^* = N_\pi[d_{yz} - \tfrac{1}{2}\Lambda_\pi(\phi_1 + \phi_3 + \phi_5 + \phi_6)] \qquad (12\text{-}54c)$$

N_π and Λ_π are factors analogous to those in Eqs. (12-52). There will be a corresponding set of bonding molecular orbitals.

Since the overlap for the σ-type is much greater than that for π-type molecular orbitals, one expects the interaction energy to be much greater in the former case. Figure 12-17 illustrates the energy-level spacing that might be expected in this simple case. The ligand electrons fill up the bonding molecular orbitals. Thus the two antibonding molecular orbitals are analogous to the t_{2g} and e_g orbitals in the crystal field model. One thus has an alternative interpretation for the ordering of energy levels in this case.

The above picture is oversimplified, as there are other central-ion orbitals (such as $4s$ and $4p$) which must be considered. One must also consider the $2s$ ligand orbitals. In certain cases, sophisticated calculations are required to obtain the correct ordering of the energy states.

One of the results of the inclusion of the ligand orbitals is a reduction in the effect of spin-orbit coupling. As an example, consider the $3d^1$ (ttdl + ttgl) case with $\alpha_t < 0$. (See Prob. 11-6.) Here the ground state in the absence of ligand participation is

$$|G\rangle = \frac{1}{\sqrt{2}}\,(|2\rangle + |-2\rangle) = d_{x^2-y^2}$$

The corresponding molecular orbital would be given in Eq. (12-52b). The Λ-tensor method of Sec. 11-6 may now be employed to compute the g

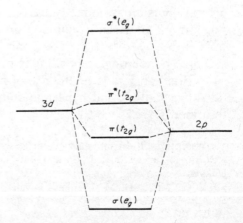

Fig. 12-17 Energy-level splitting due to interaction of central-ion $3d$ orbitals with ligand $2p$ orbitals. The bonding orbitals $\sigma(e_g)$ and $\pi(t_{2g})$ are normally filled; the antibonding orbitals $\pi^*(t_{2g})$ and $\sigma^*(e_g)$ correspond to the t_{2g} and e_g orbitals of the crystal field theory.

factors. Thus

$$\Lambda_{zz} = - \sum_{n}{}' \frac{\langle G|\hat{L}_z|n\rangle\langle n|\hat{L}_z|G\rangle}{W_n{}^{(0)} - W_G{}^{(0)}} \tag{12-55}$$

Here $|n\rangle$ will correspond to Eq. (12-54a), since $d_{xy} = (1/\sqrt{2})(|2\rangle - |-2\rangle)$.
Thus

$$\begin{aligned}
\langle G|\hat{L}_z|n\rangle &= N_\sigma N_\pi \langle d_{x^2-y^2} - \tfrac{1}{2}\Lambda_\sigma(\phi_1 - \phi_2 + \phi_3 - \phi_4)|\hat{L}_z|d_{xy} \\
&\qquad\qquad - \tfrac{1}{2}\Lambda_\pi(\phi_1 + \phi_2 + \phi_3 + \phi_4)\rangle \\
&= 2N_\sigma N_\pi
\end{aligned} \tag{12-56}$$

since there is no orbital angular momentum associated with the ligands.
Thus

$$\Lambda_{zz} = \frac{-4N_\sigma{}^2 N_\pi{}^2}{\Delta} \tag{12-57}$$

and

$$g_{\|} = g_{zz} = g_e - \frac{8\lambda N_\sigma{}^2 N_\pi{}^2}{\Delta} \tag{12-58}$$

Equation (12-58) is equivalent to Eq. (11-54a), if a reduced spin-orbit coupling constant $\lambda' = \lambda N_\sigma{}^2 N_\pi{}^2$ is used. Note that the factor $N_\sigma{}^2 N_\pi{}^2$ is less than unity.

The observation of ligand hyperfine splitting is a clear indication of ligand-orbital participation. The analysis is rather involved, but in favorable cases a quantitative measure of N_σ, N_π, Λ_σ, and Λ_π in Eqs. (12-52) to (12-54) may be obtained. The isotropic part of the ligand hyperfine splitting arises from the σ overlap, with the addition of a small core polarization contribution, whereas the anisotropic part arises from a combination of σ and π overlap. The $(NH_4)_2IrCl_6$ example is interesting in that the chlorine hyperfine structure demonstrated for the first time not only that ligand orbitals participate directly but also that π bonding with the ligands is important. This is an example of $5d^5$ $(ls)(oct)$ which corresponds to a single hole in the t_{2g} orbitals. The appropriate molecular orbitals would be Eqs. (12-54); hence only π bonding is possible.

Further details on the use of molecular orbital theory in the computation of g factors and hyperfine couplings may be found in more advanced treatments.[†,‡,§]

† J. W. Orton, "Electron Paramagnetic Resonance," chap. 8, Iliffe Books, Ltd., London, 1968.
‡ B. R. McGarvey, in R. L. Carlin (ed.), "Transition Metal Chemistry," vol. 3, p. 89, Marcel Dekker, New York, 1966.
§ J. Owen and J. H. M. Thornley, Covalent Bonding and Magnetic Properties of Transition-metal Ions. *Rept. Prog. Phys.*, **29**:675 (1966).

12-6 ELECTRON RESONANCE OF GASEOUS FREE RADICALS

Gaseous paramagnetic molecules exhibit a new feature peculiar to molecules widely separated in space. Their rotational angular momentum, which in the liquid or solid phases is quenched, couples strongly to the electronic spin and orbital angular momenta. This coupling leads to such a multitude of levels that the sensitivity is decreased because of the division of the intensity among so many lines; also, the lines are so close together that resolution suffers. For these and other reasons, the only molecules which have been studied to date are diatomic and linear triatomic radicals.

Gas-phase free radicals (e.g., NO) were first studied early in the history of ESR spectroscopy[†]; however, very few molecules other than O_2 and a number of atomic species were studied subsequently. A few short-lived diatomic species (OH, SH, SeH. TeH[‡] and SO[§]) were reported in the early 1960s; however, it is only recently that a major effort has been made to detect a wide range of gaseous free radicals.[¶,††]

A discussion of gas-phase free radicals must begin with an analysis of the coupling of the various angular momenta. For diatomic and linear polyatomic molecules, the various possibilities are classified as Hund's coupling cases.[‡‡] As an example consider Hund's case *a* illustrated in Fig. 12-18*a*. Here the electronic motion is coupled very strongly to the internuclear axis. This case applies to a molecule which has a net orbital angular momentum about the internuclear axis Z. Because of the cylindrical symmetry, the orbital angular momentum **L** precesses about Z; the Z component of **L** is quantized and is characterized by Λ ($\Lambda = 0, \pm 1, \pm 2, \ldots$). The total spin angular momentum **S** also precesses about Z because of spin-orbit coupling to **L**; the Z component of **S** is characterized by Σ. Λ and Σ add or subtract such that $\Omega = |\Lambda \pm \Sigma|$. The angular momentum **N** now couples with Ω to form the resultant total angular momentum **J**.[§§] Note that **N** must be perpendicular to Z.

As an example, consider the NO molecule. The unpaired electron resides in a π molecular orbital; hence $\Lambda = \pm 1$ and $\Sigma = \pm \frac{1}{2}$. Thus $\Omega = \frac{3}{2}, \frac{1}{2}$. These two states are characterized as $^2\Pi_{\frac{3}{2}}$ and $^2\Pi_{\frac{1}{2}}$. In NO, these two states are separated by 123 cm^{-1}, with $^2\Pi_{\frac{1}{2}}$ as the ground state. Each of these two states is doubly degenerate because $\Lambda = \pm 1$; this is commonly referred to as Λ degeneracy.

[†] R. Beringer and J. G. Castle, Jr., *Phys. Rev.*. **78**:581 (1950).

[‡] H. E. Radford, *Phys. Rev.*, **122**:114 (1960); *J. Chem. Phys.*, **40**:2732 (1964).

[§] J. M. Daniels and P. B. Dorain, *J. Chem. Phys.*, **40**:1160 (1964); **45**:26 (1966).

[¶] A. Carrington, *Proc. Roy. Soc. (London)*, **A302**:291 (1968).

[††] A. Carrington, D. H. Levy, and T. A. Miller. *Advan. Chem. Phys.*, **18**:149 (1970).

[‡‡] G. Herzberg, "Molecular Spectra and Molecular Structure," 2d ed., vol. I, pp. 219ff., D. Van Nostrand Company, Inc., Princeton, N.J., 1950.

[§§] The nuclear spin **I** may also combine with **J** to form a grand total angular momentum **F**.

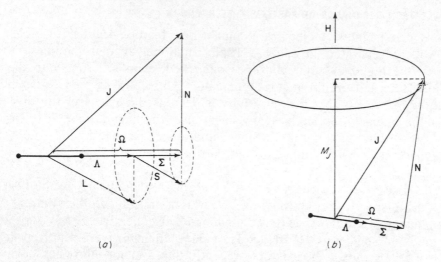

Fig. 12-18 (*a*) Coupling of angular momenta in a diatomic molecule for Hund's case *a*. The Z component (Λ) of the orbital angular momentum **L** adds to the Z component (Σ) of the spin angular momentum **S** to form a resultant Ω. Then Ω adds vectorially to **N**, the rotational angular momentum, to form the resultant total angular momentum **J**. (*b*) Vector diagram showing quantization of angular momentum in a magnetic field for a molecule corresponding to Hund's case *a*.

The application of a strong magnetic field will cause a splitting of levels given by $\hat{\mathcal{H}}_{\text{mag}}$ of Eq. (11-44). However, for the $^2\Pi_{\frac{1}{2}}$ state, the spin and orbital angular momenta virtually cancel each other, resulting in a "nonmagnetic" ground state. Now consider the lowest rotational level ($N = 0$) of the $^2\Pi_{\frac{3}{2}}$ state. Since $\Omega = \frac{3}{2}$, $J = \frac{3}{2}$ and M_J (the component of **J** along a *space-fixed* axis) $= \frac{3}{2}, \frac{1}{2}, -\frac{1}{2}, -\frac{3}{2}$. The molecular g factor[†] is given by a relation analogous to that for the Landé g factor, Eq. (11-2),

$$g = \frac{(\Lambda + 2\Sigma)(\Lambda + \Sigma)}{J(J + 1)} \tag{12-59}$$

Thus $g = \frac{4}{5}$ for the $^2\Pi_{\frac{3}{2}}$ state with $N = 0$.

The separation of the states as a function of magnetic field is illustrated in Fig. 12-19. The three allowed transitions for $J = \frac{3}{2}$ and the five for $J = \frac{5}{2}$ are those corresponding to $\Delta M_J = 1$. They occur at different magnetic fields as a result of a second-order Zeeman interaction between the $J = \frac{3}{2}$ and the $J = \frac{5}{2}$ states. For higher values of J, g becomes progressively smaller, and hence transitions within these states can only be observed at higher magnetic fields or lower microwave frequencies.

Hund's case *a* in Fig. 12-18 has been oversimplified. If $\Lambda \neq 0$, there will be a residual orbital degeneracy which has not been shown. Inter-

† G. Herzberg, "Spectra of Diatomic Molecules," 2d ed., vol. I, p. 301, D. Van Nostrand Company, Inc., Princeton, N.J., 1950.

action of the orbital and the rotational angular momenta leads to a removal of the degeneracy. The splitting ("Λ-type doubling") of the states, which will be labeled as "+" and "−," increases with increasing J. Complete removal of the Λ degeneracy makes possible four transitions for each line. Of these, two are induced by the *magnetic* component of the microwaves and two by the *electric* component. Since electric dipole transitions are approximately 100 to 1,000 times more intense than magnetic dipole transitions, it is the former which are observed. Thus the experimental arrangement must be such that the gas molecules can be exposed to regions of the cavity where the E field is large. For this purpose a cylindrical TE_{011} cavity (see Fig. 2-4) with a large sample access hole is used. We have ignored the possibility of hyperfine splitting due to a nucleus of nonzero spin. However, there are a number of illustrations of such splitting, for example, NO and ClO.

Figure 12-20 exhibits separate electron resonance spectra of $^{15}N^{16}O$ and $^{14}N^{16}O$ for both the $J = \frac{3}{2}$ and the $J = \frac{5}{2}$ states. Since $I = \frac{1}{2}$ for ^{15}N and $I = 1$ for ^{14}N, and the hyperfine splittings are large, one observes respectively two and three sets of the three transitions shown for the $J = \frac{3}{2}$ states in Fig. 12-19. Each is doubled by removal of the Λ degeneracy. Interpretation of the $J = \frac{5}{2}$ spectrum is left as Prob. 12-7.

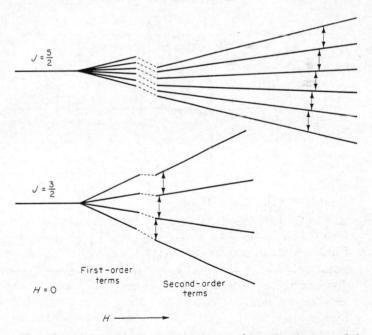

Fig. 12-19 Splitting of the $J = \frac{3}{2}$ and $J = \frac{5}{2}$ rotational states of the $^2\Pi_{3/2}$ state of NO as a function of applied magnetic field. The displacements of first-order Zeeman levels by a second-order interaction are shown by the dotted lines.

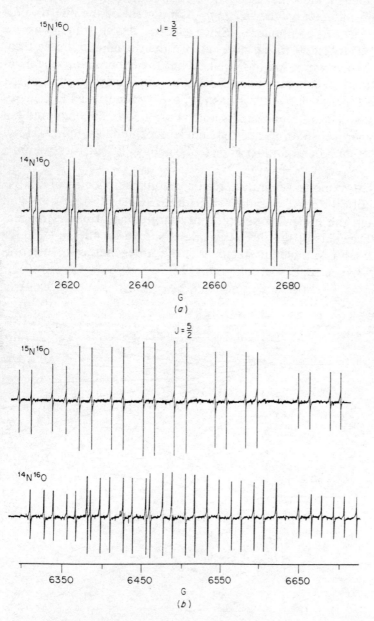

Fig. 12-20 (*a*) Electron paramagnetic resonance spectra of $^{15}N^{16}O$ and $^{14}N^{16}O$ in the $J = \frac{3}{2}$ rotational state of the $^2\pi_{3/2}$ state. $\nu = 2.8799$ GHz. The transitions correspond to $\Delta J = 0$, $\Delta M_J = 1$, $\Delta M_I = 0$ and $\pm \rightarrow \mp$. (*b*) Spectra corresponding to molecules in the $J = \frac{5}{2}$ state. [*After R. L. Brown and H. E. Radford, Phys. Rev.*, **147**:6 (1966).]

It is important to note that a molecule in the gas phase is not tumbling so as to behave isotropically. In the absence of a magnetic field and molecular collisions, **J** is *fixed in space;* in a magnetic field, **J** precesses about **H**. Thus the spectrum contains information about anisotropic interactions. In particular, the observed hyperfine interaction is a combination of isotropic and anisotropic terms. In favorable cases, these may be separated.

Another peculiarity of gas-phase electron resonance is that one cannot employ a "spin hamiltonian" as was done for the transition-metal ions. It is always necessary to consider the coupling of various angular momenta to form the total angular momentum J; the $2J + 1$ degeneracy is then removed by a magnetic field. The term "electron resonance" is employed because spectra can be observed for molecules which contain no unpaired electrons (e.g., the $^1\Delta$ state of O_2)!

A number of short-lived gas-phase radicals have now been studied; these include $(^1\Delta)SO$, $(^3\Sigma)SO$, ClO, BrO, NS, SF, $(^1\Delta)SeO$, $(^3\Sigma)SeO$, SeH, and ONS. Careful analyses of these spectra have yielded information such as bond lengths, dipole moments, quadrupole coupling constants, g factors, and zero field coupling constants.

The H_2^+ ion, which is the simplest of all molecules, represents an instructive example of the coupling of spin, nuclear, and rotational angular momenta of equivalent nuclei. Vector coupling of **I** with **S** gives a resultant F_2; the latter is then coupled with the rotational vector **K** to give the resultant **F**. However, one must take into account the fact that for ortho H_2^+, the total nuclear spin $I = 1$, and the quantum number K may take only odd values; here **S** is combined directly with **K** to give **F**. This molecule-ion has been made by irradiation of hydrogen, and transitions among its hyperfine levels were studied for the vibrational levels $\nu = 4$ to $\nu = 8$, and for rotational levels $K = 1$ and $K = 2$.†

The cases which have been studied are favorable because intense electric dipole transitions are possible. For nonlinear polyatomic molecules virtually all the orbital angular momentum is quenched; **S** is not coupled strongly to the molecular framework, and thus only magnetic dipole transitions are possible. The low intensity and the complexity of the spectra discourage the investigation of such molecules at present. For example, NO_2 in the gas phase exhibits hundreds of lines.

12-7 THE PRACTICAL INTERPRETATION OF ESR SPECTRA OF IONS IN THE SOLID STATE

The aims of an investigation of the ESR spectra of transition-metal or rare-earth systems stated in the introduction to Chap. 11 may in simple cases be achieved with little effort. In other cases, the process may be very

† K. B. Jefferts, *Phys. Rev. Lett.*, **23**:1476 (1969).

laborious if indeed an unambiguous result is achieved. There are certain questions to be answered. Some answers may be obtained in one experiment, whereas others may require a whole set of experiments. Some essential questions are the following:

1. What are the principal components of the g tensor? If the spectrum is not isotropic, does it show axial symmetry (trigonal, tetragonal, or hexagonal)? Is the symmetry orthorhombic or even lower? Are there special features such as low-field transitions or double-quantum transitions? Is there site splitting? Do the g components closely resemble those of isoelectronic ions *of the same configuration* in a similar symmetry?

2. If there is detectable hyperfine splitting, what are the principal components of the hyperfine tensor?

   a. If no splitting is obvious, are the lines too broad to permit resolution? Does it help to lower the temperature? To use a more dilute sample? To reduce the microwave power or the modulation amplitude? Are the gain and the signal-to-noise ratio sufficiently high so that contributions from nuclides of low abundance may be detected? Is there some other orientation of a single crystal which may be more favorable?

   b. If a splitting is measurable, do the relative intensities suggest one nuclide or more than one? Does the intensity ratio of lines from various nuclides correspond with the natural abundances for the suspected element? Is there any superhyperfine splitting (splitting from a nucleus other than that on which the unpaired electrons are primarily located)? Assuming the symmetry of the crystal field established from the anisotropy of g and of the hyperfine parameters, do these parameters resemble those of the suspected nuclide in similar symmetry?

3. Within what temperature ranges can an ESR spectrum be observed? How does the relative intensity of various lines change with temperature? Does the spectrum appear on lowering the temperature (suggesting a short spin-lattice relaxation time at higher temperatures)? Does it disappear on lowering the temperature (suggesting that a higher temperature is required to populate the states between which transitions are being observed)?

4. Is there a zero-field splitting? For an assumed spin in the ground state, are all possible transitions observed? If not, are more observed at a higher microwave frequency (and correspondingly higher magnetic field)? Is a "$\Delta M = 2$" (or "$\Delta M = 3$" or even "$\Delta M = 4$") line observed at low fields? Does the system show axial symmetry? What is D? If E is nonzero, what is its magnitude? Do the temperature-variation experiments in question 3 give an unambiguous determination of the sign of D and E?

5. How does the site symmetry compare with that of the host? If it is

different, what kinds of distortion would give the observed symmetry? Where would a second (or third!) defect have to be located relative to the paramagnetic center? Does the paramagnetic center correspond to a substitutional or interstitial defect?

6. Does the assumed spin hamiltonian fit the observed spectra? Are *all* of the lines, their angular dependence, and their intensity accounted for? Do the parameters still fit the spectrum if the frequency is changed?

A thorough ESR study of a relatively simple system would hopefully give satisfactory answers to all of the above questions. Many reported studies fail to give any answer to some of them. Sometimes this is due to the complexity of the system; in other cases, the experimenter has been content with a partial investigation.

BIBLIOGRAPHY

Compilations of ESR data on transition-metal and rare-earth ions

1. Kuska, H. A., and M. T. Rogers, in E. T. Kaiser and L. Kevan (eds.), "Radical Ions," pp. 579 to 745, Interscience Publishers, a division of John Wiley & Sons, Inc., New York, 1968.
2. McGarvey, B. R., in R. L. Carlin (ed.), "Transition Metal Chemistry," vol. 3, pp. 89 to 201, Marcel Dekker, New York, 1966 (examples of $3d$, $4d$ and $5d$ ions and their complexes).
3. McGarvey, B. R., in T. F. Yen (ed.), "Electron Spin Resonance of Metal Complexes," chap. 1, Plenum Publishing Corporation, New York, 1969.
4. Bowers, K. D., and J. Owen, *Rept. Progr. Phys.*, **18**:304 (1955); J. W. Orton, *Rept. Progr. Phys.*, **22**:204 (1959).
5. Low, W., Paramagnetic Resonance in Solids, in F. Seitz and D. Turnbull (eds.), "Solid State Physics," Supplement 2, 1960.
6. Low, W., and E. L. Offenbacher, in F. Seitz and D. Turnbull (eds.), "Solid State Physics," **17**:135 (1965). This review deals with simple and complex oxides.
7. Abragam, A., and B. Bleaney, "Electron Paramagnetic Resonance of Transition Ions," Clarendon Press, Oxford, 1970.

PROBLEMS

12-1. The following data were obtained for Ti^{3+} at a tetragonal site in CaO. (See Sec. 12-1a.)

(a) With the magnetic field along a [100] direction (see Fig. 12-2b) the positions of the lines are as follows at 9.52835 GHz

$^{46,48,50}Ti$ 3,504.4 G

$^{47,49}Ti$ 3,398.6, 3,428.8, 3,458.9, 3,489.1

(next line obscured), 3,549.4, 3,579.6, and 3,609.8 G

(b) With the magnetic field along a [110] direction, the corresponding data are

$^{46,48,50}Ti$ 3,508.6 G

$^{47,49}Ti$ 3,428.3, 3,451.2, 3,474.1, 3,496.9

(next line obscured), 3,542.3, 3,565.1, 3,587.4 G

(c) From the above data, confirm the values of the g and hyperfine components given in the text (Sec. 12-1a).

12-2. Solve the secular determinants derived from Eqs. (12-18) and verify the energies in Eq. (12-19b).

12-3. Carry out the perturbation-theory calculation to obtain Eqs. (12-24). Note that the E_g state functions may be written

$$|E, \pm\tfrac{1}{2}\rangle = \frac{1}{\sqrt{2}} (|2, \pm\tfrac{1}{2}\rangle + |-2, \pm\tfrac{1}{2}\rangle)$$

and

$$|0, \pm\tfrac{1}{2}\rangle$$

12-4. For a d^5 (ls)($oct + ttgl$) ion in which $\lambda \approx \delta$ if $\delta < 0$, the order of the levels in Fig. 12-4 is inverted. · Thus since $\lambda < 0$, W_+ will correspond to the lowest level.

   (a) Show that in this case

$$g_\parallel = \frac{1}{\mathscr{S}} [3(1 + \tfrac{1}{2}\eta) - \mathscr{S}]$$

$$g_\perp = \frac{1}{\mathscr{S}} [\mathscr{S} - 1 + \tfrac{3}{2}\eta]$$

   (b) Prepare plots of $g_\parallel$ and $g_\perp$ versus η for $\eta > 0$.

   (c) Ir^{4+} in $Na_2PtBr_6 \cdot 6H_2O$ exhibits an ESR spectrum at 4 K with $g_\parallel = 0.75$ and $g_\perp = 2.23$ (average of g_{XX} and g_{YY}). Estimate a value of η from these data.

12-5. Confirm that Eqs. (12-26) are the correct eigenfunctions for W_1 of Eqs. (12-25). (Hint: Start by generating the eigenfunctions corresponding to W_3 by using the operator $\hat{J}_- = \hat{L}_- + \hat{S}_-$. Then obtain the $|2, 2\rangle$ function by orthogonality. Generate its set of functions, etc., until the final result is obtained.)

12-6. Consider a $3d^3$ (ttdl) ion.

   (a) What is the ground state in the absence of spin-orbit coupling?

   (b) Apply $\hat{\mathscr{H}}'_{so}$ of Eq. (12-7) to obtain the energies:

$$W_{\frac{5}{2}} = -\frac{9\lambda}{4} \qquad W_{\frac{3}{2}} = \frac{3\lambda}{2} \qquad W_{\frac{1}{2}} = \frac{15\lambda}{4}$$

   (c) Show that the ground-state eigenfunctions are

$$|\tfrac{5}{2}, \pm\tfrac{5}{2}\rangle \equiv |\pm 1, \pm\tfrac{3}{2}\rangle$$

$$|\tfrac{5}{2}, \pm\tfrac{3}{2}\rangle \equiv \sqrt{\tfrac{2}{5}}|0, \pm\tfrac{3}{2}\rangle + \sqrt{\tfrac{3}{5}}|\pm 1, \pm\tfrac{1}{2}\rangle$$

$$|\tfrac{5}{2}, \pm\tfrac{1}{2}\rangle \equiv \sqrt{\tfrac{1}{10}}|\mp 1, \pm\tfrac{3}{2}\rangle + \sqrt{\tfrac{3}{5}}|0, \pm\tfrac{1}{2}\rangle + \sqrt{\tfrac{3}{10}}|\pm 1, \mp\tfrac{1}{2}\rangle$$

   (d) Show that $g \approx 0.6$ for this case.

12-7. Interpret the electron resonance spectra of the $J = \tfrac{5}{2}$ state of NO. (See Fig. 12-20b.)

13
Double-resonance Techniques

It was noted in Sec. 4-5 that an ESR spectrum may be used to characterize the spin of a nucleus responsible for hyperfine splitting; however, if in the same system there are two or more nuclei of the same spin, there is ambiguity in the assignment of hyperfine multiplets. Indeed, some hyperfine lines of spectra reproduced in this book were originally assigned to the wrong nucleus. Further, if the spacing of a set of hyperfine lines does not exceed their width, one fails to detect this splitting, except perhaps for a broadening. For this reason, splittings due to nuclei in neighboring molecules or to ions in rigid systems have rarely been observed directly. It would seem that in such ESR spectra one must resign himself to the loss of details of hyperfine interaction. This indeed appeared to be the case until 1956, when Feher† proposed and demonstrated the technique of Electron-Nuclear DOuble Resonance (ENDOR).‡ His brilliant contributions make it pos-

† G. Feher, *Phys. Rev.*, **103**:834 (1956).
‡ The most detailed account of ENDOR theory, including applications, is given by A. Abragam and B. Bleaney in "Electron Paramagnetic Resonance of Transition Ions," sec. 1.13 and chap. 4, Oxford University Press, London, 1970. This reference will be shortened to AABB in this chapter.

şible in some cases to regain some missing details of hyperfine interaction. For this to be possible, the ESR line being observed must be inhomogeneously broadened. In many systems, this technique completely removes ambiguities. It may provide such a wealth of detail about the wave function of the unpaired electron as to embarrass experimenter and theoretician alike. In one favorable case, a distinctive interaction of an unpaired electron with the twenty-third nearest-neighbor set of nuclei was established by an ENDOR experiment.[†]

13-1 AN ENDOR EXPERIMENT

Before undertaking a more detailed description of ENDOR processes, we shall give a brief phenomenological account of a simple ENDOR experiment on a solid-state system with $S = \frac{1}{2}$ and $I = \frac{1}{2}$. Suppose that from the resonant field positions H_k and H_m (see Fig. 3-5b) of the two hyperfine lines we have established a g factor and have calculated the hyperfine coupling to be 20 MHz. We now undertake an ENDOR experiment as follows:

1. The sample is placed in a special cavity; one type is shown in Fig. 13-1. At low microwave power, the magnetic field is scanned through the resonant field value for one of the transitions, e.g., that at H_k. The several spectrometer parameters are now optimized to give a strong ESR signal. The field is then set at H_k.
2. The microwave power level is then increased to severalfold its level in part 1.
3. A radio-frequency generator of wide range and large power output is connected to the side coils of the cavity of Fig. 13-1 so that the sample also experiences an oscillating radio-frequency magnetic field H_{rf}. The generator is set to scan the region 2 to 30 MHz, while the strip-chart recorder of the spectrometer is operating. The base line which, apart from noise, is constant indicates a constant ESR absorption. Though it will be nonlinear, the horizontal axis will also be a measure of the frequency of the rf generator.

At two frequencies of the rf generator, which we shall call ν_{n1} and ν_{n2}, the recorder will trace out lines such as those shown in Fig. 13-2. *This plot of changes in the ESR absorption intensity is called the ENDOR spectrum.* If the frequencies of these lines are carefully measured as the peak of each line is traversed, it will be noted that the difference $\nu_{n2} - \nu_{n1}$ is numerically just equal to the hyperfine coupling, that is, 20 MHz, measured from the ESR spectrum, but now determined with greater precision. Further, the mean of the frequencies ν_{n1} and ν_{n2} will be close to $\nu_0 = g_N \beta_N H_K/h$,

[†] G. Feher, *Phys. Rev.*, **114**:1219 (1959).

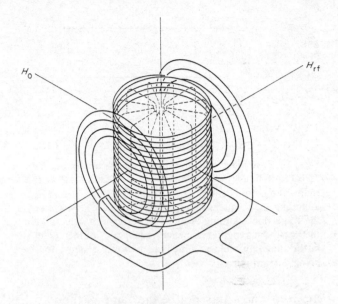

Fig. 13-1 Schematic representation of a TE_{011} cylindrical cavity designed for ENDOR studies. The sample is placed along the axis of the cavity. The side wall is a helix of spaced turns, with interstices filled by a plastic material of low dielectric loss. This design allows for penetration into the cavity of the microwave field H_{rf} and by a modulating magnetic field. The microwave magnetic field contours are shown by dotted lines. Since it is the component of the microwave or of the rf fields perpendicular to the static field H_0 which induces the electron spin and nuclear spin transitions, respectively, one seeks to keep H_0 perpendicular to both. The relative orientation of the microwave and the rf fields is in principle arbitrary; that shown here is the most efficient for a set of external coils and involves the least eddy-current loss. In other cavities, the rf field is introduced by a coil inside the cavity. To avoid coupling out microwave energy, the plane of the coil should be parallel to the microwave field. This automatically puts the microwave and the rf magnetic fields at right angles to one another. It may be of crucial importance to align the rf field appropriately with respect to a crystal axis. (See Sec. 13-3.) [*Figure after J. S. Hyde, J. Chem. Phys.*, **43**:1806 (1965).]

the NMR frequency of the nucleus in the magnetic field H_k. If the nucleus responsible for the hyperfine splitting had been uncertain, its identity would have been established from the value of g_N. If the experiment were repeated, but with the magnetic field set at H_m, the ENDOR spectrum would again consist of two lines, separated in frequency by the hyperfine coupling and symmetrically disposed (in first order) about the NMR frequency of the nucleus in the field H_m. However, the relative intensities of

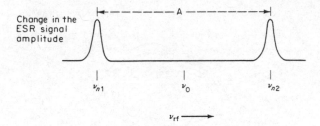

Fig. 13-2 Change in the ESR signal amplitude, i.e., "ENDOR lines" for a system with $S = \frac{1}{2}$ and $I = \frac{1}{2}$ as the radio-frequency generator is scanned through the region including the frequencies ν_{n1} and ν_{n2}; these are separated by the hyperfine coupling A and symmetrically spaced about the nuclear magnetic resonance frequency ν_0 of the nucleus in the magnetic field at which the microwave saturation is being carried out.

the two lines may not be the same in the two ENDOR spectra. (In some systems, only the ENDOR line at ν_{n1} would be observed when the magnetic field was set at H_k; further, only the ENDOR line at ν_{n2} would be observed when the magnetic field was set at H_m. We shall consider each of these cases in Sec. 13-3.)

Since the ENDOR lines typically represent a change in ESR line intensity of 1 percent of the ESR line under nonsaturated conditions, one requires a spectrometer of high sensitivity. There are also complexities of the ENDOR spectrometer which we have not enumerated. However, the method is well justified for the following cases:

1. Hyperfine lines are not resolved in the ESR spectrum.
2. Hyperfine lines are resolved, but more precise values of the hyperfine couplings are desired.
3. The identity of an interacting nucleus is to be established by a measurement of the nuclear g factor.
4. Quadrupole couplings are to be measured in a system with $I \geq 1$.

The so-called steady-state ENDOR experiment outlined briefly here will be considered in more detail in Sec. 13-3, after considering the energy levels and possible transitions of this system.

13-2 ENERGY LEVELS AND ENDOR TRANSITIONS

A description of the ENDOR lines of Fig. 13-2 (or of more complicated ENDOR spectra) requires:

1. Use of the full spin hamiltonian, including the nuclear Zeeman term (and a quadrupole term if $I \geq 1$).

2. Consideration of the populations of each state at low microwave power, under microwave saturation conditions, and during (or immediately after) passage through one of the frequencies ν_{n1} or ν_{n2} at a high rf power level. The relative populations (and thus the ENDOR line behavior) depend on the dominant relaxation mechanisms in the system. The relaxation aspects will be considered in the next section.

The spin hamiltonian for a system with $S = \frac{1}{2}, I = \frac{1}{2}$ is

$$\hat{\mathcal{H}} = \beta \hat{\mathbf{S}} \cdot \boldsymbol{g} \cdot \mathbf{H} + h\hat{\mathbf{S}} \cdot \boldsymbol{A} \cdot \hat{\mathbf{I}} - g_N\beta_N\hat{\mathbf{I}} \cdot \mathbf{H} \tag{13-1}$$

It will be convenient to begin with a fixed magnetic field and a fixed orientation of a single-crystal sample, such that the effective values g and A may be used in the simplified hamiltonian of Eq. (C-1),

$$\hat{\mathcal{H}} = g\beta\hat{\mathbf{S}} \cdot \mathbf{H} + hA\hat{\mathbf{S}} \cdot \hat{\mathbf{I}} - g_N\beta_N\hat{\mathbf{I}} \cdot \mathbf{H} \tag{13-2}$$

The first-order energy levels are [see Eqs. (C-24)]

$$W_{\frac{1}{2},\frac{1}{2}} = \tfrac{1}{2}g\beta H + \tfrac{1}{4}hA - \tfrac{1}{2}g_N\beta_N H \tag{13-3a}$$

$$W_{\frac{1}{2},-\frac{1}{2}} = \tfrac{1}{2}g\beta H - \tfrac{1}{4}hA + \tfrac{1}{2}g_N\beta_N H \tag{13-3b}$$

$$W_{-\frac{1}{2},-\frac{1}{2}} = -\tfrac{1}{2}g\beta H + \tfrac{1}{4}hA + \tfrac{1}{2}g_N\beta_N H \tag{13-3c}$$

$$W_{-\frac{1}{2},\frac{1}{2}} = -\tfrac{1}{2}g\beta H - \tfrac{1}{4}hA - \tfrac{1}{2}g_N\beta_N H \tag{13-3d}$$

These first-order levels are shown in Figs. C-1 and 13-3a and b; in the latter figure, the nuclear transitions at frequencies ν_{n1} and ν_{n2} corresponding to the selection rules $\Delta M_S = 0$, $\Delta M_I = \pm 1$ are also shown. Figure 13-3c shows the levels and transitions at constant microwave frequency. In ENDOR experiments there is no attempt to observe directly the absorption of rf power at these frequencies. Rather, one observes the enhancement of the ESR transition intensity resulting from the redistributions of populations of the various states. From Eqs. (13-3a) and (13-3b)

$$\left|W_{\frac{1}{2},\frac{1}{2}} - W_{\frac{1}{2},-\frac{1}{2}}\right| = h\nu_{n1} = \left|\frac{hA}{2} - g_N\beta_N H\right|^{\dagger} \tag{13-4}$$

or

$$\nu_{n1} = \left|\frac{A}{2} - \frac{g_N\beta_N H}{h}\right| = \left|\frac{A}{2} - \nu_0\right| \tag{13-5}$$

Likewise

$$\nu_{n2} = \left|\frac{A}{2} + \frac{g_N\beta_N H}{h}\right| = \left|\frac{A}{2} + \nu_0\right| \tag{13-6}$$

---

† The reader is reminded that the sign of A may be negative. In the normal experiment in which an oscillating radio-frequency field is used, one is unable to establish the order of energy levels; hence it is appropriate to indicate absolute magnitudes ("moduli") where differences are involved. The use of a rotating radio-frequency field will allow the order of energy levels to be determined.

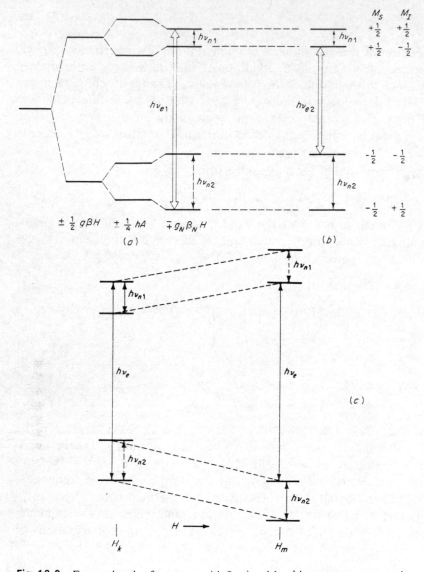

Fig. 13-3 Energy levels of a system with $S = \frac{1}{2}$ and $I = \frac{1}{2}$ in a constant magnetic field. The usual ESR transitions corresponding to the selection rules $\Delta M_S = \pm 1$, $\Delta M_I = 0$ are shown with wide arrows to symbolize the application of higher than usual microwave power. The transitions at the frequencies ν_{n1} and ν_{n2} correspond to the selection rule $\Delta M_S = 0$, $\Delta M_I = \pm 1$. The solid lines represent nuclear transitions which will give rise to ENDOR lines if there is only one cross-relaxation process represented by T_x (Sec. 13-3). The dotted transitions will also result in ENDOR lines if a second cross-relaxation process is operative. (a) Microwave saturation of the transition $M_I = +\frac{1}{2}$. ($h\nu_{e1}$). (b) Microwave saturation of the transition $M_I = -\frac{1}{2}$. ($h\nu_{e2}$). (c) Energy levels at constant microwave frequency. For the simplest assumptions about relaxation paths in steady-state ENDOR, the partially saturated transition at the field H_k will be enhanced by simultaneous irradiation with high rf power at the frequency ν_{n1}. The line at the field H_m will be enhanced if the second frequency is ν_{n2}. In some systems, precisely this behavior is observed; however, more typically, enhancement of either line will occur both at ν_{n1} and at ν_{n2}. Since one observes the enhancement as the rf field is scanned, the recorder traces out "ENDOR lines."

Here ν_0 is the magnetic resonance frequency of the nucleus in the fixed magnetic field at which the ENDOR spectrum is being observed. The two principal results derivable from the magnitudes of ν_{n1} and ν_{n2} are:

1. The determination of the hyperfine coupling A. To first order,

$$|A| = \nu_{n1} \mp \nu_{n2}\dagger$$

However, unless A is very small, one must use at least a second-order correction. The results quoted in Prob. 13-2 required fourth-order corrections to match the precision of the data. For large hyperfine couplings, one must resort to a computer solution of the spin hamiltonian.

A greatly increased precision of measurement of hyperfine couplings in inhomogeneously broadened spectra from ENDOR frequencies is possible by virtue of the usually much narrower ENDOR lines. The latter often have widths of about 10 kHz, though they have been observed to range from 3 kHz to 1 MHz or more. ESR lines in solids are considered to be narrow if their width ΔH is 1 G; if $g = 2.00$,

$$\Delta\nu = (g\beta \, \Delta H)h^{-1} \simeq 2.80 \text{ MHz}$$

If an NMR line of a nucleus with a magnetogyric ratio $\gamma = 0.63$ rad $\times 10^{+4}$ s^{-4} G^{-1} also has a width of 1 G, the corresponding frequency width is 1.13 kHz. Further, the effect of random magnetic fields upon linewidths is reduced in the ratio of the electronic to the nuclear magnetic moments. It is apparent that the smallest observed ENDOR linewidths correspond approximately to typical NMR linewidths. Hence it is not unusual to measure hyperfine couplings to about 10^{-3} percent from ENDOR frequencies.

2. The determination of an approximate value of g_N from the relation $\nu_{n1} \pm \nu_{n2} = 2\nu_0 = 2g_N\beta_N H/h.\dagger$ Even a low-precision measurement of the ENDOR frequencies permits identification of the nucleus responsible for the hyperfine splitting. In favorable cases, g_N may be determined with a precision of 0.1 percent; however, even with the use of higher-order corrections or computer solution of the spin hamiltonian, one may note a discrepancy between the calculated value of g_N and that derived from a table of nuclear moments. The discrepancy may arise from a *pseudo*nuclear Zeeman interaction: in some cases, e.g., for ions with low-lying excited states, the fractional contribution to g_N from this source is large ‡

It may occur to the reader to inquire as to why an elaborate ENDOR experimental system is used to detect transitions between nuclear spin levels instead of performing an NMR experiment at the nuclear resonance frequency. The answer is that the concentration of the nuclei present in

† The upper sign applies when $|A|/2 < \nu_0$.
‡ AABB, p. 38.

most ESR or ENDOR experiments is much too low to permit their NMR detection. The far greater ENDOR sensitivity is due to the following reasons:

1. The energy of the ESR quantum is much greater than that of the NMR quantum. Hence one may have much greater population differences for the more widely spaced levels.
2. The rate of energy absorption is far greater at the microwave frequency. (See Sec. D-2g for a discussion of sensitivity vs. frequency.)
3. The effectiveness of a nucleus in altering the intensity of an ESR line during an ENDOR experiment arises from the fact that it is acting not merely in the applied magnetic field, but in the magnetic field of the electron, which is typically of the order of 10^5 to 10^6 G at a nucleus. (See Prob. 13-4.) One may thus generate far greater population differences than would be possible if the external magnetic field governed these differences. This phenomenon is referred to as "enhancement."

13-3 RELAXATION PROCESSES IN STEADY–STATE ENDOR[†]

The usual ESR experiment involves only the one spin-lattice relaxation time T_{1e} (called T_1 in Sec. 9-2); if this is very short at 300 or 77 K, one is compelled to make ESR observations at 20 K or even at 4 K. However, even in the simplest four-level system on which ENDOR observations are to be made, there are at least three spin-lattice relaxation times which govern the distribution of population in the several levels. These not only dictate the temperature range in which ENDOR experiments may be successfully performed, but they may also dictate other experimental conditions and determine the nature of the observed spectrum. Besides T_{1e}, one is concerned with the relaxation times T_{1n} and T_x. In the absence of microwave or radio-frequency fields, the reciprocals of these times represent the rates of transition between the levels which they connect (Fig. 13-4a). T_{1n} is the nuclear spin-lattice relaxation time, i.e., that associated with the transitions $\Delta M_S = 0$, $\Delta M_I = \pm 1$. T_x is a "cross-relaxation" time associated with mutual "spin flips," i.e., processes for which $\Delta(M_S + M_I) = 0$. Usually, $T_{1e} \ll T_x \ll T_{1n}$. For most solid-state systems, one requires temperatures of the order of 4 K to do a successful ENDOR experiment. At these temperatures, one may achieve microwave saturation with modest power because T_{1e} is relatively long. A lengthened value of T_{1e} also makes it possible for the nuclear (that is, $\Delta M_I = \pm 1$) transitions to compete with the $\Delta M_S = \pm 1$ transitions. In extreme cases, e.g., phosphorus-doped silicon, T_{1e} is of the order of hours; however, more typically, it is a small frac-

---

[†] The phenomenon here referred to as steady-state ENDOR was termed "stationary" ENDOR and was first described in detail by H. Seidel, *Z. Physik*, **165**:218, 239 (1961).

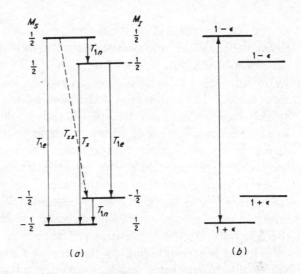

Fig. 13-4 (*a*) Relaxation paths for a system with $S = \frac{1}{2}$ and $I = \frac{1}{2}$ are indicated by arrows and are labeled by their relaxation times as follows: T_{1e}: Electron spin-lattice relaxation time; T_{1n}: Nuclear spin-lattice relaxation time; T_x: Cross-relaxation time for $\Delta(M_S + M_I) = 0$ processes; T_{xx}: Cross-relaxation time for $\Delta(M_S + M_I) = 2$ processes. (*b*) Relative state populations in the absence of a microwave magnetic field (or in the presence of a very weak microwave field).

tion of a second. Indeed, if the ENDOR linewidths are of the order of 10 kHz, corresponding to $T_2 = 10^{-5}$ s, the value of T_{1e} can be no shorter if it is not to contribute to broadening from spin-lattice relaxation. With T_{1e} and T_2 of this order of magnitude, and if T_x is not too long, one may hope to do a steady-state ENDOR experiment, i.e., to observe ENDOR lines which may be traversed at an arbitrarily slow rate and re-traversed an indefinite number of times. (By contrast, in rapid-passage types of ENDOR experiments, one observes an ENDOR signal only during a rapid traverse, and in the process, there is an equalization of populations so that a re-traverse gives no evidence of an ENDOR line.) The designation "steady state" may not be fully accurate if the relaxation time T_{1n} is very long.

It will now be profitable to consider in greater detail the steady-state ENDOR experiment which was briefly outlined in Sec. 13-1 for a system with $S = \frac{1}{2}$ and $I = \frac{1}{2}$. One commences by optimizing the intensity of an inhomogeneously broadened ESR line, after which one sets the field at the center of the line, e.g., the line at H_k in Fig. 13-3c.† The microwave power is increased somewhat beyond the value at which the intensity of a homo-

† In some types of ENDOR experiments, in which one monitors the dispersion (Sec. D-2c) instead of the absorption signal, the field is set on one side of the absorption maximum.

geneously broadened line would be a maximum. (See Fig. D-5a.) The optimum value of the microwave magnetic field H_{1e} for steady-state ENDOR observations should be such that $\gamma_e^2 H_{1e}^2 T_1 T_2 \simeq 3$.[†] Here γ_e is the magnetogyric ratio of the electron, and H_{1e} is the amplitude of the microwave magnetic field. The value of T_2 to be used here is that appropriate to the single spin packet being saturated.[‡],[§] A spin packet is a single, homogeneously broadened component of an inhomogeneously broadened line, which is the envelope of many such components. Now the power level of the radiofrequency generator—and thus H_{1n}, the amplitude of the rf magnetic field—is set sufficiently high so that the rate $\dot{n}\uparrow$ of induced upward transitions at frequency ν_{n1} is large in comparison with T_x^{-1}, i.e., $\dot{n}\uparrow T_x \gtrsim 1$. Stated another way, one requires a large value of H_{1n} because it is necessary for the $\Delta M_S = 0$, $\Delta M_I = 1$ transitions to be able to compete with the $\Delta(M_S + M_I)$ transitions corresponding to the cross-relaxation path measured by T_x. When the rf generator frequency passes through the value ν_{n1}, an ENDOR line is observed. In many four-level systems, one will also observe a second ENDOR line when the frequency ν_{n2} is traversed. If the only effective relaxation paths were those which have thus far been assumed, it would have been necessary to saturate the line at the field H_m after traversing the frequency ν_{n1} before being able to detect the ENDOR line at ν_{n2}.

We now turn to the matter of relative population of the levels under various conditions. In the absence of a magnetic field, the population of each of the four degenerate levels would be $N/4$, where N is the total number of unpaired electrons. In the presence of a magnetic field, the populations of the states will be

$$M_S = +\tfrac{1}{2}: \qquad N_{\frac{1}{2}} = \frac{N}{4} \exp\left(-\frac{g\beta H}{2kT}\right) \simeq \frac{N}{4}(1 - \epsilon) \qquad (13\text{-}7a)$$

and

$$M_S = -\tfrac{1}{2}: \qquad N_{-\frac{1}{2}} = \frac{N}{4} \exp\left(+\frac{g\beta H}{2kT}\right) \simeq \frac{N}{4}(1 + \epsilon) \qquad (13\text{-}7b)$$

Here $\epsilon = g\beta H/2kT$. [To avoid repetitious use of the factor $N/4$, we shall divide all population numbers by it; hence the relative populations in Eqs. (13-7) will be taken as $1 - \epsilon$ and $1 + \epsilon$, respectively, in the absence of a microwave field or in the presence of a very weak one. These populations are shown in Fig. 13-4b.] If it is the $M_I = \tfrac{1}{2}$ transition which is induced by the weak microwave field, the only effective relaxation path is that indi-

---

† AABB; p. 244.

‡ The value of T_2 is generally determined indirectly; saturation curves on inhomogeneously broadened lines allow $(T_1 T_2)^{\frac{1}{2}}$ to be determined. If T_1 is available from an independent measurement, then T_2 can be determined.

§ AABB, p. 264

cated by T_{1e}; the path via T_x is ineffective since the much longer T_{1n} is in series with it.

At this point, although our steady-state ENDOR experiment involves only *partial* saturation of the electron spin transition, it will simplify the discussion to assume equalization of the populations of the states $M_I = \frac{1}{2}$, as indicated in Fig. 13-5a. For complete saturation of the transition between the states $M_I = -\frac{1}{2}$, the populations will be those given in Fig. 13-5b. It is still true that for $T_{1e} << T_x$, very little cross relaxation occurs. For saturation of the $M_I = \frac{1}{2}$ transition, it is to be noted that the $|\frac{1}{2}, \frac{1}{2}\rangle$ and $|\frac{1}{2}, -\frac{1}{2}\rangle$ states differ in population by ϵ, whereas in the absence of microwave saturation they would have differed by $\epsilon_n = g_N\beta_N H/2kT$. Thus, if a short-circuiting path is provided between these two states, there can be a significant reduction in the population of the $|\frac{1}{2}, \frac{1}{2}\rangle$ state as compared with that of the $|-\frac{1}{2}, \frac{1}{2}\rangle$ state; the intense rf field at the frequency ν_{n1} will provide such a short-circuiting link. It can readily be understood that the rate of transition between the $|\frac{1}{2}, \frac{1}{2}\rangle$ and the $|\frac{1}{2}, -\frac{1}{2}\rangle$ states must at least equal T_x^{-1}. If the $M_I = -\frac{1}{2}$ transition is saturated, it is the $|-\frac{1}{2}, -\frac{1}{2}\rangle$ and the $|-\frac{1}{2}, \frac{1}{2}\rangle$ states which have the large population difference ϵ, and hence an intense rf field at frequency ν_{n2} will give rise to an ENDOR line. The equality in population of the other pair of levels—if only these relaxation processes are operative—would give no cause for expecting to see an ENDOR line at the frequency ν_{n1}. The same applies to saturation of the $M_I = \frac{1}{2}$ states, for which one should not have expected to see the ENDOR line at ν_{n2}. It is apparent that in many systems there must be at least one additional path of relaxa-

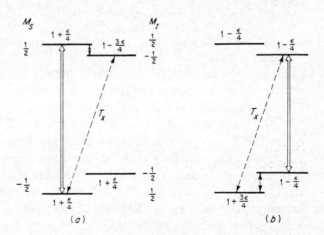

Fig. 13-5 Relative populations of levels in a system in which the ENDOR behavior is governed by the combined effects of T_{1e} and T_x. (a) Upon saturation of the low-field ESR line, only the lower-frequency ENDOR line is observed. (b) Upon saturation of the high-field ESR line, only the higher-frequency ENDOR line is observed.

tion if both ENDOR lines are to be observed on saturating either microwave transition.

We have noted repeatedly that the operators $(\hat{S}_+\hat{I}_- + \hat{S}_-\hat{I}_+)$ will admix some states $|M_S - 1, M_I + 1\rangle$ or $|M_S + 1, M_I - 1\rangle$ with a state $|M_S, M_I\rangle$. It is such mixing which makes partially allowed the $\Delta(M_S + M_I) = 0$ transitions which are associated with the relaxation path T_x. The alternative path labeled T_{xx} in Fig. 13-4a could be described as involving $\Delta(M_S + M_I) = 2$ transitions. For such transitions to be partially allowed, one requires mixing by operators of the form $(\hat{S}_+\hat{I}_+ + \hat{S}_-\hat{I}_-)$ of states $|M_S + 1, M_I + 1\rangle$ or $|M_S - 1, M_I - 1\rangle$ with the state $|M_S, M_I\rangle$. The mixing coefficient is given approximately by $(A_{XX} - A_{YY})/4h\nu_e$, where A_{XX} and A_{YY} are two principal components of the hyperfine tensor.[†] It is thus apparent that this mixing is nonzero and T_{xx} is finite only if the hyperfine interaction is not isotropic.[‡] There are now two alternative relaxation pathways (other than T_{1e}) to reach the lowest-lying $|-\frac{1}{2}, \frac{1}{2}\rangle$ state from the uppermost state $|\frac{1}{2}, \frac{1}{2}\rangle$; one involves $T_{xx} + T_{1n}$ and the other $T_{1n} + T_x$. In either case, the relaxation rate is controlled by T_{1n} (since T_{xx} is generally much smaller than T_{1n}). Application of saturating rf power at either nuclear frequency enhances the rate of the transitions $\Delta M_S = 0$, $\Delta M_I = \pm 1$ sufficiently so that the effective value of T_{1e} is reduced because of the competing relaxation path, independent of which microwave transition is saturated. *This is indeed the essential characteristic of steady-state ENDOR:*

> Having one or more cross-relaxation times and nuclear spinlattice relaxation times of favorable magnitude, the application of saturating rf power reduces the effective value of the electron spinlattice relaxation time sufficiently so that one may continuously observe ENDOR lines as long as one is within the ENDOR linewidth.

(This is to be contrasted with either "packet-shifting" or "distant" ENDOR, which constitute two other important mechanisms; these will be mentioned briefly at the end of this section.)

In many cases the intensities of pairs of ENDOR lines are similar; in others, they may be so unequal that one line is not detected. This phenomenon is particularly marked when the nuclear Zeeman and the hyperfine interactions are comparable in magnitude.[§] For systems of axial symmetry, one can calculate the differences in intensity due to differences in enhancement (see end of Sec. 13-2) of the effective rf field by the hyperfine field of the electron at the nucleus, in good agreement with observation.[§]

[†] AABB, p. 247

[‡] An alternative mechanism which may give rise to the cross-relaxation path T_{xx} and which may affect the values of the other relaxation times as well, even with isotropic hyperfine interaction, is thermal modulation of the hyperfine interaction. (See AABB, p. 248.)

[§] E. R. Davies and T. Rs. Reddy, *Phys. Letters*, **31A**:398 (1970); see also S. Geschwind, in A. J. Freeman and R. B. Frankel (eds.), "Hyperfine Interactions," p. 225, Academic Press Inc., New York, 1967, as well as AABB, p. 221.

The orientation of the rf field relative to the axis of symmetry can be extremely important in determining the intensity of a particular ENDOR line, even though the rf field is always maintained perpendicular to the static field H_0.

This discussion has thus far been limited to the response of a four-level system. In the more general case in which one has a system of spin S and nuclear spin I, the maximum possible number of ENDOR lines is $16SI$. This allows for occurrence of all of the ESR transitions which are forbidden in first order. (For the simple case $S = \frac{1}{2}$, $I = \frac{1}{2}$, saturation of the forbidden lines of Fig. 7-7 will give rise to ENDOR lines of the same frequency as the allowed lines—if indeed one could see these, since one would require a very intense rf field to compete with the very fast relaxation rate T_{1e}.) If the nucleus has spin $I \geq 1$ it is essential to add to Eqs. (13-1) or (13-2) a term to account for the quadrupole interaction. If the system has axial symmetry and the magnetic field H_0 lies along the symmetry axis, the additional term would be $Q'_\parallel [\hat{I}_z^2 - \frac{1}{3}I(I+1)]$. Even here there are so many relaxation paths that it is difficult to predict the intensity of the ENDOR lines or to specify in detail the ENDOR mechanism.

We now consider very briefly two other types of ENDOR mechanisms.[†] The first of these, "packet-shifting" ENDOR, is typified by the phosphorus-doped silicon system.[‡] Here the electron wave function of the donor extends over a large number of nuclei, some of which are ^{29}Si ($I = \frac{1}{2}$), as well, of course, as ^{31}P. Since T_{1e}, T_{1n}, and T_x are all of the order of hours for this system, one can easily "burn a hole" in one of the inhomogeneously broadened ESR lines by saturating those spin packets which experience a particular local field. Since the "hole" will recover with characteristic time T_{1e}, one may proceed in leisurely fashion with the redistribution of population of nuclear states associated with this microwave-saturated transition. The most effective means of providing a favorable redistribution of population of the nuclear levels is the "rapid-passage" technique; this allows one to invert the populations of a pair of levels by the use of an appropriate intense pulse.[§] It is expedient to observe the dispersion (rather than the absorption) signal for observation of the transient signals which arise.

A particular spin packet of the inhomogeneously broadened line (which is the envelope of all such packets) represents one particular value

---

[†] It is not to be inferred that these ENDOR mechanisms are mutually exclusive. The order of magnitude of T_{1e} for systems in which packet shifting predominates will typically be much longer than those for steady-state ENDOR; a typical value of T_{1e} for the latter case may be taken as $T_{1e} \lesssim 1$ ms.

[‡] G. Feher, *Phys. Rev.*, **114**:1219 (1959); G. Feher and E. A. Gere, *Phys. Rev.*, **114**:1245 (1959).

[§] Rapid passage represents a traverse of the resonant absorption line in a time short in comparison with T_1 and T_2, with a radio-frequency field H_{1n} large enough so that $T_2 \gg |\gamma_n H_1|^{-1}$ and $|\gamma_n H_1 T_1| \gg 1$, where γ_n is the nuclear magnetogyric ratio. See F. Bloch, *Phys. Rev.*, **70**:460 (1946).

of the local field, to which many nuclei contribute. The redistribution of populations by the rapid-passage inversion changes the local field at neighboring nuclei, which in turn may have as neighbors electrons which are not involved in the microwave saturation. The changed local field means that some spin packets are shifted to other regions of the inhomogeneously broadened line, while other packets now find themselves in just the local field which defines the region of the line which was saturated. The net changes of nuclear populations allow transient ENDOR signals to be observed both for ^{31}P and for ^{29}Si nuclear transitions. Thus the term "packet-shifting" is very appropriate for this type of ENDOR.

Another of the important ENDOR mechanisms is that of "distant" ENDOR. In the course of investigations on ruby (Cr^{3+} in Al_2O_3), ENDOR lines were observed from ^{27}Al transitions—not surprisingly—but the nuclear magnetic resonance frequency of the Al nuclei which were involved was not affected by the Cr^{3+} ions.[†] Hence these Al nuclei must have been located so far away that the dipolar interaction was negligible. It can be shown[‡] that there is marked polarization of nuclei in the vicinity of a paramagnetic ion in a magnetic field if high microwave power is applied at a frequency such that one is on the shoulder of a resonance line. "Polarization" implies preferential population of spin levels. Instead of being confined to the vicinity of the paramagnetic ion, these population differences are transmitted throughout the sample by mutual spin flips of the nuclei, at the eventual expense of the energy of the microwave field. This "spin diffusion" is thus the mode of communication of the paramagnetic ion with distant nuclei. When rf power corresponding to Al nuclear transitions is applied, the change in orientation of the distant nuclei is transmitted back to the Cr^{3+} ions; the change in their ESR signal level indicates that energy is absorbed. In consonance with this mechanism, it was noted that when the rf power was removed, the ESR signal recovered with a characteristic time of about 10 s—the nuclear spin-lattice relaxation time. For both packet shifting and for steady-state ENDOR, the recovery rate is of the order of T_{1e}, which here is about 10^{-1} s.

13-4 AN ENDOR EXAMPLE: THE F CENTER IN THE ALKALI HALIDES

Perhaps the most spectacular successes of the ENDOR method have been in its application to systems which give inhomogeneously broadened lines which are the envelope of large numbers (in some cases, literally hundreds) of overlapping hyperfine components. An example is the F center in KBr, for which the width of the gaussian ESR line is about 125 G. The six first-shell neighbors (see Fig. 8-12) are either ^{39}K (relative abundance 93.08 percent) or ^{41}K (relative abundance 6.91 percent). These nuclides also

[†] J. Lambe, N. Laurance, E. C. McIrvine, and R. W. Terhune, *Phys. Rev.*, **122**:1161 (1961).
[‡] See AABB, p. 74.

comprise the third, fifth, and ninth shells. The second, fourth, sixth, and eighth shells are composed of ^{85}Br (abundance 50.57 percent) and ^{87}Br (abundance 49.43 percent). All of these nuclei have $I = \frac{3}{2}$. Ignoring (1) the differences in their nuclear-magnetic moments, (2) anisotropic hyperfine interactions, and (3) hyperfine splittings from nuclei of shell numbers greater than 2, one may compute (see Sec. 4-7) the maximum number of hyperfine lines arising from the 6 first- and the 12 second-shell neighbors.

$$\Pi_i(2n_iI + 1) = 19 \times 37 = 703$$

To be sure, many of these lines would have very low intensity. For example, the outermost line of a set of 19 would have an intensity only $\frac{1}{580}$ that of the central line. Considering the extra lines arising from interaction with additional shells of nuclei, the nonidentity of nuclear moments, anisotropic interactions, and the effects of the nuclear quadrupole moments, it is understandable that the ESR line of the F centers in KBr gives no indication of any structure.

From Eqs. (13-3) and (13-4) one expects that the very large range of hyperfine interactions of the unpaired electron with nuclei in the various shells will insure that the ENDOR spectrum is spread over a considerable range of frequency. Looking at the ENDOR spectrum of KBr in Fig. 13-6, one sees that the frequencies at which the lines are observed varies from roughly 0.5 MHz to 26 MHz. Although there is a very great variation in line widths, the narrowest lines have a width of the order of 10 kHz. Especially in the 3- to 4-MHz region, the small width makes it possible to resolve the pairs expected from Eqs. (13-3) and (13-4), separated by $2\nu_{Br}$ for each of the bromine nuclides. Various line pairs from ^{79}Br, ^{81}Br, ^{39}K and ^{41}K are identified with brackets above the lines.

The identification of the shell numbers (indicated as a subscript to a symbol or bracketed below it) is accomplished by a study of the angular dependence of the lines. In Fig. 13-6, the field is oriented along a $\langle 100 \rangle$ axis. If the hyperfine interaction has axial symmetry, as for the first-shell nuclei, the angular dependence of a line is similar to that of Fig. 12-2c or Fig. 13-7a. For nuclei in this and other shells, one can predict the angular dependence of the dipolar hyperfine interaction. These angular dependences are given in Fig. 13-7 out to the eighth shell.† Where the angular dependence of two shells is similar, the magnitudes of the hyperfine coupling are usually very different, and the line pairs are usually assignable without ambiguity. If the hyperfine couplings are large, the second-order terms must be taken into account.

There are actually three ^{39}K lines which show the angular dependence corresponding to Fig. 13-7a. These, which also show the effects of second-

† W. C. Holton and H. Blum. *Phys. Rev.*, 125:89 (1962).

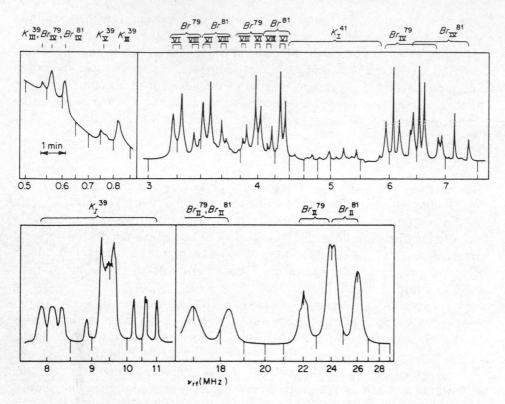

Fig. 13-6 ENDOR spectrum of the F center in KBr at 90 K for **H** ∥ ⟨100⟩. The line pairs corresponding to ^{39}K, ^{41}K, ^{79}Br, and ^{81}Br for the various shells are indicated. The triple lines illustrative of quadrupole interaction are most prominent in the 8 to 11 MHz region. [*Taken from H. Seidel, Z. Physik*, **165**:218, 239 (1961).]

order hyperfine coupling, are seen between 10 and 11 MHz. Two sets of three ENDOR lines arise from interaction of the quadrupole moment of the ^{39}K nucleus with the gradient of the electric field produced at the nucleus by the trapped electron. These lines can be understood by consideration of a spin hamiltonian, which in addition to the terms of Eq. (13-1) contains the quadrupole interaction term†

$$\hat{\mathcal{H}}_Q = h\hat{\mathbf{I}} \cdot \mathbf{Q} \cdot \hat{\mathbf{I}} \tag{13-8}$$

For systems in which the quadrupole interaction tensor $\mathbf{Q}$ has axial symmetry, this reduces to

$$\hat{\mathcal{H}}_Q = hQ'[\hat{I}_z^2 - \tfrac{1}{3}\hat{I}^2] \tag{13-9}$$

† The quadrupole interaction, measured by the quadrupole coupling tensor, is the product of the nuclear quadrupole moment and the gradient of the electric field due to all surrounding electrons. If this distribution is cubic (octahedral or tetrahedral) the electric field gradient is zero. However, if there is a concentration of charge in a bond between two atoms. e.g., a transition-metal atom and a ligand atom, the gradient is nonzero and the magnitude of the quadrupole coupling is a measure of the electron concentration in the bond. By contrast, the hyperfine coupling measures only the unpaired electron distribution.

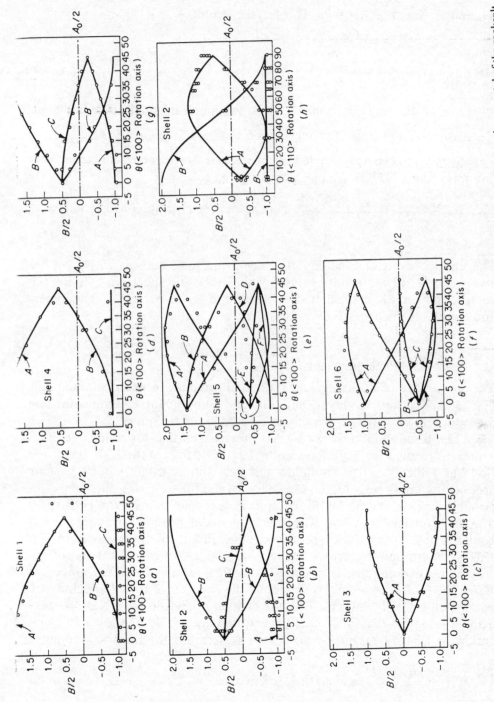

Fig. 13-7 (*a–g*) Angular dependence of $B/2$ for shells 1 to 8 of nuclei about a trapped electron of an F center in a crystal of the rocksalt structure. The rotation axis is $\langle 100 \rangle$. (*h*) $B/2$ versus θ for nuclei of shell 2, with $\langle 110 \rangle$ as the rotation axis. [*After W. C. Holton and H. Blum, Phys. Rev.*, **125**:89 (1962).] The experimental points refer to the ENDOR lines of the F center in LiF. Labels A, B, C, ⋯ refer to lines from sets of ions at equivalent positions for $\theta = 0°$.

The additive energy term $h\nu_Q$ is then (see Prob. 13-3)

$$h\nu_Q = hQ'(3\cos^2\theta - 1)(M_I - \tfrac{1}{2}) \tag{13-10}$$

Here $Q' = \dfrac{3eQ}{4I(2I-1)h}\dfrac{\partial^2 V}{\partial z^2}$

where e is the nuclear charge, Q is the nuclear quadrupole moment, $\dfrac{\partial^2 V}{\partial z^2}$ is the gradient (rate of change with distance) of the electric field $\dfrac{\partial V}{\partial z}$, and V is the potential energy. Including the nuclear Zeeman term and the hyperfine term, the ENDOR frequencies are given by

$$\nu = \frac{g_N\beta_N H'}{h} + \frac{1}{2}\left[A_{\shortparallel} + A_{\perp}(3\cos^2\theta - 1)\right] + Q'(3\cos^2\theta - 1)\left(M_I - \frac{1}{2}\right) \tag{13-11}$$

The energy levels and the ENDOR spectrum for a nucleus with $I = \tfrac{3}{2}$ are given in Fig. 13-8. From analysis of the ENDOR spectrum of KBr, the value of Q' for ^{39}K in the first shell is found to be 0.2 MHz.[†,‡] The hyperfine couplings in successive shells are given in Table 13-1.

13-5 ENDOR IN LIQUID SOLUTIONS

The possibility of detecting ENDOR of substances in liquid solution was first demonstrated by Hyde and Maki.[§] Subsequent experimental and theoretical work have shown this to be a valuable technique.[¶]

The discussion of Sec. 13-4 has emphasized that for solids one often is able to resolve far more lines in the ENDOR than in the ESR spectrum. For free radicals in liquid solution, this may still be true if inhomogeneous broadening limits resolution. However, even in these cases, the number of possible lines in the ENDOR spectrum is less than that for the ESR spectrum.[§] Consider a radical with four equivalent protons, which gives the familiar $1:4:6:4:1$ ESR spectrum. The full hamiltonian is again Eq. (13-1), where now $\hat{I}$ corresponds to $I = \Sigma I_i = 2$. In the discussion of energy levels of free radicals in Chap. 3, the final term was omitted, since just as for the hydrogen atom (Appendix C) the *energies* of ESR transitions are unaffected by its inclusion. That the number of lines in the ENDOR spectrum is less than that for the ESR spectrum may readily be shown by application of Eq. (13-2) to the four-proton system. If the hyperfine coupling

† H. Seidel, *Z. Physik*, **165**:218 (1961).
‡ W. C. Holton and H. Blum, *Phys. Rev.*, **125**:89 (1962); W. T. Doyle, *Phys. Rev.*, **126**:1421 (1962).
§ J. S. Hyde and A. H. Maki, *J. Chem. Phys.*, **40**:3117 (1964).
¶ J. S. Hyde, *J. Chem. Phys.*, **43**:1806 (1965); J. H. Freed, *J. Chem. Phys.*, **43**:2312 (1965); *J. Phys. Chem.*, **71**:38 (1967); J. H. Freed, D. S. Leniart, and J. S. Hyde, *J. Chem. Phys.*, **47**:2762 (1967); R. D. Allendoerfer and A. H. Maki, *J. Mag. Res.*, **3**:396 (1970).

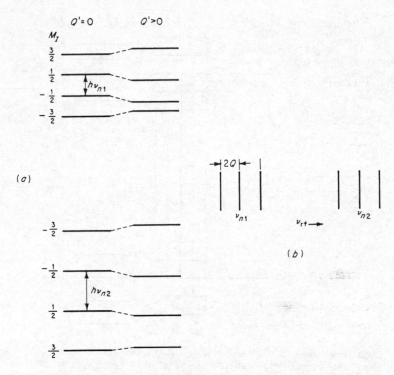

Fig. 13-8 (a) Energy-level splitting due to nuclei with $I = \frac{3}{2}$, $Q > 0$. (b) First-order ENDOR spectrum with quadrupole interaction.

constant is small enough so that second-order couplings are small, the full hamiltonian gives a set of five equally spaced levels with $M_S = +\frac{1}{2}$ (Fig. 13-9). The five levels corresponding to $M_S = -\frac{1}{2}$ have a uniform spacing greater than that of the $M_S = +\frac{1}{2}$ levels. Hence only *two* ENDOR transitions will be observed; these occur at $\nu_{\text{ENDOR}} = \nu_{\text{proton}} \pm A/2$. Here ν_{proton} is the proton NMR frequency at the *constant* magnetic field used for the ENDOR experiment. For radicals in which there are n sets of m nonequivalent pro-

Table 13-1 Hyperfine couplings in KBr at 90 K from ENDOR spectra†

Shell	Nucleus	$\dfrac{A_\parallel}{h}$	(MHz)	$\dfrac{A_\perp}{h}$
1	^{39}K	18.3		0.77
2	^{81}Br	42.8		2.8‡
3	^{39}K	0.27		0.022
4	^{81}Br	5.70		0.41
5	^{39}K	0.16		0.02‡
6	^{81}Br	0.84		0.086‡
8	^{81}Br	0.54		0.07

† H. Seidel, *Z. Physik*, **165**:218 (1961).
‡ Detectable departures from axial symmetry.

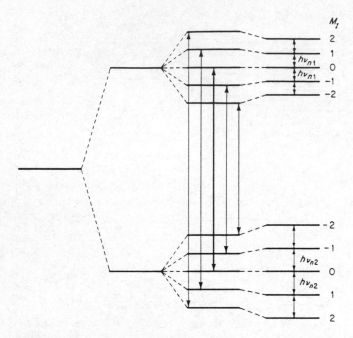

Fig. 13-9 Allowed ESR and ENDOR transitions for a system with four equivalent protons in a magnetic field. Upon saturation of any of the ESR transitions—here the central transition —one observes ENDOR transitions only at the frequencies ν_{n1} and ν_{n2}.

tons, there will be only $2n$ lines in the ENDOR spectrum, irrespective of the number of protons in any set.[†] In the ESR spectrum there will be $(m + 1)^n$ lines.

Whereas it is standard practice for a worker to build his own ENDOR system for the study of paramagnetic centers in solids, very few have undertaken to build equipment for studies of ENDOR spectra of liquids. The special difficulties in the latter case arise from the necessity of using an intense rf field, which causes heating if constantly applied. Hence a pulsed rf ENDOR system is often used. The requirement of a large rf field arises because the nuclear transitions must be saturated. T_{1e} is of the order of 10^{-5} to 10^{-6} s in a free radical at room temperatures. In this temperature region the relaxation times T_{1n} of protons are typically several orders of magnitude longer than this. The application of an intense rf field at the nuclear resonance frequency greatly reduces the lifetime of a nucleus in an excited spin state because emission is stimulated by the rf field.

[†] This is true provided that terms of the order of magnitude of $A^2/g\beta H$ can be neglected. See L. C. Kravitz and W. W. Piper, *Phys. Rev.*, **146**:322 (1966).

13-6 ENDOR IN POWDERS AND NONORIENTED SOLIDS

For many systems consisting of a paramagnetic guest in a host matrix, it is extremely difficult or impractical to grow a single crystal of size sufficient for ESR studies. We have noted earlier that for crystalline powders or glassy solid solutions one may see distinct ESR lines from those molecules which have an axis of dominant interaction at right angles to the static magnetic field. In Fig. 7-10 the predominant interaction is g-tensor anisotropy; the intense line corresponds to those molecules which have a tetragonal electric field axis perpendicular to the magnetic field. The position of this line gives $g_\perp$. Figure 7-13 represents a selection by the field of those molecules for which the field is approximately parallel to the axis of the dominant principal hyperfine component; this figure refers to a system with $S = \frac{1}{2}$ and a small anisotropy of the g tensor. Figures 10-7 to 10-10 are illustrative of triplet systems in which spin-spin interaction is dominant, with the g anisotropy small. Here the line pairs arise from molecules having principal axes of the D tensor parallel to the magnetic field. Figure 7-14 illustrates the case in which there is a marked anisotropy of both the g and the hyperfine tensors, but both tensors have the same principal axes.

These are especially favorable cases for ESR interpretation; in the most unfavorable case, all molecules would contribute equally, regardless of orientation; the detection of absorption by the sample would then be difficult. In the more typical case, molecules within a range of orientations contribute to the ESR powder line shape. If a paramagnetic molecule has several nuclides contributing a marked hyperfine anisotropy with different principal axes, the ESR powder spectrum may not be unambiguously interpretable. Provided that the signal-to-noise ratio in the ESR powder (or randomly oriented) spectrum is adequate, one may obtain ENDOR spectra and may be able to extract a number of the parameters which are obtainable from single crystals.[†,‡] These ENDOR spectra will be simpler than the ESR powder spectra and will be more readily interpretable than the latter for the following reasons:

1. The hyperfine couplings arising from each nuclide will give distinctive contributions which can be related to (or which in simple cases are centered upon) the nuclear resonance frequency ν_0 of that nuclide in the magnetic field at which the ENDOR spectrum is taken. We have noted that for nuclei with $I = \frac{1}{2}$, each hyperfine coupling gives a line pair at frequencies $\nu_0 \pm A/2$. For nuclei with $I = 1$ (for example, ^{14}N), there will be (for a small axial hyperfine interaction) a characteristic four-line pattern at frequencies $A_\perp/2 \pm \nu_0 \pm Q'$, where hQ' is the quadrupole interaction.[‡]

† W. T. Doyle, *Phys. Rev.*, **126**:1421 (1962).
‡ G. H. Rist and J. S. Hyde, *J. Chem. Phys.*, **52**:4633 (1970).

2. The ENDOR powder spectrum is the superposition of the powder spectrum of each nuclide, independent of the differences in orientations of the principal axes of the several hyperfine tensors. There is no complication of interpretation because one hyperfine splitting is equal to the sum or the difference of two others, or is a multiple of another splitting.

It seems safe to predict that there will be a marked increase in the number of ENDOR studies reported, now that it has been demonstrated that numerous nonoriented systems can profitably be studied. Up to this point, most of the single-crystal free radical studies have been carried out on compounds (such as carboxylic acids and their salts) which are easy to crystallize.

A much wider variety of paramagnetic systems including many of biological interest will doubtless be explored by the ENDOR technique. (See Sec. 14-3.)

13-7 ELECTRON–ELECTRON DOUBLE RESONANCE

In an ENDOR experiment one observes a change in intensity of a partially saturated ESR signal when one establishes a connection to an energy level belonging to a *different* hyperfine transition. A very different experiment—termed electron-electron double resonance (ELDOR)—is the observation of the reduction in the intensity of one hyperfine transition when a second hyperfine transition is simultaneously being saturated.[†] Simultaneous electron-spin resonance in one magnetic field for two different transitions re-

---

[†] J. S. Hyde, J. C. W. Chien, and J. H. Freed. *J. Chem. Phys.*, **48**:4211 (1968).

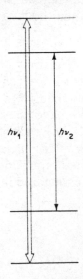

Fig. 13-10 Electron-electron double resonance experiment. The intensity of the ESR line observed in a spectrometer operating at a low microwave power level at frequency ν_2 is recorded as a function of the (high) microwave power applied to the same bimodal cavity by a separate source operating at a frequency ν_1.

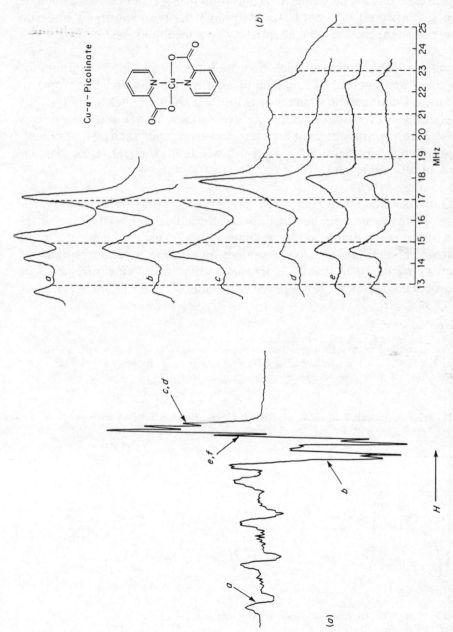

Fig. 13-11 (a) Second-derivative ESR spectrum of Cu(pic)₂ in Zn(pic)₂·4H₂O powder. The arrows indicate magnetic field values at which the ENDOR spectra in (b) are taken. (b) Powder-type nitrogen ENDOR spectra corresponding to (a). The spectrum a is a single-crystal-type spectrum, whereas the remaining spectra have contributions from molecules in many orientations. [*After G. H. Rist and J. S. Hyde, J. Chem. Phys.*, **52**:4633 (1970).]

quires irradiation at two microwave frequencies. That is, one requires a bimodal cavity tunable to two frequencies separated by a multiple of the hyperfine coupling. The simplest case in principle is that of a single nucleus of spin $\frac{1}{2}$, illustrated in Fig. 13-10. Although the two transitions have no level in common, they may be coupled by two mechanisms:

1. Rapid nuclear relaxation which may be induced by dipolar coupling of electrons and nuclei. The flipping of an electron spin under appropriate conditions can cause a simultaneous flip of a coupled nuclear spin. This mechanism is predominant at low concentrations and at low temperatures.
2. At high concentrations or at high temperatures, spin exchange or chemical exchange (see Secs. 9-5b and 9-5c) will tend to equalize the populations of all spin levels.

This technique has been used primarily to study relaxation mechanisms. It has been suggested that (as an alternative to ENDOR measurements) one might be able to discriminate between nearly identical proton and nitrogen splittings because the electron-nuclear dipolar coupling of nitrogen is greater than that of hydrogen. Very precise measurements of the coupling constants of DPPH have been made by the ELDOR technique.†

PROBLEMS

13-1. The second-derivative ESR spectrum of Cu-α-picolinate in Zn-α-picolinate powder is shown in Fig. 13-11a. The corresponding ENDOR spectrum is shown in Fig. 13-11b. Determine $A_\perp$, ν_0, and the quadrupole splitting for nitrogen.

† J. S. Hyde, R. C. Sneed, Jr., and G. H. Rist, *J. Chem. Phys.*, **51**:1404 (1969).

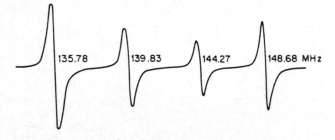

Fig. 13-12 ENDOR spectrum of Co^{++} in MgO at 4.2 K. The ENDOR frequencies given correspond to the centers of the derivative lines obtained by scanning the radio frequency while saturating the ESR transition $\Delta M_S = \pm 1$, $M_I = +\frac{1}{2}$ at a field of 1,561 G, with $\nu = 9.563$ GHz. $g = 4.280$. [*Taken from D. J. I. Fry and P. M. Llewellyn, Proc. Roy. Soc. (London)*, **A266**:84 (1962).]

13-2. The ENDOR spectrum of Co^{++} in MgO (Fig. 13-12) is observed when the transition $\Delta M_S = \pm 1$, $M_I = +\frac{1}{2}$ is partially saturated at a frequency $\nu = 9.563$ GHz, $H = 1,561$ G. From an ESR experiment, $g = 4.280$.

(a) Why are four lines observed? Assign each transition.

(b) Assuming that only second-order hyperfine corrections are necessary, estimate the hyperfine coupling A_0. (Anisotropic effects are here very small.) (The value obtained by using corrections to fourth order is 290.55 MHz.)

(c) Determine the magnetogyric ratio of ^{59}Co.

13-3. (a) Justify the number of separate curves in Figs. 13-7 a to g.

(b) Explain the angular variation for each curve for shells 1 to 4.

13-4. The magnitude of the hyperfine field H_e of the electron at the nucleus can be obtained from the expression

$$h A \hat{S}_z \hat{I}_z = -g_N \beta_N H_e \hat{I}_z$$

(a) Calculate the hyperfine field at a proton for which the hyperfine coupling is 1,420 MHz.

(b) Do the same for the $^{55}Mn^{++}$ ion ($S = \frac{1}{2}$), for which the coupling is $A/c = -9.10 \times 10^{-3}$ cm^{-1}. The magnetic moment of ^{55}Mn is 3.462 μ_N.

14
Biological Applications of Electron Spin Resonance

14-1 INTRODUCTION

The first publication on the use of ESR techniques in connection with biological systems appeared only nine years after the first demonstration of paramagnetic resonant absorption.† In that first report, a variety of biological systems, such as leaves, seeds, and tissue preparations, were shown to contain free radicals. A definite correlation was found between the concentration of the radicals and the metabolic activity of the material. This work appears to confirm earlier ideas that free radicals are involved as intermediates in metabolic processes. However, the question as to which free radicals are involved in a given metabolic process proved to be a much more difficult question to answer.

Part of the difficulty lies in the nature of the ESR spectra found for biological systems. One must realize that most biological reactions occur within an organized structure such as a membrane. Thus if radicals are produced as intermediates, they will most likely be bound or closely associated with the enzymes which catalyze the specific reaction. These immo-

† B. Commoner, J. Townsend, and G. E. Pake, *Nature*, **174**:689 (1954).

bilized free radicals give rise to ESR spectra which are similar to those which might be obtained by freezing a solution of an organic radical. Although a strong free radical signal may be seen, little or no hyperfine structure can be resolved.

The immobilization of biological paramagnetic species also presents sensitivity problems, since the "solid state" spectra generally have much broader lines than the corresponding "solution" spectra. (In Sec. D-1 it is shown that the sensitivity is inversely proportional to the inherent linewidth of the resonance signal.)

A further difficulty is presented by the fact that almost all biological materials, including *in vivo* samples, contain a high proportion of water. Because of the high dielectric loss of water, there are severe restrictions on sensitivity, sample size, and shape.

In spite of the above difficulties, impressive advances involving ESR spectroscopy have been made in certain areas. However, one must also say that the literature abounds with trivial and inconclusive reports. In this chapter we do not intend to provide a comprehensive review of this rapidly growing subject. Where they are available, we shall refer to suitable review articles or books.[†] Our intent is rather to present a few selected topics with emphasis on work in which the use of electron spin resonance has provided a major contribution to a particular biological problem.

14-2 SUBSTRATE FREE RADICALS

Ever since the work of Michaelis in 1936[‡] there has been a growing realization that many biological redox reactions proceed in distinct one-electron steps. Of the enzymes which catalyze these reactions, most are oxidative in function. One should anticipate this, since the downhill flow of energy in biological reactions proceeds from carbohydrates (reduced, energy-rich) to CO_2 (oxidized, energy-poor).

The hypothesis of one-electron steps in a redox reaction implies that free-radical intermediates are formed, though some may be too short-lived to observe. Thus it is natural that electron spin resonance should be applied to the study of these reactions. An excellent review of these applications is available.[§]

[†] M. Calvin and G. M. Androes, *Biophys. J.* **2**:Part 2, 217 (1962); M. S. Blois et al. (eds.), "Free Radicals in Biological Systems," Academic Press Inc., New York, 1961; A. Ehrenberg, B. G. Malmström and T. Vänngård (eds.), "Magnetic Resonance in Biological Systems," 2d International Conference, Stockholm, Pergamon Press, Inc., New York, 1966; G. Schoffa, "Electronspinresonanz in der Biologie," Verlag G. Braun, Karlsruhe, 1964; C. Nicolau et al., "Resonanta Paramagnetica Electronica—Applicatii in Chimie si Biologie," Editura Technica, Bucuresti, 1966; R. C. Bray, *FEBS Letters*, **5**:1 (1969); H. M. Swartz, J. R. Bolton and D. C. Borg (eds.), "Biological Applications of Electron Spin Resonance," John Wiley & Sons, Inc., New York, 1972.

[‡] L. Michaelis, *J. Biol. Chem.*, **92**:211 (1931).

[§] H. Beinert and G. Palmer, *Advan. Enzymology*, **27**:105 (1965).

Although free-radical intermediates are undoubtedly involved in *in vivo* reactions, their short lifetime and their immobilized state make detection via electron spin resonance very difficult. Thus it is not surprising that these reactions have been largely studied *in vitro* where the radical intermediates are free to tumble in solution. One of the first of these studies was the elucidation of the mechanism of reactions catalyzed by peroxidases.[†,‡,§] These enzymes catalyze the reaction

$$H_2O_2 + 2e^- + 2H^+ \rightarrow 2H_2O$$

where the two electrons must be supplied by a substrate such as ascorbic acid. By application of rapid-flow techniques,[¶] radical intermediates have been detected by electron spin resonance, and their kinetic behavior was determined. The free radicals were identified by their hyperfine structure in a steady-state ESR spectrum. (See Fig. 14-1.) In this way it was shown that each molecule of H_2O_2 produces two ascorbic acid radicals, which then decay by a second-order process.

Flow rates as high as 8 ml s^{-1} were used; these place a severe limitation on this technique, since in most cases only a limited amount of enzyme is available. Thus only enzymes which can be obtained in large quantities and which produce relatively long-lived radical intermediates can be studied by this method. Nevertheless, with the development of better mixing techniques, the use of a 35-GHz spectrometer,[††] and computer-averaging techniques, many more reactions can be studied.

† I. Yamazaki, H. S. Mason, and L. H. Piette, *Biochem. Biophys. Research Commun.*, **1**:336 (1959); *J. Biol. Chem.*, **235**:2444 (1960).
‡ I. Yamazaki and L. H. Piette, *Biochim. Biophys. Acta*, **50**:62 (1960); **77**:47 (1963).
§ L. H. Piette, G. Bulow, and I. Yamazaki, *Biochim. Biophys. Acta*, **88**:120 (1964).
¶ B. Chance, R. Eisenhardt, Q. H. Gibson and K. K. Lonberg-Holm (eds.), "Rapid Mixing and Sampling Techniques in Biochemistry," Academic Press Inc., New York, 1964.
†† D. C. Borg and J. J. Elmore, Jr., in A. Ehrenberg, B. G. Malmström, and T. Vänngård (eds.), "Magnetic Resonance in Biological Systems," p. 383, Pergamon Press, Inc., New York, 1967.

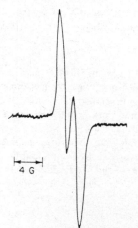

Fig. 14-1 Steady-state ESR signal obtained during continuous flow after mixing of a solution of ascorbic acid + H_2O_2 with a solution of peroxidase. (*Taken from L. H. Piette, I. Yamazaki, and H. S. Mason, in M. S. Blois et al., eds., "Free Radicals in Biological Systems," p. 195, Academic Press Inc., New York, 1961.*)

14-3 FLAVINS AND METAL–FREE FLAVOPROTEINS

Most oxidative enzymes function via one-electron redox reactions involving the production of either enzyme-bound free radicals or by a change in the valence state of a transition-metal ion.

Flavoproteins are certainly the most widely studied class of enzymes in which the free radicals produced are bound to the enzyme. These proteins all contain one or more flavin subunits which are all based on the isoalloxazine moiety

($R = CH_3$ in lumiflavin; $R = CH_2(CHOH)_3CH_2OH$ in riboflavin; and $R = CH_2(CHOH)_3CH_2OPO(OH)_2$ in flavin mononucleotide.)

The free flavin semiquinones can be generated by one-electron chemical reduction in solution; these produce spectra rich in hyperfine lines. The use of isotopic substitution techniques[†] coupled with ENDOR studies[‡] has provided a complete analysis and assignment of the hyperfine structure. Figure 14-2 shows the ESR spectra of the lumiflavin semiquinone with various isotopic substitutions. Table 14-1 gives the assigned hyperfine splittings. The spectra change drastically due to the successive protonation of the nitrogens at positions 1 and 5 as the pH is lowered.

In contrast to the rich hyperfine structure produced by flavosemiquinones in solution, the chemical reduction of metal-free flavoproteins produces a single broad line with a g factor (2.0032), the same as that of the free flavosemiquinones. There is little doubt that the signal is due to an

† L. E. G. Eriksson and A. Ehrenberg, *Acta. Chem. Scand.*, **18**:1437 (1964).

‡ L. E. G. Eriksson, J. S. Hyde, and A. Ehrenberg, *Biochim. Biophys. Acta*, **192**:211 (1969).

Table 14-1 Major isotropic hyperfine splittings for the lumiflavin semiquinone anion[†]

$a_{10}^N = 3.2$ G	$a_6^H = 3.5$ G
$a_5^N = 7.3$ G	$a_8^H(CH_3) = 4.0$ G
	$a_9^H = 0.9$ G
	$a_{10}^H(CH_3) = 3.0$ G

† [Taken from A. Ehrenberg, F. Müller, and P. Hemmerich. *European J. Biochem.*, **2**:286 (1967).]

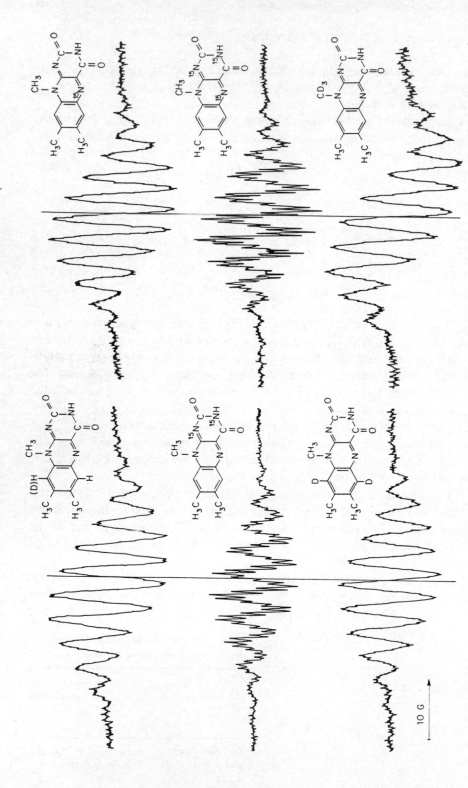

Fig. 14-2 The effect of isotopic substitution on the ESR spectra of the lumiflavin anion radical at pH 12. (*Taken from A. Ehrenberg, L. E. G. Eriksson, and F. Müller, in E. C. Slater, ed., "Flavins and Flavoproteins," p. 37, Elsevier Publishing Company, Amsterdam, 1966.*)

enzyme-bound flavosemiquinone. Thus one is forced to conclude that even a protein molecule in solution does not tumble rapidly enough to average out the anisotropic hyperfine interactions.

14-4 PHOTOSYNTHESIS

Photosynthetic systems are another example in which metabolic free-radical intermediates have been detected. Here the complexity of the system makes an identification extremely difficult. Progress in the solution of this problem provides a fascinating story in the application of electron spin resonance to biological systems.

Very early in the consideration of theories of photosynthesis it was recognized that an oxidation and a reduction must be involved. However, early theories centered around a primary act which involved a transfer of hydrogen atoms from water to carbon dioxide. In 1941 as a result of a brilliant analysis on the comparative biochemistry of the photosynthetic process in many diverse organisms, Van Niel[†] proposed that the initial photochemical act involves the production of a primary oxidant and a primary reductant. Once these are separated, the former can ultimately lead to the evolution of O_2, whereas the latter is utilized in the conversion of CO_2 to carbohydrates. Van Niel's ideas naturally led to the hypothesis that free radicals are produced as primary photoproducts. However, only with the advent of electron spin resonance was it possible to show that light-induced free radicals are indeed formed in chloroplasts,[‡] living algae,[‡] and photosynthetic bacteria.[§] However, these observations do not prove that radicals are produced as *primary* photoproducts. It is only recently that such evidence has been obtained. The following discussion represents a brief synopsis of current ideas concerning the identity and the role of free radicals found in photosynthetic systems. For further details the reader is referred to several reviews or monographs[¶],[††],[‡‡],[§§] dealing with the contributions of electron spin resonance to an understanding of the mechanisms of photosynthetic reactions.

Much of the ESR work on photosynthetic systems has been carried out with photosynthetic bacteria. These organisms may represent an earlier stage in the evolutionary development of the photosynthetic apparatus, since

[†] C. B. Van Niel, *Advan. Enzymology*, 1:263 (1941).

[‡] B. Commoner, J. J. Heise, and J. Townsend, *Proc. Natl. Acad. Sci. U.S.*, 42:710 (1956).

[§] P. B. Sogo, N. G. Pon, and M. Calvin, *Proc. Natl. Acad. Sci. U.S.*, 43:387 (1957).

[¶] E. C. Weaver, *Ann. Rev. Plant Physiol.*, 19:283 (1968).

[††] M. Calvin and G. M. Androes, in "Microalgae and Photosynthetic Bacteria," 319–335, University of Tokyo Press, Tokyo, Japan, 1963.

[‡‡] J. J. Heise and R. W. Treharne, *Develop. Appl. Spectry.*, 3:340 (1964).

[§§] A. N. Terenin and V. E. Kholmogorov, *Biochim. Biophys. Fotosintizea Akad. Nauk S. S. S. R.*, 1965:5; D. H. Kohl, in H. M. Swartz, J. R. Bolton, and D. C. Borg (eds.), "Biological Applications of Electron Spin Resonance," John Wiley & Sons, Inc., New York, 1972.

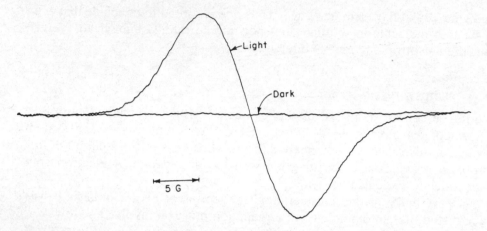

Fig. 14-3 Light-induced ESR spectrum of a preparation from the photosynthetic bacterium *Rhodopseudomonas spheroides* irradiated with red light. In the dark, no ESR signal is apparent.

they cannot liberate O_2 from H_2O. However, most can fix CO_2, and in the process they oxidize various substrates, including H_2S or molecular hydrogen.

When these bacteria are irradiated with near-infrared light (700 to 900 nm), a single-line ESR signal with $g = 2.0025$ and $\Delta H_{pp} \sim 10$ G[†] is observed. Only a weak ESR signal is seen in the dark. (See Fig. 14-3.) Considering the complexity of the system, it is surprising that any useful information can be obtained from such an unstructured spectrum! The following analysis of this problem is an excellent example of some of the approaches which are fruitful in the interpretation and assignment of ESR spectra in biological systems.

1. *Information obtained from the characteristics of the ESR spectrum.*
   The g factor (2.0025) is characteristic of an organic free radical such as might arise from a large conjugated system. The line shape most closely resembles that of a gaussian line. Thus one concludes that the line is inhomogeneously broadened. (See Sec. 9-3a.) Bacteria grown in D_2O exhibit a line width of ~ 4 G in contrast to a width of ~ 10 G for bacteria grown in H_2O. This strongly supports the idea that unresolved proton hyperfine structure accounts for a major portion of the observed line width. The oxidation of bacteriochlorophyll *in vitro* produces an ESR signal which closely resembles (both in g factor and linewidth) the *in vivo* signals.
2. *Kinetic information.* Light-induced reversible ESR signals can be seen in photosynthetic bacterial systems down to 4 K. The kinetics of forma-

---

[†] M. Calvin and G. M. Androes. *Science.* **138**:867 (1962).

tion and decay of the radicals are relatively temperature-insensitive, with half-times in the range of milliseconds. The initial rate of radical formation is directly proportional to the light intensity up to the highest light intensity used. The preceding information implies that the photoinduced free radical must result from a reaction which is either the primary photochemical step or a reaction which follows very rapidly from it.

3. *Correlations with optical measurements.* Reversible optical bleaching of a bacteriochlorophyll band centered at ~870 nm can be observed *in vivo* down to 1 K. Laser-flash studies have shown that this bleaching occurs in less than 1 μs after the onset of the flash. The quantum yield for this bleaching is between 0.9 and 1.0. This evidence strongly implicates the bacteriochlorophyll component absorbing at 870 nm as the site of the primary photochemical reaction in bacterial photosynthesis. The steady-state electron spin concentration is (within experimental error) the same as the concentration of bleached bacteriochlorophyll molecules. Further, the growth and the decay curves of the ESR signal and the optical bleaching at 870 nm are essentially superimposable. This information provides the necessary link between the ESR and the optical studies; it identifies the source of the light-induced ESR signal as a free radical derived from bacteriochlorophyll. Since the light-induced effects are reversible, and since chemical oxidants can mimic these effects, one tentatively concludes that the free radical involved is the positive ion of bacteriochlorophyll.

The light-driven reactions in green-plant photosynthesis appear to be more complex than in bacterial photosynthesis. Currently accepted hypotheses involve two photochemical systems operating in series; these are called system I and system II. System I produces a strong reductant required in the fixation of CO_2, whereas system II results in the production of O_2.

Two distinct ESR signals are seen in chloroplast suspensions extracted from green leaves or from living algae. (See Fig. 14-4.) Signal I appears rapidly in the light and decays within seconds in the dark. It has a g factor of 2.0025 and $\Delta H_{pp} \simeq 8$ G. Signal II also increases slowly in the light but requires many hours to decay in the dark. In this case $g = 2.0044$ and $\Delta H_{pp} \approx 20$G[†]; some hyperfine structure is observed.

Signal I shows a remarkable similarity to the light-induced ESR signal seen in photosynthetic bacteria. Its behavior closely parallels that of a reversible bleaching of a chlorophyll *a* component absorbing at 700 nm *in vivo*. The most likely radical giving rise to signal I is the positive ion of chlorophyll *a*, since chemical oxidation of chlorophyll *a in vitro* results in an ESR signal and optical-bleaching characteristics which are very similar to the *in vivo* light-induced responses.

[†] E. C. Weaver, *Ann. Rev. Plant Physiol.*, **19**:283 (1968).

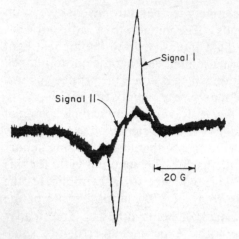

Fig. 14-4 ESR signals from living *Chlamydomonas reinhardi* algal cells. Signal I is present only when the cells are illuminated with red light. The line labeled signal II was recorded in the dark. [*From E. C. Weaver, Arch. Biochem. Biophys.*, **99**:193 (1962).]

The identity of the radical giving rise to signal II is less certain. Its g factor and linewidth are consistent with either a semiquinone-type radical or a phenoxy-type radical. It has been shown that frozen solutions of the plastochromanoxyl free radical give a signal very similar to that of signal II.

There is good evidence that signal I is associated with photosystem I and signal II with photosystem II. These findings come from studies of mutant algal strains in which one or the other photosystem is inoperative.

The detailed mechanism for the light-driven reactions of photosynthesis is by no means firmly established. However, one can say that electron spin resonance has played a significant role and will undoubtedly continue to do so.

14-5 HEME PROTEINS

The porphyrin ring system,

where M is usually a dipositive metal ion, is widely distributed in biological systems. One example has already been cited, i.e., the chlorophyll molecule in which M is Mg. The heme proteins are a very important group which have one or more porphyrin rings in which M is Fe.

One of the earliest applications of electron spin resonance to biological systems was the brilliant analysis of the ESR spectra of single crystals of ferrihemoglobin and metmyoglobin.[†] These proteins contain iron in the ferric state. Four of the six coordination positions are occupied by the nitrogen atoms of the porphyrin ring. A fifth is occupied by the nitrogen atom in a histidine ring which serves to attach the heme group to the globin part of the protein. The sixth coordination position may be occupied by a variety of ligands, for example, O_2, CO, H_2O, N_3^-, CN^-, or F^-.

The crystal field in which the high-spin Fe^{3+} atom is located has a very strong axial component; hence D is very large (often 20 to 30 cm^{-1}). The system should thus behave as one with $S' = \frac{1}{2}$, $g_\parallel = 2$, and $g_\perp = 6$. This case was discussed in Sec. 11-8. The extreme anisotropy of the g tensor provides a very sensitive method of determining the orientation of the heme planes relative to the crystal axes. This determination is made measuring the g factor as a function of angle in three crystal planes. Then using the techniques outlined in Sec. 7-4, one can obtain the principal g factors and the direction cosines for each heme plane. Figure 14-5 illustrates the orientation of the heme planes for a type A myoglobin crystal.

ESR spectra of heme proteins can also be detected in randomly oriented systems.[‡] Figure 14-6a shows the spectrum of the acid form of ferrimyoglobin taken at 77 K. This is just the pattern expected for an axial g tensor. (See Fig. 7-10.) In this case, the sixth coordination position is occupied by H_2O. However, if the pH of the aqueous myoglobin solution is

† J. E. Bennett, J. F. Gibson and D. J. E. Ingram. *Proc. Roy. Soc.* (*London*), **A240**:67 (1957).
‡ A. Ehrenberg, *Arkiv Kemi*, **19**:119 (1962).

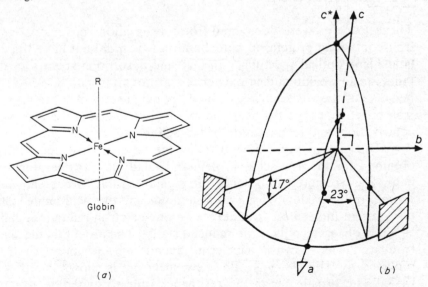

Fig. 14-5 (*a*) Structure of the "heme" portion of the hemoglobin molecule. (*b*) Orientation of heme normals and planes with respect to the crystallographic axes for type A myoglobin. (*Figure kindly supplied by Prof. D. J. E. Ingram.*)

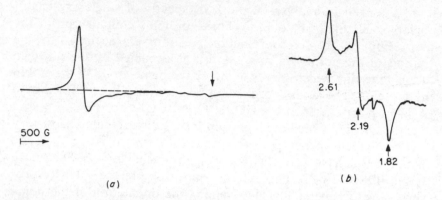

500 G

(a)   (b)

Fig. 14-6 ESR spectra of frozen aqueous solutions of ferrimyoglobin. The strong
line corresponds to $g_\perp = 6$ and the weak one (arrow) to $g_\parallel = 2$. (a) Acid form
(pH 6.0) at 167 K. (b) Basic form (pH 11.0) at 77 K. [*Spectra taken from A.
Ehrenberg, Arkiv Kemi,* **19:**119 (1962).]

~ 10.5 and it is frozen, the spectrum in Fig. 14-6b is obtained. It is clear
that the anisotropy of the g tensor is much smaller, and the crystal field is no
longer axial. OH$^-$ is a much more strongly bound ligand than H$_2$O, and
apparently this is sufficient to convert the high-spin Fe^{3+} to the low-spin
case. Studies of the magnetic susceptibility have confirmed this analysis.

14-6 IRON–SULFUR PROTEINS

These proteins were discovered from observations of residual amounts of
iron-containing proteins in mitochondria which did not have the characteristic heme optical spectrum. In fact, the absorption spectra of these proteins are so weak that their existence was not known until 1953.[†] Probably
because of the weak, poorly defined optical spectra of these proteins, electron spin resonance has been the major contributor in the discovery, char-
acterization, and identification of the active site.

Curve a in Fig. 14-7 illustrates the type of ESR spectrum which is
obtained for the iron-sulfur proteins. Usually low temperatures (77 K or
below) are required to observe the signal. Analysis has shown that this
protein (putidaredoxin) has two iron atoms and two acid-labile sulfur atoms
per protein molecule. It undergoes a one-electron reduction, and an ESR
signal is observed only in the reduced state. The proof that the signal arises
from iron was shown in an elegant experiment in which ^{56}Fe ($I = 0$) was
replaced by ^{57}Fe ($I = \frac{1}{2}$).[‡] The new spectrum is shown in curve b of Fig.
14-7. The broadening of the resonance (due to hyperfine structure from

[†] H. R. Mahler and D. G. Elowe, *J. Am. Chem. Soc.,* **75:**5769 (1953).
[‡] J. C. M. Tsibris, R. L. Tsai, I. C. Gunsalus, W. H. Orme-Johnson, R. E. Hansen, and H.
Beinert, *Proc. Natl. Acad. Sci. U.S.,* **59:**959 (1968).

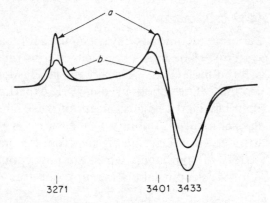

Fig. 14-7 ESR spectra of reduced puti-daredoxin (an iron-sulfur protein). (*a*) Spectrum of a protein containing ^{56}Fe. (*b*) Spectrum of a protein containing 91% ^{57}Fe ($I = \frac{1}{2}$). [*Taken from J. C. M. Tsibris, R. L. Tsai, I. C. Gunsalus, W. H. Orme-Johnson, R. E. Hansen, and H. Beinert, Proc. Natl. Acad. Sci. U.S., 59:959 (1968).*]

^{57}Fe) proves that Fe is involved in the paramagnetic center. Furthermore, the resolution of a 1 : 2 : 1 triplet pattern on the low-field peak proves that *two* iron atoms are in the paramagnetic center.

In a similar experiment in which ^{33}S ($I = \frac{3}{2}$) was substituted for ^{32}S ($I = 0$) a broadening of the resonance was observed.[†] This shows that sulfur atoms are also involved in the paramagnetic center. No resolved structure was observed; hence, the number of sulfur atoms could not be determined. However, it was possible to replace the acid-labile sulfur with selenium and still retain the activity of the protein. When the normal ^{80}Se ($I = 0$) was replaced by ^{77}Se ($I = \frac{1}{2}$), a 1 : 2 : 1 triplet structure was observed on the low-field peak, thus demonstrating the presence of two selenium atoms.[‡] The following tentative structure, which accounts for the ESR data, is attractive.

$$\begin{array}{ccc} & \diagdown \ \ S \ \ \diagup & \\ & \diagup \diagdown \diagdown \diagup & \\ Fe & & Fe \\ & \diagup \diagup \diagdown \diagdown & \\ & \diagup \ \ S \ \ \diagdown & \end{array}$$

Here each iron is tetrahedrally coordinated. For further accounts of this exciting study the reader is referred to the review literature.[§,¶,††,‡‡,§§]

† D. V. DerVartanian, W. H. Orme-Johnson, R. E. Hansen, H. Beinert, R. L. Tsai, J. C. M. Tsibris, R. C. Bartholomaus, and I. C. Gunsalus. *Biochem. Biophys. Res. Commun.*, **26**:569 (1967).

‡ W. H. Orme-Johnson, R. E. Hansen, H. Beinert, J. C. M. Tsibris, R. C. Bartholomaus, and I. C. Gunsalus, *Proc. Natl. Acad. Sci. U.S.*, **60**:368 (1968).

§ A. Ehrenberg, B. G. Malmström and T. Vänngård (eds.), "Magnetic Resonance in Biological Systems," Proc. 2d Intern. Conf. Stockholm, Pergamon Press, Inc., New York. 1966.

¶ H. Beinert and G. Palmer, *Advan. Enzymology.* **27**:105 (1965).

†† R. Malkin and J. C. Rabinowitz, *Ann. Rev. Biochem.*, **36**:113 (1967).

‡‡ A. San Pietro (ed.). "Non-heme Iron Proteins: Role in Energy Conversion," Antioch Press, Yellow Springs, Ohio, 1965.

§§ H. Beinert, in H. M. Swartz, J. R. Bolton and D. C. Borg (eds.), "Biological Applications of Electron Spin Resonance," John Wiley & Sons, Inc., New York, 1972.

14-7 SPIN LABELS

Labeling techniques have been used for many years in the study of biological systems. The labels have ranged from radioactive isotopes to fluorescent or colored molecules. Usually the label is located at or near an active site in a protein. The labels used in ESR experiments are free-radical substituents (spin labels). The advantages of these labels are (1) a sensitivity to the local environment, (2) ability to measure very rapid molecular motion, (3) the absence of interfering signals from the diamagnetic environment, and (4) a variety of spin labels which are now commercially available.†

Most spin labels are nitroxide free radicals with the general formula

$$
\underset{H_3C}{\overset{H_3C}{>}}\underset{\underset{O}{|}}{C}-\overset{R_1}{\underset{\cdot}{N}}-C\underset{CH_3}{\overset{R_2 \quad CH_3}{<}}
$$

Here R_1 and R_2 are side groups which give the nitroxide a specificity for reaction with a certain group or amino acid. Thus the choice of spin label will be dictated by the type of site which is to be labeled.

Since the first spin labeling study in 1965,‡ a great deal of interest and activity has been generated in this subject. The reader is referred to reviews for further details.§,¶,††

The type of ESR spectrum obtained for the label depends on two factors: (1) the relative ease with which the nitroxide end of the molecule can tumble or reorient and (2) the degree to which the environment is hydrophobic or hydrophilic.

We have already seen in Fig. 9-18 that the spectrum of a nitroxide changes as the reorientation rate decreases. In fact it is now possible to obtain a rough estimate of the reorientation rate from the characteristics of the nitroxide spectrum. The motion of a spin label bound at or near an active site of an enzyme is restricted: the ESR spectrum reflects the reduced rate of reorientation (see Sec. 9-7). Nanogram quantities of an opiate (or of its metabolites) in one drop of urine displace the free↔bound spin-label equilibrium sufficiently to permit ESR detection (Varian Associates).

Both the nitrogen hyperfine splitting and the g factor are solvent-dependent. The hyperfine splitting is generally 1 or 2 G smaller in hydrocarbon solvents than in polar solvents, and the g factor is about 0.0005 higher in the latter. Thus a measurement of these parameters may indicate the environment of the spin label in the biological system.

---

† Synvar Associates, 3221 Porter Dr., Palo Alto. Calif. 94303.

‡ T. J. Stone, T. Buckman, P. L. Nordio, and H. M. McConnell, *Proc. Natl. Acad. Sci. U.S.*, **54**:1010 (1965).

§ C. L. Hamilton and H. M. McConnell, in A. Rich and N. Davidson (eds.), "Structural Chemistry and Molecular Biology," p. 115, W. H. Freeman and Company, San Francisco, 1968.

¶ O. H. Griffith and A. S. Waggoner, *Accounts Chem. Res.*, **2**:17 (1969).

†† I. C. P. Smith, in H. M. Swartz, J. R. Bolton, and D. C. Borg (eds.), "Biological Applications of Electron Spin Resonance," John Wiley & Sons. Inc., New York, 1972.

Appendix A
Mathematical Operations

This appendix presents a number of mathematical techniques and equations for the convenience of the reader. Although we have attempted to summarize accurately some of the most useful relations, we make no attempt at rigor. A bibliography is included at the end of this appendix.

A-1 COMPLEX NUMBERS

A complex quantity may be represented as follows:

$$u = x + iy = re^{+i\phi} \tag{A-1}$$

where $i^2 = -1$, x, y, r, and ϕ are real numbers and $e^{i\phi} = \cos\phi + i\sin\phi$. One refers to x and y as the real and the imaginary components, respectively of u, whereas r is the absolute magnitude of u, that is, $r = |u|$. ϕ is called the phase angle. The complex conjugate of u, viz., u^*, is obtained by changing the sign of i wherever it appears; that is, $u^* = x - iy$. The relation between complex numbers and their conjugates is clarified by representing them as points in the "complex plane" (Argand diagram, see Fig. A-1). The abscissa is chosen to represent the real axis (x), and the ordinate the imaginary axis

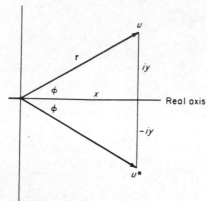

Fig. A-1   Representation of a point u and of its complex conjugate u^* in complex space (Argand diagram).

(y).   Note that the real component of u is equal to one-half the sum of u and u^*; the product of u and its complex conjugate is the square of the absolute magnitude, that is,

$$u^*u = re^{-i\phi}re^{+i\phi} = r^2 \tag{A-2}$$

A-2   OPERATOR ALGEBRA

A-2a.   Properties of operators   An operator $\hat{A}$ is a symbolic instruction to carry out a stipulated mathematical operation upon some function which is called an operand.   Unless its form is explicitly indicated, an operator will be designated by a circumflex.   One of the simplest operators is a constant multiplier; for example, $\hat{k}\alpha = k\alpha$.

An operator $\hat{\Omega}$ is said to be linear if the result of operation upon a sum of functions is the same as that obtained by operating on each function separately; i.e., if $\hat{\Omega}\alpha = \beta$, then

$$\hat{\Omega}(\alpha_1 + \alpha_2) = \hat{\Omega}\alpha_1 + \hat{\Omega}\alpha_2 = \beta_1 + \beta_2 \tag{A-3}$$

Also, if c is a constant,

$$\hat{\Omega}(c\alpha) = c\hat{\Omega}\alpha = c\beta \tag{A-4}$$

If $\alpha_i = f(q_i)$, then $\partial/\partial q_i$ is a linear operator.   An example of a nonlinear operator is "$\sqrt{\ }$".

The reader will be familiar with such operators as the summation operator Σ

$$\sum_{i=1}^{n} a_i \equiv a_1 + a_2 + a_3 + \cdots + a_n \tag{A-5}$$

Its use permits a concise representation of a series.   Frequently one wishes to summarize a set of equations with constant coefficients such as

$$\psi_1 = c_{11}\phi_1 + c_{12}\phi_2 + c_{13}\phi_3 + \cdots + c_{1n}\phi_n$$

$$\psi_2 = c_{21}\phi_1 + c_{22}\phi_2 + c_{23}\phi_3 + \cdots + c_{2n}\phi_n$$

$$\psi_3 = c_{31}\phi_1 + c_{32}\phi_2 + c_{33}\phi_3 + \cdots + c_{3n}\phi_n \tag{A-6}$$

The sum of the ψ_j can be represented by a double summation, viz.,

$$\sum_j \psi_j = \sum_j \sum_k c_{jk}\phi_k \tag{A-7}$$

Here one encounters a juxtaposition of two operators, which in the general case are represented as $\hat{A}\hat{B}$. It is understood that $\hat{A}\hat{B}$ implies operation first with $\hat{B}$ and then with $\hat{A}$. Interchange of order of the operators *may* give a different result; for example,

$$\hat{x}\frac{d}{dx}(x^2) = 2x^2 \qquad \text{but} \qquad \frac{d}{dx}\hat{x}(x^2) = 3x^2$$

If $\hat{A}\hat{B} = \hat{B}\hat{A}$, then $\hat{A}$ and $\hat{B}$ are said to be commuting operators. The difference $\hat{A}\hat{B} - \hat{B}\hat{A}$ is called the *commutator* of $\hat{A}$ and $\hat{B}$. It is represented by the symbol $[\hat{A},\hat{B}] \equiv (\hat{A}\hat{B} - \hat{B}\hat{A})$. The magnitude of the commutator of two operators is of profound significance in quantum-mechanical systems. The commutators of angular-momentum operators will be treated in Appendix B.

An operator $\hat{\Omega}$ is said to be hermitian if it obeys the following relation:

$$\int \psi_j^* \hat{\Omega}\psi_i \, d\tau = \int (\hat{\Omega}^*\psi_j^*)\psi_i \, d\tau \tag{A-8}$$

A useful aspect of the hermitian property is that (with care!) one may operate "backwards," i.e., to the left, when the operator occurs between two operands, as in Eq. (A-8). An example of operation to the left is found in Sec. B-4. Hermitian operators have the important property that if the result

Table A-1 Classical and quantum-mechanical dynamical variables

Dynamical variable	Classical quantity	Quantum-mechanical operator
Position	q	q
Time	t	t
Linear momentum	$p_q = m\dfrac{dq}{dt}$	$\hat{p}_q = -i\hbar\dfrac{d}{dq}$
Angular momentum	$\mathbf{p}_\phi = \mathbf{r} \times \mathbf{p}\ddagger$	$\hat{\mathbf{p}}_\phi = \mathbf{r} \times \hat{\mathbf{p}}$
	$p_{\phi_z} = (xp_y - yp_x)$	$\hat{p}_{\phi_z} = -i\hbar\left(x\dfrac{\partial}{\partial y} - y\dfrac{\partial}{\partial x}\right) = -i\hbar\dfrac{\partial}{\partial\phi}\,\S$
Kinetic energy associated with coordinate q	$T = \dfrac{p_q^2}{2m}$	$\hat{\mathscr{H}} = \dfrac{\hat{p}_q^2}{2m} = \dfrac{-\hbar^2}{2m}\dfrac{\partial^2}{\partial q^2}\,\P$
Potential energy	$V(q)$	$V(q)$

‡ The evaluation of vector products is described in Sec. A-4.

§ The angle ϕ measures rotation about the z axis.

¶ This form of the kinetic-energy hamiltonian is valid only for cartesian coordinates.

of operating upon a function is the function itself multiplied by a constant, one is assured that the constant is *real*. (See Sec. A-2*b*.)

Some of the most important operators of quantum mechanics are those associated with observable properties of a physical system, i.e., the "dynamical variables." A few important linear operators are listed in Table A-1. Some of these operators are identical with the variable itself, whereas others involve derivatives.

A-2b. Eigenvalues and eigenfunctions
If the result of the application of an operator $\hat{\Lambda}$ to a function ψ_n is

$$\hat{\Lambda}\psi_n = \lambda_n \psi_n \tag{A-9}$$

where λ_n is a constant, then ψ_n is said to be an *eigenfunction* of $\hat{\Lambda}$ with *eigenvalue* λ_n. (The set of functions ψ_n is often called a "basis set.") The spin functions ψ_α and ψ_β introduced in Sec. 1-5 are examples of eigenfunctions, in this case of the operator $\hat{S}_z$: that is,

$$\hat{S}_z \psi_\alpha = +\tfrac{1}{2}\psi_\alpha \tag{A-10a}$$

$$\hat{S}_z \psi_\beta = -\tfrac{1}{2}\psi_\beta \tag{A-10b}$$

(Angular momentum operator expressions are considered in detail in Sec. B-4.)

A given set of eigenfunctions ψ_n may simultaneously be eigenfunctions of several operators. Operators having the same set of eigenfunctions have the very useful property that the operators must commute. In the case of the particle in a ring considered in Sec. 1-3, the wave functions ψ are eigenfunctions both of the angular-momentum operator $\hat{p}_\phi$ and of the hamiltonian operator $\hat{\mathscr{H}}$. The eigenvalue equations are

$$\hat{p}_\phi \psi = p_\phi \psi \tag{A-11}$$

and

$$\hat{\mathscr{H}}\psi = W\psi \qquad \text{Schrödinger equation} \tag{A-12}$$

Table A-1 gives $\hat{p}_\phi = -i\hbar\, d/d\phi$, where ϕ measures the angular position of the particle. The kinetic energy of a classical particle having an angular momentum p_ϕ and moment of inertia I is

$$W = \frac{p_\phi{}^2}{2I} \tag{A-13}$$

The quantum-mechanical operator for a system with $V = 0$ is

$$\hat{\mathscr{H}} = \frac{\hat{p}_\phi{}^2}{2I} = \frac{(-i\hbar)^2}{2I}\frac{d^2}{d\phi^2} = \frac{-\hbar^2}{2I}\frac{d^2}{d\phi^2} \tag{A-14}$$

Substitution of $\hat{\mathscr{H}}$ from Eq. (A-14) into (A-12) gives

$$\frac{-\hbar^2}{2I}\frac{d^2\psi}{d\phi^2} = W\psi \tag{A-15}$$

Rearranging,

$$\frac{d^2\psi}{d\phi^2} = \frac{-2IW\psi}{\hbar^2} = -M^2\psi \tag{A-16}$$

Here the constant $2IW/\hbar^2$ has been set equal to M^2. Two solutions of (A-16) are

$$\psi_1 = Ae^{+iM\phi} \tag{A-17a}$$

and

$$\psi_2 = Ae^{-iM\phi} \tag{A-17b}$$

as is evident by substitution. From the requirement that the functions ψ be normalized, i.e., that

$$\int_0^{2\pi} \psi^*\psi \, d\phi = 1 \tag{A-18}$$

one finds that $A = (2\pi)^{-\frac{1}{2}}$. Hence

$$\psi_1 = (2\pi)^{-\frac{1}{2}}e^{iM\phi} \tag{A-19a}$$

and

$$\psi_2 = (2\pi)^{-\frac{1}{2}}e^{-iM\phi} \tag{A-19b}$$

Insertion of ψ_1 into Eq. (A-12) gives

$$\frac{-\hbar^2}{2I}\frac{d^2}{d\phi^2}\left(\frac{1}{\sqrt{2\pi}}e^{iM\phi}\right) = \frac{M^2\hbar^2}{2I}\left(\frac{1}{\sqrt{2\pi}}e^{iM\phi}\right) \tag{A-20}$$

Hence, the eigenvalue W of the operator $\hat{\mathscr{H}}$, corresponding to the eigenfunction ψ_1, is $M^2\hbar^2/2I$. Use of ψ_2 gives an identical energy value.

Operation by $\hat{p}_\phi$ on ψ_1 and ψ_2 leads to the following equations:

$$-i\hbar \frac{d}{d\phi}\left(\frac{1}{\sqrt{2\pi}}e^{iM\phi}\right) = M\hbar\left(\frac{1}{\sqrt{2\pi}}e^{iM\phi}\right) \tag{A-21a}$$

$$-i\hbar \frac{d}{d\phi}\left(\frac{1}{\sqrt{2\pi}}e^{-iM\phi}\right) = -M\hbar\left(\frac{1}{\sqrt{2\pi}}e^{-iM\phi}\right) \tag{A-21b}$$

Hence, the eigenvalues of $\hat{p}_\phi$, corresponding to the eigenfunctions ψ_1 and ψ_2, are $+M\hbar$ and $-M\hbar$, respectively.

The wave functions can be eliminated from Eq. (A-20) or (A-21) by multiplication on the left by the corresponding complex conjugate function ψ^* followed by integration. This yields expressions for the energy and for the angular momentum of a particle moving in a circle. [See Eq. (1-21) and Prob. A-3.]

A-3 DETERMINANTS

A determinant is a scalar quantity which represents a linear combination of products of terms. It may be represented by a square array, for example

$$|\mathbf{A}_2| = \begin{vmatrix} a_{11} & a_{12} \\ a_{21} & a_{22} \end{vmatrix} = a_{11}a_{22} - a_{21}a_{12} \tag{A-22a}$$

Determinants are denoted in this book by boldface type enclosed by vertical lines. More generally a determinant of order k is represented as

$$|\mathbf{A}_k| = \begin{vmatrix} a_{11} & a_{12} & \cdots & a_{1k} \\ a_{21} & a_{22} & \cdots & \cdot \\ \cdot & & & \cdot \\ \cdot & \cdots & \cdots & \cdot \\ \cdot & & & \cdot \\ a_{k1} & \cdots & \cdots & a_{kk} \end{vmatrix} \tag{A-22b}$$

A determinant may be expanded by the "method of minors." The minor of any element a_{ij} is the determinant remaining after the row and column containing the element a_{ij} are removed. The expansion is carried out by multiplying the elements of a specific row or column by their corresponding minors as follows:

$$|\mathbf{A}_k| = \sum_{i\,or\,j} (-1)^{(i+j)}\, a_{ij}\, |\mathbf{A}_{k-1}|_{ij} \tag{A-23}$$

Here $|\mathbf{A}_{k-1}|_{ij}$ is the minor corresponding to the element a_{ij}.

For a determinant of order 3 this expansion may be carried out as follows:

$$\begin{vmatrix} a_{11} & a_{12} & a_{13} \\ a_{21} & a_{22} & a_{23} \\ a_{31} & a_{32} & a_{33} \end{vmatrix} = a_{11} \begin{vmatrix} a_{22} & a_{23} \\ a_{32} & a_{33} \end{vmatrix} - a_{12} \begin{vmatrix} a_{21} & a_{23} \\ a_{31} & a_{33} \end{vmatrix} + a_{13} \begin{vmatrix} a_{21} & a_{22} \\ a_{31} & a_{32} \end{vmatrix}$$

$$= a_{11}a_{22}a_{33} - a_{11}a_{23}a_{32} - a_{12}a_{21}a_{33}$$
$$+ a_{12}a_{23}a_{31} + a_{13}a_{21}a_{32} - a_{13}a_{22}a_{31} \tag{A-24a}$$

Here the elements of the first row have been used. One could equally well have used the elements of any other row or column. The method of minors is a valuable technique for stepwise reduction of the order of a determinant. For example, a determinant of order 4 can be reduced in one step to a linear combination of four determinants of order 3.

The value of a determinant is not affected by the addition (or subtraction) of the elements of one row to those of another. This is also true for columns. Thus if any row or column is a multiple of another, the value of the determinant is zero.

A special procedure applicable only to determinants of order 3 involves a diagonal multiplication in the following manner. The sign of a term is positive if one proceeds diagonally downward, and it is negative if one proceeds diagonally upward.

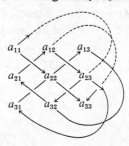

$$|A_3| = a_{11}a_{22}a_{33} + a_{12}a_{23}a_{31} + a_{13}a_{32}a_{21}$$
$$- a_{11}a_{23}a_{32} - a_{21}a_{12}a_{33} - a_{31}a_{22}a_{13} \quad (A\text{-}24b)$$

Determinants are most frequently used for obtaining solutions to sets of simultaneous equations. Consider the following set of simultaneous equations relating the dependent variables y_1, y_2, y_3 to the independent variables x_1, x_2, x_3:

$$y_1 = c_{11}x_1 + c_{12}x_2 + c_{13}x_3$$
$$y_2 = c_{21}x_1 + c_{22}x_2 + c_{23}x_3$$
$$y_3 = c_{31}x_1 + c_{32}x_2 + c_{33}x_3 \quad (A\text{-}25)$$

The solutions may be represented as follows:

$$x_1 = \frac{|\Delta_1|}{|\Delta|} \qquad x_2 = \frac{|\Delta_2|}{|\Delta|} \qquad x_3 = \frac{|\Delta_3|}{|\Delta|} \quad (A\text{-}26)$$

Here

$$|\Delta| = \begin{vmatrix} c_{11} & c_{12} & c_{13} \\ c_{21} & c_{22} & c_{23} \\ c_{31} & c_{32} & c_{33} \end{vmatrix} \quad (A\text{-}27a)$$

and

$$|\Delta_1| = \begin{vmatrix} y_1 & c_{12} & c_{13} \\ y_2 & c_{22} & c_{23} \\ y_3 & c_{32} & c_{33} \end{vmatrix} \quad (A\text{-}27b)$$

$|\Delta_2|$ or $|\Delta_3|$ is obtained in an analogous fashion by replacing the column 2 or column 3 of $|\Delta|$ by

$$\begin{bmatrix} y_1 \\ y_2 \\ y_3 \end{bmatrix}$$

If the simultaneous equations are not independent, the value of $|\Delta|$ will be zero. One of the important applications of determinants is the solution of secular equations (see Sec. 5-2) for the energies of a quantum-mechanical system.

Determinants are often used to represent antisymmetrized wave functions because interchange of two electrons corresponds to interchange of two rows of the determinant. This changes the sign of the wave function as required by the Pauli principle. For example, a two-electron wave function is written

$$\Psi = \frac{1}{\sqrt{2!}} \begin{vmatrix} \psi(1)\alpha(1) & \psi(1)\beta(1) \\ \psi(2)\alpha(2) & \psi(2)\beta(2) \end{vmatrix}$$

$$= \psi(1)\psi(2) \frac{1}{\sqrt{2}} [\alpha(1)\beta(2) - \alpha(2)\beta(1)] \tag{A-28a}$$

Equation (A-28a) is usually written in an abbreviated form

$$\Psi = \frac{1}{\sqrt{2!}} \| \psi(1)\alpha(1)\psi(2)\beta(2) \| \tag{A-28b}$$

or

$$\Psi = \frac{1}{\sqrt{2!}} \| \psi \, \bar{\psi} \| \tag{A-28c}$$

where the bar indicates β spin.

A-4 VECTORS: SCALAR, VECTOR, AND OUTER PRODUCTS

Vectors (as distinct from scalars) are quantities for which a direction is associated with a magnitude. One may add similar vectors by drawing arrows, tail-to-head, proportional to their magnitudes; the resultant is a vector drawn from the origin to the head of the last vector. It is usually more expedient to express vector quantities analytically in terms of their components. Vectors are indicated in this book by boldface type, e.g., **H**.

If the coordinates of a point with respect to fixed axes x, y, and z are 7, -3, and 4.5, the vector locating the point is written as

$$\mathbf{r} = 7\mathbf{i} - 3\mathbf{j} + 4.5\mathbf{k} \tag{A-29a}$$

Here **i**, **j**, and **k** are *unit* vectors in the x, y, and z directions, respectively. Addition of such vectors implies summation, respectively, of their x, y, and z components. Suppose a second vector is given by

$$\mathbf{s} = 3\mathbf{i} + 4\mathbf{j} - 6\mathbf{k} \tag{A-29b}$$

then the sum and difference are

$$\mathbf{r} + \mathbf{s} = (7 + 3)\mathbf{i} + (-3 + 4)\mathbf{j} + (4.5 - 6)\mathbf{k}$$

$$= 10\mathbf{i} + \mathbf{j} - 1.5\mathbf{k} \tag{A-29c}$$

$$\mathbf{r} - \mathbf{s} = 4\mathbf{i} - 7\mathbf{j} + 10.5\mathbf{k} \tag{A-29d}$$

Multiplication of two vector quantities may give a *scalar* (scalar product) *or* a *vector* (vector product). The *scalar* product $\mathbf{A} \cdot \mathbf{B}$ of vectors $\mathbf{A}$ and $\mathbf{B}$ is defined to be $AB \cos \theta_{AB}$, where θ_{AB} is the angle between the $\mathbf{A}$ and $\mathbf{B}$ vectors. If $\mathbf{A} = a_x\mathbf{i} + a_y\mathbf{j} + a_z\mathbf{k}$ and $\mathbf{B} = b_x\mathbf{i} + b_y\mathbf{j} + b_z\mathbf{k}$, then

$$\mathbf{A} \cdot \mathbf{B} = a_x b_x + a_y b_y + a_z b_z \tag{A-30}$$

since $\mathbf{i} \cdot \mathbf{i} = \mathbf{j} \cdot \mathbf{j} = \mathbf{k} \cdot \mathbf{k} = 1$ and $\mathbf{i} \cdot \mathbf{j} = \mathbf{i} \cdot \mathbf{k} = \mathbf{j} \cdot \mathbf{k} = 0$. The vectors $\mathbf{i}$, $\mathbf{j}$, and $\mathbf{k}$ are said to be *orthogonal*. If $\mathbf{A}$ and $\mathbf{B}$ are complex quantities, the scalar product is taken as $\mathbf{A}^* \cdot \mathbf{B}$.

The *vector* product $\mathbf{C} = \mathbf{A} \times \mathbf{B}$ of vectors $\mathbf{A}$ and $\mathbf{B}$ is a vector $\mathbf{C}$ perpendicular to the plane containing $\mathbf{A}$ and $\mathbf{B}$; it is drawn from the origin of $\mathbf{A}$ and $\mathbf{B}$ and is of length $AB \sin \theta_{AB}$. The sense of the vector is obtained from the right-hand rule; if the right forefinger is parallel to $\mathbf{A}$ and the middle finger parallel to $\mathbf{B}$, then the thumb indicates the direction of the vector product $\mathbf{C}$. Considering the unit vectors $\mathbf{i}$, $\mathbf{j}$, and $\mathbf{k}$, one notes that $\mathbf{i} \times \mathbf{i} = \mathbf{j} \times \mathbf{j} = \mathbf{k} \times \mathbf{k} = 0$; $\mathbf{i} \times \mathbf{j} = \mathbf{k}$, $\mathbf{j} \times \mathbf{k} = \mathbf{i}$, $\mathbf{i} \times \mathbf{k} = -\mathbf{j}$, etc. Expansion of $\mathbf{A} \times \mathbf{B}$ in terms of its components yields

$$\begin{aligned}
\mathbf{C} = \mathbf{A} \times \mathbf{B} &= (a_x\mathbf{i} + a_y\mathbf{j} + a_z\mathbf{k}) \times (b_x\mathbf{i} + b_y\mathbf{j} + b_z\mathbf{k}) \\
&= a_x b_y \mathbf{k} - a_x b_z \mathbf{j} - a_y b_x \mathbf{k} + a_y b_z \mathbf{i} + a_z b_x \mathbf{j} - a_z b_y \mathbf{i} \\
&= \mathbf{i}(a_y b_z - a_z b_y) + \mathbf{j}(a_z b_x - a_x b_z) + \mathbf{k}(a_x b_y - a_y b_x)
\end{aligned} \tag{A-31a}$$

Note that the result in Eq. (A-31a) could have been obtained directly by writing a mnemonic "determinant" with the unit coordinate vectors in the first row and the components of $\mathbf{A}$ and $\mathbf{B}$ in the second and third rows, respectively,

$$\mathbf{A} \times \mathbf{B} = \begin{vmatrix} \mathbf{i} & \mathbf{j} & \mathbf{k} \\ a_x & a_y & a_z \\ b_x & b_y & b_z \end{vmatrix} \tag{A-31b}$$

One of the important uses of vector products is in the description of the components of angular momentum. (See Table A-1 and Appendix B.)

A third type of product of two three-dimensional vectors is called an outer product. The result is a second-rank tensor. (See Sec. A-6.) The results of such a multiplication are as follows:

$$(A_1, A_2, A_3)(B_1, B_2, B_3) = \begin{bmatrix} C_{11} & C_{12} & C_{13} \\ C_{21} & C_{22} & C_{23} \\ C_{31} & C_{32} & C_{33} \end{bmatrix} \tag{A-32a}$$

or

$$\mathbf{AB} = \mathbf{C} \tag{A-32b}$$

A-5 MATRICES

A matrix is defined as any rectangular array of $n \times m$ numbers or symbols ("matrix elements") where n is the number of rows and m the number of columns. The symbol for a matrix will be a boldface capital letter, e.g., $\mathbf{B}$. If $n = m = 1$, then the matrix is a representation of a scalar quantity. If $n = 1$ and $m > 1$, then the resulting row matrix $\mathbf{R}$ may be regarded as one representation of a vector (a row vector). If $n > 1$ and $m = 1$, the column matrix $\mathbf{C}$ may similarly be regarded as a representation of a vector (a column vector).

$$\mathbf{C} = \begin{bmatrix} c_1 \\ c_2 \\ \cdot \\ \cdot \\ \cdot \\ c_n \end{bmatrix} \qquad \mathbf{R} = \begin{bmatrix} r_1 & r_2 & \cdots & r_n \end{bmatrix} \tag{A-33}$$

Such representations of vectors are a common practice. Hence the notation used for a matrix will be the same as that for a vector.

A square matrix is one in which $n = m$. This special type of matrix is said to be an nth-order matrix. The square matrix $\mathbf{B}$ may be written

$$\mathbf{B} = \begin{bmatrix} b_{11} & b_{12} & \cdots & b_{1n} \\ b_{21} & b_{22} & \cdots & b_{2n} \\ \cdot & \cdot & \cdots & \cdot \\ b_{n1} & b_{n2} & \cdots & b_{nn} \end{bmatrix} \tag{A-34}$$

If $b_{ij} = b_{ji}$, the matrix is said to be symmetric.

A-5a. Addition and subtraction of matrices The operation

$$\mathbf{D} = \mathbf{A} + \mathbf{B}$$

or

$$\mathbf{E} = \mathbf{A} - \mathbf{B} \tag{A-35}$$

is accomplished by adding or subtracting corresponding matrix elements of $\mathbf{A}$ and $\mathbf{B}$; e.g., the element d_{ij} is equal to $a_{ij} + b_{ij}$, and the element e_{ij} is equal to $a_{ij} - b_{ij}$. The following numerical examples will illustrate the procedure:

$$\begin{bmatrix} 3 & -2 & 7 \\ -2 & 5 & -4 \\ 7 & -4 & 8 \end{bmatrix} + \begin{bmatrix} 6 & 4 & -2 \\ 4 & 2 & 3 \\ -2 & 3 & -5 \end{bmatrix} = \begin{bmatrix} 9 & 2 & 5 \\ 2 & 7 & -1 \\ 5 & -1 & 3 \end{bmatrix} \tag{A-36}$$

and

$$\begin{bmatrix} 3 & -2 & 7 \\ -2 & 5 & -4 \\ 7 & -4 & 8 \end{bmatrix} - \begin{bmatrix} 6 & 4 & -2 \\ 4 & 2 & 3 \\ -2 & 3 & -5 \end{bmatrix} = \begin{bmatrix} -3 & -6 & 9 \\ -6 & 3 & -7 \\ 9 & -7 & 13 \end{bmatrix} \tag{A-37}$$

Note that only matrices of the same dimensions can be added or subtracted.

A-5b. Multiplication of matrices The multiplication of a matrix by a scalar is accomplished by multiplying each element by the scalar, for example,

$$6 \begin{bmatrix} 3 & -2 & 7 \\ -2 & 5 & -4 \\ 7 & -4 & 8 \end{bmatrix} = \begin{bmatrix} 18 & -12 & 42 \\ -12 & 30 & -24 \\ 42 & -24 & 48 \end{bmatrix} \tag{A-38}$$

The rules of matrix multiplication can be summarized as follows:

1. Two matrices can be multiplied only if the first is a $z \times n$ matrix and the second an $n \times y$ matrix, i.e., the number of columns in the first matrix must equal the number of rows in the second matrix. The resulting product matrix will have dimensions $z \times y$.
2. Each element of the product matrix is obtained as follows:

$$(ab)_{jk} = \sum_l a_{jl} b_{lk} \tag{A-39}$$

As a first example, consider the multiplication of a column matrix by a row matrix

$$[3 \quad 5 \quad -4] \begin{bmatrix} 2 \\ -1 \\ 1 \end{bmatrix} = [(6) + (-5) + (-4)] = [-3] = -3[1] = -3 \tag{A-40}$$

Note that in this case the answer is a scalar. The result of this type of multiplication is called a "scalar product" since it corresponds to the scalar product of vectors. (See Sec. A-4.) For example, the scalar product $\mathbf{H} \cdot \hat{\mathbf{S}}$ of the two vectors with components H_x, H_y, H_z and $\hat{S}_x$, $\hat{S}_y$, $\hat{S}_z$ is

$$[H_x \quad H_y \quad H_z] \begin{bmatrix} \hat{S}_x \\ \hat{S}_y \\ \hat{S}_z \end{bmatrix} = [H_x\hat{S}_x + H_y\hat{S}_y + H_z\hat{S}_z] \tag{A-41}$$

Next consider the product of a 1×3 matrix and a 3×3 matrix

$$[3 \quad 5 \quad -4] \begin{bmatrix} 3 & -2 & 7 \\ -2 & 5 & -4 \\ 7 & -4 & 8 \end{bmatrix} = [-29 \quad 35 \quad -31] \tag{A-42}$$

The product of a 3×3 matrix and a 3×3 matrix results in a 3×3 product matrix; for example,

$$\begin{bmatrix} 3 & -2 & 7 \\ (-2) & (5) & (-4) \\ 7 & -4 & 8 \end{bmatrix} \begin{bmatrix} 6 & 4 & (-2) \\ (4) & (2) & (3) \\ -2 & 3 & (-5) \end{bmatrix} = \begin{bmatrix} -4 & 29 & -47 \\ 16 & -10 & 39 \\ 10 & 44 & -66 \end{bmatrix} \tag{A-43}$$

It will perhaps be clearer if one calculates a few elements of the above product. For example, the element a_{11} in the product matrix is $(3)(6) + (-2)(4) + (7)(-2) = -4$. The element a_{23} of the product matrix is $(-2)(-2) + (5)(3) + (-4)(-5) = 39$, etc. The location of an element resulting from multiplication of a particular row and column corresponds to that obtained by mentally (or actually) drawing lines through the row and the column being multiplied. This is shown in (A-43) for the element a_{23}.

In general, $\mathbf{AB} \neq \mathbf{BA}$. If $\mathbf{AB} = \mathbf{BA}$, $\mathbf{A}$ and $\mathbf{B}$ are said to commute. For example, if the order of multiplication in (A-43) is reversed, the result is quite different

$$\begin{bmatrix} 6 & 4 & -2 \\ 4 & 2 & 3 \\ -2 & 3 & -5 \end{bmatrix} \begin{bmatrix} 3 & -2 & 7 \\ -2 & 5 & -4 \\ 7 & -4 & 8 \end{bmatrix} = \begin{bmatrix} -4 & 16 & 10 \\ 29 & -10 & 44 \\ -47 & 39 & -66 \end{bmatrix} \qquad \text{(A-44)}$$

Each *row* of the product matrix (A-43) is the same as a *column* of the product matrix (A-44). This is a result of the fact that each original matrix is symmetric. Had the original matrices not been symmetric, the product matrices would, in general, be dissimilar.

The multiplication of matrices is associative, that is,

$$\mathbf{ABC} = (\mathbf{AB})\mathbf{C} = \mathbf{A}(\mathbf{BC}) \qquad \text{(A-45)}$$

The student may satisfy himself as to the validity of this statement.

In this book there will be occasion to perform the following type of multiplication.

$$\begin{bmatrix} a_1 & a_2 & a_3 \end{bmatrix} \begin{bmatrix} g_{11} & g_{12} & g_{13} \\ g_{21} & g_{22} & g_{23} \\ g_{31} & g_{32} & g_{33} \end{bmatrix} \begin{bmatrix} b_1 \\ b_2 \\ b_3 \end{bmatrix} = ? \qquad \text{(A-46)}$$

The result is a scalar. Multiplication of the first two matrices yields a row matrix, and a row matrix times a column matrix is a scalar. [See Eq. (A-40).] It will be left to the reader to work out the result.

A further example of matrix multiplication is the operation of rotation through an arbitrary angle ϕ in the xy plane. After counterclockwise rotation of the x and y axes through the angle ϕ, the coordinates of a point in a rigid body will change from (x_1,y_1) to (x_2,y_2). Reference to Fig. A-2 yields the relations between the new and old coordinates:

$$x_2 = x_1 \cos \phi + y_1 \sin \phi \qquad \text{(A-47a)}$$
$$y_2 = -x_1 \sin \phi + y_1 \cos \phi \qquad \text{(A-47b)}$$

Alternatively, both initial and final coordinates may be expressed as column vectors:

$$\begin{bmatrix} x_2 \\ y_2 \end{bmatrix} = \begin{bmatrix} \cos \phi & \sin \phi \\ -\sin \phi & \cos \phi \end{bmatrix} \begin{bmatrix} x_1 \\ y_1 \end{bmatrix} = \begin{bmatrix} x_1 \cos \phi + y_1 \sin \phi \\ -x_1 \sin \phi + y_1 \cos \phi \end{bmatrix} \qquad \text{(A-48)}$$

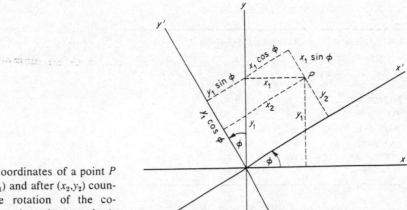

Fig. A-2 Coordinates of a point P before (x_1, y_1) and after (x_2, y_2) counterclockwise rotation of the coordinate axes through an angle ϕ.

It is clear that Eq. (A-48) is equivalent to Eqs. (A-47). The square matrix in Eq. (A-48) is called a coordinate-rotation matrix.

A-5c. Special matrices and matrix properties A given matrix may be transformed into various related matrices, some of which have especially useful properties. Table A-2 defines several matrices derived from a matrix A;

Table A-2

Matrix symbol	Components	Example
A	a_{ij}	$\begin{bmatrix} 2 & 3+i \\ 4i & 5 \end{bmatrix}$
Transpose $\tilde{A}$	$(\tilde{A})_{ij} = a_{ji}$	$\begin{bmatrix} 2 & 4i \\ 3+i & 5 \end{bmatrix}$
Complex conjugate A^*	$(A^*)_{ij} = a_{ij}^*$	$\begin{bmatrix} 2 & 3-i \\ -4i & 5 \end{bmatrix}$
Adjoint $A^\dagger$	$(A^\dagger)_{ij} = a_{ji}^*$	$\begin{bmatrix} 2 & -4i \\ 3-i & 5 \end{bmatrix}$
Inverse A^{-1}	$(A^{-1})_{ij} = \dfrac{1}{\lvert A \rvert} \dfrac{\partial \lvert A \rvert}{\partial a_{ji}}$ ‡	$\dfrac{14 + 12i}{340} \begin{bmatrix} 5 & -3-i \\ -4i & 2 \end{bmatrix}$ §

‡ $\lvert A \rvert$ refers to the determinant with elements identical with those of the matrix. The inverse exists only for a square matrix, the determinant of which is nonzero. The symbolism "$\partial \lvert A \rvert$"/∂a_{ji} implies that in taking the derivative with respect to a_{ji}, all other elements will be kept fixed. For a 2×2 determinant

$$\lvert A \rvert = \begin{vmatrix} a_{11} & a_{12} \\ a_{21} & a_{22} \end{vmatrix} = a_{11}a_{22} - a_{21}a_{12}$$

Hence

$$\frac{\partial \lvert A \rvert}{\partial a_{21}} = -a_{12}$$

§ The numerator and denominator have been multiplied by $14 + 12i$ to rationalize the denominator.

Table A-3

Matrix	Alternative definitions	
Unit	1	$a_{ii} = 1$, $a_{ij} = 0$ if $i \neq j$
Diagonal	$^d\mathbf{A}$	$a_{ij} = 0$ if $i \neq j$
Symmetric	$\tilde{\mathbf{A}} = \mathbf{A}$	$a_{ij} = a_{ji}$
Antisymmetric	$\tilde{\mathbf{A}} = -\mathbf{A}$	$a_{ij} = -a_{ji}$
Real	$\mathbf{A}^* = \mathbf{A}$	$a_{ij}^* = a_{ij}$
Orthogonal	$\mathbf{A}^{-1} = \tilde{\mathbf{A}}$	
Hermitian	$\mathbf{A}^\dagger = \mathbf{A}$	$a_{ji}^* = a_{ij}$
Unitary	$\mathbf{A}^{-1} = \mathbf{A}^\dagger$	

various examples illustrating the relations among matrix components are given in the third column.

The properties of some matrices of especial importance are defined in Table A-3.

A-5d. Dirac notation for wave functions and matrix elements A shorthand notation, introduced by Dirac[†] for wave functions, is employed in this book; for example, the wave function ψ_n is represented by $|n\rangle$, where n is an identifying label, usually a quantum number. The function $|n\rangle$ is called a "ket," For example, Eq. (A-9) can be written

$$\hat{\Lambda}|n\rangle = \lambda_n|n\rangle \qquad\qquad\qquad\qquad (A\text{-}49)$$

The ket $|n\rangle$ may be labeled with the quantum number n, since the eigenvalue λ_n is a function only of n. Spin functions corresponding to $M_S = +\frac{1}{2}$ and $M_S = -\frac{1}{2}$ are conventionally represented by $|\alpha\rangle$ and $|\beta\rangle$, respectively. Corresponding to each ket $|n\rangle$ there will be a function called a "bra," written as $\langle n|$. A bra has meaning only when combined with another ket. For the bra $\langle n|$ and the ket $|m\rangle$, the notation $\langle n|m\rangle$ implies integration over the full range of all variables, that is,

$$\langle n|m\rangle = \int_\tau \psi_m^* \psi_n \, d\tau \qquad\qquad\qquad\qquad (A\text{-}50)$$

If $m \neq n$ and $\langle n|m\rangle = 0$, the wave functions are said to be *orthogonal*. If $m = n$ and $\langle m|m\rangle = 1$, the wave functions are said to be *normalized*. It is always possible and usually convenient to choose the angular-momentum wave functions to be orthogonal and normalized, i.e., *orthonormal*. Frequently integrals of the form $\int \psi_n^* \hat{B} \psi_m \, d\tau$ are encountered as elements of a matrix [See Eq. (A-57).] In the Dirac notation such integrals are represented by

$$\int_\tau \psi_n^* \hat{B} \psi_m \, d\tau = \langle n|\hat{B}|m\rangle \qquad\qquad\qquad\qquad (A\text{-}51)$$

† P. A. M. Dirac, "The Principles of Quantum Mechanics," 3d ed., p. 18, Oxford University Press, London, 1947.

The expression $\langle n|\hat{B}|m\rangle$ is then called a "matrix element." If $n = m$, the function is called a "diagonal matrix element," and if $n \neq m$ then it is called an "off-diagonal matrix element."

The average value (or expectation value) $\langle b\rangle$ of any observable b for the state described by the orthonormal wave function ψ_n is obtained by the following operation:

$$\langle b\rangle = \int_\tau \psi_n^* \hat{B}\psi_n \, d\tau = \langle n|\hat{B}|n\rangle \tag{A-52}$$

where $\hat{B}$ is the operator corresponding to the observable b.

An important property of the bra and ket functions is given by the relation

$$\langle n|m\rangle = [\langle m|n\rangle]^* \tag{A-53}$$

In the matrix element $\langle n|\hat{B}|m\rangle$ it is assumed that $\hat{B}$ is to operate in the forward direction of the ket $|m\rangle$. To operate backwards on the bra $\langle n|$, one must take the adjoint of the matrix element, using Eq. (A-53).

$$\langle n|\hat{B}|m\rangle = [\langle n|\hat{B}|m\rangle]^\dagger = [\langle m|\hat{B}|n\rangle]^* = x_n^*[\langle m|n'\rangle]^* = x_n^*\langle n'|m\rangle \tag{A-54}$$

Here x_n can be a complex number.

If $\hat{B}$ is a nonhermitian operator (e.g., the ladder operators which will follow) the effect of $\hat{B}$ is to alter the wave function ($|n\rangle \rightarrow |n'\rangle$), as well as to multiply it by a constant x_n. If the operator $\hat{B}$ corresponds to an observable quantity, and is therefore hermitian, then x_n must be real and $n = n'$. Hence in this case it does not matter whether $\hat{B}$ operates forward or backward in the matrix element of Eq. (A-54).

A-5e. Diagonalization of matrices Many matrices of interest (for example, the energy matrix) are encountered in structural problems; they are usually hermitian but not necessarily diagonal. It is possible to transform such matrices into a diagonal form.

As an example, consider the spin operators $\hat{S}_x$, $\hat{S}_y$, and $\hat{S}_z$ for $S = \frac{1}{2}$. The eigenfunctions of $\hat{S}_z$ are usually taken as a basis set for the spin functions (see Sec. 1-5); that is,

$$\hat{S}_z|\alpha\rangle = +\tfrac{1}{2}|\alpha\rangle \tag{A-55a}$$

$$\hat{S}_z|\beta\rangle = -\tfrac{1}{2}|\beta\rangle \tag{A-55b}$$

Multiplication of Eq. (A-55a) from the left by $\langle\alpha|$ gives

$$\langle\alpha|\hat{S}_z|\alpha\rangle = +\tfrac{1}{2}\langle\alpha|\alpha\rangle = \tfrac{1}{2} \tag{A-56a}$$

Similarly

$$\langle\beta|\hat{S}_z|\alpha\rangle = +\tfrac{1}{2}\langle\beta|\alpha\rangle = 0 \tag{A-56b}$$

$$\langle\alpha|\hat{S}_z|\beta\rangle = -\tfrac{1}{2}\langle\alpha|\beta\rangle = 0 \tag{A-56c}$$

$$\langle\beta|\hat{S}_z|\beta\rangle = -\tfrac{1}{2}\langle\beta|\beta\rangle = -\tfrac{1}{2} \qquad\qquad\qquad (A\text{-}56d)$$

It will prove to be very convenient to deal with a single matrix S_z instead of numerous operator equations such as Eqs. (A-56). These equations may be combined into the matrix equation

$$\begin{bmatrix} \langle\alpha|\hat{S}_z|\alpha\rangle & \langle\alpha|\hat{S}_z|\beta\rangle \\ \langle\beta|\hat{S}_z|\alpha\rangle & \langle\beta|\hat{S}_z|\beta\rangle \end{bmatrix} = \begin{bmatrix} \tfrac{1}{2} & 0 \\ 0 & -\tfrac{1}{2} \end{bmatrix} = S_z \qquad\qquad (A\text{-}57)$$

S_z is a matrix which includes all possible matrix elements of the operator $\hat{S}_z$ between the states α and β. In general, if there are n states in the system, S_z will be a square matrix of order n. If such a matrix is diagonal, then the basis wave functions must be eigenfunctions of the operator. Thus $|\alpha\rangle$ and $|\beta\rangle$ are eigenfunctions of $\hat{S}_z$. Examples of spin operators and spin matrices are given in Sec. B-6.

Operation by $\hat{S}_x$ on $|\alpha\rangle$ and $|\beta\rangle$ gives the following results:

$$\hat{S}_x|\alpha\rangle = \tfrac{1}{2}|\beta\rangle \qquad\qquad\qquad\qquad\qquad (A\text{-}58a)$$

$$\hat{S}_x|\beta\rangle = \tfrac{1}{2}|\alpha\rangle \qquad\qquad\qquad\qquad\qquad (A\text{-}58b)$$

The corresponding matrix is then

$$\begin{bmatrix} \langle\alpha|\hat{S}_x|\alpha\rangle & \langle\alpha|\hat{S}_x|\beta\rangle \\ \langle\beta|\hat{S}_x|\alpha\rangle & \langle\beta|\hat{S}_x|\beta\rangle \end{bmatrix} = \begin{bmatrix} 0 & \tfrac{1}{2} \\ \tfrac{1}{2} & 0 \end{bmatrix} = S_x \qquad\qquad (A\text{-}59)$$

S_x is not a diagonal matrix because the basis functions are *not* eigenfunctions of $\hat{S}_x$.

Assume that the unknown eigenfunctions of $\hat{S}_x$ can be represented as linear combinations of $|\alpha\rangle$ and $|\beta\rangle$, that is,

$$|\phi_1\rangle = c_{11}|\alpha\rangle + c_{12}|\beta\rangle \qquad\qquad\qquad (A\text{-}60a)$$

$$|\phi_2\rangle = c_{21}|\alpha\rangle + c_{22}|\beta\rangle \qquad\qquad\qquad (A\text{-}60b)$$

or in matrix form

$$\begin{bmatrix} \phi_1 \\ \phi_2 \end{bmatrix} = \underset{\mathbf{C}}{\begin{bmatrix} c_{11} & c_{12} \\ c_{21} & c_{22} \end{bmatrix}} \begin{bmatrix} \alpha \\ \beta \end{bmatrix} \qquad\qquad\qquad (A\text{-}61)$$

Each row of the square matrix C in Eq. (A-61) may be considered as a row vector $\mathbf{c}$. These vectors have the property that

$$S_x\mathbf{c} = \lambda\mathbf{c} \qquad\qquad\qquad\qquad\qquad (A\text{-}62)$$

Hence, the vectors $\mathbf{c}$ are called *eigenvectors* of the matrix S_x with eigenvalues λ.

Insertion of the unit matrix 1 into Eq. (A-62) and rearrangement gives

$$(S_x - \lambda 1)\mathbf{c} = 0 \qquad\qquad\qquad\qquad (A\text{-}63)$$

Equation (A-63) represents a series of simultaneous equations (called secular equations) which are not independent if λ is an eigenvalue. If $c \neq 0$, then these equations may be solved by expansion of the following determinantal equation, often called the secular determinant†

$$|S_x - \lambda 1| = 0 \qquad (A\text{-}64)$$

Since the matrix S_x is of order 2, there will be two roots in Eq. (A-64); these are two values of λ. On expansion, Eq. (A-64) becomes

$$\begin{vmatrix} 0 - \lambda & \tfrac{1}{2} \\ \tfrac{1}{2} & 0 - \lambda \end{vmatrix} = 0 \qquad (A\text{-}65)$$

$$\lambda^2 - \tfrac{1}{4} = 0$$

$$\lambda = \pm\tfrac{1}{2} \qquad (A\text{-}66)$$

For $\lambda = \tfrac{1}{2}$, the two simultaneous equations corresponding to Eq. (A-63) are

$$(0 - \tfrac{1}{2})c_{11} + \tfrac{1}{2}c_{12} = 0$$

$$\tfrac{1}{2}c_{11} + (0 - \tfrac{1}{2})c_{12} = 0$$

or

$$c_{11} = c_{12} \qquad (A\text{-}67)$$

For $\lambda = -\tfrac{1}{2}$, $c_{21} = -c_{22}$. If $|\phi_1\rangle$ and $|\phi_2\rangle$ are to be normalized, $c_{11}^2 + c_{12}^2 = 1$ and $c_{21}^2 + c_{22}^2 = 1$. Hence, the final eigenfunctions of S_x are

$$\begin{bmatrix} \phi_1 \\ \phi_2 \end{bmatrix} = \underbrace{\begin{bmatrix} \dfrac{1}{\sqrt{2}} & \dfrac{1}{\sqrt{2}} \\ \dfrac{1}{\sqrt{2}} & -\dfrac{1}{\sqrt{2}} \end{bmatrix}}_{C} \begin{bmatrix} \alpha \\ \beta \end{bmatrix} \qquad (A\text{-}68)$$

The square matrix C in Eq. (A-68) is a unitary matrix, since $C^{-1} = C^\dagger$. It has the additional property that

$$CS_x C^{-1} = CS_x C^\dagger = \Lambda \qquad (A\text{-}69)$$

where Λ is a diagonal matrix. For S_x

$$CS_x C^\dagger = \begin{bmatrix} \dfrac{1}{\sqrt{2}} & \dfrac{1}{\sqrt{2}} \\ \dfrac{1}{\sqrt{2}} & -\dfrac{1}{\sqrt{2}} \end{bmatrix} \begin{bmatrix} 0 & \dfrac{1}{2} \\ \dfrac{1}{2} & 0 \end{bmatrix} \begin{bmatrix} \dfrac{1}{\sqrt{2}} & \dfrac{1}{\sqrt{2}} \\ \dfrac{1}{\sqrt{2}} & -\dfrac{1}{\sqrt{2}} \end{bmatrix}$$

$$= \begin{bmatrix} \dfrac{1}{2\sqrt{2}} & \dfrac{1}{2\sqrt{2}} \\ -\dfrac{1}{2\sqrt{2}} & \dfrac{1}{2\sqrt{2}} \end{bmatrix} \begin{bmatrix} \dfrac{1}{\sqrt{2}} & \dfrac{1}{\sqrt{2}} \\ \dfrac{1}{\sqrt{2}} & -\dfrac{1}{\sqrt{2}} \end{bmatrix} = \begin{bmatrix} \dfrac{1}{2} & 0 \\ 0 & -\dfrac{1}{2} \end{bmatrix} \qquad (A\text{-}70)$$

† See, e.g., G. G. Hall, "Matrices and Tensors," chap. 4, Pergamon Press, Oxford, England, 1963.

This procedure for diagonalizing a hermitian (in this case, symmetric) matrix with the use of a unitary matrix and its reciprocal is general for any hermitian matrix. The most frequent examples encountered in this book are the diagonalization of hamiltonian matrices (see Chaps. 10, 11, and 12 and Appendix C) and the diagonalization of g and of hyperfine tensors. (See Chap. 7.)

If **C** is taken to be the two-dimensional coordinate-rotation matrix [this is a unitary matrix, see Eq. (A-48)], a general method for the diagonalization of any 2×2 matrix may be developed. The appropriate diagonalization procedure is as follows:

$$\begin{bmatrix} \cos \omega & \sin \omega \\ -\sin \omega & \cos \omega \end{bmatrix} \begin{bmatrix} a & c \\ c & b \end{bmatrix} \begin{bmatrix} \cos \omega & -\sin \omega \\ \sin \omega & \cos \omega \end{bmatrix} = \begin{bmatrix} x & 0 \\ 0 & y \end{bmatrix} \qquad \text{(A-71)}$$

After matrix multiplication, the general solutions for zero off-diagonal elements are

$$\tan 2\omega = \frac{2c}{a - b} \qquad \text{(A-72}a\text{)}$$

$$\cos^2 \omega = \frac{1}{2} \left[1 + \left(1 + \frac{4c^2}{(b - a)^2} \right)^{-\frac{1}{2}} \right] \qquad \text{(A-72}b\text{)}$$

$$\sin^2 \omega = \frac{1}{2} \left[1 - \left(1 + \frac{4c^2}{(b - a)^2} \right)^{-\frac{1}{2}} \right] \qquad \text{(A-72}c\text{)}$$

$$\sin \omega \cos \omega = \frac{c}{[(b - a)^2 + 4c^2]^{\frac{1}{2}}} \qquad \text{(A-72}d\text{)}$$

$$x = a \cos^2 \omega + b \sin^2 \omega + 2c \sin \omega \cos \omega \qquad \text{(A-72}e\text{)}$$

$$y = a \sin^2 \omega + b \cos^2 \omega - 2c \sin \omega \cos \omega \qquad \text{(A-72}f\text{)}$$

The advantage of this method is that with ω given by Eqs. (A-72), the coordinate rotation matrix on the left of (A-71) become the eigenvector matrix. That is, the two elements in each row are the coefficients of one of the two eigenvectors.

A-6 TENSORS

In Chap. 1 it was explicitly indicated that the simple resonance expression $H_r = h\nu/g\beta$ is applicable only to systems which behave as if they were *isotropic* (i.e., which exhibit physical properties independent of orientation). The same restriction (isotropic behavior) applies to the expression

$$H_r = \frac{h\nu}{g\beta} - aM_I = \frac{h\nu}{g\beta} - \frac{A_0 h}{g\beta} M_I \qquad \text{(A-73)}$$

for systems showing hyperfine splitting [see Eq. (3-12)]. Essentially, g and a (or A_0) have been assumed to be scalar quantities. In an anisotropic system, the response to an external stimulus cannot be described by a single constant. Most physical properties will, in general, require *six* independent parameters to describe the response. The basic reason for requiring this large number is that for a general orientation, *the response occurs in a direction different from that of the applied stimulus.*

Except for systems of low symmetry, there are three special directions ("principal axes") along which the response will occur in the *same* direction as the stimulus. It is commonly found for anisotropic systems that the symmetry axis of highest order will be a principal axis (e.g., for a system with hexagonal symmetry, the sixfold rotation axis is a principal axis).[†]

The magnetic susceptibility of a solid illustrates many of the properties encountered in both isotropic and anisotropic systems. The magnetization $\mathcal{M}$ (response) of an isotropic solid is related to the field $\mathbf{H}$ (stimulus) by the expression

$$\mathcal{M} = \chi \mathbf{H} \tag{A-74}$$

Here χ is the magnetic susceptibility. Consider $\mathbf{H}$ applied along the z direction; Eq. (A-74) is then written as

$$\begin{bmatrix} \mathcal{M}_x \\ \mathcal{M}_y \\ \mathcal{M}_z \end{bmatrix} = \begin{bmatrix} \chi & 0 & 0 \\ 0 & \chi & 0 \\ 0 & 0 & \chi \end{bmatrix} \begin{bmatrix} 0 \\ 0 \\ H_z \end{bmatrix} = \begin{bmatrix} 0 \\ 0 \\ \chi H_z \end{bmatrix} \tag{A-75}$$

The physical property χ has been represented as a matrix. Such matrix representations of physical properties are called *tensors*. Tensors represented by a square matrix are called tensors of the second rank.[‡]

In the following discussion, "tensor" will imply a second-rank tensor. The notation for second-rank tensors will be boldface italic type, e.g., A. Note that in the above case there is a component of magnetization only along the field direction. If the field is applied successively in the x or y directions or any other arbitrary direction, an identical numerical value of $\mathcal{M}$ is obtained in the same direction as the stimulus $\mathbf{H}$.

Next consider a crystal with axial, e.g., hexagonal, symmetry, exemplified by graphite. Experiments indicate that when the field is along the hexagonal axis, the absolute magnitude of the diamagnetic susceptibility is much larger than the constant value of susceptibility for $\mathbf{H}$ in the layer plane. Taking the z direction as that of the hexagonal axis,

$$\begin{bmatrix} \mathcal{M}_x \\ \mathcal{M}_y \\ \mathcal{M}_z \end{bmatrix} = \begin{bmatrix} \chi_{xx} & 0 & 0 \\ 0 & \chi_{xx} & 0 \\ 0 & 0 & \chi_{zz} \end{bmatrix} \begin{bmatrix} H_x \\ H_y \\ H_z \end{bmatrix} = \begin{bmatrix} \chi_{xx} H_x \\ \chi_{xx} H_y \\ \chi_{zz} H_z \end{bmatrix} \tag{A-76}$$

Each element of the susceptibility tensor requires two subscripts. (Note that the number of subscripts equals the rank of the tensor.) Consider first the application of a field along one of the principal axes. The magnetization

[†] This is not *necessarily* the case for the hyperfine tensor.
[‡] Vectors are tensors of the first rank, whereas scalars are tensors of zero rank. A fuller definition of second-rank tensors is given later.

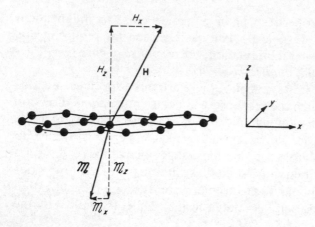

Fig. A-3 An illustration of the diamagnetic susceptibility of a graphite crystal. The z axis corresponds to the hexagonal axis of the crystal. For **H** in the xz plane, **H** can be resolved into components H_x and H_z. Upon multiplication of H_x and H_z by χ_{xx} and χ_{zz} (negative quantities), respectively, the components $\mathcal{M}_x$ and $\mathcal{M}_z$ are obtained. The resultant $\mathcal{M}$ is obviously not parallel to **H** since $\chi_{xx} \neq \chi_{zz}$.

is then along the same direction. If the crystal is rotated in the magnetic field, there will be nonzero field components H_x, H_y, and H_z along the x, y, and z axes; hence, there will be components $\mathcal{M}_x = \chi_{xx}H_x$, $\mathcal{M}_y = \chi_{xx}H_y$, and $\mathcal{M}_z = \chi_{zz}H_z$ such that $\mathcal{M}$ will not be parallel to **H**. Figure A-3 illustrates this for **H** lying in the xz plane. In this case it is clear that $\mathcal{M}$ and **H** are not parallel. For an axial system, one usually writes $\chi_\perp$ for χ_{xx} or χ_{yy} and $\chi_\parallel$ for χ_{zz} to designate the susceptibility components perpendicular or parallel to the symmetry axis. This same tensor χ is applicable to systems with three- or fourfold axes of symmetry.

The extension to rhombic or lower symmetry requires only the condition $\chi_{yy} \neq \chi_{xx}$, that is,

$$\begin{bmatrix} \mathcal{M}_x \\ \mathcal{M}_y \\ \mathcal{M}_z \end{bmatrix} = \begin{bmatrix} \chi_{xx} & 0 & 0 \\ 0 & \chi_{yy} & 0 \\ 0 & 0 & \chi_{zz} \end{bmatrix} \begin{bmatrix} H_x \\ H_y \\ H_z \end{bmatrix} = \begin{bmatrix} \chi_{xx}H_x \\ \chi_{yy}H_y \\ \chi_{zz}H_z \end{bmatrix} \qquad (\text{A-77})$$

Again, only in the case of **H** along one of the principal axes will the magnetization be parallel to **H**; an arbitrarily oriented field gives magnetization components determined by the field components H_x, H_y, H_z and the susceptibility tensor χ.

At this point it is desirable to consider a representation for χ if the axes which are used are not the principal axes of an orthorhombic crystal. Specifically, consider the use of the x', y', z' axis system, rotated through $\cos^{-1} l_{xx}$ with respect to the x axis, and having the x' axis further described by $\cos^{-1} l_{xy}$ and $\cos^{-1} l_{xz}$. These l_{ij} quantities, which measure the cosine of

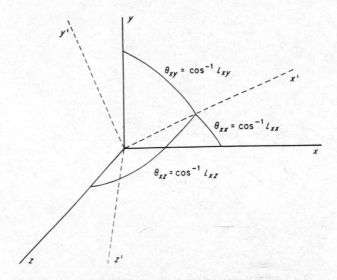

Fig. A-4 A set of new axes x', y', z' derived from the old axes x, y z by a rotation about an arbitrary direction through the origin.

the angle between axes i and j, are aptly termed direction cosines. (See Fig. A-4.) The first subscript refers to the "new" axes and the second to the "old." The "new" and "old" axes may be related by use of the direction cosines in Table A-4.

Figure A-5 illustrates the components of **H** along, say, x'. Since the old axes x, y and z are not perpendicular to x', H_x, H_y, and H_z will all contribute to the field component $H_{x'}$. The contribution from H_x will be $H_x \cos (x',x)$, that of H_y is $H_y \cos (x',y)$ and that of H_z is $H_z \cos (x',z)$. Hence

$$H_{x'} = H_x l_{xx} + H_y l_{xy} + H_z l_{xz} \qquad (\text{A-78}a)$$

Similarly

$$H_{y'} = H_x l_{yx} + H_y l_{yy} + H_z l_{yz} \qquad (\text{A-78}b)$$

and

$$H_{z'} = H_x l_{zx} + H_y l_{zy} + H_z l_{zz} \qquad (\text{A-78}c)$$

Table A-4

New \ Old	x	y	z
x'	l_{xx}	l_{xy}	l_{xz}
y'	l_{yx}	l_{yy}	l_{yz}
z'	l_{zx}	l_{zy}	l_{zz}

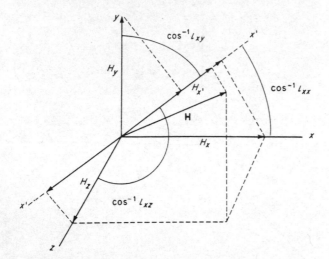

Fig. A-5 Representation of the component $H_{x'}$ and the separate contributions to it from H_x, H_y, and H_z.

The results are conveniently summarized in matrix form as follows:

$$\begin{bmatrix} H_{x'} \\ H_{y'} \\ H_{z'} \end{bmatrix} = \begin{bmatrix} l_{xx} & l_{xy} & l_{xz} \\ l_{yx} & l_{yy} & l_{yz} \\ l_{zx} & l_{zy} & l_{zz} \end{bmatrix} \begin{bmatrix} H_x \\ H_y \\ H_z \end{bmatrix} \tag{A-79a}$$

or

$$\mathbf{H'} = \mathscr{L}\mathbf{H} \tag{A-79b}$$

The corresponding transformation of $\mathscr{M}$ to the $\mathscr{M}'$ representation is given by

$$\begin{bmatrix} \mathscr{M}_{x'} \\ \mathscr{M}_{y'} \\ \mathscr{M}_{z'} \end{bmatrix} = \begin{bmatrix} l_{xx} & l_{xy} & l_{xz} \\ l_{yx} & l_{yy} & l_{yz} \\ l_{zx} & l_{zy} & l_{zz} \end{bmatrix} \begin{bmatrix} \mathscr{M}_x \\ \mathscr{M}_y \\ \mathscr{M}_z \end{bmatrix} \tag{A-80a}$$

or

$$\mathscr{M}' = \mathscr{L}\mathscr{M} \tag{A-80b}$$

The task is to relate the $\mathscr{M}_{i'}$ components to the $H_{i'}$ components. With the expectation that the appropriate susceptibility tensor is more complicated than that in (A-77), one writes

$$\begin{bmatrix} \mathscr{M}_{x'} \\ \mathscr{M}_{y'} \\ \mathscr{M}_{z'} \end{bmatrix} = \underbrace{\begin{bmatrix} \chi_{xx} & \chi_{xy} & \chi_{xz} \\ \chi_{yx} & \chi_{yy} & \chi_{yz} \\ \chi_{zx} & \chi_{zy} & \chi_{zz} \end{bmatrix}}_{\chi'} \begin{bmatrix} H_{x'} \\ H_{y'} \\ H_{z'} \end{bmatrix} \tag{A-81a}$$

or

$$\mathscr{M}' = \chi'\mathbf{H'} \tag{A-81b}$$

In general, the principal axes for a given system are not known in advance; hence one uses some convenient set of arbitrarily chosen axes x', y', z'. The tensor χ' represents the form in which the data are initially obtained. The task is to transform χ' to a diagonal form which is designated $^d\chi$. This is equivalent to the rotation of the x', y', z' axes to the *principal* axes x, y, and z.

Substitution of Eq. (A-79b) into Eq. (A-81b) relates $\mathscr{M}'$ to $\mathbf{H}$ in the principal-axis system. A similar substitution of Eq. (A-80b) into Eq. (A-81b) yields

$$\mathscr{L}\mathscr{M} = \chi'\mathscr{L}\mathbf{H} \tag{A-82a}$$

Multiplication from the left by $\mathscr{L}^{-1}$ yields

$$\mathscr{M} = \mathscr{L}^{-1}\chi'\mathscr{L}\mathbf{H} \tag{A-82b}$$

since $\mathscr{L}^{-1}\mathscr{L} = 1$. The direction-cosine matrices are unitary; hence, $\mathscr{L}^{-1} = \mathscr{L}^{\dagger}$. The direction cosines are real and thus $\mathscr{L}^{\dagger} = \tilde{\mathscr{L}}$. Equation (A-82$b$) then becomes

$$\mathscr{M} = \tilde{\mathscr{L}}\chi'\mathscr{L}\mathbf{H} \tag{A-82c}$$

or

$$\begin{bmatrix} \mathscr{M}_x \\ \mathscr{M}_y \\ \mathscr{M}_z \end{bmatrix} = \begin{bmatrix} l_{xx} & l_{yx} & l_{zx} \\ l_{xy} & l_{yy} & l_{zy} \\ l_{xz} & l_{yz} & l_{zz} \end{bmatrix} \underbrace{\begin{bmatrix} \chi_{x'x'} & \chi_{x'y'} & \chi_{x'z'} \\ \chi_{y'x'} & \chi_{y'y'} & \chi_{y'z'} \\ \chi_{z'x'} & \chi_{z'y'} & \chi_{z'z'} \end{bmatrix} \begin{bmatrix} l_{xx} & l_{xy} & l_{xz} \\ l_{yx} & l_{yy} & l_{yz} \\ l_{zx} & l_{zy} & l_{zz} \end{bmatrix}}_{^d\chi} \begin{bmatrix} H_x \\ H_y \\ H_z \end{bmatrix}$$

$$\tag{A-82d}$$

It is useful at this point to carry out the matrix multiplication indicated in Eq. (A-82d) to obtain the transformation equations relating χ' to $^d\chi$. For example, the element χ_{xx} of $^d\chi$ is

$$\chi_{xx} = \sum_{i=x,y,z} l_{xx}l_{ix}\chi_{ix'} + \sum_{i=x,y,z} l_{yx}l_{ix}\chi_{iy'} + \sum_{i=x,y,z} l_{zx}l_{ix}\chi_{iz'}$$

$$= \sum_{j=x,y,z}\sum_{i=x,y,z} l_{jx}l_{ix}\chi_{ij} \tag{A-83}$$

There will be a similar transformation equation for each element of the tensor. A tensor of second rank is then defined as any physical property which transforms according to equations such as Eq. (A-83). If the direction cosines are such that $\chi_{ij} = 0$ for $i \neq j$, then χ will be diagonal, and the direction cosine matrices are eigenvector matrices. (See Sec. A-5e.) Diagonalization of g tensors and of hyperfine tensors is accomplished in a fashion identical to that described above and is a very important aspect of the study of anisotropic systems. (See Chap. 7.) The tensors $\boldsymbol{g}$ and $\boldsymbol{A}$ must be obtained by taking the square roots of the tensors $\boldsymbol{g}^2$ and $\boldsymbol{A}^2$. The square of a tensor B is defined as: $B^2 \equiv BB^\dagger$. This definition requires B^2 to be symmetric, even if B is not. One is thus unable to distinguish between such components as $(B)_{yx}$ and $(B)_{xy}$ without additional information.[†]

A-7 PERTURBATION THEORY

In numerous structural problems, solutions are available for the major terms in the hamiltonian operator. For example, the rotational motion of a diatomic molecule is well approximated by the solutions to the quantum-mechanical rigid-rotor problem. If there is an additional term in the hamiltonian operator which is small, then perturbation theory may be applied to ascertain its effect on the wave functions and energies of the system. Centrifugal distortion in the rigid-rotor problem is an example of a small perturbation.

The hamiltonian for such systems is usually written as

$$\hat{\mathcal{H}} = \hat{\mathcal{H}}_0 + \lambda \hat{\mathcal{H}}' \tag{A-84}$$

where $\hat{\mathcal{H}}_0$ is the hamiltonian for which solutions are known and $\hat{\mathcal{H}}'$ is the perturbation operator. λ is a perturbation magnitude parameter, having values between 0 and 1.

Suppose the eigenfunctions and eigenvalues of $\hat{\mathcal{H}}_0$ are given by

$$\hat{\mathcal{H}}_0 \psi_i^0 = W_i^0 \psi_i^0 \tag{A-85a}$$

or

$$\hat{\mathcal{H}}_0 |i\rangle^0 = W_i^0 |i\rangle^0 \tag{A-85b}$$

Here the Dirac notation for wave functions has been utilized. i ranges over the full set of "zero-order" eigenfunctions $|1\rangle^0$, $|2\rangle^0 \cdots |n\rangle^0$. The W_i^0 values are called "zero-order" energies.

The unknown eigenfunctions and eigenvalues of $\hat{\mathcal{H}}$ are given by

$$\hat{\mathcal{H}}|i\rangle = W_i|i\rangle \tag{A-86}$$

Since the solutions of Eq. (A-86) must go continuously into those of Eq. (A-85b) as $\lambda \to 0$, it is assumed that $|i\rangle$ and W_i can be expanded as a power series in λ; that is,

$$|i\rangle = |i\rangle^0 + \lambda|i\rangle' + \lambda^2|i\rangle'' + \cdots \tag{A-87}$$

$$W_i = W_i^0 + \lambda W_i' + \lambda^2 W_i'' + \cdots \tag{A-88}$$

Substitution of Eqs. (A-84), (A-87). and (A-88) into Eq. (A-86) yields

---

[†] If the true spin of a system is greater than $\frac{1}{2}$, there is some imprecision in referring to "g" and "A" as tensors. (See A. Abragam and B. Bleaney, "Electron Paramagnetic Resonance of Transition Ions", pp. 650 to 653, Oxford University Press, London, 1970.)

$$(\hat{\mathcal{H}}_0 + \lambda\hat{\mathcal{H}}')(|i\rangle^0 + \lambda|i\rangle' + \lambda^2|i\rangle'' + \cdots)$$
$$= (W_i^0 + \lambda W_i' + \lambda^2 W_i'' + \cdots)(|i\rangle^0 + \lambda|i\rangle' + \lambda^2|i\rangle''$$
$$+ \cdots) \quad (A\text{-}89)$$

or

$$\hat{\mathcal{H}}_0|i\rangle^0 + \lambda(\hat{\mathcal{H}}'|i\rangle^0 + \hat{\mathcal{H}}_0|i\rangle') + \lambda^2(\hat{\mathcal{H}}'|i\rangle' + \hat{\mathcal{H}}_0|i\rangle'') + \cdots$$
$$= W_i^0|i\rangle^0 + \lambda(W_i'|i\rangle^0 + W_i^0|i\rangle')$$
$$+ \lambda^2(W_i''|i\rangle^0 + W_i'|i\rangle' + W_i^0|i\rangle'') + \cdots \quad (A\text{-}90)$$

Equation (A-90) must be valid for all possible values of λ; this is possible only if the coefficients of a given power of λ are equal on both sides of the equation. Thus one may write

$$\hat{\mathcal{H}}_0|i\rangle^0 = W_i^0|i\rangle^0 \qquad\qquad\qquad\qquad (A\text{-}91a)$$

$$\hat{\mathcal{H}}'|i\rangle^0 + \hat{\mathcal{H}}_0|i\rangle' - W_i'|i\rangle^0 + W_i^0|i\rangle' \qquad\qquad (A\text{-}91b)$$

$$\hat{\mathcal{H}}'|i\rangle' + \hat{\mathcal{H}}_0|i\rangle'' = W_i''|i\rangle^0 + W_i'|i\rangle' + W_i^0|i\rangle'' \qquad (A\text{-}91c)$$

It was assumed that the solutions to Eq. (A-91a) are known.

In Eq. (A-91b) the functions $|i\rangle'$ are unknown; similarly the effect of $\hat{\mathcal{H}}'$ on $|i\rangle^0$ is unknown. Assume that these functions of the zero-order basis set may be expanded as follows:

$$|i\rangle' = A_1|1\rangle^0 + A_2|2\rangle^0 + \cdots + A_j|j\rangle^0 + \cdots \qquad (A\text{-}92)$$

$$\hat{\mathcal{H}}'|i\rangle^0 = H_{1i}|1\rangle^0 + H_{2i}|2\rangle^0 + \cdots + H_{ji}|j\rangle^0 + \cdots \qquad (A\text{-}93)$$

Multiplication of Eq. (A-93) from the left by $^0\langle j|$ gives

$$H_{ji} = {}^0\langle j|\hat{\mathcal{H}}'|i\rangle^0 \qquad\qquad\qquad\qquad (A\text{-}94)$$

since the kets are assumed to be orthonormal.

Rearrangement of Eq. (A-91b) gives

$$(\hat{\mathcal{H}}_0 - W_i^0)|i\rangle' = (W_i' - \hat{\mathcal{H}}')|i\rangle^0 \qquad\qquad (A\text{-}95)$$

Substitution of Eqs. (A-92) and (A-93) into Eq. (A-95) gives

$$\sum_j (W_j^0 - W_i^0)A_j|j\rangle^0$$
$$= W_i'|i\rangle^0 - H_{1i}|1\rangle^0 - H_{2i}|2\rangle^0 - \cdots - H_{ji}|j\rangle^0 - \cdots \quad (A\text{-}96)$$

Multiplication from the left by $^0\langle i|$ gives

$$W_i' = H_{ii} = {}^0\langle i|\hat{\mathcal{H}}'|i\rangle^0 \qquad\qquad\qquad (A\text{-}97)$$

The quantity W_i' is the first-order correction to the energy.

Multiplication of Eq. (A-96) from the left by $^0\langle j|$ gives

$$A_j = \frac{-H_{ji}}{W_j^0 - W_i^0} \qquad i \neq j \qquad\qquad\qquad (A\text{-}98)$$

The coefficient A_i remains to be determined [since for $i = j$, Eq. (A-98) is not valid]. By requiring $|i\rangle$ to be normalized, it is readily shown that $A_i = 0$[†] Hence the first-order correction to the wave function is

$$|i\rangle' = -\sum_j{}' \frac{H_{ji}}{W_j{}^0 - W_i{}^0} |j\rangle^0 \tag{A-99}$$

Here the prime on the summation signifies that the term $i = j$ is not included.

A similar approach to the solution of Eq. (A-91c) yields the second-order corrections to the energies and wave functions

$$W_i{}'' = -\sum_j{}' \frac{H_{ij}H_{ji}}{W_j{}^0 - W_i{}^0} \tag{A-100}$$

$$|i\rangle'' = \sum_k{}' \left[\sum_j{}' \frac{H_{kj}H_{ji}}{(W_i{}^0 - W_k{}^0)(W_i{}^0 - W_j{}^0)} - \frac{H_{ii}H_{ki}}{(W_i{}^0 - W_k{}^0)^2} \right] |k\rangle^0 \tag{A-101}$$

Examples of first- and second-order corrections to energies and wave functions are given in Secs. C-7 and 11-6.

Owing to the occurrence of energy differences in the denominator of Eqs. (A-98) or (A-100), it is necessary to modify the above procedure when applying perturbation methods to systems with degenerate energy levels.[†]

A-8 EULER ANGLES

Some experimenters prefer to specify the orientation of the set of axes X, Y, and Z relative to x, y, and z by the Euler angles shown in Fig. A-6.

† H. Eyring, J. Walter, and G. Kimball, "Quantum Chemistry," pp. 94 to 101, John Wiley & Sons, Inc., New York, 1944.

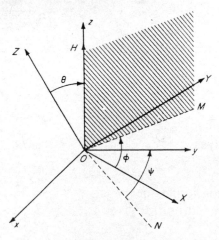

Oxyz – Laboratory axes

OXYZ – Molecular axes

Fig. A-6 Representation of the Euler angles θ, ϕ, and ψ. Line OM lies in the XY plane and represents the projection of the z axis on this plane. ON is the line of intersection of the xOY and XOY planes.

The projection of the z axis on the XY plane determines the line OM. ON represents the intersection of the xOy and XOY planes. With the magnetic field along the z direction, the energies of an electron spin system depend only upon the angles θ and ϕ, since ψ measures the rotation about z and such rotation does not alter the energy of the system.

BIBLIOGRAPHY

Anderson, J. M.: "Mathematics for Quantum Chemistry," W. A. Benjamin, Inc., New York, 1966 (good for operator algebra and perturbation theory).

Bak, T., and J. Lichtenberg: "Vectors, Tensors and Groups," W. A. Benjamin, Inc., New York, 1966 (discussion of vectors is recommended).

Hall, G. G.: "Matrices and Tensors," Pergamon Press, Oxford, England, 1963.

Nye, J. F.: "Physical Properties of Crystals," chaps. 1 and 2, Oxford University Press, Oxford, England, 1957 (excellent discussion of second-rank tensors).

PROBLEMS

A-1. Write the sum $S = A + B$, difference $D = A - B$, and products $P = AB$, $Q = BA$ of the matrices

$$A = \begin{bmatrix} 3 & 1 & -2 \\ 4 & -2 & 3 \\ -2 & 1 & -1 \end{bmatrix} \quad \text{and} \quad B = \begin{bmatrix} 2 & 0 & -1 \\ -4 & 1 & 2 \\ 1 & -1 & 0 \end{bmatrix}$$

A-2. If

$$A = \begin{bmatrix} 2 & 1 \\ -1 & 3 \end{bmatrix} \quad \text{and} \quad B = \begin{bmatrix} -1 & 2 \\ 3 & -2 \end{bmatrix}$$

show that $(A + B)(A - B) \neq A^2 - B^2$.

A-3. Multiply the wave functions in Eqs. (A-20) and (A-21) from the left by the complex-conjugate wave functions. Then integrate the resulting expressions over the appropriate range of ϕ to obtain the energy and the angular momentum for a particle moving in a circle of fixed radius.

A-4. Find $x, y,$ and z such that

$$\begin{bmatrix} 2 & -1 & 3 \\ 1 & 2 & -4 \\ -1 & 3 & -2 \end{bmatrix} \begin{bmatrix} x \\ y \\ z \end{bmatrix} = \begin{bmatrix} 1 \\ -3 \\ 6 \end{bmatrix}$$

A-5. Obtain the eigenvalues of the following matrix by the determinant method:

$$\begin{bmatrix} 5 & 0 & -2 \\ 0 & -3 & 0 \\ -2 & 0 & 2 \end{bmatrix}$$

A-6. Diagonalize the matrix

$$A = \begin{bmatrix} 5 & 1 & -1 \\ 1 & 3 & -1 \\ -1 & -1 & 3 \end{bmatrix}$$

using the eigenvector matrix

$$
C = \begin{bmatrix} 0 & \dfrac{1}{\sqrt{2}} & \dfrac{1}{\sqrt{2}} \\[2ex] \dfrac{1}{\sqrt{3}} & -\dfrac{1}{\sqrt{3}} & \dfrac{1}{\sqrt{3}} \\[2ex] \dfrac{2}{\sqrt{6}} & \dfrac{1}{\sqrt{6}} & -\dfrac{1}{\sqrt{6}} \end{bmatrix}
$$

A-7.   Perform the following matrix multiplication:

$$
\begin{bmatrix} H_x & 0 & 0 \end{bmatrix} \begin{bmatrix} g_{xx} & g_{xy} & g_{xz} \\ g_{xy} & g_{yy} & g_{yz} \\ g_{xz} & g_{yz} & g_{zz} \end{bmatrix} \begin{bmatrix} \hat{S}_x \\ \hat{S}_y \\ \hat{S}_z \end{bmatrix}
$$

A-8.   The coordinates of the corners of the rectangular base of a parallelepiped are (2.191, 0.448,0); (-1.673,1.484,0) (-2.191,-0.448,0); (1.673,-1.484,0).

(*a*) Determine the angle between the *x* axis and the long side of the rectangle.

(*b*) Using the appropriate rotation matrix (A-48), transform the coordinates to an axis system in which the long side of the rectangle is parallel to the *x* axis.

A-9.   A radical R is produced in a single crystal of RH by irradiation.   Axes *x*, *y*, and *z* of R and RH are chosen to coincide with simple directions on the faces of a single crystal.   The direction cosines of the principal axes *1*, *2*, and *3* with respect to the *x*, *y*, and *z* axes are, respectively, (0.48,0.88,0.05), (0.07,-0.10,0.99), and (0.87,-0.47,-0.11).   The elements of the *g* tensor are as follows: $g_{xx} = 2.0039$, $g_{yy} = 2.0030$, $g_{zz} = 2.0035$, $g_{xy} = -0.0007$, $g_{xz} = -0.0001$, $g_{yz} = 0$.   Find the principal elements g_{11}, g_{22}. and g_{33}.

A-10.   Prove that the coordinate-rotation matrix

$$
\begin{bmatrix} \cos\phi & \sin\phi & 0 \\ -\sin\phi & \cos\phi & 0 \\ 0 & 0 & 1 \end{bmatrix}
$$

is unitary.

A-11.   Verify by matrix multiplication of Eq. (A-71) that Eqs. (A-72) represent the requirement for diagonalization of a symmetric 2 × 2 matrix.

Appendix B
Quantum Mechanics of
Angular Momentum

B-1   INTRODUCTION

Those properties (such as the total energy or angular momentum) which are conserved in an isolated system are called "constants of motion." These constants and their corresponding operators play an important role in quantum mechanics. Consider the operator $\hat{\mathscr{H}}$ for the total energy, i.e., the hamiltonian operator. When $\hat{\mathscr{H}}$ operates on an eigenfunction $\psi_n(x,y,z)$ of a particular system the result is

$$\hat{\mathscr{H}}\psi_n(x,y,z) = W_n\psi_n(x,y,z) \tag{B-1}$$

Here W_n is the *exact* total energy of the nth state of the system. Equation (B-1) is the Schrödinger equation. Such exact values W_n are called "eigenvalues." Operators which do not correspond to constants of motion do not yield eigenvalues. Suppose now that λ_n is another constant of motion with a corresponding operator $\hat{\Lambda}$. It is an important result of quantum mechanics that $\psi_n(x,y,z)$, the eigenfunctions of $\hat{\mathscr{H}}$, can always be chosen in such a way that they are also eigenfunctions of $\hat{\Lambda}$. Angular momentum is a constant of motion for an isolated system; this implies that the operator for the total angular momentum (or more exactly the square of the total angular momen-

419

tum) should also produce an exact result when applied to an appropriately chosen $\psi_n(x,y,z)$. If this operator is designated by $\hat{J}^2$, then

$$\hat{J}^2\psi_n(x,y,z) = \lambda_j\psi_n(x,y,z) \tag{B-2}$$

Here λ_j is an eigenvalue of $\hat{J}^2$ and is expressed in units of $(h/2\pi)^2 = \hbar^2$. By an appropriate choice of an axis system, $\psi_n(x,y,z)$ can be made to be an eigenfunction of any *one* of the components of $\hat{\mathbf{J}}$, i.e., $\hat{J}_x, \hat{J}_y$, or $\hat{J}_z$, in addition to being an eigenfunction of $\hat{J}^2$ and of $\hat{\mathcal{H}}$. However, $\psi_n(x,y,z)$ can never be a simultaneous eigenfunction of two or more of the components of $\hat{\mathbf{J}}$. If $\psi_n(x,y,z)$ is to be a simultaneous eigenfunction of any group of operators, each of which corresponds to a constant of motion, all of the operators must commute among themselves. For instance, if $\psi_n(x,y,z)$ is to be simultaneously an eigenfunction of $\hat{\mathcal{H}}, \hat{J}^2$, and $\hat{J}_z$, then the following relations must hold:

$$(\hat{\mathcal{H}}\hat{J}^2 - \hat{J}^2\hat{\mathcal{H}})\psi_n(x,y,z) = 0 \tag{B-3a}$$

$$(\hat{\mathcal{H}}\hat{J}_z - \hat{J}_z\hat{\mathcal{H}})\psi_n(x,y,z) = 0 \tag{B-3b}$$

and

$$(\hat{J}^2\hat{J}_z - \hat{J}_z\hat{J}^2)\psi_n(x,y,z) = 0 \tag{B-3c}$$

λ_j, the eigenvalue of $\hat{J}^2$, has an important property. Its value is determined only by the *symmetry* properties of ψ_n. Thus, although the values of W_n vary according to the wave function used, λ_j will be the same for all wave functions which have the same symmetry characteristics. For example, all spherically symmetric wave functions have $\lambda_j = 0$. Hence, if one can obtain a set of solutions to Eq. (B-2), these eigenvalues will be generally applicable to all wave functions of the same symmetry.

$\psi_n(x,y,z)$ has been represented as a three-dimensional wave function. In this case it is found that λ_j is a function of the quantum number j, which is an integer. For instance, j could be the electron orbital angular-momentum quantum number l or the molecular rotation quantum number J. However, electrons and nuclei possess an intrinsic angular momentum (called spin angular momentum) which cannot be described in terms of the spatial wave functions $\psi_n(x,y,z)$. In order to account for this spin angular momentum, a "spin coordinate" must be included in the wave function. It is found that the quantum number j can then take on half-integral values as well (that is, $j = 0, \frac{1}{2}, 1, \frac{3}{2}, \ldots$). This spin angular momentum has no classical counterpart; however, it can be accommodated in the quantum mechanics if a generalized angular momentum is defined. (See Sec. B-3.)

B-2 ANGULAR–MOMENTUM OPERATORS

In order to obtain the quantum-mechanical operators for angular momentum, one must first consider the classical expression for the angular momentum $\mathbf{p}_\phi$ of a particle about an origin O

$$\mathbf{p}_\phi = \mathbf{r} \times \mathbf{p} \qquad (\text{B-4})$$

Here $\mathbf{r}$ represents the position vector of the particle, and $\mathbf{p}$ is its linear-momentum vector. The vector product indicated in Eq. (B-4) can be computed as in Eq. (A-31b). The components of $\mathbf{p}_\phi$ are

$$p_{\phi_x} = yp_z - zp_y \qquad (\text{B-5})$$

$$p_{\phi_y} = zp_x - xp_z \qquad (\text{B-6})$$

$$p_{\phi_z} = xp_y - yp_x \qquad (\text{B-7})$$

The operator $\hat{\mathbf{r}}$ remains the same as $\mathbf{r}$ in classical mechanics but $\mathbf{p}$ must be replaced by $-i\hbar\hat{\mathbf{\nabla}}$, where $\hat{\mathbf{\nabla}} = \mathbf{i}\partial/\partial x + \mathbf{j}\partial/\partial y + \mathbf{k}\partial/\partial z$. Thus in quantum mechanics, the components of $\hat{\mathbf{p}}_\phi$ become

$$\hat{p}_{\phi_x} = -i\hbar\left(y\frac{\partial}{\partial z} - z\frac{\partial}{\partial y}\right) \qquad (\text{B-8})$$

$$\hat{p}_{\phi_y} = -i\hbar\left(z\frac{\partial}{\partial x} - x\frac{\partial}{\partial z}\right) \qquad (\text{B-9})$$

and

$$\hat{p}_{\phi_z} = -i\hbar\left(x\frac{\partial}{\partial y} - y\frac{\partial}{\partial x}\right) \qquad (\text{B-10})$$

The angular-momentum operator is more conveniently represented in terms of the operator $\hat{\mathbf{J}}$, which is measured in units of $\hbar$, that is, $\hat{\mathbf{p}}_\phi = \hat{\mathbf{J}}\hbar$, and hence

$$\hat{J}_x = -i\left(y\frac{\partial}{\partial z} - z\frac{\partial}{\partial y}\right) \qquad (\text{B-11})$$

$$\hat{J}_y = -i\left(z\frac{\partial}{\partial x} - x\frac{\partial}{\partial z}\right) \qquad (\text{B-12})$$

and

$$\hat{J}_z = -i\left(x\frac{\partial}{\partial y} - y\frac{\partial}{\partial x}\right) \qquad (\text{B-13})$$

As in the classical mechanics of angular momentum, the square of a vector operator is equivalent to the sum of the squares of its three component operators, that is,

$$\hat{J}^2 = \hat{J}_x^2 + \hat{J}_y^2 + \hat{J}_z^2 \qquad (\text{B-14})$$

In classical mechanics the magnitude and direction of an angular-momentum vector are well defined. In quantum mechanics only the magnitude of the *total* angular momentum vector and *one* of its components may simultaneously have well-defined values. It is possible to determine two observables simultaneously only if the operators corresponding to them commute. (See Sec. B-1.)

It will be seen in Sec. B-3 that $\hat{J}^2$ commutes with $\hat{J}_z$; however, $\hat{J}_z$ does not commute with $\hat{J}_x$ or $\hat{J}_y$.

B-3 THE COMMUTATION RELATIONS FOR THE ANGULAR–MOMENTUM OPERATORS

By expanding $\hat{J}^2$, it can be shown that $\hat{J}^2$ commutes with any *one* of the components $\hat{J}_x$, $\hat{J}_y$, or $\hat{J}_z$, that is,

$$[\hat{J}^2, \hat{J}_x] = [\hat{J}^2, \hat{J}_y] = [\hat{J}^2, \hat{J}_z] = 0 \tag{B-15}$$

Here, $[\hat{J}^2, \hat{J}_x] = [\hat{J}^2\hat{J}_x - \hat{J}_x\hat{J}^2]$ etc. However, no two components commute among themselves. For example,

$$\hat{J}_x\hat{J}_y = -\left(y\frac{\partial}{\partial z} - z\frac{\partial}{\partial y}\right)\left(z\frac{\partial}{\partial x} - x\frac{\partial}{\partial z}\right)$$

$$= -\left[y\frac{\partial}{\partial x} + yz\frac{\partial^2}{\partial z\,\partial x} - xy\frac{\partial^2}{\partial z^2} - z^2\frac{\partial^2}{\partial x\,\partial y} + xz\frac{\partial^2}{\partial y\,\partial z}\right] \tag{B-16}$$

$$\hat{J}_y\hat{J}_x = -\left(z\frac{\partial}{\partial x} - x\frac{\partial}{\partial z}\right)\left(y\frac{\partial}{\partial z} - z\frac{\partial}{\partial y}\right)$$

$$= -\left[yz\frac{\partial^2}{\partial x\,\partial z} - z^2\frac{\partial^2}{\partial x\,\partial y} - xy\frac{\partial^2}{\partial z^2} + xz\frac{\partial^2}{\partial z\,\partial y} + x\frac{\partial}{\partial y}\right] \tag{B-17}$$

$$[\hat{J}_x, \hat{J}_y] = \hat{J}_x\hat{J}_y - \hat{J}_y\hat{J}_x = \left(x\frac{\partial}{\partial y} - y\frac{\partial}{\partial x}\right) = i\hat{J}_z \tag{B-18}$$

Similar expressions hold for the commutators of the other components. Consequently, it is not possible simultaneously to determine three or even two components of the angular-momentum operator. However, it is possible to determine the square of the magnitude of $\hat{\mathbf{J}}$ and *one* of the components of $\hat{\mathbf{J}}$, which is taken as $\hat{J}_z$. These important commutation relations are summarized as follows:

$$[\hat{J}_x, \hat{J}_y] = i\hat{J}_z \tag{B-19a}$$

$$[\hat{J}_y, \hat{J}_z] = i\hat{J}_x \tag{B-19b}$$

$$[\hat{J}_z, \hat{J}_x] = i\hat{J}_y \tag{B-19c}$$

and

$$[\hat{J}^2, \hat{J}_z] = [\hat{J}^2, \hat{J}_y] = [\hat{J}^2, \hat{J}_x] = 0 \tag{B-15}$$

A generalized angular momentum (i.e., one which includes spin) *is defined as any function which obeys the commutation relations of (B-19) and (B-15).*

At this point it is convenient to introduce the so-called "ladder" operators, which are linear combinations of $\hat{J}_x$ and $\hat{J}_y$:

$$\hat{J}_+ = \hat{J}_x + i\hat{J}_y$$

and

$$\tag{B-20}$$

$$\hat{J}_- = \hat{J}_x - i\hat{J}_y$$

$\hat{J}_+$ is called the "raising" operator and $\hat{J}_-$ the "lowering" operator. The significance of these operators will become apparent later. As can readily be verified by substitution of their definitions, they obey the commutation relations

$$[\hat{J}^2, \hat{J}_+] = [\hat{J}^2, \hat{J}_-] = 0$$
$$[\hat{J}_z, \hat{J}_-] = -\hat{J}_- \qquad [\hat{J}_z, \hat{J}_+] = \hat{J}_+$$

and   (B-21)

$$[\hat{J}_+, \hat{J}_-] = 2\hat{J}_z$$

B-4 THE EIGENVALUES OF $\hat{J}^2$ AND $\hat{J}_z$

Let the eigenvalues of $\hat{J}^2$ and $\hat{J}_z$ be λ_j and λ_m, respectively. The angular-momentum eigenvalues depend only on the quantum numbers j and m. j is characteristic of the total angular momentum, and m is characteristic of the z component of the angular momentum. Consequently, as far as the angular-momentum properties are concerned, the eigenfunctions are solely a function of j and m.

Since the angular-momentum properties of the wave functions depend only on j and m, the functions can be represented by the kets $|j, m\rangle$. (See Sec. A-5d.) The eigenvalue equations for $\hat{J}^2$ and $\hat{J}_z$ can then be written

$$\hat{J}^2|j, m\rangle = \lambda_j|j, m\rangle \qquad\qquad\qquad (\text{B-22})$$

and

$$\hat{J}_z|j, m\rangle = \lambda_m|j, m\rangle \qquad\qquad\qquad (\text{B-23})$$

Here λ_j and λ_m are the eigenvalues of $\hat{J}^2$ and $\hat{J}_z$, respectively. Note that the $|j, m\rangle$ kets are orthonormal. This means that

$$\langle j', m'|j, m\rangle = 1 \qquad \text{for } j' = j \text{ and } m' = m$$
$$= 0 \qquad \text{for } j' \neq j \text{ or } m' \neq m \qquad (\text{B-24})$$

The operator $\hat{J}^2$ may be expanded as

$$\hat{J}^2 = \hat{J}_x{}^2 + \hat{J}_y{}^2 + \hat{J}_z{}^2 \qquad\qquad\qquad (\text{B-14})$$
$$\hat{J}_x{}^2 + \hat{J}_y{}^2 = \hat{J}^2 - \hat{J}_z{}^2 \qquad\qquad\qquad (\text{B-25})$$

Thus the operator $(\hat{J}_x{}^2 + \hat{J}_y{}^2)$ also has the discrete eigenvalues

$$(\hat{J}_x{}^2 + \hat{J}_y{}^2)|j, m\rangle = (\hat{J}^2 - \hat{J}_z{}^2)|j, m\rangle$$
$$= (\lambda_j - \lambda_m{}^2)|j, m\rangle \qquad (\text{B-26})$$

Since the operators $\hat{J}_x{}^2$ and $\hat{J}_y{}^2$ correspond to experimental observables (and hence are hermitian), when they are applied to $|j, m\rangle$, they must give real numbers. Hence the eigenvalues of $(\hat{J}_x{}^2 + \hat{J}_y{}^2)$ must be *real and positive*, that is,

$$\lambda_j - \lambda_m{}^2 \geqslant 0 \tag{B-27}$$

To establish the exact form of the eigenvalues λ_m, it is convenient to examine the matrix elements of the commutator $[\hat{J}_z,\hat{J}_+] = \hat{J}_+$ from Eq. (B-21),

$$\langle j, m'|\hat{J}_z\hat{J}_+ - \hat{J}_+\hat{J}_z|j, m\rangle = \langle j, m'|\hat{J}_+|j, m\rangle \tag{B-28}$$

Evaluation of the left-hand side of Eq. (B-28) demonstrates the effect of $\hat{J}_+$ on the wave functions $|j, m\rangle$. The left-hand matrix element can be expanded into two matrix elements

$$\langle j, m'|\hat{J}_z\hat{J}_+|j, m\rangle - \langle j, m'|\hat{J}_+\hat{J}_z|j, m\rangle$$

Use of Eq. (B-23) allows the second matrix element to be reduced to

$$\langle j, m'|\hat{J}_+\hat{J}_z|j, m\rangle = \lambda_m\langle j, m'|\hat{J}_+|j, m\rangle$$

The first matrix element can be reduced with the help of Eq. (A-54),

$$\langle j, m'|\hat{J}_z\hat{J}_+|j, m\rangle = \lambda_{m'}^*\langle j, m'|\hat{J}_+|j, m\rangle \tag{B-29}$$

Since $\hat{J}_z$ is a hermitian operator, $\lambda_{m'}$ must be real, and $\lambda_{m'}^* = \lambda_{m'}$. Thus Eq. (B-28) reduces to

$$(\lambda_{m'} - \lambda_m)\langle j, m'|\hat{J}_+|j, m\rangle = \langle j, m'|\hat{J}_+|j, m\rangle \tag{B-30}$$

This means that the only nonzero matrix elements of $\hat{J}_+$ are those for which $(\lambda_{m'} - \lambda_m) = +1$. Hence

$$\hat{J}_+|j, m\rangle = x_m|j, m + 1\rangle \tag{B-31}$$

This is easily seen from the fact that if $\lambda_{m'} - \lambda_m \neq +1$, then Eq. (B-30) is satisfied only if $\langle j, m'|\hat{J}_+|j, m\rangle = 0$. Similarly, an examination of the matrix element of $\hat{J}_-$ shows that the only nonzero matrix elements of $\hat{J}_-$ are those for which $(\lambda_{m'} - \lambda_m) = -1$, that is,

$$\hat{J}_-|j, m\rangle = y_m|j, m - 1\rangle \tag{B-32}$$

Here x_m and y_m of (B-31) and (B-32) may be complex numbers. The factor $e^{i\phi}$, where ϕ is a phase angle, may appear in x_m and y_m. (See Sec. A-1.) It is apparent from Eqs. (B-31) and (B-32) why $\hat{J}_+$ and $\hat{J}_-$ are called raising and lowering operators, respectively.

This analysis shows that for a given value of λ_j, one may obtain a whole series of states $|j, m\rangle$ having the eigenvalues

$$\cdots \lambda_{m-2}, \lambda_{m-1}, \lambda_m, \lambda_{m+1}, \lambda_{m+2} \cdots$$

This series must terminate at both ends, since from Eq. (B-27), $\lambda_m{}^2 \leqslant \lambda_j$. Since the λ_m values differ by integers and the m quantum number is assumed to increase in integral steps for a given value of j, one may equate λ_m and m.

Within the above series, the lowest eigenvalue of $\hat{J}_z$ is designated by $\underline{m}$ and the highest eigenvalue of $\hat{J}_z$ by $\overline{m}$.
Therefore

$$\hat{J}_+ |j, \overline{m}\rangle = 0 \qquad\qquad (\text{B-33})$$

$$\hat{J}_- |j, \underline{m}\rangle = 0 \qquad\qquad (\text{B-34})$$

Otherwise, there would be a value of $\overline{m}$ higher than $\lambda_j^{\frac{1}{2}}$ and a value of $\underline{m}$ less than $-\lambda_j^{\frac{1}{2}}$; this is contrary to the limitation imposed by Eq. (B-27).

Next apply $\hat{J}_-$ to Eq. (B-33). When $\hat{J}_-\hat{J}_+$ is expanded,

$$\begin{aligned}
\hat{J}_-\hat{J}_+ &= (\hat{J}_x - i\hat{J}_y)(\hat{J}_x + i\hat{J}_y) \\
&= \hat{J}_x^2 + \hat{J}_y^2 + i[\hat{J}_x, \hat{J}_y] \\
&= \hat{J}^2 - \hat{J}_z^2 - \hat{J}_z \qquad\qquad (\text{B-35})
\end{aligned}$$

Therefore

$$\hat{J}_-\hat{J}_+|j, \overline{m}\rangle = (\lambda_j - \overline{m}^2 - \overline{m})|j, \overline{m}\rangle = 0 \qquad\qquad (\text{B-36})$$

that is,

$$\lambda_j = \overline{m}(\overline{m} + 1) \qquad\qquad (\text{B-37})$$

Similarly, by applying $\hat{J}_+$ to Eq. (B-34)

$$\lambda_j = \underline{m}(\underline{m} - 1) \qquad\qquad (\text{B-38})$$

Equations (B-37) and (B-38) are compatible only if $\overline{m} = -\underline{m}$.

Since successive values of m differ by unity, $(\overline{m} - \underline{m})$ is a positive integer which is denoted by $2j$. Hence j can have the values

$$j = 0, \tfrac{1}{2}, 1, \tfrac{3}{2}, \ldots$$

Then from $\overline{m} - \underline{m} = 2j$ and $\overline{m} = -\underline{m}$.

$$\overline{m} = j \qquad \text{and} \qquad \underline{m} = -j \qquad\qquad (\text{B-39})$$

Hence $m = j, j - 1, \ldots, -j + 1, -j$. There are thus $2j + 1$ permissible values of m for each value of j. From Eqs. (B-37) and (B-38)

$$\lambda_j = \overline{m}(\overline{m} + 1) = j(j + 1) \qquad\qquad (\text{B-40})$$

The eigenvalues of $\hat{J}^2$ and $\hat{J}_z$ are then

$$\hat{J}^2|j, m\rangle = j(j + 1)|j, m\rangle \qquad\qquad (\text{B-41})$$

and

$$\hat{J}_z|j, m\rangle = m|j, m\rangle \dagger \qquad\qquad (\text{B-42})$$

$\dagger$ Note that if the integer $(\overline{m} - \underline{m})$ had been set equal to j, then the eigenvalue of $\hat{J}^2$ would be $(j/2)[(j/2) + 1]$. Hence half-integral quantum numbers appear quite naturally in this treatment. Half-integral values of j occur for electron and nuclear spin angular momenta.

B-5 THE MATRIX ELEMENTS OF $\hat{J}_+$, $\hat{J}_-$, $\hat{J}_x$ AND $\hat{J}_y$

The quantities x_m and y_m of Eqs. (B-31) and (B-32) remain to be evaluated. Consider the matrix element

$$\langle j, m|\hat{J}_-\hat{J}_+|j, m\rangle = x_m\langle j, m|\hat{J}_-|j, m+1\rangle$$
$$= x_m y_{m+1}\langle j, m|j, m\rangle = x_m y_{m+1} \quad \text{(B-43)}$$

But from Eq. (B-36)

$$\langle j, m|\hat{J}_-\hat{J}_+|j, m\rangle = [j(j+1) - m^2 - m] \quad \text{(B-44)}$$

Thus

$$x_m y_{m+1} = [j(j+1) - m(m+1)] \quad \text{(B-45)}$$

The matrix element of Eqs. (B-43) may be evaluated in a third way:

$$\langle j, m|\hat{J}_-\hat{J}_+|j, m\rangle = x_m[\langle j, m|\hat{J}_-|j, m+1\rangle]^\dagger$$

$$\leftarrow$$

$$= x_m[\langle j, m+1|\hat{J}_+|j, m\rangle]^*$$

$$\rightarrow$$

$$= x_m x_m^*\langle j, m+1|j, m+1\rangle$$

$$= x_m x_m^* \quad \text{(B-46)}$$

Equation (A-54) has again been used together with the fact that $\hat{J}_-^* = \hat{J}_+$. Thus from Eqs. (B-43) and (B-44)

$$y_{m+1} = x_m^*$$

and

$$x_m x_m^* = |x_m|^2 = [j(j+1) - m(m+1)]$$

or

$$x_m = [j(j+1) - m(m+1)]^{\frac{1}{2}} \quad \text{(B-47)}$$

The value of x_m in Eq. (B-47) should have been multiplied by $e^{i\phi}$, where ϕ is a phase angle such that $|e^{i\phi}|^2 = 1$. By convention, ϕ is chosen to be zero. If this convention is applied consistently, then this choice has no effect on the final results since the experimental observables correspond to real numbers. Similarly, from the matrix element of $\hat{J}_+\hat{J}_-$

$$y_m = [j(j+1) - m(m-1)]^{\frac{1}{2}} \quad \text{(B-48)}$$

Hence the operation of $\hat{J}_+$ and $\hat{J}_-$ on $|j, m\rangle$ gives the results

$$\hat{J}_+|j, m\rangle = [j(j+1) - m(m+1)]^{\frac{1}{2}}|j, m+1\rangle \quad \text{(B-49)}$$

and

$$\hat{J}_-|j, m\rangle = [j(j+1) - m(m-1)]^{\frac{1}{2}}|j, m-1\rangle \quad \text{(B-50)}$$

At this point one may write the nonzero matrix elements of $\hat{J}_+, \hat{J}_-, \hat{J}_x,$ and $\hat{J}_y$ as follows:

$$\langle j, m+1|\hat{J}_+|j, m\rangle = [j(j+1) - m(m+1)]^{\frac{1}{2}} \tag{B-51}$$

$$\langle j, m-1|\hat{J}_-|j, m\rangle = [j(j+1) - m(m-1)]^{\frac{1}{2}} \tag{B-52}$$

$$\langle j, m+1|\hat{J}_x|j, m\rangle = \tfrac{1}{2}[j(j+1) - m(m+1)]^{\frac{1}{2}} \tag{B-53}$$

$$\langle j, m-1|\hat{J}_x|j, m\rangle = \tfrac{1}{2}[j(j+1) - m(m-1)]^{\frac{1}{2}} \tag{B-54}$$

$$\langle j, m+1|\hat{J}_y|j, m\rangle = \frac{-i}{2}[j(j+1) - m(m+1)]^{\frac{1}{2}} \tag{B-55}$$

and

$$\langle j, m-1|\hat{J}_y|j, m\rangle = \frac{i}{2}[j(j+1) - m(m-1)]^{\frac{1}{2}} \tag{B-56}$$

Equations (B-53) to (B-56) follow from Eq. (B-20); that is, $\hat{J}_x = \tfrac{1}{2}(\hat{J}_+ + \hat{J}_-)$ and $\hat{J}_y = (1/2i)(\hat{J}_+ - \hat{J}_-)$.

B-6 ANGULAR–MOMENTUM MATRICES

For a given value of j, the matrix elements such as those in Eqs. (B-51) to (B-56) are conveniently arrayed as a square matrix. The order of the matrix will be $(2j + 1)$, corresponding to the possible values of m. Consider the spin matrices for $j = \tfrac{1}{2}$. These will be directly applicable to the electron spin case, where $S = \tfrac{1}{2}$, and to the nuclear spin cases with $I = \tfrac{1}{2}$.

$$\mathbf{J}_x = \begin{array}{c} \\ \langle\tfrac{1}{2}, \tfrac{1}{2}| \\ \langle\tfrac{1}{2}, -\tfrac{1}{2}| \end{array} \overset{\displaystyle |\tfrac{1}{2}, \tfrac{1}{2}\rangle \quad |\tfrac{1}{2}, -\tfrac{1}{2}\rangle}{\begin{bmatrix} 0 & \tfrac{1}{2} \\ \tfrac{1}{2} & 0 \end{bmatrix}} = \tfrac{1}{2}\begin{bmatrix} 0 & 1 \\ 1 & 0 \end{bmatrix} \tag{B-57}$$

The elements appearing in the $\mathbf{J}_x$ matrix (B-57) are obtained by inserting the $\hat{J}_x$ operator between the corresponding bra to the left of a given matrix element and the ket above that element. For example, the a_{12} element of Eq. (B-57) is computed as

$$\langle \tfrac{1}{2}, \tfrac{1}{2}|\hat{J}_x|\tfrac{1}{2}, -\tfrac{1}{2}\rangle = \tfrac{1}{2}[\tfrac{1}{2}(\tfrac{3}{2}) - (-\tfrac{1}{2})(\tfrac{1}{2})]^{\frac{1}{2}} = \tfrac{1}{2}$$

In a similar fashion $\mathbf{J}_y$ and $\mathbf{J}_z$ are written as

$$\mathbf{J}_y = \begin{bmatrix} 0 & -\dfrac{i}{2} \\ \dfrac{i}{2} & 0 \end{bmatrix} = \frac{1}{2}\begin{bmatrix} 0 & -i \\ i & 0 \end{bmatrix} \tag{B-58}$$

$$\mathbf{J}_z = \begin{bmatrix} \tfrac{1}{2} & 0 \\ 0 & -\tfrac{1}{2} \end{bmatrix} = \tfrac{1}{2}\begin{bmatrix} 1 & 0 \\ 0 & -1 \end{bmatrix} \tag{B 59}$$

The matrices on the right of Eqs. (B-57) to (B-59) are often called the Pauli spin matrices, symbolized by $\boldsymbol{\sigma}_x, \boldsymbol{\sigma}_y,$ and $\boldsymbol{\sigma}_z$. Hence

$$\mathbf{J}_i = \tfrac{1}{2}\boldsymbol{\sigma}_i \qquad i = x, y, \text{ or } z \tag{B-60}$$

One can obtain the J_+ and J_- matrices either from Eqs. (B-51) and (B-52) or from matrix addition, that is,

$$J_+ = J_x + iJ_y = \begin{bmatrix} 0 & 1 \\ 0 & 0 \end{bmatrix} \qquad (B-61)$$

and

$$J_- = J_x - iJ_y = \begin{bmatrix} 0 & 0 \\ 1 & 0 \end{bmatrix} \qquad (B-62)$$

Since the eigenvalue of $\hat{J}^2$ for each spin function in the case of $j = \frac{1}{2}$ must be $\frac{1}{2}(\frac{1}{2} + 1) = \frac{3}{4}$, the matrix J^2 is

$$J^2 = \begin{bmatrix} \frac{3}{4} & 0 \\ 0 & \frac{3}{4} \end{bmatrix} \qquad (B-63)$$

This can be verified by computing

$$J^2 = J_x{}^2 + J_y{}^2 + J_z{}^2 \qquad (B-64)$$

For instance, matrix multiplication of J_x by itself gives

$$\begin{bmatrix} 0 & \frac{1}{2} \\ \frac{1}{2} & 0 \end{bmatrix} \begin{bmatrix} 0 & \frac{1}{2} \\ \frac{1}{2} & 0 \end{bmatrix} = \begin{bmatrix} \frac{1}{4} & 0 \\ 0 & \frac{1}{4} \end{bmatrix}$$

with identical results for $J_y{}^2$ and $J_z{}^2$. Addition of the matrices in (B-64) yields the desired result (B-63).

B-7 ADDITION OF ANGULAR MOMENTA

One often encounters problems in which there are two angular momenta which may or may not be coupled by an interaction. The necessity for considering interaction of angular momenta arises in the following cases:

1. Coupling of electron spin and orbital angular momenta.
2. Coupling of the angular momenta of two different particles.

We shall begin by considering two angular momenta J_1 and J_2 which initially are not coupled. The eigenfunctions for J_1 and J_2 will be taken as $|j_1, m_1\rangle$ and $|j_2, m_2\rangle$, respectively. Thus

$$\hat{J}_1{}^2|j_1, m_1\rangle = j_1(j_1 + 1)|j_1, m_1\rangle \qquad \hat{J}_2{}^2|j_2, m_2\rangle = j_2(j_2 + 1)|j_2, m_2\rangle$$

$$\hat{J}_{1z}|j_1, m_1\rangle = m_1|j_1, m_1\rangle \qquad \hat{J}_{2z}|j_2, m_2\rangle = m_2|j_2, m_2\rangle \qquad (B-65)$$

The direct-product representation $|j_1, m_1\rangle|j_2, m_2\rangle \equiv |j_1, j_2, m_1, m_2\rangle$ will be called the *uncoupled* representation.

The total angular momentum J is defined by

$$J = J_1 + J_2 \qquad (B-66)$$

If **J** is to be an angular momentum, its components must satisfy the commutation relations Eqs. (B-19). For example,

$$[\hat{J}_x, \hat{J}_y] = [\hat{J}_{1x} + \hat{J}_{2x}, \hat{J}_{1y} + \hat{J}_{2y}]$$
$$= [\hat{J}_{1x}, \hat{J}_{1y}] + [\hat{J}_{1x}, \hat{J}_{2y}] + [\hat{J}_{2x}, \hat{J}_{1y}] + [\hat{J}_{2x}, \hat{J}_{2y}] \tag{B-67}$$

Each of the middle two commutators is zero, since angular momenta in different spaces commute. Thus

$$[\hat{J}_x, \hat{J}_y] = i\hat{J}_{1z} + i\hat{J}_{2z} = i\hat{J}_z \tag{B-68}$$

The representation $|j_1, j_2, j, m\rangle$ of the eigenfunctions of $\hat{J}_1{}^2$, $\hat{J}_2{}^2$, $\hat{J}^2$, and $\hat{J}_z$ is called the *coupled* representation. Thus

$$\hat{J}_1{}^2 |j_1, j_2, j, m\rangle = j_1(j_1 + 1)|j_1, j_2, j, m\rangle$$
$$\hat{J}^2 |j_1, j_2, j, m\rangle = j(j + 1)|j_1, j_2, j, m\rangle$$
$$\hat{J}_2{}^2 |j_1, j_2, j, m\rangle = j_2(j_2 + 1)|j_1, j_2, j, m\rangle$$
$$\hat{J}_z |j_1, j_2, j, m\rangle = m|j_1, j_2, j, m\rangle \tag{B-69}$$

The coupled and uncoupled representations are connected by the transformation

$$|j_1, j_2, j, m\rangle = \sum_{m_1, m_2} C(j_1 j_2 j; m_1 m_2 m)|j_1, j_2, m_1, m_2\rangle \dagger \tag{B-70}$$

The factors $C(j_1 j_2 j; m_1 m_2 m)$ are variously called vector-coupling, Clebsch-Gordan, or Wigner coefficients.

If the operator $\hat{J}_z = \hat{J}_{1z} + \hat{J}_{2z}$ is applied to (B-70) one gets

$$m|j_1, j_2, j, m\rangle = \sum_{m_1, m_2} (m_1 + m_2) C(j_1 j_2 j; m_1 m_2 m)|j_1, j_2, m_1, m_2\rangle$$

or

$$\sum_{m_1, m_2} (m - m_1 - m_2) C(j_1 j_2 j; m_1 m_2 m)|j_1, j_2, m_1, m_2\rangle = 0 \tag{B-71}$$

Since the $|j_1, j_2, m_1, m_2\rangle$ functions are linearly independent, the above sum can vanish only if the coefficient of each term is identically zero; hence

$$(m - m_1 - m_2) C(j_1 j_2 j; m_1 m_2 m) = 0$$

Thus $C(j_1 j_2 j; m_1 m_2 m) = 0$ unless $m_1 + m_2 = m$. \tag{B-72}

Thus m, m_1, and m_2 are not independent, and the sum in Eq. (B-70) can be replaced by a sum over m_1, since $m_2 = m - m_1$; that is,

$$|j_1, j_2, j, m\rangle = \sum_{m_1} C(j_1 j_2 j; m_1, m - m_1)|j_1, j_2, m_1, m - m_1\rangle \tag{B-73}$$

---

† The function on the left-hand side of this equation is in the coupled representation, whereas the functions on the right hand side are in the uncoupled representation. The "equals" sign here should be taken to mean equivalence.

Further restrictions on the vector-coupling (VC) coefficients can be derived from the orthonormal properties of the $|j_1, j_2, j, m\rangle$ eigenfunctions. That is,

$$\langle j_1 j_2, j', m' | j_1, j_2, j, m \rangle = \delta_{jj'} \delta_{mm'}$$

$$= \sum_{m'_1} \sum_{m_1} C(j_1 j_2 j; m_1, m - m_1) C(j_1 j_2 j'; m'_1, m' - m'_1)$$

$$\times \langle j_1, j_2, m_1, m - m_1 | j_1, j_2, m'_1, m' - m'_1 \rangle \quad \text{(B-74)}$$

The VC coefficients have been assumed to be real in Eq. (B-74). Note that j and j' are obtained from the same values of j_1 and j_2.

Equation (B-74) thus restricts the sum in Eq. (B-73) to functions that have the same values of j and m. Since the values of j_1 and j_2 must also be the same, the notation may be simplified to

$$|j, m\rangle = \sum_{m_1} C(j_1 j_2 j; m_1, m - m_1) |m_1, m_2\rangle \quad \text{(B-75)}$$

where $m = m_1 + m_2$.

Nothing has yet been said about the ranges of j and m. Since $\mathbf{J}$ is a generalized angular momentum, the restriction found in Eq. (B-39) will apply; that is,

$$\overline{m} = j \quad \text{and} \quad \underline{m} = -j \quad \text{(B-39)}$$

Here $\overline{m}$ and $\underline{m}$ are, respectively, the maximum and minimum values of m. Since $m = m_1 + m_2$, the maximum value of m for all values of j is $j_1 + j_2$. This must also be the maximum value of j; otherwise there would exist a larger value of m. Thus

$$\overline{j} = j_1 + j_2 \quad \text{(B-76)}$$

where $\overline{j}$ is the maximum value of j. When j and m have their maximum values, the relation between the coupled and uncoupled representations is especially simple, since there is only one permissible value of m_1 (and hence of m_2). Thus from Eq. (B-75)

$$|\overline{j}, \overline{m}\rangle = C(j_1 j_2, j_1 + j_2; j_1 j_2) |\overline{m_1}, \overline{m_2}\rangle \quad \text{(B-77a)}$$

$$= |\overline{m_1}, \overline{m_2}\rangle$$

The standard convention takes $C(j_1 j_2, j_1 + j_2; j_1 j_2) = 1$.

Similarly

$$|\overline{j}, \underline{m}\rangle = |\underline{m_1}, \underline{m_2}\rangle \quad \text{(B-77b)}$$

(B-77a) and (B-77b) are two of the $2j + 1$ functions in the set with $j = \overline{j}$. The other members of this set can be obtained by applying $\hat{J}_- = \hat{J}_{1-} + \hat{J}_{2-}$ to (B-77a) or by applying $\hat{J}_+ = \hat{J}_{1+} + \hat{J}_{2+}$ to (B-77b); for example,

$$\hat{J}_- |\overline{j}, \overline{m}\rangle = (\hat{J}_{1-} + \hat{J}_{2-}) |\overline{m_1}, \overline{m_2}\rangle$$

or

$$y_{\overline{m}} \, |\overline{j}, \overline{m} - 1\rangle = y_{\overline{m}_1} |\overline{m}_1 - 1, \overline{m}_2\rangle + y_{\overline{m}_2} |\overline{m}_1, \overline{m}_2 - 1\rangle \qquad (B\text{-}78)$$

The y coefficients are obtained from Eq. (B-48). Hence

$$C(j_1 j_2, j_1 + j_2; \overline{m}_1 - 1, \overline{m}_2) = \frac{y_{\overline{m}_1}}{y_{\overline{m}}} \qquad (B\text{-}79a)$$

and

$$C(j_1 j_2, j_1 + j_2; \overline{m}_1, \overline{m}_2 - 1) = \frac{y_{\overline{m}_2}}{y_{\overline{m}}} \qquad (B\text{-}79b)$$

Sequential application of $\hat{J}_-$ to Eq. (B-78) will generate all of the $2j + 1$ functions in the set with $j = \overline{j}$.

The set of functions corresponding to $j = \overline{j} - 1$ will be fewer in number by two than the set with $j = \overline{j}$ and will be bounded by the functions $|\overline{j} - 1, \overline{m} - 1\rangle$ and $|\overline{j} - 1, \underline{m} + 1\rangle$. The use of Eq. (B-75) demonstrates that $|\overline{j} - 1, \overline{m} - 1\rangle$ must be related to the same uncoupled functions as $|\overline{j}, \overline{m} - 1\rangle$. Further, there cannot be any other functions with $m = \overline{m} - 1$. If one writes

$$|\overline{j}, \overline{m} - 1\rangle = C_1 |\overline{m}_1 - 1, \overline{m}_2\rangle + C_2 |\overline{m}_1, \overline{m}_2 - 1\rangle \qquad (B\text{-}80a)$$

and

$$|\overline{j} - 1, \overline{m} - 1\rangle = C_1' |\overline{m}_1 - 1, \overline{m}_2\rangle + C_2' |\overline{m}_1, \overline{m}_2 - 1\rangle \qquad (B\text{-}80b)$$

then the orthonormal properties of the functions require that $C_1' = C_2$ and $C_2' = -C_1$.

Now that the function $|\overline{j} - 1, \overline{m} - 1\rangle$ has been defined, all the other members of that set can be obtained by the application of $\hat{J}_-$. The function $|\overline{j} - 2, \overline{m} - 2\rangle$ is obtained from the condition that it must be orthogonal to $|\overline{j} - 1, \overline{m} - 2\rangle$ and to $|\overline{j}, \overline{m} - 2\rangle$. The above sequence of processes is continued until all of the functions have been generated.

The number of uncoupled states must be the same as the number of coupled states, that is,

$$\sum_{\underline{j}}^{\overline{j}} (2j + 1) = (2j_1 + 1)(2j_2 + 1) \qquad (B\text{-}81)$$

This counting procedure will determine $\underline{j}$, since $\overline{j} = j_1 + j_2$. The left-hand side of (B-81) can be evaluated using

$$\sum_{\alpha}^{\beta} j = \frac{1}{2} [\beta(\beta + 1) - \alpha(\alpha - 1)] \qquad (B\text{-}82)$$

where j, α, and β must be integers or half-integers. Thus

$$(j_1 + j_2)(j_1 + j_2 + 1) - \underline{j}(\underline{j} - 1) + j_1 + j_2 - \underline{j} = (2j_1 + 1)(2j_2 + 1)$$

or

$$\underline{j}^2 = (j_1 - j_2)^2$$

Since $j \geq 0$,

$$\underline{j} = |j_1 - j_2| \tag{B-83}$$

Thus j is restricted to the values

$$j = j_1 + j_2, j_1 + j_2 - 1, \ldots, |j_1 - j_2| \tag{B-84}$$

A number of symmetry relations exist among the VC coefficients; also a general relation may be derived for these coefficients†; however, it is often easier to evaluate the VC coefficients from relations such as Eqs. (B-79). The following example will illustrate the method:

Consider two angular momenta such that $j_1 = j_2 = 1$. From Eq. (B-77a)

$$|2, 2\rangle = |1, 1\rangle \tag{B-85a}$$

Application of $\hat{J}_-$ gives

$$|2, 1\rangle = \frac{1}{\sqrt{2}}(|1, 0\rangle + |0, 1\rangle) \tag{B-85b}$$

Equations (B-79) have been used to obtain the VC coefficients. A second application of $\hat{J}_-$ gives

$$|2, 0\rangle = \frac{1}{\sqrt{6}}(2|0, 0\rangle + |1, -1\rangle + |-1, 1\rangle) \tag{B-85c}$$

Further application of $\hat{J}_-$ or the use of $\hat{J}_+$ with $|2, -2\rangle$ gives

$$|2, -1\rangle = \frac{1}{\sqrt{2}}(|-1, 0\rangle + |0, -1\rangle) \tag{B-85d}$$

and

$$|2, -2\rangle = |-1, -1\rangle \tag{B-85e}$$

In general

$$|j, \pm m\rangle = \sum_{m_1} C(j_1 j_2 j; m_1, m_2)|\pm m_1, \pm m_2\rangle \tag{B-86}$$

Thus only the first $j + 1$ members of a j set need be evaluated.

The members of the set with $j = 1$ are evaluated by using the condition that $|2, 1\rangle$ and $|1, 1\rangle$ must be orthogonal. This requires that

$$|1, 1\rangle = \frac{1}{\sqrt{2}}(|1, 0\rangle - |0, 1\rangle) \tag{B-87a}$$

† See Bibliography on page 434.

Application of $\hat{J}_-$ to (B-87a) gives

$$|1, 0\rangle = \frac{1}{\sqrt{2}} \left(|1, -1\rangle - |-1, 1\rangle\right) \tag{B-87b}$$

The third member of the set is

$$|1, -1\rangle = \frac{1}{\sqrt{2}} \left(|-1, 0\rangle - |0, -1\rangle\right) \tag{B-87c}$$

The single function $|0, 0\rangle$ of the set with $j = 0$ is obtained from the orthogonality with $|2, 0\rangle$ and $|1, 0\rangle$. The result is

$$|0, 0\rangle = \frac{1}{\sqrt{3}} \left(|0, 0\rangle - |1, -1\rangle - |-1, 1\rangle\right) \tag{B-88}$$

The use of the coupled representation becomes especially convenient when the angular momenta $\mathbf{J}_1$ and $\mathbf{J}_2$ are coupled by an interaction term in the hamiltonian; for example $\hat{\mathbf{L}}$ and $\hat{\mathbf{S}}$ are coupled through the hamiltonian term $\lambda \hat{\mathbf{L}} \cdot \hat{\mathbf{S}}$. The coupled representation functions $|j, m\rangle$ are still eigenfunctions of $\hat{J}^2, \hat{J}_z, \hat{J}_1{}^2$, and $\hat{J}_2{}^2$; however, the uncoupled representation functions $|m_1, m_2\rangle$ are *no longer* eigenfunctions of $\hat{J}_{1z}$ and $\hat{J}_{2z}$. Examples of the application of the methods outlined in this section are found in Chap. 12 and Appendix C.

B-8 SUMMARY

For convenience in reference, the essential results of this chapter are summarized below. For orbital, spin, or nuclear angular momenta, the appropriate expressions are obtained from those below by substituting L, S, or I, respectively, for j. The quantum number expressed here as m is then M_L, M_S, or M_I, respectively.

1. Operations giving nonzero values

$$\hat{J}^2|j, m\rangle = j(j + 1)|j, m\rangle \tag{B-41}$$

$$\hat{J}_z|j, m\rangle = m|j, m\rangle \tag{B-42}$$

$$\hat{J}_+|j, m\rangle = [j(j + 1) - m(m + 1)]^{\frac{1}{2}}|j, m + 1\rangle \tag{B-49}$$

$$\hat{J}_-|j, m\rangle = [j(j + 1) - m(m - 1)]^{\frac{1}{2}}|j, m - 1\rangle \tag{B-50}$$

2. Matrix elements

$$\langle j', m'|j, m\rangle = 1 \quad \text{if } j' = j \text{ and } m' = m$$
$$= 0 \quad \text{if } j' \neq j \text{ or } m' \neq m \tag{B-24}$$

$$\langle j, m|\hat{J}_z|j, m\rangle = m \quad \text{from (B-42)} \tag{B-89}$$

$$\langle j, m|\hat{J}^2|j, m\rangle = j(j + 1) \quad \text{from (B-41)} \tag{B-90}$$

$$\langle j, m + 1|\hat{J}_+|j, m\rangle = [j(j + 1) - m(m + 1)]^{\frac{1}{2}} \tag{B-51}$$

$$\langle j, m - 1|\hat{J}_-|j, m\rangle = [j(j + 1) - m(m - 1)]^{\frac{1}{2}} \tag{B-52}$$

Matrix elements of $\hat{J}_x$ and $\hat{J}_y$, less often used, are given as Eqs. (B-53) to (B-56).

3. Angular momentum matrices

$$j = \tfrac{1}{2}: \quad \mathbf{J}_x = \frac{1}{2}\begin{bmatrix} 0 & 1 \\ 1 & 0 \end{bmatrix} \quad \mathbf{J}_y = \frac{1}{2}\begin{bmatrix} 0 & -i \\ i & 0 \end{bmatrix} \quad \mathbf{J}_z = \frac{1}{2}\begin{bmatrix} 1 & 0 \\ 0 & -1 \end{bmatrix}$$

$$\mathbf{J}_+ = \begin{bmatrix} 0 & 1 \\ 0 & 0 \end{bmatrix} \quad \mathbf{J}_- = \begin{bmatrix} 0 & 0 \\ 1 & 0 \end{bmatrix} \tag{B-91}$$

$$j = 1: \quad \mathbf{J}_x = \frac{1}{\sqrt{2}}\begin{bmatrix} 0 & 1 & 0 \\ 1 & 0 & 1 \\ 0 & 1 & 0 \end{bmatrix} \quad \mathbf{J}_y = \frac{1}{\sqrt{2}}\begin{bmatrix} 0 & -i & 0 \\ i & 0 & -i \\ 0 & i & 0 \end{bmatrix}$$

$$\mathbf{J}_z = \begin{bmatrix} 1 & 0 & 0 \\ 0 & 0 & 0 \\ 0 & 0 & -1 \end{bmatrix}$$

$$\mathbf{J}_+ = \sqrt{2}\begin{bmatrix} 0 & 1 & 0 \\ 0 & 0 & 1 \\ 0 & 0 & 0 \end{bmatrix} \quad \mathbf{J}_- = \sqrt{2}\begin{bmatrix} 0 & 0 & 0 \\ 1 & 0 & 0 \\ 0 & 1 & 0 \end{bmatrix} \tag{B-92}$$

BIBLIOGRAPHY

Edmonds, A. R.: "Angular Momentum in Quantum Mechanics," Princeton University Press, Princeton, N.J., 1960.

Rose, M. E.: "Elementary Theory of Angular Momentum," John Wiley & Sons, Inc., New York, 1957.

PROBLEMS

B-1. Show that $[\hat{J}_y,\hat{J}_z] = i\hat{J}_x$.

B-2. Derive the commutation relations in (B-21).

B-3. Show that the matrix element $\langle j, m'|\hat{J}_-|j, m\rangle$ is nonzero only if $m' = m - 1$.

B-4. Establish the angular-momentum matrices $\mathbf{J}_x$, $\mathbf{J}_y$, $\mathbf{J}_z$, $\mathbf{J}_+$, $\mathbf{J}_-$, and $\mathbf{J}^2$ for $j = \tfrac{3}{2}$.

B-5. For $j = \tfrac{3}{2}$, show that $\mathbf{J}_x^2 + \mathbf{J}_y^2 + \mathbf{J}_z^2 = \mathbf{J}^2$.

B-6. By matrix addition and multiplication, find the commutators $[\mathbf{J}_+,\mathbf{J}_-]$, $[\mathbf{J}_+,\mathbf{J}^2]$, and $[\mathbf{J}_-,\mathbf{J}^2]$ for $j = 1$.

B-7. For $j = \tfrac{1}{2}$, $\mathbf{J}_x^2 = \mathbf{J}_y^2 = \mathbf{J}_z^2$. By calculation of the appropriate matrices for $j = 1$ and $j = \tfrac{3}{2}$, show that these relations are not satisfied for $j \geq 1$. On the other hand, show that the trace (sum of diagonal elements) is the same for these matrices for a given value of j.

B-8. Verify Eq. (10-9b) by matrix multiplication.

B-9. For $j_1 = 2$, $j_2 = 1$, show that

$$|3, \pm 3\rangle = |\pm 2, \pm 1\rangle$$

$$|3, \pm 2\rangle = \frac{1}{\sqrt{6}} (2|\pm 1, \pm 1\rangle + \sqrt{2}|\pm 2, 0\rangle)$$

$$|3, \pm 1\rangle = \frac{1}{\sqrt{30}} (2\sqrt{3}|0, \pm 1\rangle + 4|\pm 1, 0\rangle + \sqrt{2}|\pm 2, \mp 1\rangle)$$

$$|3, 0\rangle = \frac{1}{\sqrt{15}} (\sqrt{3}|1, -1\rangle + 3|0, 0\rangle + \sqrt{3}|-1, 1\rangle)$$

$$|2, \pm 2\rangle = \frac{1}{\sqrt{6}} (\sqrt{2}|\pm 1, \pm 1\rangle - 2|\pm 2, 0\rangle)$$

$$|2, \pm 1\rangle = \frac{1}{\sqrt{6}} (\sqrt{3}|0, \pm 1\rangle - |\pm 1, 0\rangle - \sqrt{2}|\pm 2, \mp 1\rangle)$$

$$|2, 0\rangle = \frac{1}{\sqrt{2}} (|-1, +1\rangle - |+1, -1\rangle)$$

$$|1, \pm 1\rangle = \frac{1}{\sqrt{10}} (|0, \pm 1\rangle - \sqrt{3}|\pm 1, 0\rangle + \sqrt{6}|\pm 2, \mp 1\rangle)$$

$$|1, 0\rangle = \frac{1}{\sqrt{10}} (\sqrt{3}|1, -1\rangle - 2|0, 0\rangle + \sqrt{3}|-1, 1\rangle)$$

Appendix C
Calculation of the Hyperfine
Interaction in the Hydrogen Atom
and in an ·RH₂ Radical

The hyperfine interaction in the hydrogen atom was treated in an approximate manner in Chap. 3. An exact calculation will be presented in this appendix.

C-1 THE HAMILTONIAN FOR THE HYDROGEN ATOM

A more exact spin hamiltonian than that of Eq. (3-8) for an isotropic system of one electron ($S = \frac{1}{2}$) and one proton ($I = \frac{1}{2}$) in a magnetic field $\mathbf{H}$ is

$$\hat{\mathscr{H}} = g\beta\mathbf{H} \cdot \hat{\mathbf{S}} + hA_0\hat{\mathbf{S}} \cdot \hat{\mathbf{I}} - g_N\beta_N\mathbf{H} \cdot \hat{\mathbf{I}} \tag{C-1}$$

For $\mathbf{H}$ along the z axis, Eq. (C-1) becomes

$$\hat{\mathscr{H}} = g\beta H\hat{S}_z + hA_0(\hat{S}_z\hat{I}_z + \hat{S}_x\hat{I}_x + \hat{S}_y\hat{I}_y) - g_N\beta_N H\hat{I}_z \tag{C-2}$$

Using the operators $\hat{S}_+$, $\hat{S}_-$, $\hat{I}_+$, and $\hat{I}_-$, defined by Eq. (B-20), one finds the quantity $\hat{S}_+\hat{I}_- + \hat{S}_-\hat{I}_+$ to be $2(\hat{S}_x\hat{I}_x + \hat{S}_y\hat{I}_y)$. Hence Eq. (C-2) can be rearranged to

$$\hat{\mathscr{H}} = g\beta H\hat{S}_z + hA_0[\hat{S}_z\hat{I}_z + \frac{1}{2}(\hat{S}_+\hat{I}_- + \hat{S}_-\hat{I}_+)] - g_N\beta_N H\hat{I}_z \tag{C-3}$$

C-2   THE SPIN EIGENFUNCTIONS AND THE ENERGY MATRIX FOR THE HYDROGEN ATOM

The bra and ket notation (see Sec. A-5d) will be used for the spin eigenfunctions, that is, $|M_S, M_I\rangle$. There will be four independent spin eigenfunctions as given in Sec. 3-3.

The energy matrix consists of the matrix elements of the spin hamiltonian between all the spin eigenfunctions (i.e., $\langle M_S, M_I|\hat{\mathcal{H}}|M_S', M_I'\rangle$). It will thus be a 4×4 matrix. Use of the angular-momentum matrices for $S = \frac{1}{2}$ (computed in Sec. B-6),

$$
\mathbf{S}_z = \begin{bmatrix} \frac{1}{2} & 0 \\ 0 & -\frac{1}{2} \end{bmatrix} \quad \mathbf{S}_x = \begin{bmatrix} 0 & \frac{1}{2} \\ \frac{1}{2} & 0 \end{bmatrix}
$$

and

$$
\mathbf{S}_y = \begin{bmatrix} 0 & -\frac{i}{2} \\ \frac{i}{2} & 0 \end{bmatrix} \quad \mathbf{S}_+ = \begin{bmatrix} 0 & 1 \\ 0 & 0 \end{bmatrix} \quad \mathbf{S}_- = \begin{bmatrix} 0 & 0 \\ 1 & 0 \end{bmatrix}
$$

permits computation of these matrix elements. The matrices $\mathbf{I}_z$, $\mathbf{I}_+$, and $\mathbf{I}_-$ are exactly the same as the corresponding electron spin matrices, since these apply to any system with *total angular momentum* $\langle J \rangle = \sqrt{J(J+1)}$, with $J = \frac{1}{2}$.

The matrix elements are divided into two classes:

1. *Diagonal matrix elements.* A diagonal matrix element is one in which the bra and the ket have the same labels. Inspection of the spin matrices shows that only $\mathbf{S}_z$ and $\mathbf{I}_z$ have nonzero diagonal elements; hence the only nonzero diagonal matrix elements will be

$$
\langle M_S, M_I|\hat{S}_z\hat{I}_z|M_S, M_I\rangle \qquad \langle M_S, M_I|\hat{S}_z|M_S, M_I\rangle
$$

and

$$
\langle M_S, M_I|\hat{I}_z|M_S, M_I\rangle
$$

A typical diagonal matrix element is

$$
\langle \alpha_e, \alpha_n|g\beta H\hat{S}_z + hA_0\hat{S}_z\hat{I}_z - g_N\beta_N H\hat{I}_z|\alpha_e, \alpha_n\rangle
$$
$$
= \tfrac{1}{2}g\beta H + \tfrac{1}{4}hA_0 - \tfrac{1}{2}g_N\beta_N H \quad \text{(C-4)}
$$

α_e and α_n correspond, respectively, to $M_S = \frac{1}{2}$ and $M_I = \frac{1}{2}$; β_e and β_n correspond, respectively, to $M_S = -\frac{1}{2}$ and $M_I = -\frac{1}{2}$.

2. *Off-diagonal matrix elements.* Inspection of the spin matrices shows that $\mathbf{S}_+$, $\mathbf{S}_-$, $\mathbf{I}_+$ and $\mathbf{I}_-$ have only off-diagonal nonzero elements. Hence for the operators $\hat{S}_+\hat{I}_-$ and $\hat{S}_-\hat{I}_+$, the nonzero off-diagonal matrix elements of the hamiltonian in Eq. (C-3) will be of the type

$$\langle (M_S + 1), (M_I - 1)|\hat{S}_+\hat{I}_-|M_S, M_I\rangle$$

and

$$\langle (M_S - 1), (M_I + 1)|\hat{S}_-\hat{I}_+|M_S, M_I\rangle$$

For example

$$\langle \alpha_e, \beta_n|\hat{S}_+\hat{I}_-|\beta_e, \alpha_n\rangle = 1$$

The energy matrix is then constructed as follows:

$$
\begin{array}{c|cccc}
 & |\alpha_e,\alpha_n\rangle & |\alpha_e,\beta_n\rangle & |\beta_e,\alpha_n\rangle & |\beta_e,\beta_n\rangle \\
\hline
\langle \alpha_e,\alpha_n| & \left(\tfrac{1}{2}g\beta H + \tfrac{1}{4}hA_0 - \tfrac{1}{2}g_N\beta_N H\right) & 0 & 0 & 0 \\
\langle \alpha_e,\beta_n| & 0 & \left(\tfrac{1}{2}g\beta H - \tfrac{1}{4}hA_0 + \tfrac{1}{2}g_N\beta_N H\right) & \tfrac{1}{2}hA_0 & 0 \\
\langle \beta_e,\alpha_n| & 0 & \tfrac{1}{2}hA_0 & \left(-\tfrac{1}{2}g\beta H - \tfrac{1}{4}hA_0 - \tfrac{1}{2}g_N\beta_N H\right) & 0 \\
\langle \beta_e,\beta_n| & 0 & 0 & 0 & \left(-\tfrac{1}{2}g\beta H + \tfrac{1}{4}hA_0 + \tfrac{1}{2}g_N\beta_N H\right)
\end{array}
$$

(C-

Note that this matrix is factorable into blocks along the principal diagonal, with all other elements zero. When this factorization is possible, one may deal in succession with each of the submatrices in turn. This procedure results in a considerable simplification of the calculations. Similar considerations apply to determinants.

C-3 EXACT SOLUTION OF THE DETERMINANT OF THE ENERGY MATRIX (SECULAR DETERMINANT)

To obtain the energies of the four states, one must diagonalize the energy matrix (C-5). Diagonalization may be accomplished by subtracting a variable (say W) from each diagonal element and setting the resulting determinant (secular determinant) equal to zero. The four roots of the quartic equation in W will be the state energies. By inspection one can see that two of the state energies will be

$$W_{\alpha_e\alpha_n} = \tfrac{1}{2}g\beta H + \tfrac{1}{4}hA_0 - \tfrac{1}{2}g_N\beta_N H \tag{C-6a}$$

and

$$W_{\beta_e\beta_n} = -\tfrac{1}{2}g\beta H + \tfrac{1}{4}hA_0 + \tfrac{1}{2}g_N\beta_N H \tag{C-6b}$$

The other two state energies can be obtained by expanding the remaining 2×2 determinant

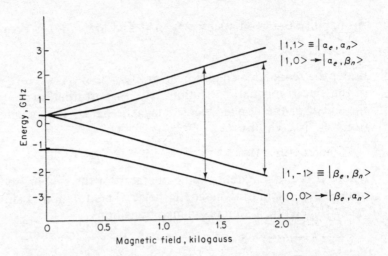

Fig. C-1 Energy levels of the hydrogen atom at low and moderate magnetic fields (Breit-Rabi diagram); allowed transitions at moderate magnetic fields are shown.

$$\begin{vmatrix} \frac{1}{2}g\beta H - \frac{1}{4}hA_0 + \frac{1}{2}g_N\beta_N H - W & \frac{1}{2}hA_0 \\ \frac{1}{2}hA_0 & -\frac{1}{2}g\beta H - \frac{1}{4}hA_0 - \frac{1}{2}g_N\beta_N H - W \end{vmatrix} = 0$$
$$(C\text{-}7)$$

One obtains the energies

$$W_{(\alpha_e\beta_n)} = \frac{1}{2}[(g\beta + g_N\beta_N)^2 H^2 + h^2 A_0^2]^{\frac{1}{2}} - \frac{1}{4}hA_0 \qquad (C\text{-}8a)$$

and

$$W_{(\beta_e\alpha_n)} = -\frac{1}{2}[(g\beta + g_N\beta_N)^2 H^2 + h^2 A_0^2]^{\frac{1}{2}} - \frac{1}{4}hA_0 \qquad (C\text{-}8b)$$

The parentheses on $\alpha_e\beta_n$ and $\beta_e\alpha_n$ are meant to give notice that the corresponding states are *mixtures* of $\alpha_e\beta_n$ and $\beta_e\alpha_n$. The eigenvalues $W_{(\alpha_e\beta_n)}$ and $W_{(\beta_e\alpha_n)}$ are subscripted as shown, since $\alpha_e\beta_n$ and $\beta_e\alpha_n$ *are* the respective correct eigenstates in the limit of very high magnetic field.

Equations (C-6) and (C-8) are called the Breit-Rabi formulas.[†]   The Breit-Rabi energies are plotted as a function of magnetic field in Fig. C-1. The unusual aspects of this diagram at low magnetic fields are considered in Sec. C-8.

C-4 SELECTION RULES FOR HIGH–FIELD MAGNETIC DIPOLE TRANSITIONS IN THE HYDROGEN ATOM

The interaction of electromagnetic radiation with the hydrogen-atom system can lead to transitions between certain energy levels.   The transition

† G. Breit and I. I. Rabi, *Phys. Rev.*, **38**:2082L (1931); J. E. Nafe and E. B. Nelson, *Phys. Rev.*, **73**:718 (1948).

probability between states $|M_S, M_I\rangle$ and $|M_S', M_I'\rangle$ is proportional to[†]

$$|\langle M_S', M_I'|\hat{\mathcal{H}}'|M_S, M_I\rangle|^2$$

under the resonance condition $h\nu = W_{M_S' M_I'} - W_{M_S M_I}$. Here ν is the frequency of the electromagnetic radiation. $\hat{\mathcal{H}}'$ is a perturbation operator representing the effect of the applied electromagnetic field. For magnetic-dipole transitions $\hat{\mathcal{H}}'$ is given by

$$\hat{\mathcal{H}}' = -\hat{\boldsymbol{\mu}} \cdot \mathbf{H}_1 \tag{C-9}$$

where $\hat{\boldsymbol{\mu}}$ is the magnetic-dipole operator for the system and H_1 is the amplitude of the oscillating magnetic field. This relation is similar to Eq. (1-1).

For electron spin resonance transitions

$$\hat{\mathcal{H}}' = g\beta\mathbf{H}_1 \cdot \hat{\mathbf{S}} \tag{C-10}$$

With $\mathbf{H}_1$ along z, the direction of the static magnetic field $\mathbf{H}$, Eq. (C-10) becomes

$$\hat{\mathcal{H}}' = g\beta H_1 \hat{S}_z \tag{C-11}$$

A general matrix element of $\hat{\mathcal{H}}'$ will be

$$\langle M_S', M_I'|\hat{\mathcal{H}}'|M_S, M_I\rangle = g\beta H_1 \langle M_S', M_I'|\hat{S}_z|M_S, M_I\rangle$$
$$= g\beta H_1 \langle M_S'|\hat{S}_z|M_S\rangle\langle M_I'|M_I\rangle$$
$$= g\beta H_1 M_S \langle M_S'|M_S\rangle\langle M_I'|M_I\rangle \tag{C-12}$$

From the orthogonality of the wave functions this matrix element will be nonzero only if $M_S' = M_S$ and $M_I' = M_I$. Hence the selection rules are

$$\Delta M_S = 0 \qquad \Delta M_I = 0 \qquad \mathbf{H}_1 \parallel z \tag{C-13}$$

Under these conditions, one cannot expect to see absorption, since no transitions are allowed.

With $\mathbf{H}_1 \parallel x$

$$\hat{\mathcal{H}}' = g\beta H_1 \hat{S}_x \tag{C-14}$$

The general matrix element of $\hat{\mathcal{H}}'$ will be

$$\langle M_S', M_I'|\hat{\mathcal{H}}'|M_S, M_I\rangle = g\beta H_1 \langle M_S', M_I'|\hat{S}_x|M_S, M_I\rangle$$
$$= g\beta H_1 \langle M_S'|\hat{S}_x|M_S\rangle\langle M_I'|M_I\rangle \tag{C-15}$$

From Eqs. (B-53) and (B-54) one notes that the matrix element of $\hat{S}_x$ is zero unless $M_S' = M_S \pm 1$. Hence, the selection rules are

$$\Delta M_S = \pm 1 \qquad \Delta M_I = 0 \qquad \mathbf{H}_1 \parallel x \tag{C-16}$$

Identical results are obtained with $\mathbf{H}_1 \parallel y$.

[†] H. Eyring, J. Walter, and G. Kimball, "Quantum Chemistry," chap. 8, John Wiley & Sons, Inc., New York, 1944.

These selection rules are valid only when the kets $|M_S, M_I\rangle$ are eigen-functions of the spin hamiltonian, Eq. (C-3), that is, when the term

$$\frac{hA_0}{2} (\hat{S}_+\hat{I}_- + \hat{S}_-\hat{I}_+)$$

is neglected. This approximation is valid when $g\beta H \gg hA_0$. The case of very low magnetic fields will be treated in Sec. C-8.

It is desirable to be able to calculate relative intensities of possible transitions between energy levels. Equation (C-15) demonstrates that H_1 must be perpendicular to z for transitions to occur. The transition prob-ability is proportional to the square of the matrix element of $\mathcal{H}'$, between the initial and final states. Hence, relative intensities of different transi-tions can be obtained by computing the square of the corresponding matrix element of $\hat{S}_x$; that is, the intensities will be proportional to

$$|\langle M_S', M_I'|\hat{S}_x|M_S, M_I\rangle|^2$$

Since $\hat{S}_x = \frac{1}{2}[\hat{S}_+ + \hat{S}_-]$, the matrix elements of $\hat{S}_+$ and $\hat{S}_-$ can also be used. In a specific instance, only one of these will be effective in causing a tran-sition; if $\Delta M_S = +1$, then

$$|\langle (M_S + 1), M_I|\hat{S}_+|M_S, M_I\rangle|^2$$

governs the intensity. If $\Delta M_S = -1$, then

$$|\langle (M_S - 1), M_I|\hat{S}_-|M_S, M_I\rangle|^2$$

determines the intensity.

C-5 THE TRANSITION FREQUENCIES IN CONSTANT MAGNETIC FIELD WITH A VARYING MICROWAVE FREQUENCY

When H is constant, the separation of the energy levels is fixed. Upon scanning the microwave frequency, one will observe resonance when $\Delta W = h\nu$. With the selection rules $\Delta M_S = \pm 1$, $\Delta M_I = 0$, resonance will be observed at the following two frequencies [see Eqs. (C-6) and (C-8)]:

$$\nu_k = \frac{W_{\alpha_r\alpha_n} - W_{(\beta_r\alpha_n)}}{h} = \frac{1}{2}(g\beta - g_N\beta_N)h^{-1}H + \frac{1}{2}A_0$$
$$+ \frac{1}{2}[(g\beta + g_N\beta_N)^2h^{-2}H^2 + A_0^2]^{\frac{1}{2}} \quad (C\text{-}17a)$$

$$\nu_m = \frac{W_{(\alpha_r\beta_n)} - W_{\beta_r\beta_n}}{h} = \frac{1}{2}(g\beta - g_N\beta_N)h^{-1}H - \frac{1}{2}A_0$$
$$+ \frac{1}{2}[(g\beta + g_N\beta_N)^2h^{-2}H^2 + A_0^2]^{\frac{1}{2}} \quad (C\text{-}17b)$$

The difference $(\nu_k - \nu_m)$ is exactly A_0, the hyperfine *coupling constant*.

C-6 THE RESONANT MAGNETIC FIELDS AT CONSTANT MICROWAVE FREQUENCY

Use of a constant microwave frequency and a varying magnetic field is the typical experimental arrangement. This situation is more complicated, since the magnetic field will *not* be the same for the two transitions. H_k will be taken as the resonant field for the transition $|\beta_e, \alpha_n\rangle \to |\alpha_e, \alpha_n\rangle$ and H_m the resonant field for the transition $|\beta_e, \beta_n\rangle \to |\alpha_e, \beta_n\rangle$. If $A_0^2 \ll (g\beta + g_N\beta_N)^2 h^{-2}H^2$, one can carry out a binomial expansion of the square root term, that is,[†]

$$[(g\beta + g_N\beta_N)^2 h^{-2}H^2 + A_0^2]^{\frac{1}{2}} = h^{-1}(g\beta + g_N\beta_N)H$$
$$+ \frac{1}{2}\frac{A_0^2 h}{(g\beta + g_N\beta_N)H} - \frac{1}{8}\frac{A_0^4 h^3}{(g\beta + g_N\beta_N)^3 H^3} + \cdots \quad \text{(C-18)}$$

Only the first two terms will be retained. For $\nu_k = \nu_m = \nu_0$ (the fixed microwave frequency), Eqs. (C-17), can be transformed by substitution of (C-18) and multiplication by H_k or H_m, respectively, to

$$h^{-1}g\beta H_k^2 - (\nu_0 - \tfrac{1}{2}A_0)H_k + \frac{1}{4}\frac{A_0^2 h}{g\beta} = 0 \quad \text{(C-19a)}$$

and

$$h^{-1}g\beta H_m^2 - (\nu_0 + \tfrac{1}{2}A_0)H_m + \frac{1}{4}\frac{A_0^2 h}{g\beta} = 0 \quad \text{(C-19b)}$$

Since $g_N\beta_N \sim 10^{-3}g\beta$, $g_N\beta_N$ is neglected compared to $g\beta$ in Eqs. (C-19). Solution of Eqs. (C-19a) and (C-19b) gives

$$H_k = \frac{h}{4g\beta}\left(2\nu_0 - A_0 + \sqrt{4\nu_0^2 - 4A_0\nu_0 - 3A_0^2}\right) \quad \text{(C-20a)}$$

and

$$H_m = \frac{h}{4g\beta}\left(2\nu_0 + A_0 + \sqrt{4\nu_0^2 + 4A_0\nu_0 - 3A_0^2}\right) \quad \text{(C-20b)}$$

It is clear that $a = H_m - H_k \neq hA_0/g\beta$. a has been called the hyperfine *splitting* constant (i.e., the separation in gauss between the experimentally observed lines on an ESR spectrum), as opposed to A_0, the hyperfine *coupling* constant (in megahertz).

For the H atom, $A_0 = 1{,}420.40573$ MHz and $g = 2.002256$.[‡] With $\nu_0 = 9{,}500$ MHz, substitution in Eqs. (C-20) gives

$$H_k = 3{,}115.93 \text{ G} \qquad \text{and} \qquad H_m = 3{,}625.67 \text{ G}$$

[†] $(a + b)^{\frac{1}{2}} = a^{\frac{1}{2}} + \frac{1}{2}a^{-\frac{1}{2}}b - \frac{1}{8}a^{-\frac{3}{2}}b^2 + \cdots\cdot$

[‡] R. Beringer and M. A. Heald, *Phys. Rev.*, **95**:1474 (1954); P. Kusch, *Phys. Rev.*, **100**:1188 (1955).

Hence $a = H_m - H_k = 509.74$ G. Since $hA_0/g\beta = 506.86$ G, one sees that a differs from $hA_0/g\beta$ by 2.88 G. The difference is not large, but it is significant. It is of interest to note the average value $(H_m + H_k)/2 = 3,370.80$ G. This is 19.18 G lower than the field $H_l = h\nu_0/g\beta$. (See Fig. 3-5b.) There would thus be a considerable error in determining the g factor from the mean position. For π-electron radicals this effect is much smaller.

C-7 CALCULATION OF THE ENERGY LEVELS OF THE HYDROGEN ATOM BY PERTURBATION THEORY

The hamiltonian [Eq. (C-3)] may be separated into two parts

$$\hat{\mathcal{H}} = \hat{\mathcal{H}}_0 + \hat{\mathcal{H}}' \tag{C-21}$$

where

$$\hat{\mathcal{H}}_0 = g\beta H \hat{S}_z + hA_0 \hat{S}_z \hat{I}_z - g_N \beta_N H \hat{I}_z \tag{C-22}$$

and

$$\hat{\mathcal{H}}' = \frac{hA_0}{2} (\hat{S}_+ \hat{I}_- + \hat{S}_- \hat{I}_+) \tag{C-23}$$

If $\hat{\mathcal{H}}' \ll \hat{\mathcal{H}}_0$, then one may use the eigenfunctions of $\hat{\mathcal{H}}_0$ as a basis set for determining the energy corrections due to $\hat{\mathcal{H}}'$.

The zero-order energies $W^{(0)}$ are just the matrix elements of $\hat{\mathcal{H}}_0$ (see Sec. A-7), namely,

$$W^{(0)}_{\alpha_e \alpha_n} = \tfrac{1}{2}g\beta H + \tfrac{1}{4}hA_0 - \tfrac{1}{2}g_N \beta_N H \tag{C-24a}$$

$$W^{(0)}_{\alpha_e \beta_n} = \tfrac{1}{2}g\beta H - \tfrac{1}{4}hA_0 + \tfrac{1}{2}g_N \beta_N H \tag{C-24b}$$

$$W^{(0)}_{\beta_e \beta_n} = -\tfrac{1}{2}g\beta H + \tfrac{1}{4}hA_0 + \tfrac{1}{2}g_N \beta_N H \tag{C-24c}$$

and

$$W^{(0)}_{\beta_e \alpha_n} = -\tfrac{1}{2}g\beta H - \tfrac{1}{4}hA_0 - \tfrac{1}{2}g_N \beta_N H \tag{C-24d}$$

Note that these are just the diagonal elements of $\hat{\mathcal{H}}_0$ in the energy matrix (C-5). The effects of each of the terms in Eqs. (C-24) are indicated sequentially in Fig. C-2. Note that the addition of the nuclear Zeeman interaction in the hydrogen atom does *not* affect the zero-order transition energies.

Regardless of the form of $\hat{\mathcal{H}}'$, one may use the general expression Eq. (A-88) with (A-97) and (A-100) for the energy due to the perturbation as

$$W = W^{(0)} + \langle i|\hat{\mathcal{H}}'|i\rangle + \sum_n{}' \frac{\langle i|\hat{\mathcal{H}}'|n\rangle \langle n|\hat{\mathcal{H}}'|i\rangle}{W_i^{(0)} - W_n^{(0)}} + \cdots \tag{C-25}$$

The second term in Eq. (C-25) is the first-order correction $W^{(1)}$ to the energy; it is given by the diagonal matrix elements of $\hat{\mathcal{H}}'$, taken over the zero-order wave functions. However, since $\hat{\mathcal{H}}'$ involves only the raising and

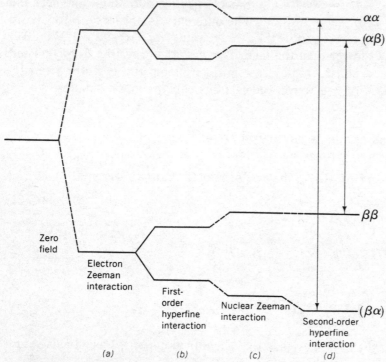

Fig. C-2 Energy levels and allowed transitions for the hydrogen atom (at constant magnetic field) showing effects of successive terms in the spin hamiltonian [Eq. (C-3)]. (a) Electron Zeeman interaction $g\beta H \hat{S}_z$. (b) Addition of the first-order proton hyperfine interaction $hA_n \hat{S}_z \hat{I}_z$. (c) Addition of the nuclear Zeeman interaction $-g_N \beta_N H \hat{I}_z$. (d) Addition of the second-order proton hyperfine interaction derived from $\frac{1}{2}(\hat{S}_+ \hat{I}_- + \hat{S}_- \hat{I}_+)$.

lowering operators, all diagonal matrix elements of $\hat{\mathcal{H}}'$ will be zero. Thus the state energies previously calculated with the use of $\hat{\mathcal{H}}_0$ are correct to first order. Hence, it is necessary to utilize the third term of Eq. (C-25) to obtain additional (second-order) corrections $W^{(2)}$. The prime on Σ implies that the summation extends over all the zero-order states $|n\rangle$ except for the state $|i\rangle$. $W_i^{(0)}$ and $W_n^{(0)}$ are the zero-order energies for the states $|i\rangle$ and $|n\rangle$, respectively.

Writing the states in Eq. (C-25) in terms of the M_S and M_I values, one finds the second-order correction to be given by

$$W^{(2)}_{M_S M_I} = \sum_{M_S' M_I'}{}' \frac{\langle M_S, M_I | \hat{\mathcal{H}}' | M_S', M_I' \rangle \langle M_S', M_I' | \hat{\mathcal{H}}' | M_S, M_I \rangle}{W^{(0)}_{M_S M_I} - W^{(0)}_{M_S' M_I'}} \tag{C-26}$$

As in the exact treatment (see Sec. C-3), the only nonzero off-diagonal matrix elements of $\hat{\mathcal{H}}'$ are

$$\langle \beta_e, \alpha_n | \hat{S}_- \hat{I}_+ | \alpha_e, \beta_n \rangle = 1$$

and

$$\langle \alpha_e, \beta_n | \hat{S}_+ \hat{I}_- | \beta_e, \alpha_n \rangle = 1$$

Therefore only the energies of the states $|\alpha_e, \beta_n\rangle$ and $|\beta_e, \alpha_n\rangle$ will be affected; the second-order energy corrections to these states are

$$W^{(2)}_{\alpha_e\beta_n} = \frac{h^2 A_0^2}{4} \frac{1}{g\beta H} \qquad \text{and} \qquad W^{(2)}_{\beta_e\alpha_n} = -\frac{h^2 A_0^2}{4} \frac{1}{g\beta H} \tag{C-27}$$

The energies to second order are shown in Fig. C-2. The transition frequencies at *constant field* will be

$$\nu_k = h^{-1} g\beta H + \frac{1}{2} A_0 + \frac{h A_0^2}{4 g\beta H} \tag{C-28a}$$

and

$$\nu_m = h^{-1} g\beta H - \frac{1}{2} A_0 + \frac{h A_0^2}{4 g\beta H} \tag{C-28b}$$

The term $g_N \beta_N H$ has been neglected in comparison with $g\beta H$. As before, $\nu_k - \nu_m = A_0$; when ν is held constant at ν_0 and the field is scanned, the expressions obtained on solving Eqs. (C-28) for H_k and H_m are exactly the same as Eqs. (C-20).

C-8 WAVE FUNCTIONS AND ALLOWED TRANSITIONS FOR THE HYDROGEN ATOM AT LOW MAGNETIC FIELDS

The fact that off-diagonal elements appear in the energy matrix (C-5) means that the basis spin functions are not all eigenfunctions of the hamiltonian equation (C-3). It is desirable to find a set of spin functions which are eigenfunctions of $\hat{\mathcal{H}}$. One first notes that $|\alpha_e, \alpha_n\rangle$ and $|\beta_e, \beta_n\rangle$ are already eigenfunctions of $\hat{\mathcal{H}}$; thus one need only diagonalize a 2×2 matrix.

The four eigenfunctions of $\hat{\mathcal{H}}$ may be expressed in the coupled representation (see Sec. B-7) $|F, M_F\rangle$, where

$$F = |S + I|, |S + I - 1|, \ldots, |S - I| \tag{C-29}$$

For the hydrogen atom, $F = 0, 1$. If $F = 1$, $M_F = 0, \pm 1$; if $F = 0$, $M_F = 0$. The eigenfunctions $|\alpha_e, \alpha_n\rangle$ and $|\beta_e, \beta_n\rangle$ then become $|1, 1\rangle$ and $|1, -1\rangle$, respectively.

The remaining two eigenfunctions $|1, 0\rangle$ and $|0, 0\rangle$ are obtained by diagonalization of the 2×2 matrix in (C-5). This is best accomplished using the coordinate-rotation matrix method outlined in Sec. A-5e. The two eigenfunctions are expressed as linear combinations, that is,

$$|1, 0\rangle = \cos \omega |\alpha_e, \beta_n\rangle + \sin \omega |\beta_e, \alpha_n\rangle \tag{C-30}$$

$$|0, 0\rangle = -\sin \omega |\alpha_e, \beta_n\rangle + \cos \omega |\beta_e, \alpha_n\rangle \tag{C-31}$$

The use of Eqs. (A-72) gives

$$\cos^2 \omega = \frac{1}{2} \left[1 + \left(1 + \frac{2 h^2 A_0^2}{g^2 \beta^2 H^2} \right)^{-\frac{1}{2}} \right] \tag{C-32a}$$

$$\sin^2 \omega = \frac{1}{2}\left[1 - \left(1 + \frac{2h^2 A_0^2}{g^2 \beta^2 H^2}\right)^{-\frac{1}{2}}\right] \qquad \text{(C-32b)}$$

Now as $H \to \infty$, $|1, 0\rangle \to |\alpha_e, \beta_n\rangle$, and $|0, 0\rangle \to |\beta_e, \alpha_n\rangle$. However, at $H = 0$

$$|1, 0\rangle = \frac{1}{\sqrt{2}}\left[|\alpha_e, \beta_n\rangle + |\beta_e, \alpha_n\rangle\right] \qquad \text{(C-33)}$$

$$|0, 0\rangle = \frac{1}{\sqrt{2}}\left[|\alpha_e, \beta_n\rangle - |\beta_e, \alpha_n\rangle\right] \qquad \text{(C-34)}$$

Because of the mixing of states, four transitions are possible at low magnetic fields. The relative intensities can be computed by evaluating the matrix elements of $\hat{S}_x$. For example, the $\hat{S}_x$ matrix element for the $|1, -1\rangle \to |1, 0\rangle$ transition is

$$\cos \omega \; \langle \beta_e, \beta_n | \hat{S}_x | \alpha_e, \beta_n \rangle = \tfrac{1}{2}\cos \omega \qquad \text{(C-35)}$$

The relative intensities of the four transitions computed in the above manner are given as follows

$$
\begin{aligned}
&|0, 0\rangle \to |1, -1\rangle \quad \text{intensity} \propto \sin^2 \omega \\
&|0, 0\rangle \to |1, 1\rangle \quad \text{intensity} \propto \cos^2 \omega \\
&|1, -1\rangle \to |1, 0\rangle \quad \text{intensity} \propto \cos^2 \omega \\
&|1, 0\rangle \to |1, 1\rangle \quad \text{intensity} \propto \sin^2 \omega
\end{aligned}
\qquad \text{(C-36)}
$$

It is clear from (C-36) that one should be able to detect resonance at zero magnetic field. These measurements have been carried out with extremely high precision in atomic beams of hydrogen atoms.[†] Zero field measurements have also been reported for Cr^{3+} in MgO.[‡]

C-9 THE ENERGY LEVELS OF AN ·RH₂ RADICAL

When more than one magnetic nucleus is interacting with the electron, the calculation of the state energies must be preceded by a careful inspection of the spin wave functions. If the nuclei are equivalent, it is usually convenient to use a "coupled" representation for the nuclear spin states. (See Sec. B-7.) The nuclear spins of the two protons of ·RH₂ are added vectorially to obtain one set of wave functions with total nuclear spin $J = 1$ and another set with $J = 0$. The new nuclear spin functions will be represented by $|J, M_J\rangle$; these are related to the wave functions $|M_{I_1}, M_{I_2}\rangle$ by

† N. F. Ramsey, Jr., "Molecular Beams," pp. 263ff., Oxford University Press, London, 1956.
‡ T. Cole, T. Kushida, and H. C. Heller, J. Chem. Phys., 38:2915 (1963).

$$|1, 1\rangle \equiv |\tfrac{1}{2}, \tfrac{1}{2}\rangle \tag{C-37a}$$

$$|1, 0\rangle \equiv \frac{1}{\sqrt{2}} \left[|\tfrac{1}{2}, -\tfrac{1}{2}\rangle + |-\tfrac{1}{2}, \tfrac{1}{2}\rangle \right] \quad J = 1 \tag{C-37b}$$

$$|1, -1\rangle \equiv |-\tfrac{1}{2}, -\tfrac{1}{2}\rangle \tag{C-37c}$$

and

$$|0, 0\rangle \equiv \frac{1}{\sqrt{2}} \left[|\tfrac{1}{2}, -\tfrac{1}{2}\rangle - |-\tfrac{1}{2}, \tfrac{1}{2}\rangle \right] \quad J = 0 \tag{C-37d}$$

Because of the fact that the *square* of the total nuclear angular momentum [characterized by $J(J + 1)$] is a constant of the motion, mixture of states with different values of J is not allowed: that is, there are no nonzero matrix elements between $J = 0$ and $J = 1$ states. Including the electron spin, the total spin wave functions will be designated by $|M_S, J, M_J\rangle$.

The spin hamiltonian for the ·RH₂ fragment will again be separated into two parts so that perturbation theory may be used, i.e.,

$$\hat{\mathscr{H}}_0 = g\beta H \hat{S}_z + hA_0 \hat{S}_z \hat{J}_z \tag{C-38a}$$

and

$$\hat{\mathscr{H}}' = \frac{hA_0}{2} (\hat{S}_+ \hat{J}_- + \hat{S}_- \hat{J}_+) \tag{C-38b}$$

The nuclear Zeeman terms have been omitted since they do not affect the transition energies. These expressions are analogous to Eqs. (C-22) and (C-23), with the total nuclear spin operator $\hat{J}$ replacing the individual nuclear spin operators $\hat{I}$.

The zero-order energies are again the diagonal matrix elements of $\hat{\mathscr{H}}_0$,

$$
\begin{aligned}
W^{(0)}_{\frac{1}{2},1,1} &= \tfrac{1}{2}g\beta H + \tfrac{1}{2}hA_0 & W^{(0)}_{-\frac{1}{2},1,-1} &= -\tfrac{1}{2}g\beta H + \tfrac{1}{2}hA_0 \\
W^{(0)}_{\frac{1}{2},1,0} &= \tfrac{1}{2}g\beta H & W^{(0)}_{-\frac{1}{2},0,0} &= -\tfrac{1}{2}g\beta H \\
W^{(0)}_{\frac{1}{2},0,0} &= \tfrac{1}{2}g\beta H & W^{(0)}_{-\frac{1}{2},1,0} &= -\tfrac{1}{2}g\beta H \\
W^{(0)}_{\frac{1}{2},1,-1} &= \tfrac{1}{2}g\beta H - \tfrac{1}{2}hA_0 & W^{(0)}_{-\frac{1}{2},1,1} &= -\tfrac{1}{2}g\beta H - \tfrac{1}{2}hA_0
\end{aligned}
\tag{C-39}
$$

Since the selection rules are $\Delta M_S = \pm 1$, $\Delta M_J = 0$, the spectrum at constant microwave frequency will consist of three lines occurring at the resonant fields,

$$H_k = \frac{h\nu_0}{g\beta} - \frac{hA_0}{g\beta}$$

$$H_l = \frac{h\nu_0}{g\beta} \tag{C-40}$$

$$H_m = \frac{h\nu_0}{g\beta} + \frac{hA_0}{g\beta}$$

The line at H_l is twice as intense as the lines at H_k or H_m because the states contributing to the line at H_l are doubly degenerate. (See Fig. 4-1a.)

As before [see Eq. (C-27)], the second-order energy corrections involve only off-diagonal matrix elements; from Eqs. (B-51) and (B-52), only the following four are nonzero:

$$\langle +\tfrac{1}{2}, 1, -1|\hat{S}_+\hat{J}_-|-\tfrac{1}{2}, 1, 0\rangle = \sqrt{2}$$
$$\langle +\tfrac{1}{2}, 1, 0|\hat{S}_+\hat{J}_-|-\tfrac{1}{2}, 1, 1\rangle = \sqrt{2}$$
$$\langle -\tfrac{1}{2}, 1, 0|\hat{S}_-\hat{J}_+|\tfrac{1}{2}, 1, -1\rangle = \sqrt{2} \qquad (\text{C-41})$$

and

$$\langle -\tfrac{1}{2}, 1, 1|\hat{S}_-\hat{J}_+|\tfrac{1}{2}, 1, 0\rangle = \sqrt{2}$$

The energies to second order are given in Fig. (C-3). The transition frequencies at *constant field* will be

$$\nu_k = h^{-1}g\beta H + A_0 + \frac{1}{2}\frac{hA_0{}^2}{g\beta H} \qquad (\text{C-42}a)$$

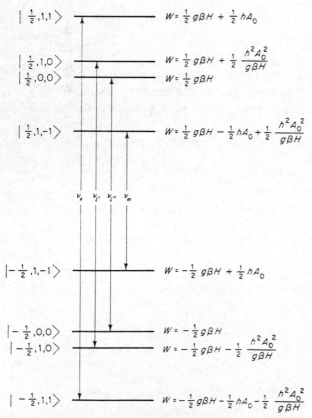

Fig. C-3 Energy levels of the ·RH₂ fragment to second order. (hA_0 has been neglected in comparison with $g\beta H$ in the second-order correction terms.)

$$\nu'_l = h^{-1}g\beta H + \frac{hA_0^2}{g\beta H} \tag{C-42b}$$

$$\nu''_l = h^{-1}g\beta H \tag{C-42c}$$

$$\nu_m = h^{-1}g\beta H - A_0 + \frac{1}{2}\frac{hA_0^2}{g\beta H} \tag{C-42d}$$

If hA_0 is small compared to $g\beta H$, the latter can be set equal to $h\nu_0$ in the correction terms, and hence the resonant fields at constant ν_0 are

$$H_k = \frac{h\nu_0}{g\beta} - \frac{hA_0}{g\beta} - \frac{1}{2}\frac{hA_0^2}{g\beta\nu_0} \tag{C-43a}$$

$$H'_l = \frac{h\nu_0}{g\beta} - \frac{hA_0^2}{g\beta\nu_0} \tag{C-43b}$$

$$H''_l = \frac{h\nu_0}{g\beta} \tag{C-43c}$$

and

$$H_m = \frac{h\nu_0}{g\beta} + \frac{hA_0}{g\beta} - \frac{1}{2}\frac{hA_0^2}{g\beta\nu_0} \tag{C-43d}$$

Thus the spectrum will now consist of four lines, with all lines except that at H''_l shifted downfield from the zero-order positions. These are illustrated in Fig. 4-29a.

PROBLEMS

C-1. Consider the spectrum of the hydrogen atom shown in Fig. 3-1. Use the expressions developed in this appendix to compute an accurate value of the hyperfine coupling constant A_0 (megahertz) and the g factor. Comment on the differences from the corresponding values for the free hydrogen atom.

C-2. Use the methods developed for ·RH₂ to calculate to second order the energies of the states for a radical containing three equivalent protons. For $A_0 = 100$ MHz, $g = 2.00232$, and microwave frequency 9.500 GHz, calculate the field positions and relative intensities of all allowed transitions.

C-3. The effects of the nuclear quadrupole moment upon the energy levels of a system with $I = 1$ in an inhomogeneous electric field may be seen by examining the shifts of levels of the deuterium atom. These levels are given by Eq. (3-13). The quadrupole moment operator for an axial electric field is

$$\hat{\mathscr{H}}_{quad} = Q'\left[\hat{I}_z^2 - \frac{I(I+1)}{3}\right] \qquad Q' = \frac{3eQ}{4I(2I-1)}\frac{\partial^2 V}{\partial Z^2}$$

where e is the nuclear charge, Q is the quadrupole moment, and $\partial^2 V/\partial Z^2$ is the gradient of the electric field $\partial V/\partial Z$ in which the nuclear quadrupole is located. Here V is the potential energy.

    (a) Plot the energy of the six levels of Eq. (3-13).

    (b) Show the shift of each of these levels by $\hat{\mathscr{H}}_{quad}$, expressing the shift in terms of Q'.

    (c) Show the allowed transitions (still $\Delta M_S = \pm 1$, $\Delta M_I = 0$).

    (d) Can one detect a quadrupole interaction from the ESR spectrum? [These shifts must be taken into account when interpreting electron-nuclear double resonance (ENDOR) spectra; see Chap. 13.]

Appendix D
Experimental Methods;
Spectrometer Performance

This appendix is designed to acquaint the reader with some experimental methods and procedures which may be used to obtain the optimum performance from an ESR spectrometer. The reader is referred to several books[†],[‡],[§] which provide a wealth of detail on many aspects of ESR measurement.

D-1 SENSITIVITY

The optimization of the signal-to-noise ratio of a spectrometer is frequently a prerequisite to the successful performance of an ESR experiment. Such optimization requires familiarity with the various factors which affect either the *noise* level or the *signal* level. The minimum detectable number of

---

[†] C. P. Poole, Jr., "Electron Spin Resonance: A Comprehensive Treatise on Experimental Techniques," Interscience Publishers, a division of John Wiley & Sons, Inc., New York, 1967.
[‡] T. H. Wilmshurst, "Electron Spin Resonance Spectrometers," Hilger, London, 1967.
[§] R. S. Alger, "Electron Paramagnetic Resonance Techniques and Applications," Interscience Publishers, a division of John Wiley & Sons. Inc., New York, 1968.

paramagnetic centers N_{min} in an ESR cavity with a signal-to-noise ratio of unity is given by

$$N_{min} = \frac{3V_C k T_s \Gamma}{2\pi g^2 \beta^2 S(S+1)H_r Q'_u}\left(\frac{FkT_d b}{P_0}\right)^{\frac{1}{2}} \tag{D-1}$$

Here

V_C = the volume of the cavity (assumed to be operated in the TE_{102} mode)

k = Boltzmann's constant

T_s = sample temperature

Γ = half-half-width (gauss) of the absorption line

H_r = magnetic field at the center of the absorption line

Q'_u = the effective unloaded Q factor of the cavity (see Sec. D-2e)

T_d = detector temperature

b = bandwidth in s^{-1} of the entire detecting and amplifying system

P_0 = microwave power (erg s^{-1}) incident on the cavity

F = a noise figure (>1) attributable to sources other than thermal detector noise. An ideal spectrometer would have $F = 1$.

The derivation† of Eq. (D-1) assumes that the absorption shape is lorentzian, that the Curie law applies, and that microwave saturation does not occur. All quantities are in cgs units.

An estimate of N_{min} is obtained by inserting the following typical values:

$Q'_u = 5,000$

$T_s = T_d = 300$ K

$H_r = 3,400$ G

$\Gamma = 1$ G

$g = 2.00$

$S = \frac{1}{2}$

$V_C = 11$ cm³   (for a TE_{102} cavity at X band)

$F = 100$

$b = 1$ s⁻¹

$P_0 = 10^6$ erg s⁻¹ $= 100$ mW

These factors give $N_{min} \approx 10^{11}$. For a typical sample, the minimum detectable concentration of paramagnetic centers is $\approx 10^{-9}M$. Figures such as these are typically quoted by manufacturers of ESR spectrometers. However, unless the conditions of measurement are given, a quoted N_{min} value may be misleading. In particular, many samples are so readily saturated

---

† G. Feher, *Bell System Tech. J.*, **36**:449 (1957); C. P. Poole, Jr., "Electron Spin Resonance: A Comprehensive Treatise on Experimental Techniques," Interscience Publishers, a division of John Wiley & Sons, Inc., New York, 1967, pp. 554ff.

that power levels in excess of 1 mW are out of the question. Hence, N_{min} may be effectively larger by a factor of 10.

When an ESR spectrum contains hyperfine lines, the intensity of a given line is only a fraction of the total intensity. Hyperfine splitting increases N_{min} by the factor

$$\mathscr{R} = \frac{\sum\limits_{j} D_j}{D_k} \tag{D-2}$$

where D_k is the degeneracy of the most intense line and $\sum\limits_{j} D_j$ is the sum of the degeneracies of all the lines in the spectrum. For the p-benzosemiquinone anion (Fig. 4-6) $\mathscr{R} = 2.67$, and for the naphthalene anion (Fig. 4-16) $\mathscr{R} = 7.11$.

D-2 FACTORS AFFECTING SENSITIVITY AND RESOLUTION

The preceding section dealt primarily with sensitivity. However, many spectra may be so rich in hyperfine components that resolution becomes an additional factor to be optimized. Increased resolution often results in a decreased sensitivity. It may then be necessary to sacrifice some sensitivity in order to gain the requisite resolution. Six factors affecting sensitivity and resolution will be considered in this section.

D-2a. Modulation amplitude It was noted in Chap. 2 that small-amplitude field modulation techniques may be employed to improve the sensitivity

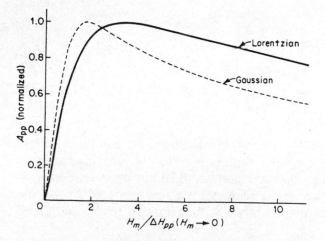

Fig. D-1 Normalized peak-to-peak amplitude (A_{pp}) for first derivatives of lorentzian and gaussian lines as a function of modulation amplitude (H_m). $\Delta H_{pp}(H_m \rightarrow 0)$ is the peak-to-peak derivative amplitude as H_m goes to zero.

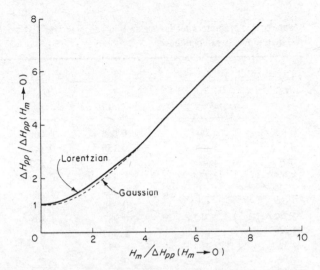

Fig. D-2 Relative first-derivative linewidths at increasing values of the relative modulation amplitude $H_m/\Delta H_{pp}$ ($H_m \rightarrow 0$).

of a spectrometer. However, an excessive modulation amplitude or an excessively high modulation frequency can lead to line distortion.

The modulation amplitude H_m† should be a small fraction of the peak-to-peak derivative linewidth ΔH_{pp}. The portion of an absorption line scanned during a half-cycle of field modulation must be nearly linear in order to obtain an output which is essentially the first derivative of the absorption line. (See Fig. 2-8.) As H_m approaches and exceeds ΔH_{pp}, the derivative line amplitude first increases linearly with H_m, then reaches a maximum, and finally decreases slowly. (See Fig. D-1.) However, long before the line amplitude reaches a maximum, the linewidth ΔH_{pp} *is significantly increased.* (See Fig. D-2.) This phenomenon has been analyzed for lorentzian‡ and for gaussian lines.§ The results are given in Table D-1 and displayed in Figs. D-1 and D-2. It is apparent that a maximum derivative amplitude A_{pp} is obtained when $H_m \cong 3.5\,\Delta H_{pp}$ for lorentzian lines and $H_m \cong 1.8\,\Delta H_{pp}$ for gaussian lines. At these settings the lines are considerably broadened (by a factor of 3 for lorentzian lines or of 1.6 for gaussian lines).

The optimum setting of H_m will depend on how much sensitivity can be sacrificed for faithfulness of line shape or vice versa. If resolution and true line shape are important, then the modulation amplitude should satisfy the

† The modulation amplitude H_m at the modulation frequency ω_m is defined by
$$H = H_0 + \tfrac{1}{2}H_m \sin \omega_m t$$
‡ H. Walhurst, *J. Chem. Phys.*, **35**:1708 (1961).
§ G. W. Smith, *J. Appl. Phys.*, **35**:1217 (1964).

Table D-1 Parameters for lorentzian and gaussian first-derivative absorption lines as a function of the relative modulation amplitude

	Lorentzian line			Gaussian line	
$\dfrac{H_m}{\Delta H_{pp}}$	$\dfrac{\Delta H_{pp}(obs.)}{\Delta H_{pp}(H_m \to 0)}$	A_{pp} (normalized)	$\dfrac{H_m}{\Delta H_{pp}}$	$\dfrac{\Delta H_{pp}(obs.)}{\Delta H_{pp}(H_m \to 0)}$	A_{pp} (normalized)
0.000	1.000	0.000	0.000	1.000	0.000
0.173	1.006	0.130	0.141	1.001	0.148
0.346	1.029	0.248	0.282	1.007	0.291
0.694	1.114	0.478	0.564	1.039	0.551
1.388	1.432	0.784	1.128	1.178	0.887
2.08	1.903	0.930	1.692	1.454	0.993
2.78	2.387	0.987	1.848	1.560	1.000
3.46	3.000	1.000	1.974	1.645	0.995
4.16	3.564	0.992	2.26	1.862	0.983
4.86	4.221	0.974	2.82	2.343	0.943
5.56	4.884	0.952	3.38	2.856	0.898
6.24	5.537	0.929	3.94	3.384	0.857
6.94	6.288	0.905	4.52	3.922	0.819
10.40	9.55	0.800	5.08	4.465	0.785
13.84	13.0	0.721	5.64	5.013	0.755
17.34	16.4	0.659	8.46	7.786	0.639
27.72	26.5	0.541	11.28	10.6	0.564
34.64	33.7	0.488	14.10	13.5	0.497
69.4	68.2	0.353	∞	∞	0.000
∞	∞	0.000			

condition $H_m \lesssim 0.2\,\Delta H_{pp}$. However, if sensitivity is the prime concern and some line distortion can be tolerated, then H_m should be increased until a maximum derivative amplitude is obtained. A reasonable compromise between sensitivity and resolution is then achieved by reducing H_m by a factor of 4 to 5 from the value which makes A_{pp} a maximum. Thus for $\Delta H_{pp} < 0.1$ G, line distortion will occur unless H_m is kept very small. The effect of increasing H_m for a very narrow line is seen in Fig. D-3. If $H_m \gg \Delta H_{pp}$ (true width at low modulation amplitude), then H_m can be determined directly from the peak separation in Fig. D-3.

D-2b. Modulation frequency The observed line will also be distorted if the modulation frequency approaches the magnitude of the linewidth in Hz, i.e., if $\omega_m = (g\beta/h)\Delta H_{pp}$. Since the crystal detector is a nonlinear device, its output contains the sum and the difference of the microwave and the modulation frequencies. This results in the production of side-band resonance lines spaced ω_m/γ_e apart and extending over a range of H_m gauss. For a modulation frequency of 100 kHz, $\omega_m/\gamma_e = 36$ mG. The development of these side bands as the amplitude of the modulation is increased is shown in Fig. D-4 for a line having a width less than 20 mG (the F center in CaO—see Sec. 8-5c). The side bands have a phase difference of 180° with respect to

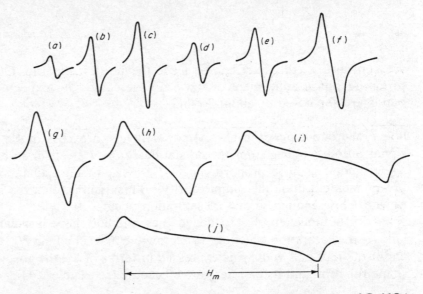

Fig. D-3 Wide-line NMR signals of protons in an aqueous solution of Cr(NO$_3$)$_3$ ($\Delta H_{pp} = 0.19$ G) as a function of modulation amplitude H_m. The field scan is the same for each trace. Values of $H_m/\Delta H_{pp}(H_m \to 0)$ are as follows: (a) 0.150; (b) 0.398; (c) 0.552; (d) 0.552; (e) 1.052; (f) 2.28; (g) 4.94; (h) 10.14; (i) 20.6; (j) 28.8. The gain for traces (a) to (c) is twice that for traces (d) to (j). [*From G. W. Smith, J. Appl. Phys.,* **35**:1217 (1964).]

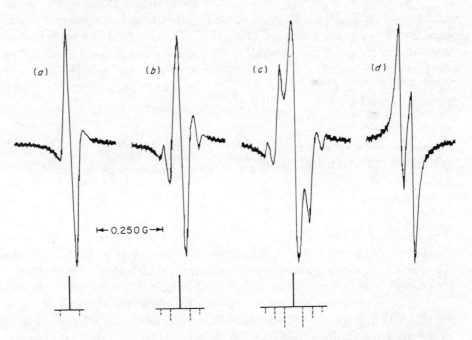

Fig. D-4 Modulation side bands on the ESR spectrum of the F center in CaO, for which the linewidth is less than 20 mG. (a) Modulation amplitude 4 mG. (b) 20 mG. (c) 50 mG. (d) The phase is adjusted so that the center line is not seen. The two lines are the first modulation side bands. The modulation side bands are opposite in phase to the central line; their positions are indicated by dotted lines in (a), (b), and (c). The separation of the side bands corresponds to the $\frac{\omega_m}{\gamma_e}$, which for 100 kHz is 36 mG.

the central line; the latter can be made to vanish (as in Fig. D-4d) by appropriate phase adjustment of the phase detector. As expected, the side bands are separated by about 36 mG.

D-2c. Microwave power level   At power levels in excess of 10^{-4} watt, the signal output voltage from the crystal detector of a reflection-cavity ESR spectrometer will be proportional to $P_0^{\frac{1}{2}}$, viz., to the square root of the power incident upon the sample cavity. This assumes that the microwave power is low enough so that no saturation occurs.

For homogeneously broadened lines, the line shape is usually lorentzian. If the saturation effect of the microwave magnetic field H_1 is included† and if the half-half width Γ is expressed in terms of T_2, the absorption lineshape function and its first derivative become (see Table 2-1)

$$Y = \frac{1}{\pi} \frac{H_1 T_2}{1 + (H - H_r)^2 \gamma^2 T_2^2 + H_1^2 \gamma^2 T_1 T_2} \tag{D-3}$$

$$Y' = -\frac{2}{\pi} \frac{H_1 T_2^3 \gamma^2 (H - H_r)}{[1 + (H - H_r)^2 \gamma^2 T_2^2 + H_1^2 \gamma^2 T_1 T_2]^2} \tag{D-4}$$

As long as $H_1^2 \gamma^2 T_1 T_2 << 1$, this "saturation" term can be neglected, and both Y and Y' will be proportional to H_1 (or to $P_0^{\frac{1}{2}}$). When the absorption line is strongly saturated ($H_1^2 \gamma^2 T_1 T_2 >> 1$), Y' will decrease with increasing microwave power (see Fig. D-5a). By computing the peak-to-peak derivative amplitude $2Y'_{max}$ and differentiating, one finds that for a value of P_0 which gives a maximum derivative amplitude the spin-lattice relaxation time is given by

$$T_1 = \frac{1}{2 H_1^2 \gamma^2 T_2} \tag{D-5}$$

Further, by computing the derivative peak-to-peak width,

$$(\Delta H_{pp})^2 = \frac{4}{3 \gamma^2 T_2^2} + \frac{4 H_1^2 T_1}{3 T_2} \tag{D-6}$$

The increase in linewidth as saturation sets in can be explained in terms of the uncertainty principle. A higher microwave power produces spin transitions at a faster rate and hence decreases the spin lifetime. This results in an increased uncertainty in the energy and hence to an increased linewidth.

It is instructive to note that when the derivative amplitude is at its maximum (Fig. D-5a), the linewidth has risen to only ~ 1.2 times the width in the absence of saturation (see Fig. D-5b). Hence, at the maximum deriv-

---

† H_1 is defined by the expression $H(\omega) = 2H_1 \cos \omega t$ where ω is 2π times the microwave frequency. Eqs. (D-3) and (D-4) are derived from the Bloch equations; a particularly good treatment is given in J. A. Pople, W. G. Schneider, and H. J. Bernstein, "High Resolution Nuclear Magnetic Resonance," pp. 31ff., McGraw-Hill Book Company, New York, 1959.

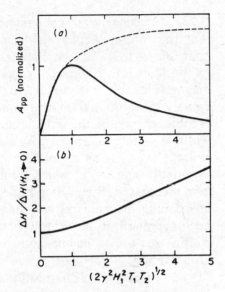

Fig. D-5 (a) The normalized derivative amplitude as a function of H_1^2 (proportional to $P_0^{\frac{1}{2}}$) for a homogeneously broadened ESR line. The dotted line refers to an inhomogeneously broadened line. (b) The normalized peak-to-peak linewidth as a function of H_1 for a homogeneously broadened line.

ative amplitude, microwave saturation does not increase linewidth as much as does excessive modulation amplitude. It is good practice first to obtain a maximum derivative amplitude. If resolution is important, then P_0 should be reduced to about 75 percent of its value for maximum line amplitude.

For inhomogeneously broadened lines (usually gaussian in shape), the derivative amplitude theoretically increases monotonically to a limiting value with increasing power. This behavior is indicated by the dotted line in Fig. D-5a. In practice, it is observed that even for lines which one classifies as inhomogeneously broadened, the amplitude goes through a maximum with increasing power. This implies some measure of homogeneous broadening, arising, perhaps, from mutual spin flips with the surroundings.

Some spectrometers are capable of detecting the dispersion† (i.e., the real part of the rf magnetic susceptibility) which accompanies absorption. The dispersion signal does not saturate as readily as the absorption signal. Hence it is desirable to detect dispersion when dealing with signals which saturate readily. This is especially important when working at liquid helium temperatures, where T_1 may be very long.

---

† The phenomenon of dispersion always accompanies the resonant absorption of energy from the microwave field—indeed, dispersion always accompanies absorption in any region of the spectrum. Dispersion represents the real part of the microwave magnetic susceptibility, whereas absorption is a measure of the imaginary part. It manifests itself in a shift in the resonant frequency of the cavity. As the magnetic field approaches the region of an absorption line, the frequency shift of the cavity first is negative, then increases rapidly through zero at the center of an absorption line to a maximum and finally decreases asymptotically to zero. As most spectrometers have the klystron frequency locked to the cavity resonant frequency, any device which will detect this frequency shift may be used to display the dispersion signal.

D-2d. The concentration of paramagnetic centers. It was noted in Chap. 9 that the intermolecular electron spin–electron spin exchange interaction contributes to $1/T_2$. At moderate concentration levels this contribution to the linewidth is directly proportional to the concentration of the paramagnetic center. However, the peak-to-peak derivative amplitude is inversely proportional to the *square* of the linewidth. (See Table 2-1.) Hence, as long as the modulation amplitude is kept constant and the linewidth is mainly determined by spin exchange, the derivative amplitude will actually *increase* as the concentration is *decreased*. Of course, a point is eventually reached at which a further decrease in concentration will result in a decreased signal amplitude. Other concentration-independent linebroadening mechanisms are then important. Linewidths of ~ 50 mG are not uncommon in the ESR spectra of free radicals in solution; these linewidths require concentrations of less than $10^{-4}M$ to avoid exchange broadening.

For free radicals in solution it is often convenient to have a concentration gradient within the sample tube. Then the concentration in the cavity may easily be adjusted by moving the sample tube up or down.

If electron transfer with a diamagnetic molecule contributes to the linewidth, a decrease in the overall concentration of all species will have the same effect as if the broadening were due to electron spin exchange.

In solids, a major source of line-broadening is the dipole-dipole interactions between neighboring electron spins. To minimize this problem, it is necessary to dilute the paramagnetic centers by incorporating them in a diamagnetic host. Since limiting linewidths in solids are usually ≥ 1G, concentrations of 10^{-2} to $10^{-3}M$ can be tolerated if there is no tendency toward pairing or aggregation.

D-2e. Temperature Even if the sample temperature does not govern the linewidth, for maximum sensitivity, one should work at as low a temperature as feasible, since the signal amplitude is inversely proportional to the absolute temperature (Curie's law). However, in many cases, the temperature also has an effect on the linewidth. For each system there will usually be an optimum temperature at which the linewidth is a minimum. (See Fig. 9-20.) This temperature is usually below room temperature; hence, a decrease in sample temperature often produces improved spectra.

If the linewidth is determined by a short value of T_1 as in transition-metal ions, a significant decrease in sample temperature can have a dramatic effect on the ESR spectrum. The reason is that T_1 is a strong function of the sample temperature (e.g., in some cases T_1 is proportional to the inverse seventh power of the sample temperature or even increases exponentially as the temperature is lowered). For some samples liquid helium temperatures (4 K or less) are necessary to obtain sufficiently narrow lines. This is especially true of many of the rare-earth and actinide ions.

D-2f. *Q* **Factor of the cavity** The *Q* factor was defined in Chap. 2 as

$$Q = \frac{2\pi \text{ (maximum microwave energy stored in the cavity)}}{\text{energy dissipated per cycle}} \qquad (\text{D-7})$$

The value of *Q* under the condition that only losses within the cavity are considered (i.e., resistive losses in the walls) is called the *unloaded Q factor* Q_u. However, a cavity must be coupled by an iris to the waveguide system. Since this entails additional losses, there will be a further lowering of *Q*. This coupling loss is measured by

$$\frac{1}{Q_r} = \frac{\text{energy lost through coupling holes per cycle}}{2\pi \text{ (stored energy)}}$$

The ratio

$$\beta = \frac{Q_u}{Q_r} \qquad (\text{D-8})$$

is called the coupling parameter.

For optimum coupling (i.e., maximum transmission of power into the cavity) $\beta = 1$. The overall or loaded quality factor Q_L is then defined by

$$\frac{1}{Q_L} = \frac{1}{Q_r} + \frac{1}{Q_u} \qquad (\text{D-9})$$

Hence, when $\beta = 1$, $Q_L = \frac{1}{2}Q_u$.

If within the cavity there are materials which have a nonvanishing imaginary part of the dielectric constant, additional losses can occur. To account for these losses, the dielectric *Q* factor is defined as

$$Q_\epsilon = \frac{2\pi \text{ (maximum microwave energy stored in the cavity)}}{\text{energy lost per cycle in dielectric loss}} \qquad (\text{D-10})$$

A factor Q_u' is defined such that

$$\frac{1}{Q_u'} = \frac{1}{Q_\epsilon} + \frac{1}{Q_u} \qquad (\text{D-11})$$

Q_u' is the factor which should be used in sensitivity calculations. Most of the dielectric losses usually occur within the sample or the sample tube. Thus it is important to position the sample in a region of the cavity for which the microwave electric field is a minimum. Positioning is extremely critical for samples (such as aqueous solutions) which exhibit a high dielectric loss. For these, the most satisfactory container is a flat high-purity silica cell which can be accurately oriented along the nodal plane of the **E** field. It is found that optimum sensitivity occurs when $Q_u = Q_\epsilon$, i.e., for a reduction in Q_L to one-half of its value in the absence of dielectric loss.†

† Charles P. Poole, Jr., "Electron Spin Resonance: A Comprehensive Treatise on Experimental Techniques," p. 587, Interscience Publishers, a division of John Wiley & Sons, Inc., New York, 1967.

For aqueous solutions this requires that the silica plates be separated by ~ 0.3 mm for X-band frequencies. At higher frequencies, the dielectric loss for water is even more serious. For organic solvents with $\epsilon < 10$, cylindrical sample tubes with an internal diameter of ≈ 3 mm are permissible.

Since most glass and silica materials give strong ESR signals, they are not suitable as sample-tube materials. Fused silica of high purity avoids these difficulties. This material is also advantageous since it has a very low dielectric loss and hence is a desirable material for the fabrication of dewar inserts. One should be forewarned that uv, γ, or x irradiation of many materials, including fused silica, generally gives rise to defects which show ESR signals.

Cylindrical cavities operated in the TE_{011} mode (Fig. 2-4) generally have a significantly larger Q factor than does a rectangular cavity in the TE_{102} mode. Thus if the sample has a low dielectric loss, use of such a cylindrical cavity may be advantageous.

D-2g. Microwave frequency The microwave frequency is a parameter which is varied but little in most ESR work. The principal reason is that most spectrometers permit frequency variations of no more than ± 10 percent of the center frequency of the klystron. However, there are numerous situations for which the inconvenience (and expense!) of introducing a major change—usually an increase—in the microwave frequency can result in a very significant improvement in sensitivity. Several cases will be considered.

1. *Constant factors: filling factor, microwave power, and dielectric loss.* As the microwave frequency increases, the size of the cavity (for the same mode) must decrease. If the sample size is scaled in the same proportion as the cavity dimension, then the filling factor remains constant. If in addition, the sample has a low dielectric loss, then the sensitivity will increase as $\nu_0^{\frac{3}{2}}$, where ν_0 is the microwave frequency. If the dielectric loss is significant, then this factor will usually be smaller, and it may even have a negative exponent. Since the manipulation of a sample in a small cavity is difficult, there would be little advantage of an increase in frequency *in this case.* Indeed, if the sample is readily saturated, then the improvement is minimal; for constant H_1 at the sample, the sensitivity increases only as $\nu_0^{\frac{3}{2}}$.

2. *Constant factors: sample size and microwave power, with negligible microwave loss.* If the sample size is limited, as is often true for single crystals, then an increase in the microwave frequency may result in a dramatic improvement in sensitivity. In this case, the sensitivity increases as $\nu_0^{\frac{9}{2}}$. The important factor here is that for constant sample volume, the filling factor increases as ν_0^3. All other factors being equal, a change from X band to Q band would result in a 500-fold increase in sensitivity!

3. *Aqueous samples*. Unfortunately, the dielectric loss of water increases with frequency from X band to Q band. The resultant reduction in the Q factor largely cancels any gain in sensitivity which would otherwise be achieved. It is for this reason that almost all aqueous solution studies are carried out at X band or lower.

D-2h. Signal averaging When the signal-to-noise ratio is very low, there are two techniques for improving this ratio by signal averaging. The first involves filtering the output of the phase-sensitive detector by the use of a resistive-capacitive filter (see Fig. D-6).

The product RC has units of seconds if R is expressed in megohms and C in microfarads. The filter attenuates noise components with a frequency greater than $\sim (RC)^{-1}$. However, the use of this filter restricts the rate at which a line may be scanned. A good working rule is to adjust the scan rate so that the time τ required to go from peak to peak on a derivative line is such that $\tau > 10RC$. Scan times less than this will distort the line. However, if sensitivity is the prime factor, then $\tau \sim RC$ gives the best sensitivity. It is desirable to have a large range of RC combinations (e.g., 10^{-3} to 100 s) depending on the signal-to-noise ratio required and the rate at which the spectrum is to be scanned.

The use of an RC filter improves the signal-to-noise ratio in proportion to $\sqrt{RC}$, since the effective bandwidth of the spectrometer is usually governed by the time constant of the output filter, i.e., $b = (RC)^{-1}$. However, there are practical considerations which limit the magnitude of the RC product which can be used. These usually include: (1) a limited lifetime of the paramagnetic species and (2) instrumental instability (baseline drift, drift of klystron frequency or power output).

A second technique for signal averaging involves addition of spectra obtained by repetitive scans. This is accomplished by dividing the spectrum into equal intervals (typically 512 or 1,024). Each portion is then stored in a separate channel of a time-averaging computer. The coherent signal in each channel will rise in proportion to n, the number of scans. However, due to its randomness, the noise signal will tend to cancel; in fact, it will increase in proportion to $\sqrt{n}$. Thus there will be an improvement in the signal-to-noise ratio in proportion to $\sqrt{n}$.

If both the sample and the spectrometer are sufficiently stable, the time-averaging computer technique has little advantage over RC filtering. However, for short-lived paramagnetic species or for rapid kinetic studies, the improved bandwidth of the time-averaging computer technique makes possible studies which might otherwise be hopeless.

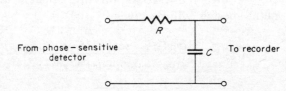

From phase – sensitive   To recorder
detector

Fig. D-6   An *RC* filter.

D-3 ABSOLUTE INTENSITY MEASUREMENTS

In any analytical applications, one desires a knowledge of the number of paramagnetic centers giving rise to the observed signal. The following factors determine the absolute intensity of the signal:

1. The area under the absorption curve. For derivative spectra this requires a double integration of the first-derivative curve.
2. The modulation amplitude at the sample.
3. The amplitude of the microwave magnetic field H_1 at the sample. This requires a knowledge of P_0, Q'_u, and the distribution of H_1 within the cavity.
4. The overall spectrometer gain.
5. The sample temperature.
6. The spin of the paramagnetic species.
7. The g factor.
8. The microwave frequency.
9. The filling factor η. For a small sample in a TE_{102} cavity, $\eta \sim 2V_s/V_C$, where V_s is the sample volume and V_C the cavity volume.

In practice, absolute determinations on a single sample are rarely carried out since so many errors can enter in. Standards are employed to minimize some of these errors. Such standards are of two types: (1) concentration standards and (2) absolute spin number standards.

If one requires only the concentration of paramagnetic species in a liquid solution or solid, then a concentration standard can be employed. The following conditions should apply to both standard and unknown:

1. The same solvent or host and the same sample geometry should be employed in order to ensure that the microwave magnetic field H_1 is the same for the unknown and the standard samples.
2. The amplitude of the ESR signals should be proportional to $(P_0)^{\frac{1}{2}}$; i.e., there should be no saturation for either sample or standard. Ideally, P_0 should be the same for sample and standard.
3. The modulation amplitude can be large [that is, $H_m \sim (2-4)\,\Delta H_{pp}$] providing that the *area* under the absorption curve is determined. This is especially important for weak signals where overmodulation must be employed to achieve adequate sensitivity.
4. The unknown and standard samples should be at the same temperature.

Under these conditions the concentration of the unknown paramagnetic species will be given by

$$[X] = \frac{[\text{std}]A_x \mathscr{R}_x (\text{scan}_x)^2 G_{\text{std}} M_{\text{std}} (g_{\text{std}})^2 [S(S+1)]_{\text{std}}}{A_{\text{std}} \mathscr{R}_{\text{std}} (\text{scan}_{\text{std}})^2 G_x M_x (g_x)^2 [S(S+1)]_x} \qquad \text{(D-12)}$$

Here A is the measured area under the absorption curve. This may be in arbitrary units as long as they are the same for unknown and standard. Scan is the horizontal scale in gauss per unit length on the chart paper. G is the relative gain of the signal amplifier. M is the modulation amplitude in gauss, and $\mathscr{R}$ is defined by Eq. (D-2).

The area under the absorption curve can be obtained by computing the first moment[†],[‡] by electronic double integration,[§] digital integration, or by weighing a cut-out absorption curve. This procedure requires a constant base line or a base-line correction after each integration.

One must use extreme care in evaluating the area under an absorption curve. Serious errors may result from failure to extend measurements sufficiently far from the center of the line.[¶] The percentage error resulting from finite truncation of the first-derivative curve is shown in Fig. D-7, which may be used to apply corrections. The errors are especially large for lorentzian lines.

The following concentration standards have proved useful:

1. *α,α'-diphenyl-β-picrylhydrazyl* (*DPPH*). This substance can be weighed out and it dissolves readily in benzene; however its solutions are not stable over a long period of time.
2. *Potassium peroxylamine disulfonate* $[K_2NO(SO_3)_2]$. This is a good standard for aqueous solutions, since concentrations can be determined optically.[††] Aqueous solutions should be prepared in 10% Na_2CO_3 but these are stable for only about one day.
3. $MnSO_4 \cdot H_2O$ *and* $CuSO_4 \cdot 5H_2O$. These are good intensity standards since they are readily available in pure form; however, their lines are rather broad. (Note that for Mn^{++}, $S = \frac{5}{2}$.)
4. The following nitroxide compounds

[†] The first moment is computed by evaluating the integral $\int_{-\infty}^{\infty} Y'(H - H')\,dH$, where Y' is the amplitude of the first-derivative curve, H is an arbitrary value of the magnetic field, and H' is the field at the center of the derivative curve.

[‡] C. P. Poole, Jr., "Electron Spin Resonance: A Comprehensive Treatise on Experimental Techniques," p. 784, Interscience Publishers, a division of John Wiley & Sons, Inc., New York, 1967.

[§] M. L. Randolph, *Rev. Sci. Instr.*, **31**:949 (1960).

[¶] M. L. Randolph, in H. M. Swartz, J. R. Bolton, and D. C. Borg (eds.), "Biological Applications of Electron Spin Resonance," John Wiley & Sons, Inc., New York, 1972.

[††] M. T. Jones, *J. Chem. Phys.*, **38**:2592 (1963).

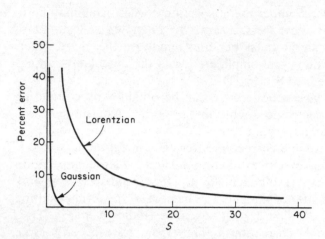

Fig. D-7 Percent error in determining the area under an absorption line when the first-derivative lorentzian or gaussian curve is truncated at the finite limits $\pm S$, where S is measured in units of ΔH_{pp}. (*Data taken from M. L. Randolph, "Quantitative Considerations in the ESR Studies of Biological Materials," in H. M. Swartz, J. R. Bolton, and D. C. Borg, eds., "Biological Applications of Electron Spin Resonance," John Wiley & Sons, Inc., New York, 1972.*)

have proved to be very versatile standards. They can be obtained in a pure form and dissolve readily in a variety of solvents, including water. In addition, the solutions, stored in a refrigerator, are stable over periods of months.

The determination of the absolute number of spins is more difficult, since the spatial variations of both the modulation field and the microwave magnetic field can cause serious errors. The unknown and the standard must be placed in equivalent positions and must have a small sample volume. Weighed samples of DPPH, $CuSO_4 \cdot 5H_2O$, or the nitroxide compounds mentioned above serve as satisfactory standards.

The use of a dual-sample cavity (TE_{104}) is advantageous when making absolute-intensity measurements since both standard and unknown can be run simultaneously. The standard and unknown samples should be interchanged to ensure that differences between the two sample positions are accounted for.

D-4 MEASUREMENT OF g FACTORS AND HYPERFINE SPLITTINGS

The absolute determination of g factors and hyperfine splittings requires accurate measurements of the external magnetic field at the sample. Such measurements are usually made by determining the proton magnetic reso-

Table D-2   Radicals for which the g factors are accurately known

Radical	Solvent†	g factor‡	Reference
Naphthalene⁻	DME/Na −58°C	2.002743 ± 0.000006	¶
Perylene⁻	DME/Na	2.002657 ± 0.000003	¶
Perylene⁺	Conc. H_2SO_4	2.002569 ± 0.000006	¶
Tetracene⁺	Conc. H_2SO_4	2.002590 ± 0.000007	¶
p-benzosemiquinone⁻	Butanol with KOH at 23°C	2.004665 ± 0.000006§	¶
Würster's blue cation	Absolute ethanol	2.003037 ± 0.000012	††
DPPH (Sec. D-3)	None (powder)	2.0037 ± 0.0002	

† DME is dimethoxyethane.
‡ Not corrected for second-order shifts.
§ Temperature dependent.
¶ B. G. Segal, M. Kaplan, and G. K. Fraenkel. *J. Chem. Phys.*, **43**:4191 (1965) as corrected by R. D. Allendoerfer. *J. Chem. Phys.* **55**:3615 (1971).
†† W. R. Knolle, Ph.D. thesis, University of Minnesota, 1970.

nance frequency of a water sample doped with a paramagnetic salt such as $FeCl_3$. However, precautions must be taken to correct for field variations between the positions of the ESR and NMR samples.†

The use of a dual-sample cavity considerably simplifies the measurement of g factors if a suitable secondary standard is employed. Table D-2 lists some secondary standards useful for this purpose. Matched standard samples are inserted into the dual-sample cavity, and their spectra are run on a dual-channel recorder. The separation of the centers of the two spectra is a measure of the field difference at the two sample positions. One of the standard samples is then replaced by the unknown sample. The field difference ΔH (corrected for the standard-sample field difference) between the spectral centers is then measured. As long as ΔH is small (less than 1 percent) compared with the magnetic field H_s at the center of the standard ESR spectrum, the g factor g_x for the unknown may be determined from

$$g_x - g_s = \frac{-\Delta H}{H_s} g_s \tag{D-13}$$

Here g_s is the g factor of the standard.

The magnetic-field sweep may also be calibrated conveniently by the use of a dual-sample cavity. Here it is convenient to use as a standard a substance giving a many-line spectrum for which an accurate determination of the hyperfine splittings has been made. One useful standard is Würster's blue perchlorate.‡

Table D-3 lists the line positions and relative intensities for some of the strong lines in this spectrum (some of the weak outermost lines are also included).

† B. G. Segal, M. Kaplan, and G. K. Fraenkel. *J. Chem. Phys.*, **43**:4191 (1965).
‡ W. R. Knolle, Ph.D. thesis, University of Minnesota, 1970.

Table D-3   Field positions of the strong lines in half the ESR spectrum of Würster's blue perchlorate in (degassed) absolute ethanol at 23°C[†]

$$a_{CH}^{H} = 1.989 \pm 0.009 \text{ G}$$
$$a_{CH_3}^{H} = 6.773 \pm 0.005 \text{ G}$$
$$a^{N} = 7.051 \pm 0.007 \text{ G}$$

$g = 2.003015 \pm 0.000012$ (corrected for second-order shifts)
$ = 2.003051 \pm 0.000012$ (uncorrected)

Line position, G, relative to the center	Relative intensity	$\tilde{M}_{CH}^{H}$	$\tilde{M}_{CH_3}^{H}$	$\tilde{M}^{N}$
0.000	16,632	0	0	0
0.278	9,504	0	−1	1
1.711	6,336	1	1	−1
1.989	11,088	1	0	0
2.267	6,336	1	−1	1
2.796	2,376	−2	1	0
3.978	2,772	2	0	0
4.785	9,504	−1	1	0
5.062	7,392	−1	0	1
6.496	5,940	0	2	−1
6.773	14,256	0	1	0
7.051	11,088	0	0	1
7.329	4,752	0	−1	2
8.762	9,504	1	1	0
9.040	7,392	1	0	1
11.558	5,940	−1	2	0
11.836	6,336	−1	1	1
13.547	8,910	0	2	0
13.824	9,504	0	1	1
14.102	5,544	0	0	2
15.536	5,940	1	2	0
15.813	6,336	1	1	1
18.609	3,960	−1	2	1
18.887	3,168	−1	1	2
20.598	5,940	0	2	1
20.875	4,752	0	1	2
22.587	3,960	1	2	1
22.864	3,168	1	1	2
25.382	1,760	−1	3	1
25.660	1,980	−1	2	2
27.371	2,640	0	3	1
27.649	2,790	0	2	2
29.360	1,760	1	3	1
29.638	1,980	1	2	2
32.433	880	−1	3	2
34.422	1,320	0	3	2
36.411	880	1	3	2
41.195	396	0	4	2
43.184	264	1	4	2

[†] W. R. Knolle, Ph.D. thesis, University of Minnesota, 1970.

PROBLEMS

D-1. Suppose that kinetic studies are to be conducted on a radical for which the half-life is $\approx 10^{-3}$ s at 0°C. With an active sample volume of ≈ 0.5 cm³, Q'_u of the cavity is $\sim 3,000$. The radical gives rise to a single lorentzian line with a peak-to-peak linewidth of 5 G. The derivative signal amplitude reaches a maximum when the microwave power is ≈ 10 mW. The measurements are made at X band and a TE_{102} cavity of volume 11 cm³ is used. The g factor is 2.01. Assume $F = 200$, $T_d = 300$ K, and $b = 10^{-4}$ s.

   With the conditions as given above, determine the appropriate minimum radical concentration detectable in this system for a signal-to-noise ratio of 10.

D-2. Determine the value of $\mathscr{R}$ in Eq. (D-2) for the following radicals:
1. Anthracene anion
2. Pyrazine anion
3. $^{13}CD_2H$

D-3. For a certain radical $T_1 = 10^{-5}$ s. Determine the value of H_1 which would give a maximum derivative amplitude. The peak-to-peak linewidth is 1 G.

D-4. Calculate the relative intensities of the two lines of Fig. 2-11, using the method of moments.

Table of Symbols

a	designation of an orbitally nondegenerate level.
a, b, c	crystallographic axes; also unit-cell dimensions.
a'	zero field splitting parameter for an S-state ion in an octahedral field.
a_i	isotropic hyperfine splitting due to the ith nucleus, G.
A	designation of an orbitally nondegenerate state.
A	area.
A_{eff}	effective hyperfine coupling. [See Eq. (7-37).]
A_{ij}	i, j component of the hyperfine tensor.
A_0	isotropic hyperfine coupling, MHz.
A_{pp}	peak-to-peak amplitude of the first-derivative curve.
A_t	parameter of the tetragonal component of the crystal field.
$A_{\parallel}$	hyperfine coupling parallel to a symmetry axis.
$A_{\perp}$	hyperfine coupling perpendicular to a symmetry axis.
(A)	antisymmetric benzene orbital. (See Sec. 5-4)
$\mathbf{A}$	hyperfine tensor.
$^d\mathbf{A}$	diagonal form of the hyperfine tensor.
$\mathscr{A}$	surface area on a sphere.
b	bandwidth.
B	purely anisotropic part of the hyperfine coupling. [See Eq. (7-28).]
B	designation of an orbitally nondegenerate state.
B	reference energy in Tanabe-Sugano diagram. (See Sec. 11-7.)

B_c	parameter of the octahedral component of the crystal field potential.
c	speed of light.
c_{ij}	coefficient of the ith atom in the jth molecular orbital.
C	capacitance.
d	states of electron orbital angular momentum with $l = 2$.
D	states of total orbital angular momentum with $L = 2$.
D	zero field parameter defined in Eq. (10-24a).
D'	zero field splitting parameter in magnetic field units, that is, $D/g\beta$.
Dq	crystal field splitting parameter. (See footnote p. 272.)
D_{XX}, D_{YY}, D_{ZZ}	diagonal elements of the D tensor.
$\mathbf{D}$	zero field splitting tensor.
$^d\mathbf{D}$	diagonal form of D tensor.
e	designation of a twofold orbitally degenerate level.
e	electronic charge.
eQ	quadrupole coupling.
E	zero field splitting parameter defined in Eq. (10-24b).
E'	$E/g\beta$
E	designation of a state with twofold orbital degeneracy.
E_1	amplitude of a microwave electric field.
f	states of electron orbital angular momentum with $l = 3$.
F	states of total orbital angular momentum with $L = 3$.
F	quantum number for combined electron and nuclear angular momenta.
F	noise figure. (See Sec. D-1.)
F center	electron trapped at a negative ion vacancy in a crystal.
g	g factor.
g'	Landé g factor.
g_e	g factor of the free electron.
g_{eff}	effective g factor defined in Eq. (7-3).
g_{ij}	i, j component of the g tensor.
g_N	nuclear g factor.
g_{XX}, g_{YY}, g_{ZZ}	diagonal components of g tensor.
$g_{\parallel}$	g component for $\mathbf{H}$ parallel to axis of symmetry.
$g_{\perp}$	g component for $\mathbf{H}$ perpendicular to axis of symmetry.
$\mathbf{g}$	g tensor.
$^d\mathbf{g}$	diagonal form of the g tensor.
G	gauss.
G	giga $= 10^9$.
$|G\rangle$	ground state.
h	Planck's constant.
$\hbar$	Planck's constant divided by 2π.
H_d	hyperfine dipolar field.
H_{eff}	effective magnetic field experienced by an electron. [See Eq. (7-26).]
H_{hf}	hyperfine field at the electron. [See Eq. (7-26).]
H_i	resonance field for a shifted line. [See Eq. (6-3).]
H_{ij}	$i = j$—coulomb integral; $i \neq j$, j adjacent to i—resonance integral.
H_k	field corresponding to the kth line in a spectrum.
H_{local}	local magnetic field. [See Eq. (3-2b).]
H_m	amplitude of modulation field.
H_{min}	minimum resonant magnetic field.
H_r	resonant magnetic field.
H_x, H_y, H_z	magnetic fields in specified directions.
H_0	external static magnetic field.
H_1	rf or microwave magnetic field.
$H_{\parallel}$	magnetic field parallel to the symmetry axis.

$H_\perp$	magnetic field perpendicular to the symmetry axis.
H'	resonant magnetic field in the absence of local fields.
$\mathbf{H}$	magnetic field vector.
Hz	hertz = 1 cycle per second.
δH_e	separation of two lines in the presence of interconversion of species.
δH_0	separation of two lines in the absence of interconversion of species.
ΔH	separation in field.
ΔH_{pp}	separation in field of extrema for a first-derivative ESR line.
$\hat{\mathcal{H}}$	hamiltonian operator.
$\hat{\mathcal{H}}'$	perturbation hamiltonian.
$\hat{\mathcal{H}}_{cr}$	hamiltonian for a crystalline electric field.
$\hat{\mathcal{H}}_{dipolar}$	dipolar hamiltonian defined in Eq. (7-23).
$\hat{\mathcal{H}}_{mag}$	Zeeman hamiltonian defined in Eq. (11-33).
$\hat{\mathcal{H}}_{oct}$	hamiltonian for an ion in an octahedral electric field. [See Eq. (11-20).]
$\hat{\mathcal{H}}_s$	general spin hamiltonian.
$\hat{\mathcal{H}}_{so}$	hamiltonian operator for spin-orbit coupling. [See Eq. (11-32).]
$\hat{\mathcal{H}}_{ss}$	spin-spin hamiltonian operator.
$\hat{\mathcal{H}}_{ttgl}$	hamiltonian for an ion in a tetragonal electric field. [See Eq. (11-21).]
$\hat{\mathcal{H}}_0$	zero-order hamiltonian.
$\mathcal{H}$	hamiltonian matrix.
i	current.
i	$\sqrt{-1}$.
I	quantum number for nuclear spin angular momentum.
I	moment of inertia.
$\hat{I}_\pm$	raising and lowering nuclear spin angular momentum operators.
$\mathscr{I}$	intensity.
J	electron exchange interaction constant, MHz.
J	vector sum of angular momenta.
J'	fictitious angular momentum defined in Sec. 12-1.
J_i	ith component of the spin-plus-orbital angular momenta; also for general angular momentum.
$\hat{J}_\pm$	raising and lowering spin-plus-orbital angular momentum operators.
k	Boltzmann's constant.
k	rate constant.
k'	orbital reduction factor. [See Eq. (12-27).]
K	hyperfine parameter defined in Eqs. (11-49).
K	constant associated with excess charge. [See Eq. (6-20).]
K band	12 to 35 GHz microwave region.
l	quantum number for the electron orbital angular momentum.
l_i	direction cosine.
l_{ij}	the i, j element of the $\mathscr{L}$ matrix.
L	total orbital angular-momentum quantum number.
L	inductance.
L'	fictitious orbital angular momentum quantum number defined in Sec. 12-1.
$\hat{\mathbf{L}}$	operator for orbital angular momentum.
$\hat{L}_i$	component of the orbital angular-momentum operator.
$\hat{L}_\pm$	raising and lowering orbital angular-momentum operators.
$\mathscr{L}$	matrix of direction cosines.
m	electron rest mass.
m_N	proton rest mass.
M	quantum number for the z component of an angular momentum.
$\tilde{M}$	M_I value for a set of equivalent nuclei such that high-field lines have $\tilde{M}_I > 0$.
M_I	quantum number for the z component of the nuclear spin angular momentum.

M_J	quantum number for the z component of the spin-plus-orbital angular momentum.
M_l	projection of orbital angular momentum on a space-fixed direction.
M_L	quantum number for the z component of the orbital angular momentum.
M_S	quantum number for the z component of the electron spin angular momentum.
$\mathcal{M}$	magnetization (magnetic moment per unit volume).
n	population difference of two levels.
$\vert n \rangle$	ket representation of the function ψ_n
$\langle n \vert$	bra representation of the function ψ_n.
n_0	equilibrium population difference of two levels.
N	number of magnetic dipoles per unit volume.
N	quantum number associated with the rotational angular momentum of a diatomic molecule.
N_{min}	minimum detectable number of spins.
oct	octahedral.
p	states of electron orbital angular momentum with $l = 1$.
$\mathbf{p}$	linear momentum.
$\mathbf{p}_\phi$	angular momentum.
P	states of total angular momentum with $L = 1$.
P_i	probability for species i.
P_0	incident microwave power.
q	charge.
q_i	total electron charge density at atom i. (See Sec. 6-5.)
Q	merit factor. [See Eq. (2-2).]
Q	proportionality constant between a and ρ.
Q band	33 to 50 GHz microwave region.
Q_L	Q factor of loaded cavity.
Q_u	Q factor of unloaded cavity.
Q'	quadrupole interaction parameter, Hz.
Q_ϵ	Q factor assignable to dielectric losses.
r	radius.
r_0	Bohr radius.
R_j	vector locating an ion (Sec. 11-3).
R	resistance in ohms.
$\mathcal{R}$	hyperfine multiplicity factor in Eq. (D-2).
s	states of electron orbital angular momentum with $l = 0$.
S	states of total orbital angular momentum with $L = 0$.
(S)	symmetric benzene orbital. (See Sec. 5-4.)
$\mathbf{S}_i$	matrix composed of the elements of $\hat{S}_i$.
$\hat{S}_i$	operator for the ith component of the spin angular momentum.
S_{ij}	overlap integral.
S'	fictitious spin such that $(2S' + 1)$ is the multiplicity of the ground state.
$\hat{S}_\pm$	raising and lowering electron spin angular momentum operators.
t	time.
t	designation of a threefold orbitally degenerate level.
trgl	trigonal.
ttdl	tetrahedral.
ttgl	tetragonal.
T	absolute temperature (Kelvin).
T	designation for states with threefold orbital degeneracy.
$\mathbf{T}$	purely anisotropic tensor (traceless).
$\vert T_i \rangle$	zero field eigenfunction of the zero field operator $\hat{\mathcal{H}}_{ss}$.
T_x	cross-relaxation time. (See Sec. 13-3.)

T_1	spin-lattice relaxation time.
T_{1e}	electron spin-lattice relaxation time.
T_{1n}	nuclear spin-lattice relaxation time.
T_2	inverse linewidth, Hz^{-1}.
T_2'	spin-spin relaxation time.
TE_{ijk}	transverse electric designation of a cavity mode. (See Sec. 2-3a.)
Tr	trace of a matrix.
u, U	spin hamiltonian parameters defined in Eq. (11-50).
v	velocity.
V	potential; voltage.
V_C	volume of a cavity.
V_j	electric-field potential ($j = x, y, z$). [See Eq. (11-5).]
V_{oct}	octahedral potential. [See Eq. (11-9).]
V_s	sample volume.
V_{ttgl}	tetragonal potential. [See Eq. (11-10).]
V_1 center	defect illustrated in Fig. 7-1.
W	energy.
$W_{dipolar}$	dipolar energy. [See Eqs. (3-2) and (7-22).]
W_G	energy of a ground state.
W_i	energy of the ith level.
x, y, z	laboratory-fixed axes.
X band	8.2 to 12.4 GHz microwave region.
X, Y, Z	molecule-fixed axes.
$\mathscr{X}, \mathscr{Y}, \mathscr{Z}$	zero field splitting parameters. [See Eqs. (10-15) and (10-23).]
Y	amplitude of an absorption line.
Y'	amplitude of first-derivative of an absorption line.
Y''	amplitude of second-derivative of an absorption line.
α	coulomb integral. (See Sec. 5-2.)
α	coefficient of $\hat{L}'$. [See Eq. (12-1).]
$|\alpha\rangle$	ket representation of the state corresponding to $M_S = +\frac{1}{2}$ or $M_I = +\frac{1}{2}$.
α proton	proton attached to the carbon atom on which the unpaired electron is primarily localized in an alkyl radical.
α_t	parameter of the tetragonal component of the crystal-field hamiltonian.
β	Bohr magneton.
β	resonance integral. (See Sec. 5-2.)
β	cavity coupling parameter defined in Eq. (D-8).
β_c	parameter of the octahedral component of the crystal-field hamiltonian.
β_N	nuclear magneton.
β proton	proton on a carbon atom adjacent to the carbon atom on which the unpaired electron is primarily localized in an alkyl radical.
γ	magnetogyric ratio.
γ_e	magnetogyric ratio of the electron.
γ_p	magnetogyric ratio of the proton.
Γ	half the linewidth at half-height in the absence of microwave saturation.
Γ_0	linewidth in the absence of exchange processes.
δ	separation of orbital energy levels due to the tetragonal component of the crystal field.
Δ	separation of orbital energy levels in an octahedral or tetrahedral crystal field.
Δ	splitting parameter defined in Eqs. (12-49) and (12-50).
Δ	state of a diatomic molecule in which $\Lambda = \pm 2$.
ϵ	mixing coefficient. [See Eq. (6-14).]
ϵ	population difference. (See Sec. 13-3.)

ϵ_i	excess charge. [See Eq. (6-20).]
η	viscosity.
η	filling factor of a cavity.
θ	angle between $\mathbf{H}$ and $\mathbf{r}$.
θ	dihedral angle.
κ	proportionality factor between T_2^{-1} and linewidth.
λ	deBroglie wavelength.
λ	spin-orbit coupling parameter.
λ	perturbation mixing coefficient.
λ_i	eigenvalue.
$\hat{\Lambda}$	general operator.
Λ	tensor defined in Eqs. (11-37) and (11-38).
Λ	quantum number corresponding to the projection of $\mathbf{L}$ on the internuclear axis in a diatomic molecule.
Λ_{ij}	i, j component of the Λ tensor.
$\boldsymbol{\mu}$	magnetic moment.
$\boldsymbol{\mu}_e$	magnetic moment of the electron.
$\boldsymbol{\mu}_N$	magnetic moment of nucleus N.
ν	frequency.
ν_e	electron-resonance frequency.
ν_n	nuclear-resonance frequency.
ν_0	fixed microwave frequency.
ν_p	proton-resonance frequency.
π	molecular orbital comprised of atomic $2p_z$ orbitals.
π_{rt}	atom-atom polarizability between atoms r and t.
Π	state of a diatomic molecule in which $\Lambda = \pm 1$.
ρ	spin density; also unpaired-electron density.
ρ_i	spin density in the ith $2p_z$ atomic orbital.
σ	orbital which is cylindrically symmetric about a bond.
Σ	state of a diatomic molecule in which $\Lambda = 0$.
Σ	quantum number corresponding to the projection of $\mathbf{S}$ on the internuclear axis of a diatomic molecule.
τ	lifetime in a state.
τ	time constant of an RC circuit.
τ	general variable of integration.
$\int_\tau$	integration over the full range of variables.
ϕ	atomic wave function.
ϕ	polar angle.
ϕ	Euler angle. (See Fig. A-6.)
χ	magnetic susceptibility.
$\boldsymbol{\chi}$	magnetic susceptibility tensor.
ψ	wave function.
ω	angular frequency.
ω_m	modulation frequency.
Ω	solid angle.
Ω	quantum number corresponding to $\Lambda + \Sigma$ for Hund's case a.

Name Index

Subject Index

Page numbers in *italics* indicate main coverage; numbers in **boldface** indicate figures.

Table C Nuclear spins, abundances, moments, and hyperfine couplings for some common magnetic nuclei[†]

Nucleus	Spin	% Natural abundance	Magnetogyric ratio[‡] $(rad\ G^{-1}s^{-1} \times 10^{-4})$	Anisotropic hyperfine coupling B, MHz[§]	Isotropic hyperfine coupling A_0, MHz[¶]
^{1}H	$\frac{1}{2}$	99.985	2.67510	—	1,420
^{2}H	1	0.015	0.41064	—	218
^{6}Li	1	7.42	0.39366	—	152*
^{7}Li	$\frac{3}{2}$	92.58	1.03964	—	402* (291 calc)
^{9}Be	$\frac{3}{2}$	100	−0.37594	—	−358
^{10}B	3	19.58	0.28748	17.8	672
^{11}B	$\frac{3}{2}$	80.42	0.85828	53.1	2,020
^{13}C	$\frac{1}{2}$	1.108	0.67263	90.8	3,110
^{14}N	1	99.63	0.19324	47.8	1,540
^{15}N	$\frac{1}{2}$	0.37	−0.27107	−67.1	−2,160
^{17}O	$\frac{5}{2}$	0.037	−0.36266	−144	−4,628
^{19}F	$\frac{1}{2}$	100	2.51665	1515	47,910
^{23}Na	$\frac{3}{2}$	100	0.70760	—	886*
^{25}Mg	$\frac{5}{2}$	10.13	−0.16370	—	—
^{27}Al	$\frac{5}{2}$	100	0.69706	59	2,746
^{29}Si	$\frac{1}{2}$	4.70	−0.53141	−86.6	−3,381
^{31}P	$\frac{1}{2}$	100	1.08290	287	10,178
^{33}S	$\frac{3}{2}$	0.76	0.20517	78	2,715
^{35}Cl	$\frac{3}{2}$	75.53	0.26212	137	4,664
^{37}Cl	$\frac{3}{2}$	24.47	0.21818	117	3,880
^{39}K	$\frac{3}{2}$	93.10	0.12484	—	231*
^{43}Ca	$\frac{7}{2}$	0.145	−0.17999	—	—
^{45}Sc	$\frac{7}{2}$	100	0.64989	—	1,833
^{47}Ti	$\frac{5}{2}$	7.28	−0.15079	—	−492
^{49}Ti	$\frac{7}{2}$	5.51	−0.15083	—	−492
^{51}V	$\frac{7}{2}$	99.76	0.70323	—	2,613
^{53}Cr	$\frac{3}{2}$	9.55	−0.15120	—	−630
^{55}Mn	$\frac{5}{2}$	100	0.65980	—	3,063
^{57}Fe	$\frac{1}{2}$	2.19	0.08644	—	450
^{59}Co	$\frac{7}{2}$	100	0.63171	—	3,666
^{61}Ni	$\frac{3}{2}$	1.19	−0.23905	—	1,512
^{63}Cu	$\frac{3}{2}$	69.09	0.70904	—	4,952
^{65}Cu	$\frac{3}{2}$	30.91	0.75958	—	5,305
^{67}Zn	$\frac{5}{2}$	4.11	0.16731	—	1,251
^{75}As	$\frac{3}{2}$	100	0.45816	255	9,582
^{77}Se	$\frac{1}{2}$	7.58	0.51008	376	13,468
^{79}Br	$\frac{3}{2}$	50.54	0.67021	646	21,738
^{81}Br	$\frac{3}{2}$	49.46	0.72245	696	23,432
^{83}Kr	$\frac{9}{2}$	11.55	−0.10293	—	—
^{85}Rb	$\frac{5}{2}$	72.15	0.25829	—	1,012*
^{87}Rb	$\frac{3}{2}$	27.85	0.87533	—	3,417*
^{95}Mo	$\frac{5}{2}$	15.72	0.17428	—	−3,528
^{97}Mo	$\frac{5}{2}$	9.46	−0.17796	—	−3,601
^{107}Ag	$\frac{1}{2}$	51.82	−0.10825	—	−3,520
^{109}Ag	$\frac{1}{2}$	48.18	−0.12445	—	−4,044
^{127}I	$\frac{5}{2}$	100	0.53522	—	—
^{129}Xe	$\frac{1}{2}$	26.44	−0.73995	1,052	33,030
^{131}Xe	$\frac{3}{2}$	21.18	0.21935	—	—
^{133}Cs	$\frac{7}{2}$	100	0.35089	—	2,298*
^{207}Pb	$\frac{1}{2}$	22.6	0.55968	—	—

† Compiled from data in the following references:

1. J. R. Morton, J. R. Rowlands, and D. H. Whiffen, *National Physical Laboratory Bulletin*, No. BPR 13, 1962.
2. S. Flügge (ed.), "Handbook of Chemistry and Physics," 50th ed., p. E75, 1969.
3. J. R. Morton, *Chem. Rev.*, **64**:453 (1964).
4. D. H. Whiffen, *J. Chim. Phys.*, **61**:1589 (1964).
5. B. A. Goodman and J. B. Raynor, *J. Inorg, Nucl. Chem.*, **32**:3406 (1970).

‡ The magnetic moment (erg G^{-1}) can be obtained from the magnetogyric ratio by using the relation

$$\mu_N = \hbar I \gamma_N$$

§ The anisotropic hyperfine couplings are tabulated as

$$B = \tfrac{2}{5} h^{-1} g_N \beta_N g \beta \langle r^{-3} \rangle$$

where $\langle r^{-3} \rangle$ is computed for a valence p electron from self-consistent-field wave functions. The couplings are such that the principal values of the traceless tensor are respectively -1, -1, and $+2$ times the number quoted.

¶ The isotropic hyperfine couplings are tabulated as

$$A_0 = \frac{8\pi}{3} h^{-1} g_N \beta_N g \beta |\psi_S(0)|^2$$

where $\psi_S(0)$ is the value of the valence-shell, self-consistent-field S wave function at the nucleus of the neutral atom. Values indicated with an asterisk are the experimental atomic hyperfine couplings as measured using the atomic-beam technique. [See P. Kusch and H. Taub, *Phys. Rev.*, **75**:1477 (1949).]